D0087553

THE
CAMBODIAN
AGONY

M. E. SHARPE, INC.
ARMONK, NEW YORK
LONDON, ENGLAND

To Our Parents

80 Business Park Drive, Armonk, New York 10504

Available in the United Kingdom and Europe from M. E. Sharpe, Publishers, 3 Henrietta Street, London WC2E 8LU.

Library of Congress Cataloging in Publication Data

The Cambodian agony.

1. Cambodia—History—1975- . I. Ablin, David A.
II. Hood, Marlowe.
DS554.8.C358 1986 959.6'04 86-17900
ISBN 0-87332-421-8

Printed in the United States of America

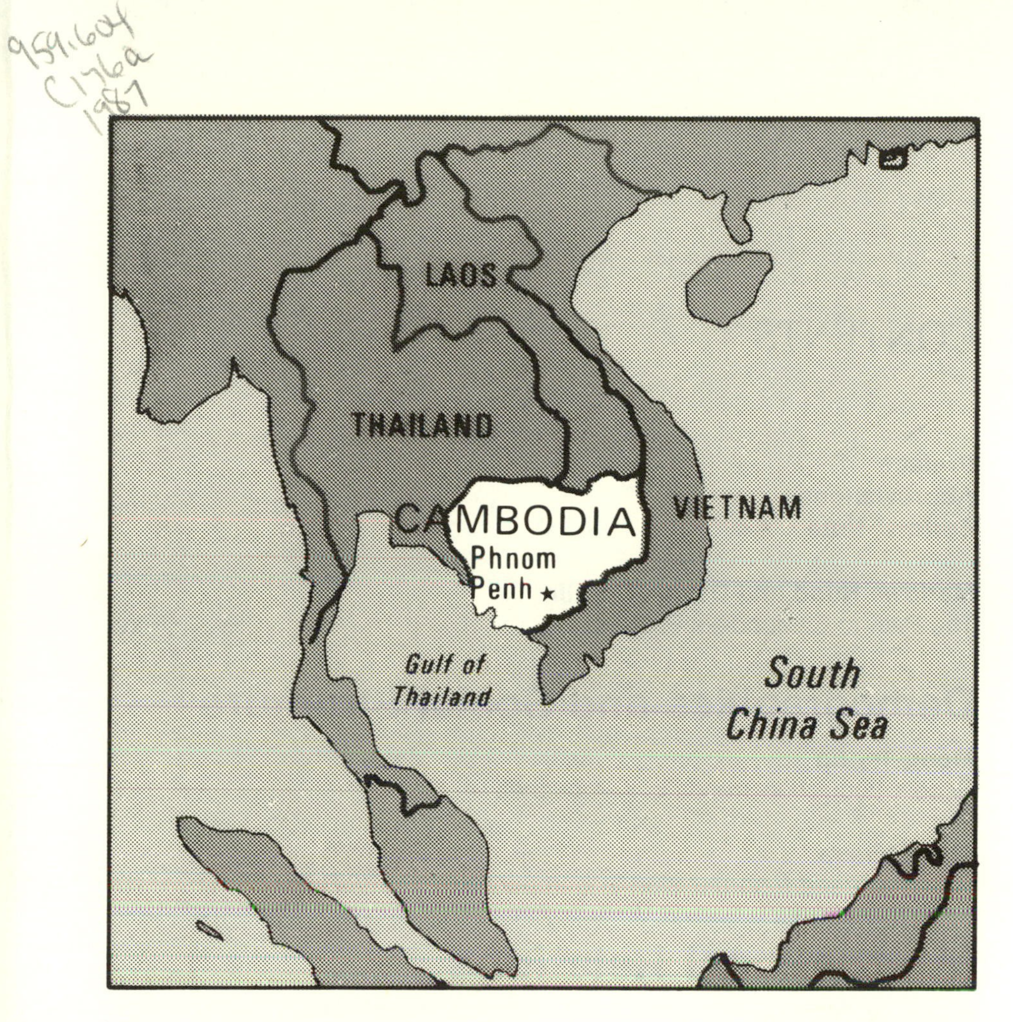

Source: United States Department of State.

CONTENTS

Agriculture and Economic Development

Refugees

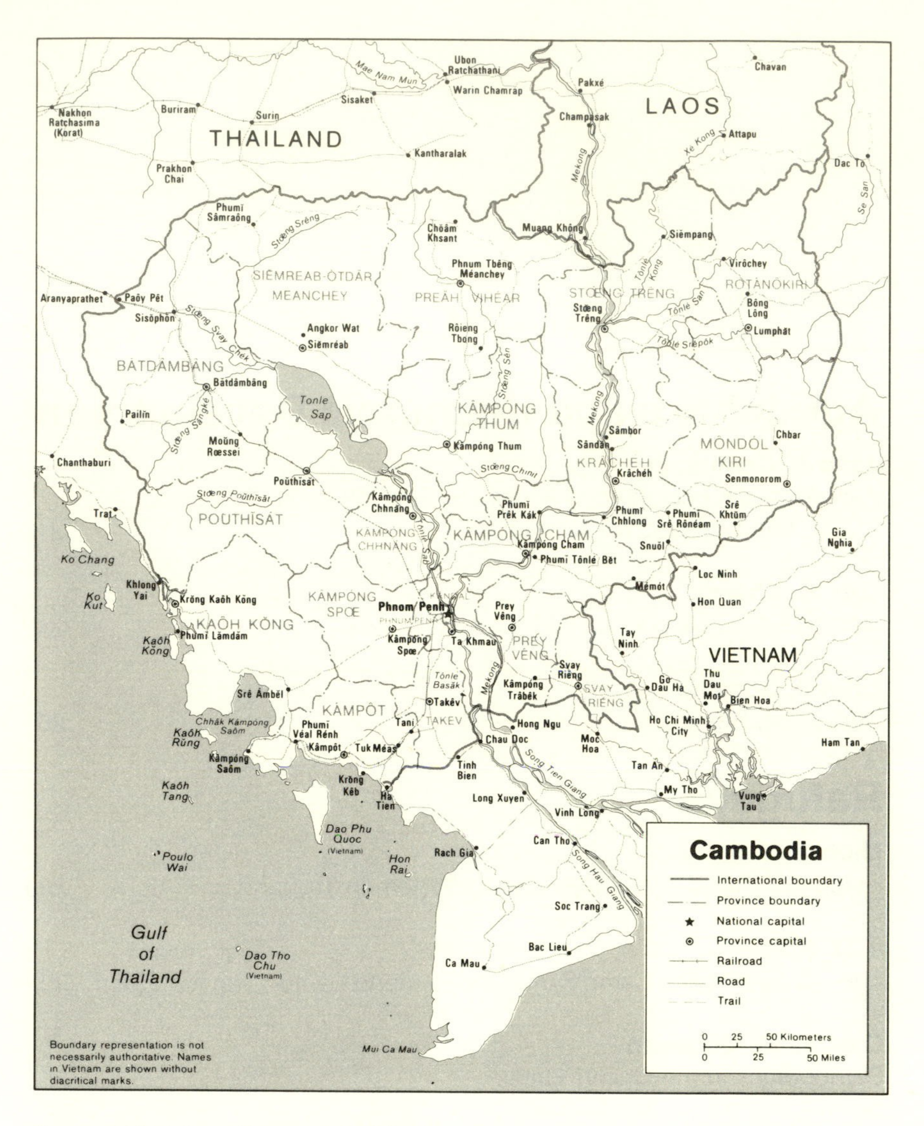

Source: United States Department of State.

PREFACE

This book emanated from an international conference, "Kampuchea in the 1980s: Prospects and Problems," organized by the editors and held in November 1982 at the Woodrow Wilson School of Public and International Affairs, Princeton University. The conference brought together more than 200 scholars, policy makers, relief organization officials, journalists, and observers from all over the world. Some of the papers included here were solicited for the conference, others after it had adjourned. Those originally presented at the conference have been updated and considerably expanded for this publication.

This book was a collective enterprise. Taking into consideration the conference from which it emanated, several dozens of people were involved in its creation.

A staff of student volunteers logged hundreds of hours of work during the months preceding the conference. Our coworkers included Peter Benda, Hwa Soo Chung, Richard Hoffman, Ronald Lillejord, John Mosler, Long Nguyen, April Oliver, Page Pulver, Jim Reynolds, Lucy Swift, and David Williamson. We are most grateful to Mark Steiner, without whose enormous contribution things would have been quite desperate indeed.

Many of the faculty at Princeton University gave freely of their time and wisdom in helping us prepare for the conference and edit the book. We were both moved by the generous help and support provided by our faculty advisers, Professors Leon Gordenker, Richard Ullman, and Lynn White. Professor Miles Kahler and Dr. David Morell were very helpful in getting the project off the ground.

Many of the participants in the conference and contributors to the book were also invaluable counselors: John McAuliff, Murray and Linda Hiebert, Robert Porter, Sina Than, and Joel Charny.

Funding for the conference and this book came from several sources. We would like to thank Dr. Enid C. B. Schoettle, Dr. Paul Balaran, and The Ford Foundation; Dr. Laurence Stifel and The Rockefeller Brothers Fund; Mrs. Geraldine Kunstadter and The Albert K. Kunstadter Family Foundation; President William G. Bowen and The President's Fund at Princeton University; and Dean Donald Stokes, Dean Ingrid Reed, and the Woodrow Wilson School of Public and International Affairs.

We are indebted to the entire staff of the Woodrow Wilson School for their technical and logistical support. We think especially of Suzanne Cox, Sally

Coyle, Monica Hamilton, Edna Lloyd, Joyce Mix, Agnes Pearson, Audrey Pitman, and Ginie Reynolds.

We would also like to acknowledge the tireless and consistently cheerful help of Elizabeth Speir and Melanie Thompson.

Myron Sharpe and the editors at M. E. Sharpe, Douglas Merwin and Arnold Tovell, were unfailingly helpful and patient. Thank you.

Finally, Marlowe extends a personal apology to Cathy Yeh for having burdened her with so many of the problems stemming from the conference and this book.

ABBREVIATIONS

ACVA	American Council of Voluntary Agencies
AFSC	American Friends Service Committee
ANS	Armée Nationale Sihanoukienne (National Army of Sihanouk)
ARRK	Consortium for Agricultural Relief and Rehabilitation in Kampuchea
ASEAN	Association of South East Asian Nations
BP	*Bangkok Post*
CCSDPT	Committee for the Coordination of Services to Displaced Persons in Thailand
CGDK	Coalition Government of Democratic Kampuchea
CIA	Central Intelligence Agency (U.S.)
CIDSE	Cooperation Internationale pour le Developpement et la Solidarité (International Cooperation for Development and Solidarity)
COMECON	Council for Mutual Economic Assistance
CPK	Communist Party of Kampuchea
CWS	Church World Service
DK	Democratic Kampuchea
DSM III	*Diagnostic and Statistical Manual of Mental Disorders*, 3d ed.
FANK	Forces Armées Nationales Khmeres (Khmer National Armed Forces)
FAO	Food and Agriculture Organization (UN)
FBIS	Foreign Broadcast Information Service (U.S.)
FEER	*Far Eastern Economic Review*
FUNCINPEC	Front Uni National Pour un Cambodge Indepéndant, Neutre, Pacifique, et Cooperatif (National United Front for an Independent, Neutral, Peaceful and Cooperative Cambodia)
FUNK	Front Uni National du Kampuchea (National United Front of Kampuchea)
GRUNK	Gouvernement Royal d'Union Nationale du Kampuchea (Royal Government of National Union of Kampuchea)

HEW	Department of Health, Education and Welfare (U.S.)
HHS	Department of Health and Human Services (U.S.)
ICK	International Conference on Kampuchea (UN)
ICP	Indo-China Communist Party
ICRC	International Committee of the Red Cross
ICSC	International Commission of Supervision and Control
INS	Immigration and Naturalization Service (U.S.)
IRC	International Rescue Committee
IRRI	International Rice Research Institute
KCP	Kampuchean Communist Party (*same as* Communist Party of Kampuchea)
KGP	Khmer Guided Placement Project (U.S.)
KID	Khao-I-Dang (*displaced persons camp*)
KK	*Khmer Krom*
KNAF	Khmer National Armed Forces
KNUFNS	Kampuchean National United Front for National Salvation (*same as* Kampuchean United Front for National Salvation)
KPNLA	Khmer People's National Liberation Army
KPNLAF	Kampuchean People's National Liberation Armed Forces [1972–1975] *or* Khmer People's National Liberation Armed Forces [1979-]
KPNLF	Khmer People's National Liberation Front
KPRAF	Kampuchean People's Revolutionary Armed Forces
KPRP	Khmer People's Revolutionary Party [1951–1960] *or* Kampuchean People's Revolutionary Party [1979-]
KR	Khmer Rouge (Red Cambodians)
KUFNCD	Kampuchean United Front for National Construction and Defense
KUFNS	Kampuchean United Front for National Salvation (*same as* Kampuchean National United Front for National Salvation)
MOULINAKA	Mouvement pour la Liberation Nationale du Kampuchea (Movement for the National Liberation of Kampuchea)
NADK	National Army of Democratic Kampuchea (*same as* National Army of Kampuchea)
NAK	National Army of Kampuchea (*same as* National Army of Democratic Kampuchea)
NGO	Nongovernmental organization
NIMH	National Institute of Mental Health (U.S.)
NR	*Nation Review*
NUFK	National United Front of Kampuchea
OECD	Organization for Economic Cooperation and Development
ORR	Office of Refugee Resettlement (U.S.)

PAFNLK	People's Armed Forces for the National Liberation of Kampuchea
PAVN	People's Army of Viet Nam
PCF	Parti Communiste Français (Communist Party of France)
PDFGNUK	Patriotic and Democratic Front of Great National Union of Kampuchea
PDR	People's Democratic Republic
PLO	Palestine Liberation Organization
PRC	People's Republic of China
PRK	People's Republic of Kampuchea
PRKAF	People's Republic of Kampuchea Armed Forces
PRPK	People's Revolutionary Party of Kampuchea (*same as* Kampuchean People's Revolutionary Party)
PTSS	Post-traumatic Stress Syndrome
RAK	Revolutionary Army of Kampuchea
RCGNU	Royal Cambodian Government of National Union (*same as* Royal Government of National Union of Kampuchea)
RGNUK	Royal Government of National Union of Kampuchea
SEATO	South East Asia Treaty Organization
SF	Salvation Front (*same as* Kampuchean National United Front for National Salvation)
SPK	Saporamean Kampuchea News Agency
SRV	Socialist Republic of Vietnam
UIF	United Issarak Front
UN	United Nations
UNBRO	United Nations Border Relief Operations
UNHCR	United Nations High Commissioner for Refugees
UNICEF	United Nations International Children's Emergency Fund
USG	United States Government
USSR	Union of Soviet Socialist Republics
VCP	Vietnam Communist Party
VNA	Vietnam News Agency
VWP	Vietnam Worker's Party
WCC	World Council of Churches
WFP	World Food Program (UN)
WPK	Worker's Party of Kampuchea
WVI	World Vision International
YMCA	Young Men's Christian Association

EDITORS' NOTE

''Cambodia'' and ''Kampuchea'' refer to the same place. ''Cambodia'' is an Anglicized variant of the French ''Cambodge,'' whereas ''Kampuchea'' is a more direct and accurate transliteration of the Khmer language word. Both are used in this book. Each has a number of connotations. Most strongly, the former connotes colonial domination and the latter brings to mind, for non-Khmer, the Khmer Rouge. All things considered, we have used ''Cambodia'' in the title and in our introduction, except when referring to the political entities ''Democratic Kampuchea'' and the ''People's Republic of Kampuchea.'' Because of the various connotations, we did not standardize the use of these terms throughout the book. We have tried to standardize the Romanized spellings of other Khmer words, especially important proper names. Some inconsistency, however, is bound to remain.

Marlowe Hood and David A. Ablin

THE PATH TO CAMBODIA'S PRESENT

Cambodia Rising From the Ashes

Cambodia has risen from the ashes, but the embers still smolder.[1] The survivors cling to a precarious stability. What has been their ordeal? How have they recovered? What prospects do Cambodians have for a peaceful, prosperous existence? These are some of the questions addressed in this volume.

The relative calm since 1979 cannot compensate for the preceding decade of terror. It began with peasant protests and government suppression during the waning years of Prince Sihanouk's Royal Kingdom of Cambodia, which fell in 1970. The next five years saw a full-scale civil war accompanied by massive U.S. bombing, and ended with the defeat of Lon Nol's American-supported Khmer Republic. The triumphant Khmer Rouge (Red [Communist] Cambodians)* sustained a climax of terror which did not abate until the Vietnamese invasion of December 1978. The denouement of this decade of tragedy was the following year of chaos and starvation.

Some of the wounds have healed. Others have been left open to fester. The total damage is incalculable, and may only become apparent over several generations. The imponderability of Cambodia's trauma is even more vexing because the pain continues today. Occupied by a traditional rival, Vietnam, threatened by the still formidable Khmer Rouge, and caught in a struggle among the world's most powerful nations, Cambodia remains an active, if exhausted, battlefield.

A Geographic and Demographic Profile

Cambodia is a small, tropical country, approximately the size of England and Wales combined. It is ringed by rugged, mountainous jungle along its western and northern borders with Thailand, its northwestern border with Laos, and

*We use ''Khmer Rouge'' throughout this introduction to refer to the Cambodian communist movement from 1930 until 1979, when, for the first time, the movement becomes formally disunited. The communist party went through several different name changes and to use each name where appropriate, although more precise, would create a virtually impenetrable narrative.

much of its eastern border with Vietnam. To the southeast, Cambodia faces the South China Sea, which once covered the great central plain that dominates the country's interior. These natural boundaries are by no means impregnable. Since the decline of the Khmer kingdom of Angkor in the fifteenth century, Cambodia has repeatedly been a victim of the expansionist tendencies of Thailand and Vietnam. It has proven most vulnerable along its eastern border, where the broad and fertile plains of the Mekong River Delta, which had once belonged exclusively to the Khmer empire, are now shared with Vietnam. Threats to Cambodia's lowland rulers have also emanated from the Cardamom Mountains to the west, historically a sanctuary for foreign enemies and domestic rebels.

The foundation of Cambodia's well-being has always been its land. Agricultural development has been the domestic cornerstone of every Khmer regime since the third century, except for Lon Nol's hapless Khmer Republic. In a world of apparently inexorable urbanization and modernization, Cambodia, after the destruction of the 1970s, is more of a peasant society today than it was twenty years ago.

Tonle Sap (The Great Lake), near the center of the country, is the pulse that regulates the ebb and flow of agricultural production. During the dry season, from November until May, the lake covers some two thousand square kilometers of Cambodia's interior. But when monsoon rains and melting Himalayan snows swell the Siem Reap River, and reverse the flow of the Mekong River, Tonle Sap covers up to an additional 10,000 square kilometers of forests and fields. Fish become so abundant that one can simply scoop them out of the shallow waters with a hand-held net. These annual flood waters coat the otherwise poor soil with a layer of rich silt and, if stored, provide a source for irrigation during the dry season. John Dennis, in his chapter in this book, "Kampuchea's Ecology and Resource Base: Natural Limitations on Food Production Strategies," explores the importance of this hydrological cycle for Cambodia's development.

There are three immediate observations to make about Cambodia's demography. First, Cambodia has a very small population compared to those of its neighbors, except for Laos. Given an estimated current population of 6.5 to 7.5 million, Thailand and Vietnam each have approximately nine times as many inhabitants. Second, Cambodia has a relatively favorable land/labor ratio. On average, there are fewer than two people per hectare of arable land, in contrast to nine in Vietnam and more than three in Thailand. These two facts combined help explain the constant pressure that Cambodia has experienced on both its eastern and western borders. Finally, Cambodia's population has actually declined since 1970, and has major distortions in age and sex groupings. In the absence of crisis, there would have been 10 million people in Cambodia in 1980. As a result of war, the willful extermination of parts of the population by the Khmer Rouge government, and famine during the 1970s, however, the actual figure in 1980, according to optimistic estimates, was 7 million. Few infants and elderly people survived the Khmer Rouge rule from 1975 through 1978; this absence distorts social patterns and will affect production in the future. In addition, the fact that nearly seventy percent of the adults in the countryside are women reduces the likelihood of returning to a surplus-producing agricultural base anytime soon. Meng-Try Ea, in his paper here, "Recent Population Trends in Kampuchea," analyses the demographic upheavals of the 1970s by dividing the decade into relevant periods and the population into political and sociological groups.

These geographic and demographic features are the nonpolitical infrastructure of a country and a people. They will at least partially determine the future course of events in Cambodia.

Cambodia Before 1945

In touching upon a few points in Cambodian history, we concentrate on those that still resonate today.

The area known today as Cambodia has been occupied by people since at least 4000 B.C., according to archaeological findings. Whether these people came from the Malay Archipelago or the area of present-day India or China is uncertain. Another group of people, who spoke the Khmer language, migrated to Cambodia, apparently from Thailand, in the first millennium B.C. Their ultimate place of origin is also unclear. The regions surrounding Cambodia were gradually inhabited by other peoples: the Vietnamese, who migrated from southern China into the Mekong Delta in the first millennium B.C.; the Burmese, who came from Tibet during the eighth and ninth centuries A.D.; and the Thai, who did not descend from southwest China into present-day Thailand until the thirteenth century. In the fifteenth century, the Vietnamese, moving south, drove many of the Cham, a Malayo-Polynesian group living in the lower Mekong area since at least the first millennium B.C., into present-day Cambodia. Finally, in the eighteenth and nineteenth centuries, a large number of ethnic Chinese immigrated to Southeast Asia, where they typically made their living through commerce. Cambodia was exceptional in this respect only in that the Chinese took root in the countryside as well as in cities and towns.

Little is known about how Cambodians lived before the oldest surviving written records were made, in the third century A.D. Envoys of the Han dynasty in China visited a civilization they called "Funan," centered near present-day Phnom Penh. They reported that the people of Funan

> lived in walled cities, palaces and houses. . . . They devote themselves to agriculture. They sow one year and harvest for three. . . . Taxes are paid in gold, silver, pearls and perfumes. . . . There are books and depositories of archives.[2]

The basis of Funan's prosperity was twofold: trade and agriculture. As piracy and competition ruined their maritime trade, the Funanese attempted to maintain their wealth solely through agriculture. They constructed a system of drainage canals in order to control the annual flood waters and improve crop yields. But despite its technological innovations, which were the foundation for later empires, Funan fell prey to invasions by its vassal state, Chenla, which subsumed Funan at the end of the sixth century.

Since the 1800s, scholars have portrayed the kingdoms of Funan and Chenla as unified states with control over much of the Southeast Asian mainland, with contacts stretching as far as the Mediterranean. Recent research has cast doubt on their grandeur, seeing instead in each period a group of small kingdoms which would occasionally collaborate with one another.

The Funanese were descendents of the first settlers in Cambodia, with perhaps some intermingling with the Khmer people. Chenla was a distinctly Khmer state. The oldest Khmer language inscriptions that have been found were written in the seventh century, while Chenla was the dominant kingdom. Some of these inscriptions tell us of the founder of the Khmer people, "Kambu," who named his country "Kambuja." Kambuja, actually a Sanskrit word taken from India, is the source for the national term in Khmer, "Kampuchea," as well as the French "Cambodge" and, from that, "Cambodia."

Chenla was torn by internal disputes and, according to Chinese chronicles, split into "Land Chenla" on the north and "Water Chenla" on the south. Water Chenla was forced into submission by a Malaysian trading state, until a Khmer prince reestablished Khmer sovereignty and built the first great civilization of Southeast Asia—Angkor.

When a French naturalist rediscovered the massive Angkor Wat (a *wat* is a Buddhist temple) complex in the middle of the nineteenth century, the splendor of ancient Khmer civilization had been all but engulfed by the jungle. Archaeologists and historians have since revealed an empire that at its height stretched across most of the land now occupied by south Vietnam, Laos, Thailand, Burma, Malaysia and, of course, Cambodia.

The founding Deva-Raja (God-King) of Angkor was Jayavarman II, who ruled from A.D. 802 to 850. From his reign onward it became the sacred duty of every king to erect a stone *linga*, housed in a great temple mountain, as an architectural embodiment of his persona. His son established another important precedent. He constructed the first in a network of massive *barays*, or water reservoirs, the largest of which held more than 30 million tons. A significant advance on the hydrological technology of Funan, these reservoirs curbed flooding during the rainy season and assured adequate water for irrigation when the weather was dry, which allowed for three and sometimes four crops each year. This prodigious agricultural productivity was the economic base on which the mammoth temples were constructed.

After recovering from an invasion by the neighboring state of Champa, Angkor reached its zenith in the twelfth and thirteenth centuries. But even as it extended its influence across most of Southeast Asia, a series of inept rulers had to confront problems that even their most able predecessors would have found difficult to manage. The costs of temple construction, maintaining the empire, and fighting evermore powerful enemies drained the royal coffers. To raise income, the Khmer kings increased the taxes and labor obligations of a people already chafing under the oppressiveness of a society in which much of the population was classified as one or another type of slave. Amid these pressures there was a subjective change which contributed as much as any single factor to the empire's decline: the rise of Theravada Buddhism.

The early Angkor kings had promoted various Hindu sects, especially those dedicated to the gods Siva and Vishnu, until Jayavarman introduced Mahayana Buddhism as the official court religion in the twelfth century. Although all three of these faiths were supported by elaborate ecclesiastical hierarchies, their influence did not extend far beyond the capital of Angkor Thom. The common people practiced an amalgam of ancestor worship and animism, and associated the state-promoted religions with slave labor and high taxes. In the late twelfth century, however, Theravada, or Lesser Vehicle Buddhism, swept through Southeast Asia and took root among the Cambodian population. Propagated by mendicant monks who lived in austerity and humility, Theravada did not require vast wealth to maintain its sacred symbols, thus eroding the raison d'etre of an extractive and corrupt state. Theravada remained the dominant and unchallenged belief system of the Khmer people until 1975, when the Khmer Rouge nearly destroyed not only the monkhood and temples, but the people's faith as well. Whether Buddhism can or will be revitalized is as yet unclear.

During the thirteenth and early fourteenth centuries, the expanding kingdom of Siam (Thailand) launched repeated attacks against Cambodia, finally capturing the capital, Angkor Thom, in 1431. The routed Khmers tried to reestablish the royal court on the site of the modern capital, Phnom Penh, but were forced to relocate several times as they retreated from further military incursions. At the end of the sixteenth century, with most of the royal family in exile, the Siamese enthroned a Khmer prince as their vassal.

Thus began the humiliating practice whereby Khmer monarchs were installed by a foreign power—either Siam, Vietnam, or France. In 1794, for example, the Khmer kingship was actually bestowed in Bangkok. Forty years later, the next Khmer king, Ang Chan, ruled at the behest of the Vietnamese, to whom he had turned in the wake of Thai annexation of three Cambodian provinces. Caught between the expansion of these two powers, Cambodia's territory and population dwindled until the French established a Protectorate in 1864. Though Cambodia never disappeared as an entity, by the middle of the nineteenth century, only a sliver of the once great empire remained. In recent years Khmers have been acutely aware of this history. Indeed, their fear of being extinguished altogether as a people and a culture is not unfounded.

The French Protectorate froze regional antagonisms. In fact, the French reversed Cambodia's decline by insisting that Thailand return Battambang and

Siem Reap provinces. The French established the boundaries that exist—and are still disputed—today. Cambodia did not figure prominently in French colonial policy. Unlike in Vietnam, there was minimal French capital investment, and little economic development occurred. The colonial administration in Cambodia was staffed with Vietnamese civil servants—a practice which did nothing to improve Khmer-Vietnamese relations.

For the French, as for the Thai and Vietnamese before them, it was a matter of great concern that they select a Cambodian king who would be amenable to their desires and helpful in legitimating their rule. In selecting, they alternated between the two main branches of the royal family, the Norodoms and the Sisowaths. In 1941 they enthroned eighteen-year-old Norodom Sihanouk, whose youth and pliant nature seemed to make him an ideal guardian of colonial interests. For a while this was so.

During the Second World War, most of Southeast Asia was occupied by the Japanese, who left the Vichy administration in Cambodia largely undisturbed until the beginning of 1945. When the war turned against them, however, the Japanese ousted the French and prompted King Sihanouk to declare independence. They also installed a new government headed by Son Ngoc Thanh, a bona fide Cambodian nationalist. Although Thanh was soon arrested and exiled by the returning French authorities, his short-lived regime gave form to the sense of nationalism that had been smoldering in Cambodia for several years.

The Sihanouk Years

When the French returned in the fall of 1945, things were not as they had been before the war. Bands of armed guerrillas were organizing throughout the countryside to fight for independence. Although they were only loosely connected, and did not share any other ideological commitments, these bands made the countryside ungovernable for the French. In addition, the French faced pressure from the Cambodian elite in Phnom Penh. As an initial concession, France granted Cambodia autonomy within the French Union in 1946. Though they ceded little power in doing so—the army and police remained under French control—there were important changes involved: the absolute monarchy was abolished and a constituent assembly was elected for the purpose of drafting a constitution, which was adopted the following year.

The new political arena was dominated by the Democratic Party, which sought full independence and a European-style democracy. However, the next several years were rife with political infighting, and despite a majority in the National Assembly, the Democrats faltered in the face of concerted conservative opposition. Young King Sihanouk, who had been testing the waters of party politics, was among their opponents.

In January 1953, Sihanouk dissolved the squabbling National Assembly. He then embarked on his famous "*Croisade Royale pour L'Indèpendance*" to drum up international support for Cambodian independence. The French, demoralized by a war in Vietnam which was turning against them, and unwilling to duplicate troubles in Cambodia, transferred virtually all the remaining attributes of state power to Sihanouk's Royal Government of Cambodia in November 1953. The King became a hero.

Writer William Shawcross has described Norodom Sihanouk as follows:

> [He] presided feudally over Cambodia from 1941 to 1970, as King, Chief of State, Prince, Prime Minister, head of the main political movement, jazz band leader, magazine editor, film director and gambling concessionaire, attempting to unite in his rule the unfamiliar concepts of Buddhism, socialism and democracy. His exercise of power was so astonishing and so individual that he came to personify his country and its policies, abroad as well as at home.[3]

Sihanouk's rule was marked by contradictory domestic policies and continually shifting domestic and international alliances. Underlying these apparent inconsistencies, however, was Sihanouk's conscious attempt to balance opposing powers, both for Cambodia's sake and his own. Internally, he played off social revolutionaries, conservative aristocrats, and members of an emerging educated class seeking personal advancement and national modernization. Internationally, he tried to keep Cambodia independent by pursuing a policy of neutrality vis-à-vis the superpowers and China. He was successful in the sense that he maintained relative social peace with less oppression than was used in other Southeast Asian states and he kept Cambodia within the eye of the political hurricane that was swirling around it.

In his foreign policy, Sihanouk courted all, but took vows with none. The Geneva Agreements of 1954 brought an end to the First Indochina War after the stunning military defeat of the French by the Vietnamese communists at Dien Bien Phu. The signatories, including France, the Soviet Union, the People's Republic of China and North Vietnam, also formally recognized Cambodia's international position as one of neutrality.

In contrast, the United States declined to sign the agreements, extended large amounts of military and economic aid to Cambodia, and pressured Sihanouk to join the South East Asia Treaty Organization (SEATO), an anticommunist security alliance which included the United States, Thailand, the Philippines, Australia, France, and Pakistan. When Sihanouk refused to join SEATO and insisted on Cambodia's neutrality, the United States added a second track to its policy. Thailand and South Vietnam were encouraged to maintain pressure on the Prince by harassing the small Cambodian army. The U.S. Central Intelligence Agency (CIA) gave support to Son Ngoc Thanh, who had returned from France and formed a right wing guerrilla force dedicated to Sihanouk's ouster, the

Khmer Serei (Free Khmer), which operated along Cambodia's borders.

Sihanouk resented American attempts to bully him into joining their side of the Cold War. In a conspicuous manner he accepted aid from China and openly endorsed the struggle of the Viet Minh against South Vietnam. He extended diplomatic recognition to the Soviet Union and the People's Republic of China in the late 1950s.

Despite these actions, Sihanouk's cooperation with regional communist powers was neither ideologically motivated nor unqualified. Sihanouk harshly suppressed the tiny communist movement inside Cambodia, and complained bitterly of Viet Minh intrusions onto Cambodian soil. But he recognized the fact that Vietnam and China were regional superpowers that would continue to exert their influence long after the Americans had gone home.

On the domestic front Sihanouk achieved an unbroken series of political victories in the years following independence. Because of constitutional restrictions on the monarch, Sihanouk's first step, in 1955, was to abdicate the throne in favor of his father. Later that year, he formed the Sangkum Reaster Niyum (People's Socialist Community), which absorbed nearly all opposition parties and won every seat in the first election of the new National Assembly. Sihanouk became prime minister. Most of his government appointees were politically and socially conservative, such as the new minister of defense, Lon Nol. Both the Khmer Rouge, who had formed a legal political party, Pracheachon (People's Group), to pursue their struggle within the electoral system, and the Democratic Party faded into the background. In 1960, Sihanouk's father died and Sihanouk assumed the new position of head of state, which he held for the next ten years. A new king was not selected, but neither was the monarchy abolished.

In the early 1960s, Sihanouk felt increasingly that his ties with the United States were doing him no good. He believed (correctly, as Shawcross later documented) that the United States was trying to undermine his authority. American aid programs were creating a military class dependent on the United States. And it seemed as if Cambodia would soon be faced with a unified communist Vietnam. Sihanouk renounced all U.S. aid in 1963. In 1965 he severed diplomatic relations.

In order to cope with the loss of American assistance, Sihanouk nationalized major sectors of the economy under a policy he called "Buddhist Socialism." Ultimately, however, he proved unable or unwilling to confront the country's established economic interests or the corruption involved in every important national project.

By 1966, when Sihanouk ensured the election of a more conservative National Assembly, the economic situation had become, in his own words, "alarming." American aid, which had been equivalent to sixteen percent of Cambodia's GNP, was gone. Foreign investment had plummeted. Nationalization was stifling private enterprise without providing the benefits of central control. Large portions of the main export commodity, rice, began to be sold to communist troops in Vietnam, depriving the state of tax revenue. Prices were rising, and the value of the currency was falling.

At the same time, Cambodia's foreign relations had become newly unstable. The Cultural Revolution in China was interfering with the realpolitik cooperation that had existed between the two countries. (Cambodia had been the first noncommunist country to receive aid from the People's Republic.) Cambodian territory was increasingly being used as a supply route and base for Vietnamese communist troops. This, in turn, brought cross-border raids by the United States and then secret American bombing of Cambodian territory.

Domestic and international problems gave rise to increasing discontent and polarization within Cambodia. The late 1960s saw a rapid succession of governments of the Right and the moderate Left.

Sihanouk was accumulating enemies along a broad spectrum of Cambodian society. The rising urban middle class resented his imperious and whimsical rule. The cessation of American assistance added material privation to indignation. The military budget was cut by one third, and civil servants could no longer count on bribes to supplement their meager incomes. A growing mass of high school and university graduates could not find jobs commensurate with their education. The commercial classes were upset by the restraints on private enterprise. Intellectuals chafed under censorship and suppression, many of them fleeing to France or, alternatively, into the jungle to join the Khmer Rouge. There was an increasingly common conviction—on both the Left and Right—that Sihanouk's influence on the country was malevolent.

Even the Prince's immense popularity with the Cambodian peasantry seemed threatened. In 1967 and 1968 peasant revolts erupted in Battambang Province. An alarmed Sihanouk publicly blamed China and authorized the army to quell the uprisings. This was done with ruthless efficiency.

In June 1969, Lon Nol returned to the government as head of a new "Government of National Salvation." Prominent in the new cabinet was Sirik Matak, Sihanouk's cousin and most outspoken critic. That fall, the government began dismantling what remained of Sihanouk's economic program. They also tried to vie with Sihanouk for control of the country's foreign policy.

In January 1970, Sihanouk went to France for a vacation. While he was gone, Matak lodged official protests and organized mass demonstrations against the use of Cambodia's eastern border region by North Vietnamese and Viet Cong troops. In what may have been a decisive error, Sihanouk, from France, threatened to remove Lon Nol and his associates from power for trying to destroy Cambodia's peaceful relations with the communist powers. On March 18, while Sihanouk was in Moscow seeking Soviet help in controlling the Vietnamese presence, Lon Nol and Sirik Matak staged a coup d'etat. They had the support of the Cambodian army and the American CIA. Sihanouk was removed as head of state and prohibited from returning to Cambodia.

Sihanouk proceeded to the next stop in his itinerary, Beijing. The response there was cautious. China and North Vietnam began trying to persuade the Lon Nol regime to continue Sihanouk's policy of acquiescing in the use of Cambodian territory for the war effort against South Vietnam. In effect, they offered to recognize Lon Nol's government in exchange for its cooperation.

The new government in Phnom Penh might have insured its survival had it made this concession. However, the men in Phnom Penh were flush with a new sense of freedom and power. They rejected Chinese entreaties, even repeating a meaningless demand that Vietnamese communist troops be withdrawn within forty-eight hours.

In Beijing Sihanouk decided to join forces with the miniscule Khmer Rouge, led by Pol Pot, to overthrow the Lon Nol regime. He announced the formation of the National United Front of Kampuchea (NUFK), and a government-in-exile, the Royal Government of National Union of Kampuchea, serving as titular head of both. Having failed to receive cooperation from the new government in Phnom Penh, China and North Vietnam decided that their interests were best served by actively backing a revolutionary movement, NUFK, in Cambodia.

This was a major turning point in Cambodian history. Not only was Sihanouk deposed, but the Sihanouk state was destroyed. The careful balance among internal interests which Sihanouk had maintained for more than fifteen years gave way to a civil war between Cambodians who had vastly different plans for their country. At the same time, the exclusion of international rivalries was replaced by the intense involvement of other countries which pursued their goals using Cambodian lives. In the span of ten years the Khmer people would experience four radically different systems of government: a constitutional monarchy, a right wing military dictatorship, a totalitarian communist state, and an occupation regime. The next three chapters of Cambodia's history were to be sad ones.

The Khmer Republic and Civil War

The Lon Nol regime came to power at a time when authoritarian regimes backed by the United States in the Third World seemed destined to give way to leftist revolutions. To many, the demise of the Lon Nol government seemed to be an inevitable occurrence, the details of which were merely local variations on a global theme. However, as similar regimes have reformed and stabilized in Southeast Asia and elsewhere, the particular events in Cambodia between 1970 and 1975 take on new significance. Also, it has become clear that the particular events of the Lon Nol years left indelible marks on the memories of many Cambodians and helped to shape the horrific regime that followed.

The coup d'etat against Sihanouk was met by enthusiasm in the cities and discontent in the countryside. Numerous sources reported that the urban populations of Cambodia were happy to be rid of Sihanouk, the one-man rule, the censorship of speech, the corruption of his court, and above all, his constant presence, in the press, the arts—seemingly everywhere. In the countryside, however, Sihanouk was still regarded as the God-King, the protector of the nation and the faith, and the defender of the rural "poor folk" against the powerful city dwellers. Demonstrations demanding Sihanouk's return occurred in several rural areas, resulting in the deaths of Lon Nol's brother and a few local government officials, but they were easily suppressed by the army.

The first months of the new regime were turbulent. Hundreds, perhaps thousands, of innocent Vietnamese living in Cambodia were killed by mobs incited by government anti-Vietnamese propaganda. More than 200,000 ethnic Vietnamese are estimated to have fled Cambodia by August. Tens of thousands of young Khmer men enlisted in the army to drive out the 60,000 or so Vietnamese communist troops encamped in Cambodia's eastern provinces.

But the confrontation with North Vietnam was extremely ill-considered. The poorly equipped Cambodian army was hardly prepared to battle experienced Vietnamese units. Within weeks of the coup, the Lon Nol government had lost major portions of Cambodian territory.

Saving the endangered Lon Nol regime was one of U.S. President Nixon's motivations for ordering the invasion of Cambodia on April 30, 1970. The primary objectives of the incursion, however, had nothing to do with Cambodia. Nixon hoped to destroy the communist military headquarters for South Vietnam, thought to be located inside Cambodia, and to rout the Vietnamese sanctuaries so that the withdrawal of American troops from Vietnam could proceed without threatening the stability of the Saigon regime.

These objectives were partially met, but there were severe, unintended consequences. The invasion caused an uproar in the United States, leading to the Kent State deaths and spurring the revival of the antiwar movement. The introduction of 31,000 American and 43,000 South Vietnamese troops into Cambodia also stripped the neutralist veneer from the Lon Nol regime. (The United States took Lon Nol's approval for granted and did not inform him of the invasion until after it had begun.) China, until then restrained in its support for Sihanouk, gave its unqualified backing to the resistance effort.

Invading South Vietnamese troops soon antagonized the local population, "indulging in a frenzy of raping and looting in the Cambodian villages they overran."[4] By contrast, North Vietnamese troops in Cambodia were known for their discipline and decorum, indeed "exemplary" behavior.[5]

Although temporarily stabilizing the military situation, in the long run the invasion was a disaster for the Lon Nol regime. North Vietnamese troops were driven into the center of Cambodia, which proved to be the critical factor in the development of an indigenous Cambodian insurgency. Shielded from the Lon Nol army by the experienced Vietnamese troops, the Khmer Rouge safely organized administrative and military structures. In the name of NUFK, and using taped exhortations from Sihanouk, they recruited from the Cambodian peasantry. The Khmer Rouge had an estimated 800 men under arms in early 1970; 12,000 at the end of that year; 18,000 in 1971; 35,000 in 1972; and 40,000 by 1973, when the Vietnamese withdrew and the Cambodians began fighting entirely on their own.[6] As scholar Laura Summers put it,

> It was not by any means historically necessary that Cambodia should experience widespread revolt or revolution. Although there were conflicts in the precoup period, no national uprising on the scale observed since 1971 was foreseen. The Cambodian liberation forces seem to have been created by the military logic of foreign intervention.[7]

American troops withdrew from Cambodia two months after they invaded. However, North and South Vietnamese battalions remained for two years, turning Cambodia into a major battleground of the Vietnam war. The impact of the war was devastating. The population of Phnom Penh doubled to 1,200,000 people within a few months of the coup. Within a year the capital would be home to more than 1 million refugees who had fled the fighting. Most of the roads and railways were impassable. By November 1970, half the country's schools were closed.

The economy went haywire. Commodity prices doubled in the first year of the war. The official unemployment rate rose from twenty to forty percent during the same period. Rubber exports, which accounted for nearly half of all export earnings in 1969, virtually ceased after April 1970. By November, the five largest rubber estates stopped functioning, not to resume production again during the war. Rice output dropped from 3.8 million tons in 1969 to 2.7 in 1970. By 1971, Cambodia, formerly a major rice exporter, was importing rice. The fiscal and trade deficits were only partially offset by large grants of U.S. military and nonmilitary aid, which continued until nearly the end of the war.

The American response to postcoup Cambodia was confused. From the beginning of the war, Nixon was under pressure from the Congress and the antiwar movement not to involve the United States in Cambodia, lest the United States become committed to another indefensible ally in Southeast Asia. Nixon himself reflected this sentiment in his declaration of the ''Nixon Doctrine,'' which called for U.S. support for friendly third-world governments without direct military intervention and the concomitant commitment of American prestige. However, this was not an easy policy to put into practice. The immensity of U.S. military assistance to the Lon Nol regime led most people in Phnom Penh to assume that the United States would do whatever was necessary to defeat the communist adversary. The feeble Cambodian elite was overwhelmed by the U.S. presence, and a psychology of dependence set in.

The attempt at state-building in Phnom Penh became one of the spectacles of the war. In October 1970, the two-thousand-year-old monarchy was abolished and the Khmer Republic was established. A constitution was finally adopted in 1972 and legislative and presidential elections were held that year. In practice,

Lon Nol, a sickly man who frequently sought the counsel of astrologers, held the reigns of power until nearly the end of the war. The elections were rigged. The American embassy was afraid that Lon Nol's departure might bring chaos and/or collapse. There were innumerable cabinet shake-ups, a dizzying game of musical chairs involving every important Cambodian political leader. (One of the players was Son Ngoc Thanh, who ended his armed resistance and returned to Phnom Penh as a consequence of Sihanouk's overthrow.) The government was continually immobilized by crises that revealed its weaknesses without prompting any significant reform. To a large degree, the country was run by Lon Nol, his brother, several of their cohorts, and the American embassy.

The most important military engagement of the war, Chenla II, occurred in the latter half of 1971, when the Khmer Republic undertook its last major offensive against the Khmer Rouge. The campaign was a disaster; the army's overextended lines were broken by communist troops. For the remainder of the war, the fighting was a slow grind. NUFK forces employed a Maoist guerrilla strategy of consolidating rural areas, cutting off roads, and eventually surrounding the capital. Chenla II also revealed to those living in areas controlled by the Khmer Republic the new nature of the war: North Vietnamese troops had largely been replaced by a Cambodian insurgency. Now, for the first time, it was Khmers killing Khmers.

The defeat of Chenla II greatly discredited the government. The enthusiasm which had originally greeted the Lon Nol regime was dissipated. Enlistment in the army, which had grown from 30,000 in Sihanouk's day to 200,000 under the new regime, ground to a halt. Students who had applauded the regime's "republicanism" in January 1971 were enraged by its "authoritarianism" one year later. There were large demonstrations against the government. The government's critics charged that it offered "Sihanoukism Without Sihanouk," referring to its inefficiency and corruption. Open nostalgia for Sihanouk, a sentiment whose expression would have resulted in a jail sentence only a year earlier, was common in 1972. Only two years after Sihanouk was overthrown, there was great pessimism about the future of the regime.

Corruption was endemic. Whole companies of phantom soldiers were created by top military officers who would then pocket their salaries. A significant proportion of the arms used by the Khmer Rouge were purchased from the armed forces of the Khmer Republic. The Nixon and Ford administrations chose to ignore the corruption, fearing that giving voice to the problem would make Congress and the country less sympathetic to further aid. In any case, there was doubt about what could be done, since many of the most corrupt officers were personal friends of Lon Nol. There were long stretches when those who were actually fighting went without pay or provisions. In Cambodia, unlike in South Vietnam, U.S. aid was not sufficient for both the corruption and the war effort.

The Khmer Rouge would probably have defeated the Khmer Republic in 1973 if the United States had not been freed to allocate enormous resources to

Cambodia that year. The Paris Peace Agreements, finalized in January 1973 by U.S. National Security Advisor Henry Kissinger and Vietnamese communist party politburo member Le Duc Tho, ended direct American involvement in Vietnam. No agreements were reached on Cambodia because the Khmer Rouge refused to negotiate, believing that a compromise would rob them of their impending victory. Prohibited from further bombing in Vietnam, the United States diverted the bombers to Cambodia, which was now "the only game in town," as William Colby, second in command at the CIA, put it.[8] American bombers pounded away at the Cambodian countryside for eight months in search of the Khmer Rouge. In all, 539,129 tons of bombs were dropped on Cambodia, more than three times the amount of explosives that had been dropped on Japan during World War II.

The destruction was unimaginable, and the effects still unknown. William Shawcross has said that even after the nightmarish Khmer Rouge period, some of the peasants he interviewed spoke of the U.S. bombing seven years earlier as their most memorable and frightening experience. The bombers hit wedding parties and funerals, rice fields and water buffaloes, villages, hospitals and monasteries. Of course, the bombing did as much to alienate the population from the Khmer Republic as to kill the Khmer Rouge. American journalist Richard Dudman reported that

> The bombing and shooting was radicalizing the people of rural Cambodia and was turning the countryside into a massive, dedicated, and effective revolutionary base. . . . American shells and bombs are proving to the Cambodians beyond any doubt that the United States is waging an unprovoked colonialist war against the Cambodian people.[9]

In recent years it has been said that the bombing created a crazed and angry psyche among the Khmer Rouge, accounting in part for their later savagery. Whether or not this is true, scholar Michael Vickery expresses what must have been a view from behind Khmer Rouge lines.

> During the severe U.S. bombing of the first eight months of 1973, there was no reaction in Phnom Penh other than relief and it must have seemed to NUFK that their urban compatriots were quite willing to see the entire countryside destroyed as long as they could enjoy a parasitical existence as U.S. clients.[10]

The U.S. Congress ended the bombing on August 15, 1973.

The year 1973 was important in other ways as well. The Khmer Rouge had diligently recruited among the peasantry for two years, and their forces now numbered an estimated 40–50,000 men. The following account, given by a Khmer Rouge cadre to scholar Serge Thion while he was behind Khmer Rouge lines, provides a flavor of their rural organizing during the war:

> "We sent out many invitations and we expected about 200 [Buddhist] Superiors. Three hundred and fifty came and with those accompanying them there were 1500 people who had to be worthily fed, decently lodged with respect for the hierarchy and, above all, protected and camouflaged for many days. An exhausting task, and I don't know how we managed it. But in the end the Congress was an enormous political success. That is to be expected. They wanted to know what we were worth. Now they are firmly with us."[11]

With a halt to the American bombing in sight, and victory only a matter of time, the Khmer Rouge started moving against their "enemies" both external and internal. Sometime during 1973, the Khmer Rouge leadership defined Vietnam as the "long-term acute enemy." Already operating with minimal Vietnamese support, the Khmer Rouge started expelling the remaining communist Vietnamese troops and advisors. On several occasions, the Khmer Rouge attacked communist Vietnamese units. The Khmer Rouge also moved to rid the anti-Lon Nol front, NUFK, of its Sihanoukist elements. Where before the Prince's voice was used to exhort peasants, the propaganda was now aimed at discrediting him and radicalizing the insurgency. Sihanouk loyalists were removed from positions of authority and often killed. Finally, internal Khmer Rouge purges were underway, with the Pol Pot leadership assassinating Cambodian communists thought to be sympathetic to Vietnam.

It must be emphasized, however, how obscure these events were at the time—and still are—to anyone outside inner communist circles. The man thought to be the leader of the communist movement (Tou Samouth) had actually been dead for a dozen years. Three men long presumed dead (Khieu Samphan, Hu Nim, Hou Yuon) were in fact key figures in the communist party. Until 1975, some scholars persisted in referring to the insurgents as Sihanoukists, when in reality Sihanouk's power in the movement was just sufficient to prevent his own execution. The Front's political program was vaguely progressive, and, as became clear later, thoroughly misleading.

Nineteen seventy-three was also the year in which the Khmer Rouge made their first major attempts to reshape Cambodian life, especially in rural areas. In sections of the country under their control, the Khmer Rouge began implementing a "Democratic Revolution," a term probably taken from Mao which indicates the stage in the revolutionary consecution in which the broad popular front against imperialism is abandoned and struggle begins against internal economic classes that are antagonistic to the revolution.[12] With variation according to which faction was in control in a particular area, the Khmer Rouge abolished private landownership, confiscated private property, outlawed trade, demonetized the economy, collectivized dining, separated children from parents, disallowed religious practices and holidays, prohibited jewelry and colorful dress, imposed harsh communal work schedules, and uprooted and relocated some villages. Worse than any of the rules was the severity of the enforcement: complaints or transgressions could bring death.

An irony of the 1970–75 period is that one of the reasons for the relative popularity of the resistance forces was their conservative appeal. Making extensive use of radio appeals by Prince Sihanouk, NUFK reminded peasants that the disruption of their lives was caused by "non-Khmer" city dwellers, their South Vietnamese friends, and American bombs. These appeals to tradition became hollow, however, when Khmer Rouge ideology came into play.

> In early 1973 when the KK [Khmer Rouge] entered the new harsh phase of their campaign in which all rules were strictly enforced and unpopular programs were carried out, with stiff penalties for noncompliance, almost all popular feeling turned against them.[13]

In the early part of 1973, thousands of Cambodians had fled into Khmer Republic-controlled areas to escape American bombs. From the latter part of 1973 onward, thousands of Cambodians fled to escape Khmer Rouge policies.

The Khmer Republic was expected to fall soon after the bombing halt, but the war continued for another twenty months. In part this was because Khmer Rouge forces lacked sufficient supplies, a consequence of disputes with North Vietnam. But the question for these months was not who would win, but how and when. The leaders of the Khmer Republic did not offer to negotiate with the Khmer Rouge until they were so weak that they had nothing to offer. Sihanouk and the Chinese leadership were eager for negotiations (so as to maximize their influence in the eventual peace relative to the unpredictable Khmer Rouge), but the United States did not respond. William Shawcross has explained this as a kind of criminal indifference on the part of then Secretary of State Henry Kissinger, one of many for whom Cambodia was a "sideshow" to Vietnam. Michael Vickery has a different, though equally disturbing, view.

> A major mystery over the last two years of the war was what U.S. goals really were in Cambodia. In spite of repeated protests by their own subordinates and responsible Cambodians, the American leadership insisted on preserving Lon Nol, even when it was certain that a change of regime offered some possibility of renewing morale and turning the war around. Of course, turning the war around could only

> have meant a temporary advantage, and perhaps a position strong enough to force the other side to compromise, for any leadership capable of rallying strong support would have seen that the ultimate solution lay in a coalition between the best elements of Phnom Penh and NUFK and the establishment of a moderate socialist regime. This was probably the last thing the U.S. government wanted.[14]

As the Khmer Republic lost more and more territory (Lon Nol was jokingly referred to as the Mayor of Phnom Penh), its international standing declined. The Sihanouk/Khmer Rouge government-in-exile was recognized by 22 countries in 1971, 28 in 1972, 44 in 1973, and 62 in 1974. The Khmer Republic was just barely able to hold onto its seat at the United Nations in credentials votes at the end of 1973 and 1974. One of the oddities of the period was the stance of the Soviet Union, which maintained relations with the Khmer Republic until the end of the war, reportedly even supplying the regime with trucks. NUFK issued vitriolic attacks on the Soviets in Sihanouk's name.

The Khmer Republic became, by the end of the war, a sickly dependent of the United States. American aid in 1974 exceeded the total Cambodian national budget for 1969. Two percent of the Khmer Republic's 1974 income came from domestic sources. Rice production was at one-quarter of prewar levels. Prices had risen five thousand percent in five years. Over the course of the war, the United States had provided the Khmer Republic with $1.18 billion in military aid and $503 million in humanitarian aid. The United States had spent $7 billion bombing the country. Cut off from most of the countryside, the Khmer Republic survived for several weeks near the end only because of a U.S. airlift of food.[15]

On the first of January 1975, the Khmer Rouge launched their final offensive. Newly rearmed, they faced a Khmer Republic which had lost its primary backer, Richard Nixon, and was out of favor with the U.S. Congress. The Khmer Republic ran out of money. Late in January, 70,000 hardened Khmer Rouge troops cut the Mekong and isolated Phnom Penh. The last fighting was ferocious, as the Khmer Republic finally rallied on its own behalf. But the effort came too late. The capital fell on April 17, 1975.

As the war ended virtually no one fled to the United States, as would many Vietnamese before the fall of Saigon a few weeks later. Cambodians believed that their traditional peaceableness would prevail.

Some of the effects of the war can be calculated. Between 500,000 and 1 million people were killed. Approximately 3,400,000 had become refugees. The population of the capital had grown from 600,000 to 3 million. Of fourteen hundred rice mills, eleven hundred were destroyed. Seventy-five percent of the draft animals were killed. And so on. By any measure, the result was horrifying.

The Rise of Communism in Cambodia, 1930–1973

On April 17, 1975 the Khmer Rouge entered Phnom Penh and created a state of extremes: xenophobia, chauvinism, autarky, oppression. It was not until the Vietnamese routed Pol Pot nearly four years later that a coherent, if incomplete, picture of the regime and the previous development of Cambodian communism began to emerge. Before 1970, the Khmer Rouge were insignificant by almost any measure. But the history of Cambodian communism assumes importance because of Pol Pot's ultimate victory; if the Khmer Rouge had never prevailed, or if Pol Pot had never consolidated power, then many of the episodes described here would have remained in blessed obscurity.

The two most important structures for the development of radical activity in Cambodia were both established in 1930. The Buddhist Institute in Phnom Penh became one of the centers of organized anticolonial activity. Its ranks included Son Ngoc Thanh, who later served briefly as prime minister, and Son Ngoc Minh and Tou Samouth, later key figures in the communist party. Also in 1930, the Indochina Communist Party (ICP) was formed in Hong Kong under the leadership of Ho Chi Minh. The first ICP cell in Cambodia was established in the French rubber plantations of the southwest. All of the cell's founding members were ethnic Vietnamese. Revolutionary activity in Cambodia was not significant until the 1940s.

During World War II there were several demonstrations, encouraged by the Japanese, against French colonial rule. Armed Khmer Issarak (Khmer Independence) groups sprung up throughout the countryside. Several of the leaders of these groups joined the ICP.

While the postwar political order was taking shape in Phnom Penh, armed resistance to the colonial administration gained momentum in the countryside. In 1946 a Khmer Issarak group took and briefly held the provincial capital of Siem Reap. French colonial records indicate a growing concern, and document a sizeable resistance: they reported 500 Khmer Issarak casualties in May 1947 alone.

After repeated efforts to consolidate the fractured resistance movement, the United Issarak Front was formed under the leadership of ICP member Son Ngoc Minh in 1948. But the return from exile in 1951 of Son Ngoc Thanh, who refused to join the communist-influenced front, and the defection of several key figures to the Phnom Penh government, weakened the already loose coalition.

That same year, for their own tactical reasons, the Vietnamese disbanded the ICP and regrouped as the Vietnam Worker's Party, leaving the Cambodian and Lao communists to form their own small parties. On September 30, 1951, the Khmer People's Revolutionary Party (KPRP) held its inaugural meeting. Its politburo was dominated by "Khmer Viet Minh," so-called because of their close association with the Vietnamese communists.

By 1953, the United Issarak Front was experiencing a resurgence. They had a guerrilla force of some 3,500 and controlled large portions of rural Cambodia. Partly because the Khmer Issarak prevented them from subduing the countryside, the French granted independence to Cambodia that year. Importantly, though, Cambodia achieved independence before its anticolonial movement controlled the country or became committed to the goals of social revolution.

The 1954 Geneva Agreements set the stage for the virtual collapse of the communist movement in Cambodia. Because Cambodia had already achieved formal independence, the Issarak movement and the Cambodian communists were denied official representation at the Geneva Conference, despite entreaties by Ho Chi Minh's delegation, which then spoke on behalf of these groups. China and the Soviet Union pressured North Vietnam to agree that Cambodia's armed insurgency would surrender their weapons to the Royal Government of Cambodia and demobilize. Following the Conference, most of the Cambodian insurgency gave up revolutionary activity and reintegrated with mainstream Cambodian society. A few others went into hiding in Cambodia to continue the work of the KPRP. Between one and two thousand insurgents, including most of the members of the KPRP, resettled in Hanoi. They were afraid or unwilling to remain in Sihanouk's Cambodia. They stayed in Vietnam for sixteen years.

It was about this time that a group of Cambodian students, including Pol Pot, Ieng Sary and Hu Nim, returned from France, where many of them had become members of the French Communist Party. The students joined forces with the few members of the KPRP who had remained in Cambodia after the Geneva Conference.

Communist political activity took two forms. Some of the KPRP members in Cambodia formed a legal political party, Pracheachon (People's Group), to participate in the 1955 general elections. Others went underground to organize. The entire KPRP politburo remained in Hanoi, except for Tou Samouth and Sieu Heng, who had returned to Cambodia to lead the urban and rural committees, respectively.

The late 1950s was a bleak period for the Cambodian communists, and revolutionary activity was minimal. The young king turned head-of-government was immensely popular, particularly among the section of the population that would be most receptive to a radical message, the rural poor. The standard of living of the peasantry was improving, and Sihanouk's foreign policy was congruent with the goals of the Left. The country was at peace for the first time in memory. Sieu Heng, secretary-general of the KPRP, openly defected to the Sihanouk government in 1959 and revealed that he had been subverting the Party from within since 1955. By the time of his defection, the rural organization under his command had lost most of its membership. Finally, Sihanouk launched a concerted offensive against the political Left in 1959.

The KPRP's second General Congress was held in Phnom Penh in 1960. Pol Pot was elected to the politburo, all of whose other members, except for secretary-general Tou Samouth, were still in Hanoi. At this point, the KPRP also changed its name to the Worker's Party of Kampuchea, signaling that the Party now led a class struggle rather than an anticolonial revolution.

The direction of the communist movement began to shift. The communist leadership in Vietnam, and the Cambodian communists who accepted guidance from them, had led the Cambodian communist movement to a state of near ruin. Nor were they going to change direction. They believed that the revolution in Vietnam took priority, and that once that was successful, a revolution in Cambodia would follow naturally. As long as Prince Sihanouk maintained his neutrality, and because he in fact provided minor assistance to the revolutionary movement in South Vietnam, the Vietnamese leadership did not want to antagonize him by supporting revolutionary activity in Cambodia. Pol Pot and his young allies in the Party saw this as selling out Cambodia to aid Vietnam. They went to work to takeover, rebuild, and redirect the communist party. Over the next ten years, they attempted, with little success, to organize a revolution against Sihanouk based primarily on the peasantry. The Party was so skeletal in 1960, and the direction it had taken so abhorrent to the Pol Pot group, that some later Party histories would identify the 1960 meeting as the Party's founding congress.

Tou Samouth was assassinated in 1962, perhaps by Pol Pot, who became acting secretary-general. Pot's position was made permanent at the third General Congress in 1963, and his allies joined the politburo. A few months later, to avoid government harassment, Pot fled to the rugged northeast of Cambodia, which became his base for the next seven or eight years. In 1966, the Party changed its name again, from the Worker's Party of Kampuchea to the Communist Party of Kampuchea, reflecting its increasingly radical posture. When Sihanouk, in one of his periodic shifts, moved to the Right in 1966–67, some of the prominent leftists in the government, including Khieu Samphan, Hu Nim, and Hou Yuon, fled and joined forces with the Khmer Rouge in the countryside.

In 1970, following Sihanouk's overthrow, many of the Cambodian cadres who had gone to Hanoi in 1954 returned to Cambodia to participate in the revolution against the Lon Nol regime. Pol Pot saw these "regroupees," as they came to be known, as traitors who, in the 1950s, had sacrificed the Cambodian revolution in order to benefit Vietnam. Of the 1,500 or so returned cadres, Pol Pot killed half by 1975 and most of the rest by 1978.

According to one analysis, three factions emerged within the communist movement after 1970.[16] Pol Pot's group, based in the northeast and north, was violently anti-Vietnamese and favored rapid social change once state power was attained. The veteran communist leadership, based in the east and the southwest, aligned itself with the Vietnamese communist movement and advocated less chauvinistic and radical policies. Better living conditions for the population in the areas they controlled reflected their willingness to accept a slower rate of social and economic change. A third group drew its ideological and perhaps some material support from the Cultural Revolution leadership in China. Pol Pot triumphed over all intraparty opposition, but only at great human cost, and not fully until 1978.

Democratic Kampuchea

Democratic Kampuchea—the name the Khmer Rouge chose for Cambodia after they took power in 1975 and which has become synonymous with their regime—was probably the most radical effort at social transformation ever undertaken by national leaders.[17] Beginning with the evacuation of Phnom Penh, the Khmer Rouge directed an orgy of iconoclasm and destruction even as they sought to rebuild the country. They were much more successful at the one than the other.

Within hours of their victory march into the capital, the Khmer Rouge began the forced exodus of its 3 million residents. Many reasons for this astounding event have been advanced. At the time, the Khmer Rouge told the population that evacuation was a temporary move, necessary because of the possibility of an American bombing attack. Later, they claimed humanitarian motives. The only way to avoid massive famine in a city cut off from food, they suggested, was to take people to where the food was. Perhaps the Khmer Rouge were desperate to begin the task of economic reconstruction, which they believed should be based almost entirely on agriculture. The evacuees were, indeed, immediately herded into the countryside to begin farm work. But the primary motivation of the Khmer Rouge appears to have been political. In a 1977 interview with Chinese reporters Pol Pot said, "The decision to evacuate was made two months ahead because we knew that until the smashing of all sorts of enemy organizations, our strength was not adequate to defend the revolutionary regime."

The dismantling of Cambodian society was seen by the Khmer Rouge as an integral part of the revolutionary process. This consisted of many things. Most of the population was relocated, not only the people in the capital, but those in other cities and towns, and in many villages as well. The Khmer Rouge established social categories that determined not just the quality of one's life, but often if one lived at all. They directed most aspects of daily life, from marriage to eating. They defrocked Buddhist monks and forced them into hard labor, and destroyed temples or converted them to other uses. They abolished financial institutions, all currency, and the most trivial forms of private property. Even barter was forbidden. They reformed the language and closed most schools. They sought to create a rural utopia, "one big work camp," as Pol Pot once boasted. At the individual level they sought nothing less than the complete transformation of Cambodian character: "life-style meetings" were held to explain the proper mode of "drinking, sleeping, walking, and talking."

Khmer Rouge analyses found nearly twenty different economic classes in the Cambodian population. In theory, the economic class designation combined with a political attitude rating to place each individual within one of three political categories: full rights, candidate, depositee. The Marxist underpinning of Khmer Rouge ideology was tenuous. While they had a maniacal animus toward "private property" and a Marxist vocabulary ("capitalist" vs. "proletariat"), their use of these terms seemed largely metaphorical and without clear referent. The Khmer Rouge never translated a single text of Marx, Lenin, or Mao. The existence of the communist party and the composition of its leadership were not even publicly revealed by the regime until September 1977. David Chandler, in his article here, "A Revolution in Full Spate: Communist Party Policy in Democratic Kampuchea, December 1976," analyses an internal Khmer Rouge document and considers this obsession with political enemies and secrecy.

If the Khmer Rouge's Marxist distinctions were somewhat academic, the social divisions that came into use were not. In practice, most people thought of themselves as belonging to one of two categories. "Base People" were those living in areas under Khmer Rouge control in April 1975, and politically suspect "New People" were those residing in Phnom Penh when it was taken by Khmer Rouge forces. This was not, however, a simple rural-urban dichotomy. Half of the population in Phnom Penh in April 1975 were peasants—refugees from the war, the American bombing, and Khmer Rouge policies. Despite their "correct class origin," they too were seen by the Khmer Rouge as politically tainted. The treatment individuals received once they had been relocated in the countryside depended in large measure on their classification. While overgeneralizing, it can fairly be said that new people were "last on distribution lists, first on execution lists, and had no political rights." Approximately half of the population fell in each category.[18]

Other kinds of groups suffered disproportionately under Democratic Kampuchean rule. The harsh routine of physical labor that the Khmer Rouge demanded took a particularly high toll on intellectuals, professionals, other people from urban backgrounds, including most of the ethnic Chinese, and the defrocked Buddhist clergy. Members of the Cham ethnic minority were often killed when they resisted "Khmerization," the Khmer Rouge's program of forced cultural assimilation. While there is no evidence that Pol Pot singled out these groups for collective extermination, they suffered so enormously under Democratic Kampuchean rule that they might as well have been the object of a pogrom. David Hawk, in his article "International Human Rights Law and Democratic Kampuchea," marshalls the available data on the persecution of these groups and the treatment of the Cambodian population as a whole and presents it within the framework of international human rights law. He points out that Pol Pot and his colleagues could still be brought to trial today in an international court of law on charges of genocide.

Invoking the great Angkor civilization in a chauvinistic appeal to rebuild the country, the Khmer Rouge instituted a massive system of collectives and work camps. The primary emphasis, as it had been five hundred years earlier, was on rice: "With rice we can have everything!" The conditions in the countryside varied greatly at the outset, becoming more uniform as Pol Pot consolidated power. Even from the beginning, however, there was a general disregard for the physical well-being of most toilers: inadequate food, nonexistent or primitive health care, many hours and days of labor with little rest. The majority of those who died in Democratic Kampuchea—probably between 1.2 and 2 million people—fell victim to some combination of exhaustion, starvation, and disease. Though rice production improved from 1975 to 1978, exporting and stockpiling meant that less was distributed to the people at the end of this period than at the beginning. The psychological scars from the murderous atmosphere and the labor camp conditions are the subject of David Kinzie's chapter, "The 'Concentration Camp Syndrome' Among Cambodian Refugees."

Democratic Kampuchea never attained the uniformity suggested by Pol Pot's image of a work camp. Until they emerged victorious from internecine struggle in mid-1978, Pot and his immediate coterie of lieutenants constituted but one of several factions within the Khmer Rouge. Regional differences in food distribution, work discipline, the quantity of executions, and administrative structure reflected these divisions within the Khmer Rouge.

Pol Pot saw enemies everywhere—not only within the communist party, but also among the professional and educated elite, ethnic and religious minorities, and urban dwellers. The nearly four years of Democratic Kampuchean rule is in large part the story of his effort to eliminate perceived and real sources of opposition.

Pol Pot was actually ousted as prime minister in September 1976, only to be reinstated in late October, perhaps a reflection of power struggles among the leadership in China. In 1977 he crushed a coup attempt in Prey Veng. The final and most serious challenge to his leadership came from the Eastern Zone, where the old communist vanguard, still sympathetic to Vietnam, led a rebellion in mid-1978. When this uprising had failed and most of its leaders were executed, Pol Pot no longer faced any threatening pockets of resistance within the Khmer Rouge. By that time, his troubles were across the border in Vietnam.

There are three salient features of Democratic Kampuchean foreign policy: (1) close cooperation with the People's Republic of China; (2) hostility toward, and eventually full-scale war with, Vietnam; (3) isolation from the rest of the world. Despite its self-imposed autarky, by the end of 1978, Cambodia was at the center of an enormously complicated web of regional and international antagonisms.

The conflict in Indochina from 1975 to 1978 is difficult to comprehend because, though the evidence suggests that Cambodia was aggressive and belligerent toward Vietnam, it seems ridiculous and implausible that the leaders of a weary country with 6 million people would confront a neighboring country with 50 million people and one of the largest armies in the world. However, this characterization, supported by virtually all accounts, begins to seem possible when we remember that the internal policies of the Pol Pot government are equally incredible. Indeed the evidence suggests that China and Vietnam, now ferocious enemies, would have been able to continue managing their own differences without a full-scale confrontation had it not been for the demands that Cambodia made on both countries.

As soon as the Khmer Rouge government took power in April 1975, Khmer Rouge soldiers moved to occupy territory historically claimed by both Cambodia and Vietnam. North Vietnam defeated the Saigon regime the next month, and North Vietnamese troops attempted to move into all the territories claimed by Vietnam. A series of clashes between Cambodia and Vietnam ensued. That summer, high level talks between the two countries resulted in a cooling of hostilities and an agreement to procede with discussions to resolve conflicting border claims. The border was quiet for the remainder of 1975 and during 1976.

In May 1976 Democratic Kampuchea and Vietnam held their first and last talks on the border dispute. Both sides agreed that the Brevie line, drawn by the French governor-general in 1939, was the legitimate delineation of the border. In contention was where exactly the Brevie line fell on the ground and whether it should govern offshore navigational rights as well. At stake was no more than a few hundred kilometers of territory. Both countries occupied territory that was jointly claimed. The talks broke down because there was an irreconcilable conflict over how the differing claims could be resolved.

The Democratic Kampuchea regime insisted that the conflicting claims were "nonnegotiable" and could only be decided unilaterally by their government. The basis for this claim was two letters sent by North Vietnam and the Viet Cong to Prince Sihanouk in 1967 which recognized Cambodia's "existing borders." Sihanouk publicly claimed that the letters signified Vietnam's acceptance of the long-held Cambodian position that any future disputes over the border were nonnegotiable, that is, it could only be resolved in favor of the position held by the Cambodian government. As the Cambodians saw it, they would only renounce their historical claims to Kampuchea Krom (lower Kampuchea)—once part of the Khmer empire, now the bulk of southern Vietnam, including Ho Chi Minh City—in exchange for a guarantee that there could be no further Vietnamese erosion of Cambodian territory. The North Vietnamese and Viet Cong publicly implied that they agreed with Sihanouk's interpretations of their letters.[19] In 1976, the Vietnamese rejected this position and wanted to conduct negotiations along more conventional lines, resolving disputes to the satisfaction of both sides. The Cambodian government suspended the talks.

In January 1977 there was a new series of major clashes on both the Cambodian-Thai and Cambodian-Vietnamese borders. The hostilities with Thailand tapered off by the fall of that year, apparently because the Pol Pot government decided it could not handle two combatants at the same time. The appearance of the more belligerent Cambodian posture in January 1977 may have been a consequence of Pol Pot's consolidation of power at the expense of other, less vigilantly nationalistic and anti-Vietnamese, factions within the Khmer Rouge.

Starting in January 1977, Democratic Kampuchea tried to force Vietnamese troops out of tracts of land that they considered to be Cambodian territory. In April, in order to convince Vietnam that it would not tolerate continued Vietnamese occupation of these lands, Democratic Kampuchea began to launch military forays into territory that was indisputably Vietnamese. That summer, Vietnam retaliated with forays into Cambodian territory. In September, Democratic Kampuchea launched a major counterattack into Vietnam. In December, Vietnam sent 60,000 men into Cambodia's eastern provinces in an invasion apparently aimed at subverting Khmer Rouge intransigence. This aim was not achieved at all, indeed Democratic Kampuchea expelled the Vietnamese diplomatic mission from Phnom Penh and suspended relations on January 1, 1978.

In January, Cambodia launched a successful counteroffensive against Vietnam. Vietnam brought foreign journalists to record the aftermath of Kampuchean atrocities against the civilian populations of several villages in Vietnamese territory. Their reports were treated skeptically by many until Cambodia was opened to the world ten months later and it became clear that the Khmer Rouge had behaved comparably in their own country.

With neither Vietnam nor Kampuchea foreseeing an easy victory, China tried to mediate between the two countries, even hinting that it might end its substantial aid to Cambodia if the leadership there was not more flexible in trying to settle the dispute. Democratic Kampuchea firmly rejected China's offers of mediation, claimed to be the sole aggrieved party, and dismissed the threats as an affront to their national sovereignty.

In February 1978, Vietnam made their last serious offer to resolve the conflict with Cambodia. They proposed an end to hostilities, troop withdrawal from both sides of the border, and international supervision of any settlements. China put forth a proposal similar to Vietnam's.[20] Democratic Kampuchea flatly rejected the offers. The Pol Pot government reiterated the position it had held since it reinaugurated hostilities in early 1977. Democratic Kampuchea said that before it would participate in any talks, Vietnam would have to withdraw from all the territories that Kampuchea considered its own and accept Kampuchea's rendering of the land borders and navigational rights. Once seven months had passed without any fighting, Democratic Kampuchea would come back to the negotiating table. Vietnam found the terms unacceptable.

We know little about the thinking of the Democratic Kampuchean leadership as regards its relations with Vietnam. Vietnam analyst Douglas Pike offers a comparatively sympathetic characterization.

> In this context [of Cambodia's long history of territorial decline] then it can be seen that the current ''border'' war has deeper psychological implications, and is not simply a matter of arguments over boundaries. The genesis of the war may be found in questions of demarcation. Boundaries were only

> crudely fixed by the French and usually at Cambodian expense. But the war touches fundamental passions and fears that cannot be treated merely by adjudication that draws some new lines on a map.
>
> Also in this context it can be seen that Cambodia's hostile, if not aggressive, behavior toward Vietnam and Thailand is not entirely irrational. Cambodia has tried various means [over the centuries] to fend off its enemies. Nothing has worked well. What is left is irrational, or seemingly irrational, behavior. This might be called the bristly dog gambit. The rule—as it is for a small dog surrounded by bigger, stronger dogs—is to bristle, assuming an aggressive posture and appearing so fearfully troublesome, so indifferent to consequences, as to convince others to leave you well enough alone. It may not work, but it holds as much promise for the Cambodians as any other.[21]

Scholar Ben Kiernan also thinks that a rational calculation was an important part of Cambodian thinking. He suggests that the Pol Pot group kept the conflict with Vietnam active specifically for the purpose of discrediting and defeating internal communist party rivals who advocated a more accommodating policy toward Vietnam.[22] Other interpretations emphasize the racial chauvinism and millennialism of the Pol Pot leadership. Cambodian government propaganda included allusions to returning to the strength and grandeur of Angkor; Cambodians were frequently exhorted to recapture all of southern Vietnam for the Khmer nation. Every Cambodian was equivalent to and would kill thirty Vietnamese, Democratic Kampuchea radio repeatedly claimed. The "barbarians" would be defeated "even if this fight lasts seven hundred years or more."[23] Sheer paranoia and hysteria were also apparent at times. Deputy Prime Minister Ieng Sary, Pol Pot's brother-in-law and apparently the second most powerful Khmer Rouge official, claimed more than once that Vietnam, which had "a dark scheme to conquer our land and destroy the Khmer race," was collaborating with the Soviet Union and the United States in its attacks on Cambodia.

The thinking of the Vietnamese is only slightly more accessible. The Vietnamese were accused by Democratic Kampuchea of attempting to force Cambodia and Laos into an "Indochina Federation," which, according to the Kampucheans, would be just a vehicle for eventually subsuming these countries into an enlarged Vietnam. The Vietnamese claim that this was never their aim (though the term was used in the 1930s and 1940s). They have called instead for a "special relationship" between the three Indochinese countries, which they sometimes suggest means nothing more than the economic and political cooperation that would "naturally" follow friendly relations. (Even this term, however, was dropped, at least temporarily, by the Vietnamese because of the sinister connotations it held for the Pol Pot government.) It seems clear that the Vietnamese leadership feels that Vietnam is owed more from Laos and Cambodia than the minimally correct relations that are normally claimed by states to be sufficient for peaceful coexistance between neighboring countries. On the other hand, it also seems clear that when Cambodia was protesting so loudly against any cooperation with Vietnam, the Vietnamese leadership was willing to settle for merely correct relations rather than go to war. They were not willing to unilaterally concede all disputed territory, perhaps because they felt Democratic Kampuchea's insistence that they do so itself forebode future belligerency, perhaps simply because of their own national pride. In the event, it seems that the Kampuchean fear of Vietnamese domination seemed so real to them that they did everything possible to bring it about.

In February, when it became clear that the Pol Pot government could be neither persuaded nor forced to even begin negotiations to resolve the border dispute, Vietnam gave up hope of dealing with it. That month, the Vietnamese communist party central committee decided to provide support to Cambodian communists in Kampuchea's Eastern Zone for a revolt intended to overthrow Pol Pot. In May, however, Pol Pot preempted the insurrection and massacred most of the Eastern Zone cadres. Following this, in June, apparently, the Vietnamese leadership decided to use their own forces to overthrow the Pol Pot regime.[24]

Vietnam's relations with China were deteriorating at the same time as the final break with Cambodia. In the spring of 1978, thousands of ethnic Chinese were fleeing Vietnam for China and the high seas as a consequence of Vietnam's promulgation of several new harsh socialization measures. China took the plight of its expatriate nationals to heart, and accused Vietnam of "racial discrimination." China discontinued its aid to Vietnam, which had amounted to nearly 1 billion dollars a year.

But the decisive factor in the break between China and Vietnam was Cambodia. Unable to persuade Democratic Kampuchea to deescalate the conflict, China was forced to choose between backing and abandoning its closest ally in the world, Cambodia. It chose the former. China already had ten to twenty thousand advisors in Cambodia. It now increased its arms shipments and aid. Pressure was put on Vietnam's northern border to warn it against taking action against Cambodia.

Facing the prospect of a two-front war, Vietnam took three steps to consolidate its international position. In June 1978, Vietnam joined the Soviet economic union, COMECON (The Council for Mutual Economic Assistance), a step which Vietnam had been resisting for several years, and which, when it came, greatly antagonized China. In September, Vietnam formally dropped its demand for reconstruction aid as a necessary precondition to diplomatic normalization with the United States, though the United States now declined to establish relations for other reasons, and thus Vietnam's gesture was to no effect. Third, and most significantly, in November Vietnam signed a mutual security pact with the Soviet Union.

On December 25, 1978, Vietnam launched a full-scale invasion of Cambodia. Within three weeks, the Vietnamese army had taken Phnom Penh and secured all but the region bordering on Thailand. On January 10, Vietnam oversaw the proclamation of the People's Republic of Kampuchea (PRK) with Heng Samrin as head of state. The Khmer Rouge established bases in the mountainous jungle along the Thai border and immediately began armed resistance to the PRK.

Would Democratic Kampuchea have collapsed on its own if the Vietnamese invasion had not occurred? The answer may be so grim that it is rarely considered. But one must note that Pol Pot had eliminated his political enemies internally by the time he had provoked them internationally. Whether he could have continued to rule with an administration ravaged by his own paranoia over a population seething with discontent is something we will fortunately never know for certain.

The People's Republic of Kampuchea

Cambodia's history since the Vietnamese invasion can be fairly neatly divided into two periods, a turbulent time of population movement, food shortage, international relief, refugee creation, and political regrouping, and a much longer period—undoubtedly extending well into the future—of regime building, economic reconstruction, simmering guerrilla warfare, and international political stalemate. All the elements of the relatively static current period were in place by mid-1981.

Destruction, Chaos, Reconfiguration, 1979–1981

One hundred and twenty thousand Vietnamese troops began pouring into Cambodia on December 25, 1978. The Vietnamese leadership was surprised by the positive reception they received from the population, and by the rapid collapse of the Democratic Kampuchea (DK) army.* Because their troops moved faster than they had anticipated, Vietnam's supply lines were overextended and the advance slowed. The Vietnamese may have thus missed their best opportunity to completely eliminate the DK forces, which retreated into the hills bordering Thailand.

Even most of those who are highly critical of Vietnam agree that the Cambodian population was quite happy to see Vietnam invade. Cambodians will

---

*The Khmer Rouge, which had remained organizationally and formally unified for nearly fifty years, divided into two distinct groups after the 1978 Vietnamese invasion. Thus, it is least confusing, when referring to the communist movement after 1978, to identify its main branch, headed by Pol Pot and operating as an anti-PRK resistance movement, as the "Democratic Kampuchea forces," which is the name

commonly tell you that they hoped only for death to end their suffering under Pol Pot. In the fall of 1978, Vietnam prompted a group of Cambodians who had fled to Vietnam from Pol Pot's Cambodia to form an organization, the Kampuchean National United Front for National Salvation, which then "invited" Vietnamese troops to enter Cambodia to help them overthrow the Pol Pot regime. Salvation Front propaganda attempted to calm popular fears that the new communist regime would continue Democratic Kampuchean policies. They promised freedom and respect for religion, restoration of the family unit, an end to communal eating, the reopening of schools, and an end to the wanton brutality. Although the regime has largely kept its promises, Cambodians even today remain fearful of a return to earlier policies. Scholar Stephen Heder found that substantial support for the DK forces persisted among members of the poorest stratum of the peasantry, who had advanced socially and economically under Pol Pot. The greatest enthusiasm for the Vietnamese invasion, of course, came from former members of the bourgeoisie, who had suffered the most under Democratic Kampuchea.[25]

The international response to the invasion was very different. Sihanouk, who had been under house arrest for much of the DK period, was flown out of Phnom Penh to Beijing and then to New York. There he appealed to the United Nations to condemn the Vietnamese invasion. The Security Council passed a resolution to this effect, but it was vetoed by the Soviet Union. The noncommunist nations of ASEAN (the Association of South East Asian Nations: Indonesia, Malaysia, the Philippines, Singapore, and Thailand, now including Brunei) also quickly condemned the invasion, but in mild tones. Few people thought at the time that Vietnam would keep troops in Cambodia for an extended period. During the coming months, Australia, Denmark, England, Japan, Sweden, and other countries suspended, reduced, or cancelled their aid programs to Vietnam to protest the invasion and occupation.

China was ferocious in its criticism of Vietnam. And it went beyond criticism. On February 17, 1979, China sent 320,000 troops into northern Vietnam. The exact purpose of the invasion was unclear. China's leader, Deng Xiaoping, said the intent was to "teach Vietnam a lesson." Some have seen the invasion as an attempt to restore Chinese prestige after the rout of their ally in Cambodia, as a warning to Vietnam not to disregard China's interests in the future, and as an effort to persuade the United States that a stronger anti-Soviet policy would be firmly backed by China.

The attack was humiliating for China, which performed badly on the battlefield and was unable to penetrate very deeply into Vietnam. However, it did

---

of the regime they ran for four years and remains the name of their government-in-exile. The Cambodian communists installed in power in Phnom Penh by the Vietnamese in 1979 were also originally part of the Khmer Rouge, though they individually broke from the movement at various points between 1970 and 1979. We identify them by the name of the regime that they now lead, the People's Republic of Kampuchea.

show off China's destructive potential. Hanoi reported that four provincial capitals were decimated, 320 villages were damaged, 300,000 people were made homeless, and many schools, hospitals, and factories were destroyed.[26] The best estimates are that approximately 50,000 people were killed in total in the four weeks before China withdrew, roughly equivalent to the number of American casualties in Vietnam from 1961 to 1975.

During the spring and summer of 1979 millions of Cambodians left the communes to which they had been assigned in Democratic Kampuchea. They went to search for relatives and to return to native villages or towns. Large-scale migration to the Thai border also began during this period. The reasons for the exodus were varied. Many people of bourgeois background felt that they could make a better life abroad. They were disillusioned when it became clear that the Salvation Front government would not return them to what they perceived as their rightful place at the top of Cambodian society. Many poor peasants had no food in the chaotic months after the invasion. Others left to avoid the residual fighting between the PRK and the DK forces.

In June 1979, Thai soldiers forcibly repatriated more than 40,000 refugees by pushing them through a ravine in a remote and deserted area of Cambodia without food. Hundreds were killed by exploding land mines or were executed by Thai soldiers. Thousands wandered for days before reaching food or water. There is no doubt that many thousands died, though the exact number is unknown.

No government came to the rescue of these Cambodians, but the episode did serve to focus international attention on the refugee problem. Thailand informally agreed not to forcibly repatriate any more refugees as long as "third countries," primarily the United States, France and Australia, continually took steps to ensure that Thailand would not be left with a permanent refugee population. That has been its policy since.

In general, the tremendous self-congratulations of the world's governments and the United Nations for their humanitarian response to the Indochinese refugee problem is in no way warranted. Approximately half of the people who have left Vietnam by boat are thought to have drowned at sea, a number in the tens of thousands. In every city in the Western world one can find numerous Vietnamese who were raped and/or robbed by Thai pirates, yet there have been few arrests or convictions. The refugee camps are often run by paramilitary thugs, who routinely rape and rob, and sometimes kill, Cambodians. Vietnamese, Khmers, Laotians and Hmong are crowded like animals for months, sometimes years, in transit camps.

In the fall of that year, the flow of Cambodians to the border increased, prompting the United Nations to set up several massive refugee camps. The largest and most famous of these, Khao-I-Dang, sheltered some 200,000 Cambodians. By the end of 1979, there were 600–700,000 Cambodians along the Thai border. That population is thought to have declined to approximately 350,000 in 1985. An estimated 250,000 Cambodians have resettled abroad since 1979, including 120,000 to the United States. Michael Vickery in his study in this book, "Refugee Politics: The Khmer Camp System in Thailand," examines how the border population was used, and to some extent created, as an instrument of Thai and American policy toward the PRK and Vietnam.

The mass population movement in Cambodia in the spring had devastating repercussions in the fall. The main rice crop for the year was not planted. A U.S. satellite survey showed that five percent of the fields were under cultivation. A large number of emaciated Cambodians started making their way to the border. Officials from international relief agencies and the United Nations predicted a famine in Cambodia; some declared that it had already begun.

A major international response was forthcoming. In mid-October, relief agencies started sending food and agricultural equipment up the Mekong in barges. On October 24, President Carter announced a $70 million grant for humanitarian assistance. On November 5, at a special conference, twenty-four nations pledged $600 million for United Nations (UN)/Red Cross relief operations in Cambodia. When it became clear that not enough rice seed could be delivered by boat in time to plant the spring crop, an expensive airlift to Phnom Penh was begun. Food and rice seed were distributed to 100,000 Cambodians each week across a "land bridge" from Thailand to Cambodia. Western societies were mobilized to a rare extent. Elementary school children learned about what was happening to their peers in Cambodia. Colleges students sponsored fundraising drives, labor unions exhorted their members, corporate executives lobbied their colleagues, rock bands donated their ticket sales. There were newspaper appeals and telethons, charity balls and church suppers.

By 1982, roughly 1 billion dollars had been donated by the nonsocialist world for Cambodian relief. An additional $250 million came from Vietnam and the Soviet Union. Ninety-five voluntary agencies had been involved. One hundred trucks were originally thought necessary to transport supplies; sixteen hundred were eventually used. The Cambodia operation far exceeded all previous relief efforts on the globe.

There were also problems. Western political objectives dictated that most of the aid be targeted for "relief" rather than "development" or even "rehabilitation" to the country's pre–1970 level of functioning. In practice, this meant that hoes were acceptable, but tractors were not. Fertilizer was permissible, but a fertilizer factory was not. Cambodia's infrastructure, decimated after five years of civil war and bombing, and four years of totalitarianism without running trains, was virtually unimproved.

Another problem with the relief effort was the disproportionate amount of money spent on the Cambodians near the border as opposed to the vast majority of the population in the country's interior. The governments with the greatest influence over the relief operations, especially the United States and Thailand, preferred to see the aid distributed at the Thai border, where it would not help to support or legitimize the Vietnamese-installed government in Cambodia. It was also simply easier to truck the food to the border than to ship it to Phnom Penh. William Shawcross has estimated that approximately $50 was spent per Cambodian in the interior, $440 in the border region, and $1120 per person in the refugee camps.[27]

There was also a fair share of corruption, the major beneficiaries being members of the Thai military. Poor judgment resulted in some conspicuous waste. The International Committee of the Red Cross earned a well-deserved reputation in this regard.

The relief effort was thoroughly infused with politics from the start. The Thai government insisted that a portion of the food aid be given to the remnants of the DK regime and army along the Thai border. This saved the DK forces from starvation and enabled them to rebuild their army. (It is most likely, of course, that China would have fed the Pol Pot forces itself had relief supplies not been available.) Similarly, some of the food delivered to Phnom Penh was used to pay the employees of and build the PRK regime. (Likewise, if relief money could not have been diverted, Vietnam would have used other money to the same effect.) The Vietnamese came in for heavy criticism during this period. Among other things, they were wrongly accused of diverting food to Vietnam. In fact, the border operation was the far more wasteful and political of the relief programs. On the other hand, Vietnamese and PRK waffling and paranoia wasted important time and resulted in unnecessary suffering.

When the initial famine scare—"2 million dead by Christmas"—had passed, many wondered why more Cambodians had not starved. Only a small proportion of the food that the UN and the relief agencies had said was necessary to prevent famine had actually been delivered in time. Yet Cambodian peasants interviewed since have said that despite widespread hunger, there was little, if any, actual starvation. In part, it seems that the threat was exaggerated. For instance, the photos of starving Cambodians coming to the border, which were beamed to television screens across the world, captured an unrepresentative population: peasants who had accompanied the Pol Pot forces into the jungle and emerged half-dead after months in isolation from food supplies. In part, a small rice crop, harvested at the end of 1979, helped tide the population over until the main harvest the next fall. In part, in the absence of rice and fish (the richly endowed Tonle Sap could barely be exploited that year because of the absence of nets and boats), Cambodians substituted traditionally scorned edibles, such as wild potatoes. In sum, the press and governments of the West sounded an exaggerated call for emergency aid, to which they then failed to respond. The aid that they did confer was important in preventing catastrophic shortages. Orlin Scoville and Joel Charny coordinated their papers for this book to look at different aspects of the relief effort and to consider what kind of long-term development plan would be appropriate for Cambodia. Orlin Scoville's "Rebuilding Kampuchea's Food Supply" considers the food production situation in 1979 and examines the choices that were made and the alternatives that were eschewed in implementing a relief and rehabilitation program. Joel Charny's "Appropriate Development Aid for Kampuchea" looks at the production economy at the end of the emergency period and asks what kind of development strategy and international aid are appropriate for a country that must restore its traditions while it builds its future.

There were two other important developments in the period immediately after the Vietnamese invasion. One was the creation of the People's Republic of Kampuchea, and its associated communist party, army, constitution, and popular fronts. The other was the refiguration of the DK forces along the Thai border and the formation of two other resistance organizations, both noncommunist.

As noted above, the Vietnamese invasion occurred in the name of the Kampuchean National United Front for National Salvation, which has remained the main mass organization of the PRK ever since. The Front has been dominated by Cambodian communists, but its leadership includes many noncommunists, some in high posts. A particular effort is made to place Buddhist monks in prominent positions.

The Front "established" the People's Republic of Kampuchea and the Kampuchean People's Revolutionary Council, which was the highest governing body in the PRK until May 1981. At that time, a National Assembly was elected. Although there was a choice of candidates in each constituency, the government made clear which candidates it favored and all of them won overwhelmingly.

The National Assembly immediately adopted a constitution, the country's fourth since independence. The new document specifies three main state organs for the PRK. The members of the National Assembly elect from among themselves a Council of State, which acts as a kind of executive committee when the Assembly is in session and functions as the premier body when the Assembly is in recess. The Council of Ministers is the executive of the government.

The constitution guarantees the standard panoply of individual rights found in most Western nations. In addition, it declares that the "good customs" of national minority groups are to be respected. The foreign policy of the state is to be "nonalignment," but there is to be "solidarity" with "Vietnam, Laos, the Soviet Union, and other fraternal socialist countries." The sections on the economy protect personal property and the right to "make commercial deals on agricultural produce and family handicrafts." The nation's most important economic commodity, land, may be passed on to one's children, but not sold. The constitution deals with many issues rarely found in such documents in the West, from guarantees of paid prenatal leave for women, to the state's obligation to protect ancient temples and art, to the encouragement of gymnastic activities. An interesting provision, perhaps aimed at Vietnamese soldiers or settlers, insists that foreigners in Cambodia must respect PRK laws.[28]

Immediately after they invaded in December 1978, Vietnam put into power in Phnom Penh a disparate group of Cambodian communists comprised of individuals who had quit the Khmer Rouge and fled to Vietnam at various times between 1970 and 1978. Once in power, they began to build a new Cambodian communist party. It was first publicly unveiled, however, in May 1981, at the same time as the adoption of the constitution and the first national elections. The Kampuchean People's Revolutionary Party considers itself the legitimate descendent of the Cambodian communist movement, and its founding is celebrated as the date in 1951 when the Khmer People's Revolutionary Party was formed in the wake of the dissolution of Ho Chi Minh's Indochina Communist Party. According to official rendering, the revolutionary process was on track until 1975, when the "Pol Pot-Ieng Sary-Khieu Samphan clique plundered the revolutionary gains and sold themselves as lackeys" to the "reactionaries of the Chinese leading circle" who, in league with the United States, were trying to roll back the victories against capitalist neo-imperialism. This "perversion" on the part of Pol Pot and his coterie was the consequence of a variety of ideological maladies, including Maoism and Trotskyism. But the true Cambodian communism reasserted itself when the Salvation Front "abolished the dictatorial, fascist and genocidal regime of the Pol Pot-Ieng Sary clique" on January 7, 1979.

As the PRK organized itself, several different kinds of Cambodian opposition to the regime coalesced. When Vietnamese troops approached, Pol Pot and the other leaders of the DK government fled Phnom Penh. Some went to China, where they asked for and received assurances of financial and diplomatic support. Others led the retreating DK army into the jungle, destroying whatever they could in the way of factories, food stocks, and transports as they went.

The DK strategy since regrouping on the Thai border has been simple. Within Cambodia, they have tried to recast themselves as a nationalist force seeking to protect the Khmer nation against the barbaric Vietnamese. Internationally, they have emphasized the illegality of their overthrow and their usefulness in resisting the expansion of Soviet power. Publicly, the DK has declared that their guerrilla struggle will lead to the mobilization of the Cambodian people, and eventually to the defeat of the Vietnamese on the battlefield. Little else is known about the group's understanding of its time in power or its plans for the future.

The DK group has also announced a number of structural changes, aimed at improving their image inside Cambodia and internationally. On August 21, 1979 they formed a new organization, the Patriotic and Democratic Front of the Great National Union of Kampuchea. In December of that year, Pol Pot formally gave up the prime ministership of the DK government-in-exile to Khieu Samphan, who is considered the least powerful of the top DK leadership, but is also the individual who is best-received in the diplomatic community. Pol Pot later also officially retired as secretary of the Communist Party of Kampuchea and commander of the DK military, though no one doubts that he remains in de facto control of the DK resistance movement. In announcing their goals for the nation in July 1980, the DK leadership declared that they no longer sought a communist state. They claimed they wanted to install a "liberal democratic system." In accord with this, in December 1981, they announced the dissolution of the Communist Party of

Kampuchea. Few believe that this was actually done, but no concrete evidence has yet surfaced to document that a communist party, as such, is functioning in the DK-controlled border areas.

While the DK forces were regrouping along the border, a second anti-Vietnamese force was taking shape, the Khmer People's National Liberation Front, proclaimed in October 1979. The group's sixty-eight-year-old founder and president, Son Sann, had held numerous top positions when Sihanouk was in power. He lived in Paris during the Lon Nol and Pol Pot periods. Sann is an affable man with a reputation for personal integrity and a style reminiscent of nineteenth century French diplomats. Despite his lifetime of high government service, many Cambodians have no idea of who he is.

Sann established control over a number of small militias headed by various Cambodian warlords strung out along the Thai border and began to meld them into a resistance army. He also has a relatively simple strategy: to offer the Cambodian people an alternative to both the DK and the Vietnamese-supported PRK. Unlike Pol Pot, Son Sann is totally credible when he says that he desires a Western-style democracy with a capitalist economy.

Immediately after the Vietnamese invasion, Prince Sihanouk pleaded the case of the DK regime at the United Nations. After learning more about what had gone on in Democratic Kampuchea while he was under house arrest, Sihanouk became a strident critic of the Khmer Rouge, calling their regime "Hitlerian." He credited the Vietnamese with saving his people from extinction. Nonetheless, as time passed, Sihanouk concluded that continuing Vietnamese control over Cambodia might mean the eventual disappearance of the nation and Khmer culture. Sihanouk's opposition to the PRK may also have been motivated by the fact that the new order did not have a substantive role for him.

Between the middle of 1980 and early 1981, Sihanouk formed three organizations: the Movement for the National Liberation of Kampuchea, the National Army of Sihanouk, and, encompassing both, the National United Front for a Cooperative, Independent, Neutral and Peaceful Cambodia. Sihanouk's organizations are based in Paris and on the Thai border, where they have their own camps and population base. (Sihanouk himself, when not traveling, divides his time between Beijing and Pyongyang.) Sihanouk envisions an independent, nonaligned Cambodia with himself at the top, probably similar in essence to the regime he headed for sixteen years. It remains unclear, however, what he would be willing to settle for, an ambiguity which makes his international backers, primarily China and the ASEAN states, quite nervous.

Sihanouk's greatest asset continues to be the affection with which he is regarded by so many Cambodians. He still carries the prestige of the monarchy, particularly for peasants, and, in retrospect, his time in power seems idyllic. The Prince is also well-liked in the international diplomatic community, especially among third world nations. Working against him are memories of the corruption in his regime, resentment of the part he played in bringing the Khmer Rouge to power, and his lack of stature among Cambodians too young to remember events before 1970.

Most significantly, working against the idea of resistance in general, is the reluctance of many Cambodians to risk their lives, to bring their exhausted country again to combat, to fight a powerful Vietnamese army, while the DK forces wait ominously in the background.

By the middle of 1981, all of the elements of the present political situation were in place. Vietnam had a large occupation army in Cambodia. A group of Cambodian communists who had split from the main communist movement were in charge of a weak, pro-Vietnamese socialist government which controlled most of the country's land and population. They were trying to build a new state, governmental structure, communist party, and army. Most Cambodians were where they wanted to be: at home, at the border, on their way abroad, or already there. The economy was stable, but output was low. The government was isolated from all the nations of the world except for a few allies of the Soviet Union. Three resistance groups were operating from bases near the Thai border. China, the United States, and the ASEAN states were giving these resistance groups financial and diplomatic support in an effort to raise the cost of Vietnam's occupation and prevent the consolidation of the government Vietnam had created. And people were waiting.

Progress On All Fronts, 1981–

Since 1981, change has been mainly incremental. There has been scarcely a single event of importance. In their cumulative effect, however, certain developments in Cambodia and in the international maneuvering around Cambodia are worth noting.

After nearly collapsing, the Cambodian economy is once again functioning. The PRK organized the rural population into mutual aid teams of ten to fifteen families each. There are ninety-five thousand of these "solidarity groups" (*krom samaki*) which include ninety-five percent of the rural population. There is insufficient information to draw an overall picture of these groups, which can range from fairly highly organized communes to more traditional cooperative ventures. A significant portion of agricultural land has been distributed for private cultivation by individual families.

Surplus produce is now legally sold in open markets, which have become commonplace in both rural and urban areas. Until 1982, the government was able to avoid imposing taxes because of the availability of international food aid and financing from Vietnam and the Soviet Union. The taxes levied since 1982 have been very limited. There is an abundant trade in consumer goods smuggled in from Thailand. Thai baht, Vietnamese dong, American dollars, and the Cambodian riel, reintroduced by the PRK in 1980, are all commonly used.

Exploitation of the Tonle Sap has advanced slowly. In 1984, 45,000 tons of fish were caught, in contrast to close to zero in 1979 and between 150,000 and 200,000 tons per annum prior to 1970. There is little more than anecdotal evidence concerning industry, but it appears that there has been little progress since 1979.

Few resources are available for the development of any kind of economic activity, whether it be agriculture, fishing, industry, or anything else. Fertilizer, seeds, fishing nets, boats, and machinery are all in short supply. There are approximately 1 million draft animals today, down from 2.5 million before 1970. A physical infrastructure virtually needs to be recreated; electricity is sporadically available in limited areas, roads are in terrible condition, telephones are practically nonexistent. Human resources are also depleted. The male death toll under Pol Pot was disproportionately high, and seventy percent of the adult population is now female. Thousands of engineers, teachers, nurses, and others with various professional skills have perished or been siphoned off by the refugee camp system.

Soviet aid amounts to approximately $100 million annually. Much smaller quantities of assistance come from East Germany, Czechoslovakia, Hungary, Poland, Cuba, and India. Vietnam's contributions to the Cambodian economy are hard to calculate. Vietnam's own economy has performed badly since reunification, and in some years since 1978 there have been severe food shortages. All Vietnamese aid to Cambodia thus ultimately comes from the Soviet Union. With food production in Cambodia approaching subsistence levels, Western governments have discontinued their contributions to the United Nations program in Cambodia, as well as most of their aid distributed at the border. A few private voluntary agencies in the West have maintained small assistance programs in both places. There is a small amount of export, mainly in rubber, kapok, cotton and fish, primarily to Soviet-bloc countries.

Living conditions in Cambodia are still miserable, but they are better than they have been in fifteen years. Average per capita income is less than $100 a year, which is the lowest level in the world, except for Chad. By contrast, Haiti's is three times higher; Thailand's, El Salvador's, and Nicaragua's are each eight times greater. In 1985, average life expectancy at age one was forty-five years, as contrasted with sixty-three years in Thailand and seventy-six in Sweden. The child mortality rate is the fourth highest in the world. One of every five infants die before their first birthday. A 1983 UN study concluded that more than half of the children below the age of thirteen suffer from moderate to severe malnutrition.

Medical care was virtually abolished under Pol Pot. All but fifty-five Cambodian doctors are thought to have died during his rule. There has been a gradual restoration of medical care since then. Twenty-five primitive hospitals and some 100 health centers are now in operation. A medical school has been opened in Phnom Penh. Despite some progress, a 1983 UN Food and Agriculture Organization report called the health care system "disastrous." Unfortunately, the PRK has turned down some Western offers of doctors and nurses.

During the DK period, a large proportion of the women in Cambodia stopped menstruating, owing to malnutrition and trauma. Figures range around eighty-five percent.[29] Thus one sign of the return to normality under the PRK is the current baby boom. Many women are now having children to compensate for children who died under Pol Pot or to make up for lost time. Cambodia now has one of the highest birthrates in the world.

Perhaps the most remarkable achievement of the PRK has been its creation of an educational system. Within a short time, 1.5 million children were attending elementary schools, a marked contrast to the DK period when schooling had been virtually eradicated. At higher levels, the growth has been slower. There are 60,000 children in middle school and 4,000 in high school. Although some schools of higher education have been established, most students above the high school level go to Vietnam or to Eastern Europe for training. It has been reported that approximately 1,000 Cambodian students are in the Soviet Union at any given time, with another thousand in the Eastern European countries.

The PRK has encouraged the revival of traditional art forms such as theater and dance, which were largely absent in DK. The regime also uses some of these media for propaganda. Ben Kiernan gives an evocative description of one politically inspired theater-dance performance:

> And then there is "Blood and Tears," a shocking reenactment of the final stages of a Pol Pot massacre; the semidarkness and slow-motion violence are punctuated by a moving lament of the suffering sung with tremendous sadness and beauty by a woman. In the end Heng Samrin's Salvation Front Troops arrive and apprehend the Pol Pot executioners. Their victims wake up and stand in a chorus as fists go up, all eyes turn to the massive PRK flag suddenly projected at the back of the stage, and then hands are religiously outstretched toward it. The scene ends with a eulogy of the new revolution. Despite the agonizing memories it recalls, and the ceremonial ending, "Blood and Tears" is a popular scene in Phnom Penh, no doubt because of the degree of realism. The Kampuchean tragedy has not lacked melodrama.[30]

Every year the PRK marks a "National Day of Hatred" for the DK regime.

Religion has been allowed to revive, even been encouraged, within certain constraints. It appears that most of the 70,000 to 100,000 Buddhist bonzes of pre-1975 Cambodia died in DK. More than 5,000 bonzes have been reordained since 1979. The government has been restoring temples desecrated under DK and religious ceremonies are commonplace. New "Buddha factories" churn out statuettes in Phnom Penh. However, the authorities will not allow anyone below the age of fifty to join the monkhood, and there is pressure placed on the bonzes to propagate the government's political line.

The Chams, who are Muslims, were one of the most persecuted groups under Pol Pot, and have received some of the most solicitous policies of the new government. Cham mosques have reopened, and Cham leaders have been given considerable freedom in reorganizing their communities.

Catholics and Protestants, few in number, have been harassed when they have attempted to hold religious services. Presumably, the association of these faiths with unfriendly Western governments makes the authorities wary. Similarly, there are few signs of the revival of the community practices of the once very large ethnic Chinese population. Again, the regime's hostile relations with China are probably to blame. In her study here, "Revolution and Reformulation in Kampuchean Village Culture," May Ebihara traces religion and other activities of daily significance in Cambodia's villages, home of eighty percent of the Cambodian population, as they have gone through turbulent changes under the last four political regimes.

As is evident from its economic and social policies, the regime has pursued its ideological goals timidly. Other socialist states may be tolerant of markets, but rarely are they such an important part of the official structure as they are in Cambodia. Similarly, the official enthusiasm for religion is unique. Marxism-Leninism is not mentioned in the constitution, as it is in Laos, Vietnam, and other countries. Indeed, the Soviet Union does not define Cambodia, unlike Laos and Vietnam, as a state progressing toward socialism. Particularly in light of the political and economic experimentation that is now occurring in several communist states, including Vietnam, Cambodia's ideological future is difficult to forecast.

The building of the new communist party, the Kampuchean People's Revolutionary Party, has been a more conventional process. Promising young people are identified while they attend one of the many political education courses and are groomed for Party membership. Between 1979 and 1985, the Party is reported to have grown from 66 to 7,500 members. Initially, the Party had no members outside Phnom Penh; it now has organizations down to at least the district level.

Outside observers usually refer to the party and government leadership as consisting of three main groups: cadres who lived in Vietnam from the time of the Geneva Agreements until 1970, when they returned to Cambodia to participate in the civil war, and who then fled back to Vietnam when Pol Pot's wing of the Khmer Rouge started killing these "regroupees"; cadres who first went to Vietnam between 1970 and 1978, when they became disenchanted with Pol Pot's policies or felt themselves to be a likely target of a new round of Pol Pot purges; and people without previous revolutionary experience. Although some have written of these groupings as if they were competing factions, there is little evidence to support such a view. It is said that the Vietnamese have favored the regroupees in building the government and party, but reportedly less than 100 members of this group survived to participate in the PRK. The Party's rank and file is dominated by people without a revolutionary background.

Building the state and the Party has been difficult for a number of reasons. There is a great shortage of skilled people, especially among those who might share the political goals of the regime. For example, William Shawcross has reported that in 1979, Hun Sen, the twenty-seven-year-old foreign minister who is considered one of the rising stars of the regime and Party, had to be "reminded" what the United Nations was. There have been purges of lower-level officials, both for ineptitude and subversive cooperation with the Cambodian resistance. There are frequent invectives against corruption in the media. A full year after the invasion, there were virtually no functioning ministries.

There have been reports of political struggle within the regime and Party, but not enough information gets out of the country to allow foreign observers to clearly discern what is happening. There seems to be conflict as to whether the regime should distance itself from Vietnam to a greater degree. One avenue for officials so inclined is to work with the Soviet Union, which seeks independent leverage on the regime. Also, Vietnam appears unable to persuade its "puppet regime" to sign a border treaty or to offer a high position to Prince Sihanouk as an inducement to join the government. Many of these issues and the way they are handled combine to form a distinctive Cambodian style of politics, as Serge Thion shows in his chapter, "The Pattern of Cambodian Politics," which finds surprising similarities among the last four regimes in Cambodia.

As important as establishing a viable regime, Vietnamese and PRK officials are trying to build a Cambodian army to defend it. At the height of the occupation, there were 220,000 Vietnamese troops in Cambodia. This number has gradually declined to about 140,000. Vietnam had promised to have all of its troops withdrawn by 1987. They later revised this schedule, promising full troop withdrawal by 1990. This, too, is highly unlikely. At a minimum, the PRK must be able to defend itself before Vietnam will withdraw. That prospect is not in sight. In any case, Vietnam may choose to keep some troops in Cambodia indefinitely, as they have in Laos, where there are 50,000 Vietnamese troops.

The size of the PRK armed forces hovers around 30,000 troops. A total "Khmerization" of the war is feared by the ASEAN states because they would then be in the uncomfortable position of fueling a civil war in a ravaged land, rather than supporting "liberation forces" against a foreign occupation. But the PRK has had great trouble obtaining compliance with the lengthy military service requirements, prompting Khmer soldiers to actually fight against other Khmers, and preventing desertions. The construction of an indigenous army by the PRK has not been analyzed before and is the main subject of Timothy Carney's "The Heng Samrin Armed Forces and the Military Balance in Cambodia."

While the PRK is better able to protect its territory, the resistance is now more capable of attacking it. The DK forces grew from 20,000 in 1979 to 35,000 in 1985. Son Sann had 3,000 men under arms in 1979 and 15,000 in 1985. Sihanouk forces had not yet been organized in 1979, but totaled 9,000 men in 1985. The reality, though, is not as impressive as the numbers: the noncommunists resist more than they attack and are more of a nuisance to the regime than a serious hindrance, at least militarily. The DK army is, by all accounts, a formidable force.

In June 1982, the three resistance groups formed the Coalition Government of Democratic Kampuchea. Formally, this meant that the Democratic Kampuchea government-in-exile maintained by the DK leadership from 1979 to 1982 was replaced by a tripartite government-in-exile maintained by all three resistance groups. Prince Sihanouk was made president, Son Sann was named prime minister, and Khieu Samphan became vice-president in charge of foreign affairs.

Sihanouk and Son Sann have been frank in stating that they were forced into the Coalition by the ASEAN countries and China, which threatened to discontinue military and financial aid to their forces if they did not comply. ASEAN and Chinese diplomats apparently felt that a coalition of the three groups would give the resistance a recognizable shape in the eyes of the world community and would draw attention away from the role of the notorious Pol Pot forces, thereby making it easier for other nations to continue denying recognition to the PRK. The great disadvantage of this strategy is that Sihanouk's and Son Sann's alliance with the hated Pol Pot may limit their ability to recruit fighters and garner support among the Cambodian population. When the Coalition was formed, the Soviet Union called it "'a coalition' of hangmen and traitors cobbled together by foreign hands, taking cover on a foreign territory, and serving foreign interests for foreign pieces of silver."

Although the three resistance forces have never formed a joint military command, the creation of the Coalition should have put an end to the fighting among themselves, but it has not. Armed clashes have continued, with no discernable cause other than indiscriminate DK belligerence. In 1983, noncommunist military commanders reported that the DK forces "now posed more of a day to day threat to them than the Vietnamese forces in Kampuchea." Fighting between troops loyal to Son Sann and to Sihanouk is occasionally reported, though these two coalition partners generally prefer attacking each other in print.

As promised, however, the formation of the coalition has brought additional aid to the noncommunist factions. All three forces are supplied by China. Son Sann's and Sihanouk's groups receive additional support from some of the ASEAN states, West Germany, and a few other countries. There is significant evidence that Son Sann's group has also been receiving covert financial and military backing from the American CIA. In 1985, the U.S. Congress allocated funds for "humanitarian" assistance to both noncommunist factions.

The fighting between the resistance forces and the Vietnamese and PRK armies follows a predictable cycle every year. The Vietnamese and PRK troops attack the resistance bases on the border from October to May, when their regular formations can take advantage of the dry weather. During the rest of the year, when there is frequent rain, the resistance fighters penetrate into PRK territory and launch guerrilla assaults. The Vietnamese and PRK forces have concentrated their attacks on the Son Sann and Sihanouk troops, often leaving the DK forces alone. The obvious interpretation is that they feel that the DK has no real potential to galvanize the populace and develop into a popular nationalist resistance. By keeping the noncommunist forces weak, they can present to the Cambodian population and the rest of the world a choice between continued PRK governance and a return to power of the Pol Pot forces.

The significance of the development of the resistance is not clear. They have made the interior of Cambodia less secure. However, they remain unable to hold significant amounts of territory for any period of time. From a narrow balance of power perspective, they remain in the same position as they were in when they first formed. With the availability of sanctuary in Thailand, the resistance cannot be completely defeated. Nor can they defeat the Vietnamese, who have vastly superior resources. As Prince Sihanouk has put it: "I foresee a long, long war without losers because we cannot lose, but without winners, too, because we cannot win either."

Nonetheless, resistance leaders do not hope to win an outright victory, but rather to make the costs of sustaining the PRK government in power sufficiently high that Vietnam will ultimately seek a negotiated settlement. In these terms, it is difficult to know whether, in the last five years, the resistance has become more likely to ultimately succeed. Their progress depends on changes in the subjective calculations of the Vietnamese leadership.

The strategy of imposing costs on Vietnam is shared by the ASEAN states, which, at a minimum, want to see Vietnam's troops out of Cambodia—that is, away from Thailand—and which would like, in addition, to have a friendly government in power in Phnom Penh. The ASEAN states have thus far rejected the compromise position of acquiescing in the continuation of the PRK regime in exchange for the removal of Vietnamese troops, which, among other things, they see as an unenforceable arrangement.

Besides directly supporting the resistance, the ASEAN governments have succeeded in persuading many other nations to eschew economic relations with Vietnam and have marshalled diplomatic pressures against the PRK and Vietnam. Each year between 1979 and 1982, the ASEAN states were able to block Vietnamese efforts to unseat the Democratic Kampuchea delegation at the UN. The voting margins were so decisive that Vietnam gave up the attempt after 1982. By even larger margins each year since 1979, the General Assembly has passed ASEAN-sponsored resolutions calling for a Vietnamese withdrawal from Cambodia. Only thirty-five nations recognize the PRK; one hundred and thirty-eight either recognize the Coalition Government of Democratic Kampuchea or no Cambodian government at all.[31] The ASEAN states also pressed the United Nations to convene an extraordinary meeting, the International Conference on

Kampuchea, in July 1981. The stated purpose of the Conference was to provide a forum for a negotiated settlement of the Cambodian conflict, but with Vietnam and its allies boycotting the proceedings, the meeting was used as a forum to score diplomatic points against Vietnam and the PRK.

The ASEAN states accrue benefits from their Cambodia policies quite apart from any impact on the situation in Cambodia. Thailand gets additional military aid from the United States. Thailand and Malaysia have insisted that they could only cooperate closely with China on the Cambodia question if China stopped supporting the indigenous communist insurgencies in their countries. China has promised to defend Thailand if it is attacked by Vietnam and has granted it favorable trading terms. The "external" Vietnamese threat has prompted close cooperation among the ASEAN states. Their diplomatic endeavors with regard to Cambodia have led to a dramatic increase in their international prestige. This incentive structure is one of the issues tackled by David Elliott in his article, "Deadlock Diplomacy: Thai and Vietnamese Interests in Kampuchea," which considers why years of diplomacy have failed to resolve the Cambodia conflict.

United States rhetoric has largely focused on the restoration of Cambodian sovereignty, but, as Michael Vickery has pointed out, sovereignty is not the only issue: "Had Thailand, in response to similar provocation, administered the lesson, set up *its* Cambodians, the Khmer Serei [anticommunist Cambodians], in Phnom Penh, and overseen the same progress as has occurred in the last two years, it would be hailed as a great victory for the Free World and its methods."[32] As the Reagan administration sees it, the most important goals of United States policy toward the conflict are the maintenance of Thai security and the containment of Soviet power, seen as being channeled through Vietnam. The U.S. role has been to back the resistance diplomatically and materially, and to encourage the ASEAN states to take a hardline against Vietnam. As for the ASEAN states, the consequences for Cambodia are not the only or even the main determinants of U.S. policy. U.S. analysts readily acknowledge that the United States would be extremely reluctant to do anything with respect to Cambodia that would antagonize China, even if it was believed that current U.S. policy was likely to be fruitless.

Among all the international actors, China is the most strongly committed to denying legitimacy and permanence to the PRK. Chinese officials can see no advantage to conceding predominant influence in Cambodia, which has long been viewed as a natural ally of China, to Vietnam or to the Soviet Union, even allowing for the likelihood that the PRK cannot be displaced or modified. China, which is virtually the only country to provide aid to the DK forces, plays a major role in supplying the noncommunist resistance, battles Vietnam diplomatically, and maintains military pressure on Vietnam's northern border. China also tells the Soviet Union that its support of the Vietnamese occupation of Cambodia is the single greatest obstacle to improved Soviet-Chinese relations. (The Soviet occupation of Afghanistan and the military buildup on the Soviet-Chinese border are the second and third priorities on China's agenda in bilateral talks with the Soviet Union.) The Chinese hope that Moscow will sacrifice its hegemony over Cambodia in order to achieve a friendly and stable relationship with China, and will thus ultimately induce Vietnam to make critical concessions.

Since 1979 Vietnam has shown little inclination to make any such concessions. Hanoi sees no reason to give up hard-earned security on its southern flank in order to achieve what would likely be limited economic assistance and nebulous gains in international prestige. The Vietnamese also believe that it is merely a matter of time before their opponents acknowledge the futility of their efforts and desist from actively opposing the PRK. As one Vietnamese diplomat put it, "We have already won. The other side just hasn't noticed yet." Because Vietnam has rejected the idea of free elections in Cambodia, even with a guarantee that the DK forces would not be allowed to return to power, it must be said that they are willing to sacrifice Cambodian sovereignty in order to ensure that Cambodia remains a Vietnamese ally.

Hanoi has, of course, been able to withstand the diplomatic, economic, and military pressure largely because of the Soviet Union, which supplies Vietnam with between 1 and 2 billion dollars in aid annually. In return, the Soviets have had, since 1979, the use of Cam Ranh Bay and Da Nang, two huge American-built military installations in southern Vietnam. These are now the largest Soviet bases outside of the Soviet Union. The Soviet Union is the only country that has clearly gained in power and influence as a result of these last ten years of turmoil in Asian interstate relations.

Vietnam's intransigence and its ability to withstand the economic, military and diplomatic pressures have led ASEAN strategists to begin to speak of the need to pressure Vietnam over a very long period. They envision a gradual erosion of Vietnam's comparative military strength owing to the expansion of their own economies and the stagnation of Vietnam's. Some now predict that the confrontation will continue for several decades. China's diplomats have always spoken in terms of a protracted struggle with Vietnam. Given the relative sizes of the two countries, they say China can "bleed Vietnam white," even waiting 100 years if necessary.

A crucial question for the future is whether the ASEAN states can sustain their enthusiasm for the conflict over such a long period. ASEAN leaders recognize that Vietnam provides a valuable counterweight to growing Chinese power in the region. ASEAN governments, particularly in Indonesia and Malaysia, which tend to be most wary of China, harbor no desire for a crippled Vietnam.

Another important question is whether a long-term strategy is supportable. The noncommunist resistance is led by a man already well into his seventies, and another, Prince Sihanouk, whose health is only fair. Sihanouk hints every so often that he would like to pursue an acceptable arrangement with the PRK leadership. Pol Pot is reportedly quite ill, and the cohesion of the DK resistance after his departure is not at all certain. The maintenance of the schedule of pressures clearly depends on the loyalty of all the major international actors, but this is not guaranteed. A change in leadership in the United States or China could result in a fatal defection. There are already signs of declining support for the anti-Vietnamese strategy in the international community. Small countries occasionally add their names to the list of countries that recognize the PRK. Australia and France have both backed away from their support for the ASEAN position. Anthony Barnett argues in his analysis of the structure of the international conflict, "Cambodian Possibilities," that the conflicting goals of the governments that oppose Vietnam will undermine their alliance prior to such a time when Vietnam would be forced to make major concessions.

Cynthia Coleman's chapter, "Cambodians in the United States," which closes this book, is different from the other essays in that it is in large part a personal chronicle rather than a detached analysis. We were more than happy to include it here though because it captures Cambodians in the midst of those crucial acts of adaptation that have, sometimes, ensured their survival through the last three Cambodian political regimes. The last fifteen years have been agonizing for Cambodians. One can only hope that history asks less of them in the future.

Notes

1. We thank Katherine David, Daniel Fineman, Jeremy Primer, and Adam Vital for their comments on this chapter. David Weiner read this essay with tremendous care and skill, improving it enormously, for which we are most grateful.
2. Kenneth Hall, "The 'Indianization' of Funan: An Economic History of Southeast Asia's First State," *Journal of Southeast Asian Studies* 13 (1982): 82.
3. William Shawcross, *Sideshow: Kissinger, Nixon and the Destruction of Cambodia* (New York: Simon & Schuster, 1979), 46.
4. "Cambodia," *Far Eastern Economic Review 1971 Yearbook* (Hong Kong: Far Eastern Economic Review, 1971), 116.
5. "Communist Infrastructure in Cambodia," U.S. Defense Intelligence Agency Appraisal, July 8, 1971.
6. The primary source for these numbers is Timothy Carney, Cambodia: The Unexpected Victory, MS, 25.
7. Laura Summers, "Cambodia: Model of the Nixon Doctrine," *Current History* 65 (1973): 256.
8. Shawcross, *Sideshow*, 265.
9. Richard Dudman, *Forty Days With the Enemy* (New York: Liveright, 1971), 69-70.
10. Michael Vickery, "Looking Back at Cambodia," *Westerly*, 1976, no. 4: 27.
11. Serge Thion, "With the Guerrillas in Cambodia," *Indochina Chronicle* 17 (1972): 11.
12. This is entirely different from Marx's "bourgeois democratic revolution."
13. Kenneth Quinn, "The Khmer Krahom Program to Create a Communist Society in

Southern Cambodia,'' Department of State, February 1974, 34.

14. Vickery, ''Looking Back at Cambodia,'' 27.

15. See George C. Hildebrand and Gareth Porter, *Cambodia: Starvation and Revolution* (New York: Monthly Review Press, 1976).

16. Ben Kiernan, ''Conflict in the Kampuchean Communist Movement,'' *Journal of Contemporary Asia*, 10 (1980).

17. Technically, the civil war was won by the government-in-exile, the Royal Government of National Union of Kampuchea, which had been headed by Prince Sihanouk since his overthrow in March 1970. Thus this Royal Government was officially in power through January 8, 1976, when the constitution and state of Democratic Kampuchea were unveiled. However, the Khmer Rouge exercised state power from April 17, 1975 onward, and the entire time, from April 17, 1975 to January 8, 1979, is commonly referred to as the Democratic Kampuchea period.

18. Stephen R. Heder, *Kampuchean Occupation and Resistance*, Asian Studies Monographs, No. 027 (Bangkok: Institute of Asian Studies, Chulalongkorn University, 1980), 5-7.

19. Stephen R. Heder, ''The Kampuchean-Vietnamese Conflict,'' *The Third Indochina Conflict*, ed. David W. P. Elliott (Boulder, CO: Westview, 1981).

20. Frederic A. Moritz, ''Cambodia Snubs Viet Cease-fire Bid,'' *Christian Science Monitor*, 9 February, 1978, 4.

21. U.S., Congress, House, Committee on International Relations, Subcommittee on Asian and Pacific Affairs, *Vietnam-Cambodia Conflict*. Report prepared by the Congressional Research Service, Library of Congress, 95th Cong., 2d sess., 4 October 1978, 3-4.

22. Kiernan, ''Conflict,'' 64.

23. Lowell Finley, ''The Propaganda War,'' *Southeast Asia Chronicle*, 1978, no. 64: 33.

24. Gareth Porter, ''Vietnamese Policy and the Indochina Crisis,'' *The Third Indochina Conflict*, ed. David W. P. Elliott (Boulder, CO: Westview, 1981), 105.

25. Heder, *Kampuchean Occupation and Resistance*.

26. Porter, ''Vietnamese Policy,'' 110.

27. William Shawcross, *The Quality of Mercy: Cambodia, Holocaust and Modern Conscience* (New York: Simon and Schuster, 1984), 393. The numbers are so rough that they are really only indicative of the magnitude of the differences. They refer specifically to the expenditures of the UN/International Committee of the Red Cross Joint Mission.

28. Quotations from Albert P. Blaustein and Gisbert H. Flanz, eds., *Constitutions of the Countries of the World*. Supplement for the People's Republic of Kampuchea (Dobbs Ferry, NY: Oceana Publications, 1982).

29. Murray Hiebert and Linda Gibson Hiebert, ''Famine in Kampuchea: Politics of a Tragedy,'' *Indochina Issues*, 1979, no. 4: 1.

30. Ben Kiernan, ''Kampuchea 1979-1981: National Rehabilitation in the Eye of an International Storm,'' *Southeast Asian Affairs* 1982 (Singapore: Institute of Southeast Asian Studies, 1982), 182.

31. The following countries recognize the PRK: in Asia—Afghanistan, India, Laos, Mongolia, Syria, Vietnam; in Europe—Albania, Bulgaria, Czechoslovakia, East Germany, Hungary, Poland, Soviet Union; in Africa—Angola, Benin, Cape Verde, Congo, Ethiopia, Guinea, Guinea-Bissau, Libya, Madagascar, Morocco, Mozambique, Sao Tome and Principe, Seychelles, South West African People's Organization, Democratic Yemen,

Zimbabwe; in the Americas—Cuba, Grenada, Guyana, Jamaica, Nicaragua, Panama.

Because the Coalition Government of Democratic Kampuchea is a government-in-exile, and because of its unsavory reputation, a large proportion of the countries that recognize it as the official Cambodian government do not maintain diplomatic relations with it. We were not able to obtain a list of countries that recognize the Coalition Government. The countries that maintain diplomatic relations with it are: in Asia—Bangladesh, China, Malaysia, North Korea, Pakistan, Turkey; in Europe—Yugoslavia; in Africa—Cameroon, Egypt, Gabon, Gambia, Liberia, Mauritania, Nigeria, Senegal, Sierra Leone, Somalia, Sudan, Togo.

32. Michael Vickery, *Cambodia: 1975-1982* (Boston: South End, 1984), 247.

DEMOGRAPHICS
AND
SOCIAL STRUCTURE

Meng-Try Ea

RECENT POPULATION TRENDS IN KAMPUCHEA

Since the *coup d'etat* of March 18, 1970, all conditions for demographic and socioeconomic development in Kampuchea have been thoroughly disrupted. Favorable demographic trends had prevailed without interruption up to that time, but the following decade (1970–1980) was one of serious decline: war, revolution, and the Vietnamese invasion all affected the size, dynamism, and character of the Kampuchean population. Nonetheless, signs of demographic recovery became visible around 1981.

The documentary information for a thorough study of these transformations does not exist. The dimensions of the human losses during this decade of crises are known only through indirect estimates and fragmentary data, some of which are arbitrary and some of which have been manipulated to serve political ends. This study is based on the accounts of witnesses and various available statistics.[1]

The Pattern of Decline

All firsthand accounts support an image of demographic disaster, and all are corroborated by available figures—some of which are admittedly based on such accounts. A comparison of "actual" population with hypothetical population (i.e., in the absence of conflict) affords an approximate evaluation of the human losses sustained during the years 1970–1980.

The hypothetical population is the population Kampuchea would have had in the absence of crises. Presuming an annual rate of increase of 3 percent between 1970 and 1975 and 3.2 percent between 1975 and 1980, this would have been 10 million inhabitants by 1980. Due to the severe disturbances, however, the population in 1980 had actually declined in absolute numbers. Recently published estimates for the 1980 population vary considerably. The United Nations cites 6.7 million inhabitants,[2] while Peter Donaldson, basing his calculations on documents from official U.S. intelligence sources, figures the population at about 4.8 million (table 1).[3]

My calculation, which is derived from adjusting the rate of increase in accordance with political and socioeconomic development in Kampuchea during

Table 1

Kampuchea: Population Estimates
(in thousands at mid-year)

Year	Sources United Nations[1]	 U.S. Bureau of the Census[2]	 Official
1970	7,060	7,060	
1971	7,270	7,133	
1972	7,490	7,201	
1973	7,710	7,270	
1974	7,920	7,334	
1975	7,098	6,726	
1976	7,024	6,191	7,135[3]
1977	6,919	6,012	
1978	6,812	5,899	7,800
1979	6,746	5,767	4,500[4]
1980	6,747	4,800 (?)	6,600
1981	6,828		

Notes

1. Prior to 1980, the UN publications gave population estimates in the absence of catastrophes. The 1981 publications show that the estimates have been revised back to 1975. The estimates prior to 1975 have not yet been revised.
2. The U.S. Bureau of the Census based its estimates on figures provided by the Central Intelligence Agency.
3. The 1976 and 1978 estimates were published by the Pol Pot administration.
4. The 1979 and 1980 estimates were published by the new administration of Heng Samrin and do not include the inhabitants of the zone of resistance, which could number up to one million as of the end of 1982.

the relevant period, indicates a population of approximately 6 million in 1980. This figure thus reflects a discrepancy of more than 4 million in reference to the hypothetical population for that year. My estimates for each year, starting in 1970, are given in table 2.

These figures, which describe the dimensions of the demographic catastrophe, generate several observations. First, from 1970 to 1874, the population remained quasi-stagnant.

Second, from 1975 to 1978, Kampuchea's population declined steadily, from 6.8 million to 6.2 million, resulting in a 9 percent drop during the nearly four years of the Democratic Kampuchean regime. The dramatic drop in 1975 was due to famine and the forced evacuation of the entire population of Phnom Penh (3 million) on the very day of the Khmer Rouge victory. The sharp decline in 1978 is attributable to the armed conflict with Vietnam and a series of violent purges.

Figure 1 **Kampuchea: Population Estimates, 1970–1980.**

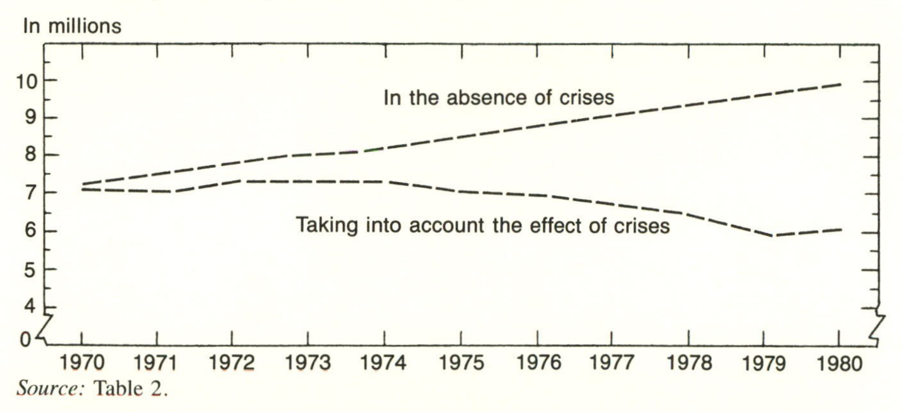

Source: Table 2.

Third, from 1979 to 1980, the population continued to decline during the first two years of the Vietnamese invasion and occupation, due to famine and the massive emigration of refugees.

Factors in the Depopulation of 1970–78

The negative demographic trends of 1970–78 are easy to account for. They essentially resulted from an increase in deaths caused directly and indirectly by war and revolution.

Mortality Rate

According to my estimates, over 600,000 people lost their lives because of the Second Indochinese War, 8 percent of the 1970 population. Specifically, these deaths during this period (1970–75) were a result of the direct combat between the U.S.-supported army of Lon Nol and the Vietnamese and Khmer Communist forces, as well as the U.S. saturation bombing from 1969 through most of 1973. Mortality among infants, young children, and the elderly rose precipitously due to extreme privation, epidemics, and sickness, especially in cities blockaded or under siege. No one will ever be able to determine the exact number of victims.

The mortality rate during the revolutionary period is much more difficult to calculate. Available figures vary widely and are occasionally ludicrous, depending on whether the source is friendly or hostile to the DK regime. The various methods of measuring loss of life during the Pol Pot revolution—notably those based on a reckoning of the number of widows[4] and the number of victims per family[5]—are to be used with considerable caution. Lacking precise information on migratory patterns differentiated according to sex, these procedures tend to exaggerate the masculine mortality rate resulting from the revolutionary context. It is illogical to generalize, extrapolating from the events observed in a few villages or even in a few provinces, on the status of Kampuchea's population as a

Table 2

Kampuchea: Differential Population Estimates

Year	In the absence of crises[1]	Taking into account the effect of crises[2]	Variation	
			Absolute number	Percentage
1970	7,300	7,160	– 140	– 1.9
1971	7,540	7,272	– 268	– 3.5
1972	7,770	7,380	– 390	– 5.0
1973	8,000	7,344	– 656	– 8.2
1974	8,240	7,365	– 875	–10.6
1975	8,490	7,095	–1,395	–16.4
1976	8,760	7,016	–1,744	–19.9
1977	9,040	6,860	–2,180	–24.1
1978	9,330	6,480	–2,850	–30.5
1979	9,630	6,000	–3,630	–37.7
1980	9,940	6,100	–3,840	–38.6

Notes:

1. The estimates in the absence of crises were calculated using a mean rate of annual growth. The rates used are 3 percent and 3.2 percent for the 1970-75 period and the 1975-80 period, respectively.

2. The estimates taking into account the effect of the crises are based on the following birth and death rates: 47 percent (births) and 17 percent (deaths) in 1970; 48 percent and 18 percent in 1971 and 1972; 46 percent and 21 percent in 1973; 44 percent and 23 percent in 1974, and 40 percent and 28 percent in 1975.

Births from 1976 to 1978 were calculated using a fluctuating rate of between 20 and 25 percent, while the deaths from all causes were estimated at 1 million. For 1979 and 1980 birth rates of 30 percent and 28 percent were adopted, whereas the death rate for both years was 22 percent. The years 1981-83 show an average annual rate of increase of 1.5 percent per year.

The mortality rate between 1970 and 1975 and 1979 and 1980 did not take into account approximately 600,000 violent deaths due to war in the first instance, and 500,000 deaths due to famine at the time of the Vietnamese invasion in the second.

The calculation did take migratory patterns into account.

whole, and especially the actual dimensions of the catastrophe.

The following is one example of many that show the danger of extrapolating information based on this period of crisis and political struggle. Some sources claim that there remained only sixty surviving doctors at the time of the Vietnamese invasion,[6] implying that all the rest had perished under the previous regime. The number of doctors (more than 200)[7] who fled the country and resettled abroad, mainly in France, has not been taken into account.

Confronted with these impressions and unverifiable estimates, I provisionally advance the figure of 600,000 dead during the war and adopt, somewhat arbitrarily, a mortality rate of 17–28 per 1,000 between 1970 and 1975.[8] In this calculation I have chosen an overall figure of 1 million deaths under the DK regime, including unnatural or violent deaths and normal mortality. In all, the death toll for the years 1970–78 figures at 2.3 million, up to 300,000 of whom were males who died in combat, purges, or political repression.[9] The remaining deceased consist mainly of infants, women, and the elderly.

Decline in the Birth Rate

The birth rate dropped during the war, as it did under the DK regime, due to several factors: the high mortality rate of men and women of reproductive age, mass military mobilization, intensive labor projects, and amenorrhoea caused by famine.[10]

The birth rate actually increased by one or two points during the first two years of the war (1970–72) due to a relaxation of social mores, parental insecurity prompting early marriage of daughters, and the decline of unemployment as a result of military recruitment. The fall in the birth rate began in 1973 and deepened during the 1975–78 period. U.S. Central Intelligence Agency figures vary according to social category: for the period July 1975 to July 1976, they note 10 to 15 births per 1,000 among the so-called "new people" and 30 per 1,000 among that portion of the population designated as "old" or "base people." Based largely on the testimony of refugees in Thailand, whose isolated accounts are not without exaggeration, the birth rate for the "new people" is no doubt underestimated. The majority of refugees fall into this category. I prefer a rate of between 20 and 25 births per 1,000 from 1976 to 1978 for the Cambodian population as a whole, which is 50 percent below what the rate would have been in the absence of conflict.

Official figures for the period from mid-1977 to mid-1978 record a birth rate of 50 per 1,000 inhabitants.[11] This is certainly too high; it can be taken, at best, as a general indication of a trend. Even if the slight political relaxation of the Pol Pot government in 1977 (an improvement in food distribution and policies encouraging marriage) allows one to consider the possibility of an augmentation in the birth rate the following year, the war with Vietnam, mass mobilization, and bloody purges all militate against so rapid an increase.

My theoretical calculation gives a figure of approximately 1.1 million births between 1970 and 1978.

Emigration

Emigration patterns were also altered drastically during this troubled period.[12] The most dramatic exodus was the expulsion of approximately 200,000 Vietnamese by the government of Lon Nol, which perceived them as a threat to Cambodian security. At least 12,000 Vietnamese were massacred as well. This forced emigration to Vietnam was the main cause of the sharp decline in population toward the end of 1970. Those Vietnamese who remained in Cambodia (some 200,000) at that time had to repatriate or were killed after the Khmer Rouge victory.

The other emigrants before 1979 were mainly Chinese and Khmer. They numbered at least 200,000, of which 100,000 were from the provinces adjoining Vietnam (Svay Rieng, Prey Veng, Kampong Cham), where they sought refuge in 1978 amidst the bloody struggles among Khmer Rouge leaders. All told, outward migration from 1970 to 1978 was on the order of 650,000 people, close to 10 percent of the population on the eve of full-scale war in 1970.

Demographic Changes Since 1979

Did the Vietnamese occupation, in deposing the Pol Pot regime on January 7, 1979, permit the Khmer population to "catch their breath," to recover without shock or obstacles? Here are some figures that indicate the dimensions of the upheaval and the demographic trends of the last five years.

The Famine

The famine of 1979–80 was the most catastrophic in Cambodian history.[13] It was caused most immediately by the devastation and destruction of crops during harvest season at the time of the Vietnamese invasion. Khmer Rouge soldiers also took a significant portion (10 to 20 percent) of the already meager 1979–80 harvest[14] as they retreated into the mountainous regions of the north.

When the new government came to power, it tried to "console" the population with a "laissez-faire" policy, including free travel within the country. As a result, the food supplies still housed in the cooperatives of the previous regime were emptied and squandered by the population. Peasants abandoned agricultural work as they returned to their villages or took up the search for family members, scattered under Democratic Kampuchean rule.

It is estimated that at the beginning of 1979 Kampuchea had only enough rice to feed its population for three months.[15] The diplomatic community issued warning signals about the critical shortage of food as early as March 1979, but political obstacles impeded a swift response. International public opinion was alerted only after July 1979, by the firsthand accounts of International Red Cross and UNICEF representatives, and by the mass media's coverage of the refugee influx along the Thai border. In September 1979, according to American sources, only 10 percent of previously cultivated land in Kampuchea was planted in seed.[16] In the winter of 1979 relief agency officials claimed that 2.25 million persons were in danger of dying of hunger.[17] According to the estimates of international relief organizations, the Khmer population needed external aid in the amount of 165,000 tons of rice, 15,000 tons of sugar, and 8,000 tons of oil for a six-month period.[18] Meanwhile, the World Food Program had provided for the distribution of 15,000 tons of rice per day to the 500,000 refugees in Thailand.[19]

The harvest of December-January 1979–80 was far too small to meet the needs of the population. It yielded only some 340,000 tons of rice.[20] Drought, scarcity of seed, and the reduction of paddy field acreage caused UNICEF and the IRC to revise their estimates of basic food aid needed for 1980 from 200,000 tons to 300,000 tons.[21]

Massive international aid did help to limit the extent of the catastrophe, but it did not avert famine. Political and logistical obstacles aggravated the difficulty of reaching regions with pressing needs, especially the countryside and the southern provinces (Svay Rieng, Prey Veng, Takeo, and Kampot). Indifference and incompetence on the part of the new regime only exacerbated conditions, particularly in 1979. The end result of all these factors was a subsistence crisis that took approximately 300,000 human lives in 1979 and another 200,000 in 1980—fully 7 percent of the 1970 population.[22]

The loss of life due to famine in 1979 and 1980 varied sharply according to region, category of population, and age. The death rate was relatively low in Phnom Penh, especially among soldiers, functionaries, the militia, and those factory workers who were better supplied and provided for. The army, police, and segments of the population sympathetic to the regime received priority in the distribution of humanitarian aid. The sixty-two remaining doctors (who in a population of 6 million made for a ratio of one doctor for every 100,000 inhabitants) were concentrated in the capital and treated, first of all, cadres and the wounded Vietnamese soldiers fighting to support the Heng Samrin regime.

The mortality rate in refugee camps in Thailand was also very high: for example, of 31,900 persons treated in hospitals during a thirty-eight-day period from October 28 to December 12, 1979, 509 died. The principle causes of death in the hospital in Sakeo camp from November 8 to December 5, 1979, to cite another example, were malaria (28 percent), pneumonia (28 percent), malnutrition (15 percent), diarrhea (dysentry) (11 percent), and premature birth (7 percent).[23] These illnesses were almost without exception the stepchildren of famine.

In the zone controlled by the resistence forces of the Khmer Rouge, according to the estimates of UNICEF and the IRC for 1979, two-thirds of the inhabitants were afflicted with malaria, over three-fourths were malnourished, and half the infants died before the age of one.

There are no figures to allow for an analysis of the impact of the famine on the population according to age and sex. One thing is certain: the famine claimed disproportionately more victims among very young and old cohorts, resulting in a relative swelling of the adult population. This was especially true in the southern provinces and in the countryside, where the catastrophe was most severe and relief efforts most inadequate. Population losses below the age of five and above the age of fifty were extreme.

Migration

The crisis in Kampuchea manifests itself in all its catastrophic magnitude in the migratory movements, which can be grouped into four overlapping categories.[24]

1) There was massive internal movement directly related to famine conditions in two directions: one group (500,000–600,000)[25] moved from the countryside toward the capital, Phnom Penh, while the other (about 1 million) moved toward the Thai border during the crisis period—the latter half of 1979—in search of help.

2) Another aspect of the migration was tied to a sense of insecurity and to the occupation. The movement toward the Thai frontier only constituted a stopover for many on the way to final emigration. These were refugees who established themselves provisionally with some family member while waiting. They numbered about 360,000 in 1981, 190,000 of whom had left Thailand for permanent resettlement in third countries—especially France, the United States, and Australia—since 1975.

3) There was also a return of about 100,000 Khmer from Vietnam, where they had taken refuge during the Pol Pot period. These refugees originated for the most part from Svay Rieng, Kompong Cham, and Prey Veng provinces. As they were supporters of the counterrevolutionary struggle of Heng Samrin, most returned to work for the new government in Phnom Penh.

4) Finally, there was a return migration of Vietnamese. These were original residents of Kampuchea who had left during the Lon Nol or Pol Pot regimes. Upon their return they generally took up their previous mode of livelihood. They became fishermen in the Great Lake region around Beng Tonle Sap, workers in rubber plantations in Kompong Cham and Kratie provinces, and farmers in Svay Rieng and Prey Veng provinces. Besides these, there were also Vietnamese cadres, technicians, experts, and advisers who filled the spaces left by the massacres under Pol Pot and the exodus under Heng Samrin. Though it cannot be verified with precision, the total number of Vietnamese immigrants as of 1980 was probably between 300,000 and 500,000.[26]

These massive migrations, which fluctuated dramatically during crisis periods, are very hard to quantify. The net migration is nonetheless certainly negative, though it probably did not exceed 200,000 if one takes into account the 100,000 Khmer returnees from Vietnam. The trend could be reversed, however, with the continuous and regular immigration of Vietnamese.

This net out-migration constituted not only a numerical decline in the population but also a qualitative impoverishment due to its selective nature. There also was a loss of capital, whether direct (gold or jewels carried by refugees or paid for the cost of passage) or indirect (the cost of lost educated human capital). The United States and those European countries accepting refugees from the camps in Thailand for resettlement preferred those with a high level of education (doctors, engineers, professors, and teachers) over poor and illiterate peasants. Many cadres, intellectuals, and youth left Kampuchea to seek their fortune abroad and had no intention of repatriating, especially during a foreign occupation. The hemorrhaging of educated talent simply adds to Kampuchea's dependence on Vietnam.

Demographic Recovery

According to the estimates of the new government, there was a spectacular rise in the population immediately after the fall of the Phnom Penh regime: from 4.5 million in July 1979, to 6.1 million toward the end of 1979, to 6.6 million a few months later. This would be an increase of more than 40 percent in six months. As the official figures only cover those inhabitants under government control, the total population would be approximately 8 million, counting the refugees and those living along the border in the resistance zone.

These official figures are suspect—for political more than technical reasons—and only indicate a trend. They are inevitably inflated at the time of census taking, in order to glorify the regime and solicit foreign aid, and could well be underestimated under the opposite circumstances.

After the establishment of the communist regime in 1975, Kampuchea did not have a civil administration that registered marriages, births, and deaths. Under the Heng Samrin regime, it is the heads of solidarity groups who keep a record of the number of households in the countryside in order to verify migratory patterns, establish a stable order, and distribute basic food supplies. It is in their interest to exaggerate the number of households, especially in the countryside, who suffered enormously during the subsistence crisis of 1979–80 due to a lack of aid. Thus the official population estimates given in table 3 are certainly inflated.

The alleged demographic explosion immediately after the fall of the Democratic Kampuchean regime can be explained, according to the new authorities, above all by the "baby boom": 48 births per 1,000 inhabitants in 1980 and 55 in 1981.[27] The former rate represents, they say, the birth rate for the entire population; the latter is the rate among 11,500 inhabitants of four villages situated outside of Phnom Penh. The rates are greater than those during "normal" times (45 to 47 per 1,000 between 1965 and 1970).

There was, without any doubt, a recovery in the birth rate beginning in 1981, following the consummation of marriages delayed or impeded by war and revolution, and following an increase in the number of households able to produce offspring (given a return to family living and the reunion of families separated under the Pol Pot regime).

The recovery in the birth rate and the increase in the number of marriages could not, however, be as rapid or prodigious as the new government would hope and seeks to demonstrate. To begin with, the number of households able to produce children was and is less than in 1970. Furthermore, the male mortality rate, especially among unmarried males, during the catastrophic decade prevented the marriage of a certain number of young women in the immediate recovery period. The deficit in the male population is consistent with a higher marriage rate among men but results in low nuptuality among women, especially those over thirty. In short, the famine, epidemics, and massive migration of 1979–80 are all

Table 3

Distribution of the Population of Cities and Provinces, 1968, 1979, and 1980

Location	Population (thousands)		
	1968[1] (year-end)	1979[2] (year-end)	1980[2] (mid-year)
City			
Phnom-Penh	570	200	
Kompong Som	15	10	
Province			
Battambang	685	800	830
Kampot	414	420	360
Kandal	805	770	730
Kampong Cham	977	400	1,090 (?)
Kompong Chhnang	331	250	250
Kompong Speu	361	350	350
Kompong Thom	379	375	390
Kratie	162	155	
Prey-veng	558	700	750
Pursat	223	200	180
Siemreap	371	400	480
Svay-rieng	346	340	310
Takeo	541	580	600
Koh Kong	45		
Mondulkiri	17		
Oddar Meanchey	50		
Preah Vihear	45	80	
Ratanakiri	56		
Stung Treng	44		
(5 preceding provinces)	(212)	(100) (?)	
Total	6,995	6,130	6,320 (c)

Notes:

1. Migozzi, 1973.
2. The 1979 and 1980 estimates are those published by the new administration of Heng Samrin and relate only to the inhabitants of the zone of resistance and refugees.
3. By adding in the population of Phnom Penh, which was at least up to 300,000 by mid-1980, the total population is easily 6.6 million according to these calculations.

countervailing forces to the growth in nuptuality and procreation.

All of these factors lead to the conclusion that the birth rate figures claimed by the pro-Vietnamese government are overestimates. A rate of 55 percent (55 per 1,000), even if the survey were accurate, would be an exception pertaining to a privileged segment of the population living near Phnom Penh. Based on my calculations for the period 1981–83, I propose a birth rate and mortality rate of 35 percent and 20 percent, respectively, which yield an average annual growth rate of 1.5 percent. These figures show a rise of 20 percent in the birth rate and a decline of 10 percent in the mortality rate, due to improvements in agricultural production beginning in 1980. The demographic resurgence is thus still weak in an underpopulated Kampuchea, which needs more people to develop economically.

Conclusions

What conclusions can be drawn from these approximate results? There was certainly a decline in the population during the 1970s in the wake of political and socioeconomic catastrophe. The demographic recovery started in 1981, but even then only slowly. The demographic situation is not yet stable—it rests on a fragile foundation of insecurity, uncertainty, emigration, and foreign occupation. Thus it is too soon to affirm that this positive surge is a reversal of the major negative trend of the 1970s and not simply a temporary aberration.

Might the imbalance of sexes in the population bring with it the possibility of changes in the behavior of Khmer women, for example, toward Vietnamese soldiers and immigrants, formerly seen primarily as foreigners and today as invaders, occupiers, and colonialists? These are social and psychological questions that require the expertise of other disciplines.

The Vietnamese occupation raises a question of fundamental importance:[28] What price will be exacted for the demand of independence, peace, and the freedom of the Khmer people to reconstruct their country and their population? The fate of several million Khmer hangs in the balance. The development problems in Kampuchea are in need, fundamentally, of a political solution. This is my opinion, but what do the politicians say?

Notes

1. Many articles and books provide eyewitness accounts of the Pol Pot revolution; they are generally of doubtful character. Here are a few: John Barron and Anthony Paul, *Murder of a Gentle Land* (New York, 1977); Boun Sokha, *Cambodge, La massue de l'Angkar* (Paris, 1980); Jean Lacouture, *Survive le peuple cambodgien!* (Paris, 1977); Jean Morice, *Cambodge, Du sourire à l'horreur* (Paris, 1977); Francois Ponchaud, *Cambodge, Année Zéro* (Paris, 1977); Pin Yathay, *L'utopie meurtrière* (Paris, 1981).

2. United Nations, *World Population Prospects as Assessed in 1980* (Department of International Economic and Social Affairs, Population Studies, no. 78) (New York, 1981).

3. Peter J. Donaldson, *New York Times* (NYT), April 22, 1980; U.S. Central Intelligence Agency, *Kampuchea: A Demographic Catastrophe* (GC80-10019U) (Washington, D.C., 1980); Bureau of the Census, U.S. Department of Commerce, *World Population 1979* (Washington, D.C., 1980).

4. Chantou Boua, *Women in Kampuchea* (Bangkok: UNICEF, 1981).

5. Honda Katuiti, *Journey to Cambodia, Investigation into Massacre by Pol Pot*

Regime (Tokyo, 1981).

6. Figures released by the Heng Samrin government.

7. Personal communication from the president of the French Medical Association.

8. Sihanouk also cites the figure of 600,000 deaths during the 1970–75 war. (See Shawcross, *Sideshow*, p. 381; Boun, *La massue de l'Angkar*, p. 208; and Morice, *Cambodge, Du Sourire a l'Honeur*, p. 361.) The 600,000 figure consists of losses due directly to the war, including military and civilian casualties. Adding indirectly caused deaths—due to poor living conditions and inadequate health facilities from 1970 to 1975—the figure, according to my estimate, increases to 700,000.

9. According to my estimates, of the 1 million deaths caused specifically by the Pol Pot revolution (above and beyond "normal" mortality), 900,000 were caused by famine, epidemics, and poor work conditions. More than 100,000 individuals were killed by violent means. Overall, 1.3 million died during 1970–75, and 1 million during 1975–78, yielding a figure of 2.3 million for the eight-year period.

10. See especially Jean-Pierre Willem, "The Population of Cambodia is in Danger of Disappearing," *Science and Life* 736 (January 1979), pp. 34–35.

11. Statement by Pol Pot on the nineteenth anniversary of the founding of the Communist Party of Kampuchea (September 12, 1978), Phnom Penh.

12. On the expulsion of Vietnamese under Lon Nol, see Joseph Pouvatchy, "The Vietnamese Exodus from Cambodia in 1970," *Asia World* 7 (Fall 1976), pp. 339–49. See also *Baltimore Sun*, August 22, 1970. For emigration during the Pol Pot period, see Milton Osborne, "The Indochinese Refugee: Cause and Effects," *International Affairs* 26 (January 1980).

13. On the famine of 1979–80, see *New York Times*, October 24 and 25, 1979; *Southeast Asia Refugee Crisis, Hearing before the Subcommittee on East and Pacific Affairs of the Committee on Foreign Relations, United States Senate*, 96th Congress, September 27, 1979 (Washington, D.C., 1980); *1979, Tragedy in Indochina: War, Refugees, and Famine, Hearings before the Subcommittee on Asian and Pacific Affairs of the Committee on Foreign Affairs, House of Representatives*, 96th Congress, 1st Session (Washington, D.C., 1980); *Cambodian Famine and U.S. Contingency Relief Plans, Hearing before the Subcommittee on Arms Control, Oceans, International Operations, and Environment of the Committee on Foreign Relations, United States Senate*, 96th Congress, 1st Session, November 1979 (Washington, D.C., 1980) *Cambodian Relief: A Report to the Committee on Foreign Relations, United States Senate, October 1980* (Washington, D.C., 1980); *Food Aid to Cambodia, Hearing before the Subcommittee on Foreign Agricultural Policy of the Committee on Agriculture, Nutrition, and Forestry, United States Senate*, 96th Congress, 1st Session, November 19, 1979 (Washington, D.C., 1980); *1980—The Tragedy in Indochina Continues: War, Refugees, and Famine, Hearings before the Subcommittee on Foreign Affairs, House of Representatives*, 96th Congress, 2d Session (Washington, D.C., 1980); International Disaster Institute (London), *Kampuchea—An Assessment of Current Information, November 5, 1979* and *Kampuchea Update, December 11, 1979*; Elizabeth Becker, "The Politics of Famine in Cambodia," *Washington Post*, November 18, 1979; William Shawcross, "The End of Cambodia?" *The New York Review of Books*, January 24, 1980; "Cambodia's Long Path Back," *Nature* 281 (October 11, 1979); "The Cambodian Picture Becomes Clearer," *Nature* 282 (November 2, 1979); "Kampuchea's Shattered Agriculture Is on the Way To Recovery," *World Agricultural Report*, June 10, 1981; Julius Holt, "The Kampuchean Emergency," *Food*

Policy (February 1980); Roger Kershaw, ''Multipolarity and Cambodia's Crisis of Survival,'' *Southeast Asian Affairs 1980* (Institute of Southeast Asian Studies, Heineman Asia); Richard Nations, ''Battle for the Hearts and Stomachs,'' *Far Eastern Economic Review*, December 7, 1979; ''Starvation, Deathwatch in Cambodia,'' *Time*, November 12, 1979; Joseph J. Zasloff, *Kampuchea: A Question of Survival* (Reports, American Universities Field Staff, Hanover, NH, 1980).

14. Zasloff, *Kampuchea*.

15. *Business Times* (Singapore), May 18, 1979.

16. ''Cambodia's Long Path Back,'' *Nature* 281 (October 11, 1979). See also Murray and Linda Gibson Hiebert, ''Famine in Kampuchea: Politics of a Tragedy,'' *The Southeast Asia Record*, December 14–20, 1979, p. 17.

17. *New York Times*, August 8–10, 1979. See also Zasloff, *Kampuchea*.

18. *Le Monde*, November 6, 1980.

19. Ibid.

20. Presuming a minimum of 420 grams of rice per individual per day, the annual consumption of a population of 6 million would be more than 900,000 tons.

21. *Le Monde*, November 20, 1980.

22. Author's estimate.

23. *The Lancet*, April 19, 1980; see also Susan E. Holck and Willard Cates Jr., ''Fertility and Population Dynamics in Two Kampuchean Refugee Camps,'' *Studies in Family Planning* 13, 4 (April 1982).

24. See especially Barry Wain, *The Asian Wall Street Journal* (Singapore), November 14 and 15, 1981.

25. According to the *New York Times*, April 29, 1980, the population of Phnom Penh grew from 100,000 at the beginning of 1979 to 500,000 at the beginning of 1980, a fivefold increase due to the influx of starving people.

26. This estimate is based primarily on refugee accounts. In 1983, the Heng Samrin government claimed there were only 50,000 new immigrants from Vietnam. Sihanouk claimed in November 1983, on the other hand, that there were over 1 million. *Far Eastern Economic Review* (Hong Kong), November 10, 1983.

27. Michael Richardson, ''Cambodia, the 10-Child Family,'' *Far Eastern Economic Review*, February 12, 1982; Frances Starner, ''Born Out of Sorrow, a National Baby Boom,'' *Asiaweek* (Hong Kong), June 19, 1981.

28. See my article ''Kampuchea: A Country Adrift,'' *Population and Development Review* 7, 3 (June 1981).

May Ebihara

REVOLUTION AND REFORMULATION IN KAMPUCHEAN VILLAGE CULTURE

Peasant communities in Cambodia/Kampuchea have undergone some extraordinary changes as the result of major sociopolitical upheavals in the nation over the past decade. This discussion examines continuities and discontinuities in local-level social organization and culture through a comparison of village culture in the "old society" prior to 1975, the reformulations of local life attempted in Democratic Kampuchea (DK), and the situation in the People's Republic of Kampuchea (PRK). The nature of village organization and culture in the PRK cannot be fully understood without reference to preceding periods.

I approach the subject matter as an anthropologist dealing most generally with culture in the broadest sense as encompassing the various material, social, and ideological components of people's lives, including both shared norms and behavior. More specifically, I am concerned with the grass-roots existence of the peasantry and shall use various anthropological concepts to discuss the sociocultural transformations that occurred in their lives. While it is a dictum in anthropology that peasant communities cannot be viewed as isolated entities but must be seen within the context of a larger society (and, indeed, global economic and political systems), my primary focus will be the local level as it was affected by and reacted to policies from above and circumstances generated by larger forces.

While my knowledge of pre-1975 Cambodian culture is based on firsthand field experience in a Khmer peasant village, my discussion of Democratic Kampuchea and the PRK must rely on the work of others who are making significant contributions to our knowledge through research using a variety of documentary materials and refugee accounts.[1] Finally, it must be emphasized that there has been variation in community social organization both regionally and through time, such that it is often difficult to make universally applicable statements about conditions. Nonetheless, with this proviso in mind, I offer a broad characterization of change and persistence in local life as abstracted from available data.

Pre-1970 Village Organization

This description of pre-1970 village life is based primarily on my own field research conducted in 1959–1960 during the Sihanouk era in a Khmer peasant village.[2] I do not pretend to claim that this particular community was typical of all villages in Cambodia because, nationwide, there were variations with regard to primary subsistence base, ethnic composition, average size of landholdings, etc.[3] Nonetheless, the villagers I knew were good representatives of the majority of Cambodian peasants who were small landholders-cum-cultivators; many fundamental aspects of peasant existence and Khmer culture were manifest in the community. Further, I believe that village life as I knew it continued basically unmodified until the late 1960s and early 1970s when conditions in the country at large became increasingly disturbed as the Vietnamese conflict spilled over into Cambodia. The following discussion will focus on certain aspects of village life that are pertinent to and provide a point of departure for consideration of changes in local-level organization that occurred in Democratic Kampuchea and the People's Republic of Kampuchea.

The community I studied, to which I gave the pseudonym Sobay, was located about thirty kilometers southwest of Phnom Penh in a region Delvert (1961: 542) characterized as "typically" Cambodian in terms of ethnic composition, ecology, and cultural practices. Its population of about 790 residents, all of whom were ethnically and culturally Khmer, lived in three named hamlets or neighborhoods; my research focused in particular on one of these sections, West Hamlet Sobay, with about 160 inhabitants living in 32 households. Physically, Sobay was a cluster of wood and/or thatch houses set amidst and interspersed with a verdant growth of palms, fruit trees, kitchen gardens, and patches of wild flora. Separate from but adjacent to this residential area of homes was a vast patchwork quilt of rice paddies, in varied shapes and sizes, which formed the basis of village economy.

The vast majority of Sobay's villagers were peasant owner-cultivators[4] of wet-rice, with one crop a year grown (during the rainy season) primarily for family consumption rather than the market. Means of production—land, draft animals, plows, etc.—were owned individually.[5] Both men and women could and did own land and transmit other kinds of property to both daughters and sons, with an ideal of equal division among offspring. Because of this bilateral inheritance pattern operating over a number of generations in a region that had been fairly densely populated since the nineteenth century, paddy lands had become heavily parceled and limited in amount of holdings per household. In West Sobay the average amount of paddies owned was about one hectare per household, with a range of .06 to, in one case, 4 hectares.[6] Absentee landlordism did not exist in Sobay, but there was tenancy between villagers; that is, those with little or no land might work someone else's fields in return for half the crop, or an individual who had inherited paddy land in another village (but moved to Sobay upon marriage) might have that land worked by a kinsman who would then turn over half the produce. Such arrangements were regarded as mutually beneficial, not as exploitative rent extracted by the rich from the poor.

The household—usually a nuclear or stem family[7]—was the basic unit of production and consumption. It cultivated its fields using primarily the labor of adult and adolescent family members, although kinsmen and neighbors commonly formed small or large cooperative work groups during the busiest periods of plowing, transplanting, and sometimes harvesting and threshing. Such cooperative efforts (*provas dai*) operated on the basic principle of reciprocal labor exchange between families/households. For example, several men might form a plowing team to work one another's fields in turn, while groups of women did the same for transplanting, thereby completing arduous tasks more quickly and easily than could family members alone. In a slightly different vein, there were also arrangements whereby, for instance, a man would agree to plow the fields of a widow without oxen or plow in return for her help in transplanting.[8]

The relatively small size of paddy holdings per household meant that Sobay villagers produced rice primarily for subsistence rather than for sale. The yield per hectare varied according to such factors as the quality of the soil, type of rice planted, and especially the timing and amount of rainfall necessary for cultivation of wet-rice in the absence of irrigation systems. Rice consumption per household also varied according to the number and ages of family members, but, generally speaking, an "average" rice crop from one hectare of paddies might or might not provide sustenance until the next harvest, depending on the size of the family and the general state of its resources (e.g., whether they had debts or major expenses).[9] While those families with two or more hectares of paddies were relatively well-off in that they could produce enough rice for food and usually some surplus for sale (except in years of very poor harvest), those with one hectare or less led a more variable and precarious existence: they might be able to sell some rice if the crop was abundant, but in bad years the granary might well be exhausted before the next harvest.

Villagers were, then, primarily subsistence cultivators, and other foods and household items were also produced at home. Families commonly owned fruit trees and palms, cultivated vegetables and herbs in kitchen gardens, caught fish in the rice paddies during the rainy season, and fabricated some material goods such as baskets and mats. However, these were not sufficient to satisfy all needs. Villagers were by no means economically self-sufficient but were tied into a national (and ultimately global) market economy that required money. Some cash income was needed to purchase certain foods and essential household goods not produced at home, to pay taxes or debts,[10] or to use for ceremonial purposes, such as giving offerings to the Buddhist temple or sponsoring weddings, funerals, and other domestic rituals.

Hence, as is common with peasantry elsewhere, many households (particularly those with one or less hectare of paddies who could not depend on marketing surplus rice) sold other produce or labor to get some cash income. A variety of part-time activities were pursued, including raising pigs and chickens for sale (not home consumption), making palm sugar for sale, [11] marketing surplus produce from gardens and trees, vending cooked food items, coolie labor on temporary construction jobs in the area, and men working as pedicab (*cyclo*) drivers in Phnom Penh during slack seasons of rice cultivation.

Sobay's economy, then, operated largely (if not totally) in terms of what Sahlins has called a "domestic mode of production" insofar as there was family control of productive resources and disposition of produce, small-scale technology with "homespun" implements handled by the family, and production geared primarily to meet household needs (with market sale of goods in order to get money for purchase of other commodities).[12]

Within the national political structure, Sobay—like other villages—constituted the lowest level of a territorial administrative hierarchy that descended from the central government down through the province, district (*srok*), and subdistrict (*khum*) to the village (*phum*), each with its own officials. Sobay's local administration had an elected village chief, deputy chief, and two assistants, whose primary responsibilities were to pass on orders and policies from above and to transmit taxes, vital statistics, and records (e.g., property transactions, marriages, divorces) from below. The *mekhum* or subdistrict chief was also a local resident and fellow villager elected by the adult population of the six communities that were included in the *khum*. He performed similar duties and served some additional functions such as occasional adjudication of disputes. He was also a critical link or mediator[13] in connecting local with national administration through his maintenance of contacts with the district office (located in a nearby market town), which was headed by a professional bureaucrat assigned to his post by the central government.

In 1959–1960 Sihanouk was in firm control of the government and greatly respected by the villagers. Sobay's residents were generally apolitical in the sense of knowing (and, seemingly, caring) little of national or international politics; they usually echoed whatever Sihanouk's latest pronouncement had been. Any sort of subversive thoughts or activity were strongly censured and guarded against by a village militia made up of able-bodied adult males who took turns patrolling the village at night. Thus, while some villagers had heard of Communist doctrine (which was, as one villager put it, "The *komuni* want to take from the rich to give to the poor"), there was no interest professed in it—or anything else contrary to Sihanouk's latest views.[14]

In addition to forming a territorial-political entity, the village also constituted a social unit, an aggregate of known and trusted kinsmen, friends, and neighbors. Although there were few communal activities as such, diverse and overlapping bonds of kinship and friendship linked various households together, and the village was a focus for identity and loyalty, contrasting "our village" to "others."

Within the realm of social relationships, family and kinship were of great importance. There were several major types of families due to variable patterns of postmarital residence: married couples could set up independent households or, because of financial necessity or filial piety, live with the family of either the husband (patrilocal residence) or wife (matrilocal residence). Hence, within Sobay one found nuclear families, stem families (in which one child remained at home upon marriage and brought his/her spouse into the household), and extended families in which relatives of one kind or another (e.g., an orphaned nephew or widowed sister) had been taken into the household because of obligations to care for needy kin. I have already noted that the family operated as a fundamental economic unit of production and consumption. It was also the locus of very strong emotional bonds, loyalty, and mutual obligations, particularly between parents and children and, ideally, between siblings as well.

There were no larger organized kin groups beyond the family, but the bilateral kinship system meant that each individual recognized a broad circle of relatives (termed a ''kindred'' by anthropologists) on both mother's and father's sides of the family and including kin by both blood and marriage. A network of kin relations linked people within Sobay itself and spread out also to other communities (including Phnom Penh) because of a high percentage of village-exogamous marriages. Within the kindred there was latitude for personal likes and dislikes to operate, such that an individual might be intimate with some relatives and rarely see others. Nonetheless, kinsmen were generally regarded as important sources of affection, support, and aid of various sorts.

Apart from the family/household, there were no organized groups in Sobay, whether formal associations, clubs, political parties, or the like. Neither were there major class strata within the community, although the question of socioeconomic levels is a complicated one with several facets. Pre-1975 Cambodian society as a whole could be roughly divided into three broad classes: an upper stratum of royalty and high government officials; a ''middle class'' of white-collar workers, professionals, minor bureaucrats, etc.; and a lower class of peasants, artisans, and urban proletarians. Sobay's residents all constituted part of this lower stratum. The villagers spoke of themselves as ''poor country folk'' in contrast to urbanites and those with nonagricultural occupations such as school teachers. Within this broad category, however, finer distinctions were drawn by the villagers themselves (and, subsequently, by analysts of rural society). People in Sobay recognized degrees of relative prosperity based primarily on how much paddy land was owned by a family. At the top were ''people who have'' (*neak min*)[15] many paddies, a fine house, nice clothes, good food (two families with over two hectares of paddies fell into this category). Next were ''people with enough'' (*neak kuésóm*) with an adequate amount of paddies, food to eat through the year, and a comfortable existence (families with two hectares of paddies). Most of the hamlet was said to be ''poor folk'' (*neak króó*), families with a hectare or less of rice fields who might barely make it through to the next harvest. Finally, about six families with less than half a hectare of land or no paddies whatsoever were characterized as ''poorer than poor'' (*neak toal*), leading a hand-to-mouth existence.[16]

Although villagers recognized degrees of relative wealth among themselves, there also existed simultaneously a general ethic of fundamental egalitarianism. This was manifest not only in the villagers' regarding themselves as "poor people of the rice fields" (even those who were relatively well-off spoke of themselves as such), but also in the fact that material wealth in and of itself was not the basis for high status among fellow villagers. Rather, individuals were given special respect or prestige on the basis of qualities such as age, religiosity, and especially "good character" (*chet l'óó*). Those who were relatively comfortable economically did not engage in conspicuous displays of wealth or condescending behavior toward their neighbors. It would have been difficult for a stranger in Sobay to tell who was relatively rich or poor from the way people dressed, behaved toward one another, or conducted their daily activities. While some might view this egalitarian ethic and behavior as a mask for underlying economic inequalities, the fact remains that such conceptions and behavior were important aspects of village life.

A final critical aspect of village life was religion, which combined Theravada Buddhism and folk beliefs and practices revolving around a variety of animistic spirits. The Buddhist temple served as a moral, social, and educational center for villagers. Monks were accorded the highest respect as living embodiments of moral values and virtues. Seventy-five percent of West Hamlet males over the age of seventeen had been monks at some point in their lives for periods ranging from a few months to several years. The conception that an individual goes through a cycle of reincarnations and that the number of meritorious deeds performed in this lifetime will affect one's position in the next life exerted a powerful influence on individual behavior. Villagers strove to earn merit in a number of ways: by becoming a monk (a path open, however, only to males); by giving offerings of food, money, goods, and services to the monks and temple; by attending ritual observances at the temple; and, importantly, by following Buddhist norms of conduct in one's daily life, particularly the major injunctions against killing living creatures, lying, theft, immoral sexual relations, and alcoholic beverages. The major annual festivals (especially New Year, *Prachum*, and *Katún*) were high points in the annual round of village social and ceremonial life. These occasions offered opportunity not only to accumulate merit through prayers and offerings, but also to socialize with kin, friends, and acquaintances from other communities.

Such features of peasant life continued, I believe, largely unmodified until the late 1960s and early 1970s when Cambodia began to crack under the pressure of numerous difficulties. During the Lon Nol regime, which deposed Sihanouk in 1970, there were increasing problems with a deteriorating economy, widespread corruption, alienation between the government and people, the growing organization and militancy of the Khmer Rouge, and, of course, the spillover from the raging conflict in neighboring Vietnam.[17] Much of the countryside experienced severe disruption and devastation in the fighting between government troops and insurgents, as well as the "strategic bombing" policy instituted by Richard

Nixon and Henry Kissinger, which dropped a staggering tonnage of bombs (more than had been used on Europe during all of World War II) on not only Communist enclaves but ordinary Cambodian folk as well.[18] Clearly, then, the fabric of customary peasant life was already being seriously torn in these years. With the remarkable growth of the Khmer Rouge in strength and numbers through the early 1970s,[19] communities in various parts of the country gradually came under their control prior to 1975 and had early experience (in more moderate form) of some of the organizational and ideological changes that would soon be imposed on the entire population. In other regions where control was contested, there was obviously great turmoil and destruction. Numerous people fled the countryside to the relative security of Phnom Penh. I suspect this is what Sobay villagers did, as they had done once before in another period of civil unrest during the Issarak movement after World War II.

While I have no firsthand information on what happened to Sobay, news items on Cambodia in the *New York Times* during the summer and early fall of 1973 periodically mentioned a market town with which I was very familiar because it was only two kilometers from Sobay.[20] This town and its environs were the scene of heavy fighting, mortar and rocket attacks, and bombing by American jets as government and rebel troops battled for control of the area. Some accounts also mentioned a massive flow of refugees out of the region to Phnom Penh in the early summer. In July, both U.S. and Cambodian planes bombarded this area daily for a week. Knowing that bombs and artillery fire never land precisely on target and that the village was only a short distance from the town, I believe it is more than likely that in the summer of 1973 Sobay as a physical entity was quite literally blown apart.[21]

Democratic Kampuchea

This discussion of Democratic Kampuchea will not consider the complex questions of how and why the Khmer Rouge came to power or the various forces, both external and internal, that shaped DK economic and political policies. The focus will be on certain aspects of local-level social organization and culture, noting in particular the dramatic and often drastic alterations in life under the DK regime as compared to that of the "old society." While our knowledge of Democratic Kampuchea is still incomplete and perspectives on this period differ, information on events and conditions during this time is expanding rapidly with the research of scholars such as Kiernan, Carney, Chandler, Vickery, Summers, and Heder, whose work forms the basis for this discussion. The existing data are sufficient to permit a broad characterization of some of the sociocultural developments instituted by DK forces in "base" regions of the country occupied before 1975 and nationwide after their victory.

The policies of Democratic Kampuchea aimed at radically reformulating Khmer society and culture into a new revolutionary order. DK ideology and rhetoric spoke of a Kampuchea purged of contaminating colonialist-imperialist domination and influence. This purification did not mean the construction of a society entirely *de novo*.[22] As has been the case in other revolutionary regimes, DK leaders used or tried to build upon selected elements of customary culture that involved links to precolonial or noncolonial tradition, thus providing symbols and models that would be familiar and hence presumably acceptable to the populace.[23] This can be seen simply and graphically in the DK flag, which used three towers (presumably of Angkor but not identified as such) to represent, in the words of the DK Constitution, "national tradition," while the red ground "symbolizes the revolutionary movement."[24] There are other examples of how traditional concepts and models can be utilized to organize and institute a new social order. The drive for economic reconstruction, maximization of production, and development of relative economic self-sufficiency all involved an ideological glorification of peasantry and manual labor as being of high social value, and a pragmatic policy of turning most of the population into peasants/workers. Prescribed garb was the checkered *kroma* scarf, a ubiquitous item of dress with multiple uses among common folk, and the plain black clothing worn by villagers when doing hard labor.[25] Further, the long-established peasant tradition of mutual aid and cooperative work teams was vastly extended and institutionalized to form the basis for a new structuring of community organization and social relations of production.

While certain elements of the old society persisted, more striking (and, from an anthropological perspective, quite significant) were the changes wrought by Democratic Kampuchea's attempt to create a new social order. Several critical cornerstones of village life were forcibly suppressed or altered: villages were reorganized into communes and work teams; household production and consumption were replaced by collectivized property, labor, and distribution; family and kin relationships were seriously weakened in favor of other associational units; Buddhism was crushed and supplanted by secular "revolutionary" norms and behavior. Whatever the official rationales were for such policies,[26] the changes seem geared toward the dissolution of certain basic institutions—the family, village, and *wat*—that would have competed with the state for people's loyalties and labor. An effort to rechannel individual sentiments and activities to the service of the revolutionary order was certainly evident in DK rhetoric and in the sometimes Draconian measures used to ensure compliance with new behaviors that were not easily or voluntarily acceptable to much of the population.

Many others have stressed the important point that local organization and conditions varied both regionally and through time, depending on such factors as ecological setting, ideological divergences and contention for power among factions of Kampuchean communists, stages of collectivization, and interpretations and implementations of policy by local cadres. Some communities had reasonable work requirements and adequate food, while others were subject to stringent discipline and starvation. After 1977, however, when Pol Pot's faction gained ascendancy, conditions generally became much harsher. Many regions experienced stricter rules of communalism and controls over behavior, cuts in food rations, and large-scale purges and executions.[27] Recognition that such variations existed is very important, but in what follows I must speak instead in more general terms of broad changes.

In the society as a whole there were several major and interrelated developments that restructured the overall sociopolitical order and formed the context for changes at the local level. First, the old system of provinces was replaced by new territorial-administrative zones (given compass designations such as Northeast Zone), which were divided into *damban* (variously translated as regions or sectors). Subdivisions of *srok* (district), *khum* (subdistrict), and *phum* (village) were retained, although the last in particular came to be modified. Each territorial level was administered by a committee, with the hierarchy reaching ultimately to the Central Committee of the Kampuchean Communist Party, that somewhat mysterious body known popularly as *Angkar*, the "Organization," which was eventually dominated by Pol Pot and his associates. As in the past, policies came down from above; but local cadres, especially before 1977, implemented such policies in different ways.[28]

Second, the DK regime initially evacuted urban areas and dispersed their populations to various rural communities. The reasons for this evacuation have been discussed and debated in various works,[29] but most likely involved a complex of factors: need for agricultural production,[30] facilitation of political control by a limited administrative cadre, fear of subversive elements in Phnom Penh, ideological conceptions of the city as symbolic of corrupt colonialist-imperialist forces, and revolutionary ideals of a communal, egalitarian society.

Whatever the causes, along with this extraordinary demographic displacement came a third and critical social transformation: the attempt to level the class structure of the old society. The old system of stratification was replaced, however, by new social distinctions that often affected an individual's placement and treatment in the new order. There were several dimensions of status. The Communist Party of Kampuchea (CPK) offered several class analyses of the old society as a whole, with different versions delineating major classes (feudal, capitalist, petty bourgeois, peasant, worker) and subclasses within each (e.g., rich, middle, and poor peasants).[31] After 1975 the CPK "formally defined only two classes: the workers and peasants plus the 'revolutionary ranks,' "[32] although former class background was a criterion used by the new regime. Obviously, in the new social order poor peasants and workers were now socially and politically favored over those who came from higher classes, and the latter were expected to assume the behavior and demeanor of the former.[33]

Another, evidently quite important, distinction was made between "base people" and "new people."[34] The former (also referred to as "old people" and by several other appellations) were those liberated by the Khmer Rouge before 1975, from DK base areas, or belonging to the "basic classes" of the poor and lower-middle peasantry. "New people" included those liberated as of 1975, primarily evacuees from Phnom Penh, other urban centers, and market towns. According to Heder (1980: 5–6), this "simplistic dichotomization" was not official DK policy and was complicated by the fact that other features such as "political attitudes" and class background were also taken into account. Various reports suggest, however, that the base people/new people categorization was widely used for social and political purposes in local-level organization, with base people generally (if not necessarily always) favored in various respects.

Another component of sociopolitical status was "political attitude" in terms of proven loyalty and adherence to the CPK.[35] There was a tripartite classification of membership in cooperatives: persons with "full rights" (*penh sith*), "candidate" (*triem*), and "depositee" (*bannhau*). The first were entitled to join the party and army and, within the cooperative, could hold any political position and receive full food rations. "Candidates" could hold some low-ranking political positions and were next in line for food distribution; "depositees" were "last on distribution lists, first on execution lists, and had no political rights."[36] Class background and base/new people designations were essentially permanent statuses, although some individuals managed to conceal higher class origins and "pass" as lower class. But one could change statuses in the depositee/candidate/full rights classification, moving upward if one proved worthy, or downward if one fell from grace for any reason.[37] Heder says that this categorization was not widely known among the populace, but Summers suggests that in the immediate post-1975 period it was more important than class background because, at that time, the CPK's class analysis was "either unknown or interpreted freely . . . in a manner that reinforced the judgment of 'political attitude.'" This system of classification was in effect until the Central Committee suddenly abolished it in September 1978 when Full Membership (in cooperatives) status was conferred on everyone.[38]

An individual's status was, then, a complex mix of several variables (class background, base/new, full rights, etc.) in which one or another factor might take precedence in defining one's general social and political position.[39] A further complication was regional and even local variability as to how people of different categories were treated, depending on available food resources, policies of local administrative cadres, and attitudes of local populations, etc.[40]

Economy

Where the village or hamlet had constituted a fundamental sociopolitical unit in the old society, the cooperative became "the organizational foundation of Cambodia" from 1975 to 1978.[41] In some places, cooperatives were contained within or coincident with villages; that is, existing communities (albeit with changed or expanded populations after 1975 when urban evacuees were dispersed through the countryside) were collectivized. Other cooperatives were evidently new creations formed when people were sent or relocated into previously unsettled areas. At the time of the DK victory and in the initial period of the regime, cooperative organization varied from place to place with respect to degree of collectivization, age of the cooperative, implementation of directives by local cadres, and ecological resources. From the end of 1976 into 1977 there was a move toward increasing collectivization of various aspects of life and consolidation of smaller cooperatives under subdistrict (*khum*) administration.[42] Whatever the particular genesis or size of a cooperative, its basic structure was quite different from traditional village organization in a number of respects.

In old-society villages, the means of production (land, oxen, plows, other agricultural implements) were individually owned, and the family served as the primary unit of production and consumption. In Democratic Kampuchea the ownership of resources, production, and consumption all became increasingly communal through time.[43] The DK Constitution (Article 2) explicitly states that "All important means of production are the collective property of the people's State and the common property of the people's communities."[44] The Constitution also states that "property for everyday use remains in private hands," but it is not clear to what this refers. Some private plots around a family home, evidently comparable to the kitchen gardens that existed in old-society villages, were permitted in pre-1975 base area communities to grow vegetables, fruit, and tobacco for private consumption or trade and are mentioned also in a 1978 radio broadcast.[45] But I do not know whether such family plots were widespread during 1975–78.

With Democratic Kampuchea's deep concern for restoring and increasing agricultural productivity after the ravages of war, people were set to work not only on cultivation but also on construction of large-scale irrigation projects. A system of reservoirs, dams, and canals was intended to eliminate dependence on monsoon rains and enable more than one harvest a year or higher yields per hectare.[46] Such irrigation systems would also rationalize the layout of paddy fields (formerly broken up into parcels of varying shapes and sizes) into regularized, quadrangular plots.[47] (Note that the DK national emblem depicts "a network of dikes and canals which symbolize modern agriculture," along with "a factory symbolizing industry, framed by an oval garland of rice ears."[48])

In addition to rice, cooperatives were also encouraged to cultivate other crops: vegetables and fruits (bananas, corn, beans, tubers) as well as nonfood crops (cotton, rubber trees, jute). Fishing and the raising of pigs, chickens, and ducks provided additional food resources in various areas.[49]

DK cooperatives combined old and new models of organization: mutual aid and military structure. In the old society, villagers had formed ad hoc, temporary work crews during the busiest periods of cultivation. Now the populace was formally organized into permanent labor teams, and cooperative effort became the norm for virtually all activities. The tradition of *provas dai*, mutual aid, was institutionalized on a vast scale to become a cornerstone of collectivization.

Joined to this was the model of military structure extended from the armed forces to civilians. Work in Democratic Kampuchea became regimented both literally and figuratively as labor teams were organized on the basis of a military hierarchy of unit and behavior was subject to order, uniformity, and discipline.[50]

A fundamental unit of organization was the *krom*, a group variously reported to have consisted of ten to fifteen "families."[51] "Nuclear households" were used as "accounting units."[52] In pre-1975 cooperatives in base areas, these *krom* were organized into larger "production teams" of twenty or more families (perhaps 100–200 people), several of which formed a cooperative.[53] Toward the end of 1976 and into 1977, according to Summers (1981: 23), the government instituted "a new 'socialist revolution'" with "further rationalization and centralization of the cooperative management system as a whole." Cooperative units were grouped together at the *khum* (subdistrict) level of management in most cases; and the basic *krom* were organized into progressively larger units using a military system of designation into platoons, companies, battalions, regiments, brigades, and divisions, each composed of three units of the lower level (e.g., three companies making up one battalion).[54]

While some work teams were evidently more or less sedentary (i.e., based in certain communities), there were also mobile labor teams composed primarily of unmarried people over age fourteen, although they sometimes also included married men. These so-called shock troops, which were of varying sizes and divided into male and female groups, might work for their own cooperatives or be sent elsewhere to construction projects, state farms, or on special tasks such as harvesting in different areas.[55] In addition, children and the elderly were not excluded from the labor system but organized into their respective groups and assigned various tasks commensurate with their ages and abilities.[56] The harshness of work requirements in Democratic Kampuchea is a subject that has been debated. This is a difficult question because several variables are involved: local variations in work loads, changes through time, and differing perceptions of labor depending on the backgrounds of individuals. Obviously an upper or middle class urbanite would find manual labor more taxing than a peasant. On the one hand, work hours reported for some areas were quite similar to the work-day schedule I observed in Sobay during much of the year. Work loads as well were not unusual in some places. On the other hand, there is also evidence that work in many regions was indeed severe in terms of both hours and amount of labor required, and that conditions became generally harsher in a number of respects after 1977–78.[57] Furthermore, work in old-society villages such as Sobay had been regulated primarily by the ecological demands of cultivation and, within that framework, carried out according to familial and individual predilections and with various periods of relative leisure.[58] Labor in Democratic Kampuchea, however, was regimented and dictated by official authorities, and failure to fulfill work requirements could be punished by cuts in food rations, beatings, or even death.

Accompanying the collectivization of resource ownership and labor, the collection and distribution of produce and goods also became centralized. Rice, other crops, and foods went to cooperative storehouses (which existed also for storage of tools and supplies). From the common granary, some portion of rice was appropriated by the state; some might be used to barter for other food or goods for the cooperative;[59] and some was redistributed to cooperative members. Reports indicate that in the early years of Democratic Kampuchea and for the mobile work teams, individuals were given full to partial food rations according to whether they filled daily work requirements.[60] Children, the elderly, and disabled presumably received lesser rations than did able-bodied workers; and I presume further that in this early period, individual rations were eaten in family meals (or, in the case of the mobile work teams, by some unit of the group). But even food consumption became collectivized with the institution of communal dining halls, which appeared in some regions at the end of 1975 and became widespread in 1977.[61] Such communal dining, which was coincident with cuts in food rations and generally harsher living conditions in most areas, served several functions. By controlling distribution down to bedrock level, the state could extract more produce, which, Summers suggests, was used to support industrial development and military defense. It also enabled greater control over the population by preventing storage of food for escapees and by allowing a means of daily ''rewards and punishments'' for individual conduct. Finally, it also undermined the family as a social unit.[62]

Individuals were dependent on the cooperative for other items as well, such as clothing and other necessities, although how exactly these were acquired and allocated is not clear.[63] Additionally, cooperatives were supposed to provide social services such as health care and schools, but there have been reports that both were very limited in nature.[64]

Family and Kinship

In its attempt to control various aspects of life and to transfer authority and loyalty from local foci to the central state, Democratic Kampuchea undermined the solidarity of what was perhaps the most important grass-roots social unit: the family.[65] It is true that numerous individual families had already been physically torn apart or decimated by separations and deaths in the turmoil of the early 1970s, but the concept of the family remained. In Democratic Kampuchea, however, various measures, new forms of organization, and revolutionary ideology served to weaken greatly the family and, by extension, broader kinship bonds. Traditional family structure, sentiments, and activities were remolded or suppressed by a variety of means.

The village family in the old society was a unit of parents and children (and often other relatives as well) who lived and worked together, shared resources, and were emotionally attached to one another. In Democratic Kampuchea, the family was changed in all these respects, first by physical separation of family members. While husbands and wives evidently still resided together, they might see little of one another because of long hours of labor in separate work teams; able-bodied married men could also be put into mobile teams and sent away from home for extended periods of time.[66] More significant and striking was the rupture of the strongest affective tie in village life, the relationship between parents and children. Children past the age of about seven were separated from their families, housed separately in dormitory-like arrangements, and mobilized into youth labor teams working in the community or, in the case of teenagers, often sent elsewhere.[67] Henceforth they saw parents only infrequently or, according to one report, not at all.[68]

This separation of children from parents had implications beyond the family itself. Youth were a special target for indoctrination into revolutionary ideology;[69] they would be, of course, more amenable to socialization (or resocialization) into new forms of thought and behavior. Such indoctrination, combined with actual organization into distinct work teams and youth associations, as well as physical separation from home, would produce alienation from family ties and development of primary loyalties to other groups such as the association, the army, the party, and the revolutionary state in general.[70] As a corollary, children past a certain age were also no longer under the authority of their families but rather of various representatives of the government and party who supervised the groups just noted. Hence, in Democratic Kampuchea the customary relationship between the old and young was reversed. The parents and elders who were once accorded respect could now be viewed as survivors of the reactionary old regime while, conversely, youth were given special attention and respect as bearers of the new revolutionary culture.[71]

State control over what had once been parental or individual prerogative was manifest also in another critical sphere, marriage. In the old society, free choice of mates was the ideal. A girl was presumably free to accept or reject a marriage proposal brought by a go-between from a young man and his parents to her parents, although in practice the parents themselves often exerted a strong or subtle control over a child's marriage.[72] In Democratic Kampuchea, however, the state, through its local officials, replaced parents in the negotiation and regulation of marriages. Ponchaud notes that refugee accounts regarding marriage were quite diverse, possibly reflecting regional variation, differential treatment of civilians and soldiers, and change through time.[73] Basically, however, there appear to have been both free choice and forced marriages.[74] In the first instance, both men and women were "equal and free" to initiate marriage proposals;[75] and, in an interesting mutation of an old custom, the chairpersons of the young men's and young women's groups acted as go-betweens.[76] Forced marriages occurred when soldiers or disabled veterans were allowed to marry any woman of their choosing. Ordinary men and women were sometimes arbitrarily married off to one another in hasty mass weddings.[77]

While it may seem paradoxical that marriage was supported while the family was undermined, its maintenance as an institution served several important purposes. First, by encouraging and even imposing marriage, the government could exercise some control over sexuality. Democratic Kampuchea, like some other revolutionry regimes, seemed to have a puritanical streak: men and women were strictly segregated in various spheres of life such as work teams, associations, and dining halls;[78] illicit affairs were punished; and there were rules of conduct such as women having to keep their shirts buttoned to the neck.[79] By marrying people off, possible problems of promiscuity (both moral and pragmatic) would be lessened. Second, marriages were obviously a crucial means of sustaining reproduction of children, who would become important additions to the labor force and revolutionary ranks. Third, forced marriages made it likelier that couples would not have strong emotional commitment to one another but, theoretically at least, would direct their loyalties and energies to serving the regime rather than the family.[80]

In addition, the solidarity and functions of the family were further reduced by collectivization. A family no longer formed a discrete economic unit with its own property, managing its day-to-day production and consumption; it became submerged into larger work teams whose efforts and produce were strictly regulated by the state. The institution of communal dining halls was particularly significant and symbolic in this regard, because commensality (the sharing and eating of food together) is an important means of expressing and reinforcing social ties in all cultures. The imposition of communal dining was not simply a means whereby the state controlled distribution; it further demonstrated that the work team or cooperative had superseded the family as the basic social unit in Democratic Kampuchea. Boua makes an interesting point that women have spoken of communal dining as the thing they hated most in Democratic Kampuchea. Not only did all produce have to be turned over to the commune; in times of meager rations, individuals found themselves hiding whatever food they could find—wild plants, for example—instead of sharing it with family members. Such acts "destroyed family confidence and solidarity" and sometimes led to feelings of distrust, hostility, and even vengeance between spouses.[81] It is possible that women had particularly strong negative reactions to communal dining because it symbolized the shattering of the domestic domain in which the wife/mother was a key figure exercising considerable authority over familial matters.[82]

The same forces that broke up families also had repercussions on the sphere of wider kin relationships that had been important in the old society. People obviously lost many relatives through separation and death in the upheavals both preceding and following 1975. When the cities were emptied, some urban evacuees tried to get to villages where they had relatives,[83] but this was usually not possible because DK cadres often dictated where new people should settle. There were also subsequent relocations of the population that broke up many kindreds. In any case, the ties among family, kinsmen, friends, and neighbors that had formed the basis for mutual aid and cooperation in the old society were no longer necessary for the communal organization imposed by fiat in Democratic Kampuchea.[84]

Additionally, Democratic Kampuchea favored people addressing one another as "*mit*," usually translated as "comrade" or "friend," in place of the custom widespread among villagers of using kin terms appropriate to sex and relative age (e.g., "grandmother" or "uncle"), regardless of whether a person was actually related.[85] The attempt to minimize use of kin terms (as well as other vocabulary that connoted status differences between social superiors and inferiors) was part of the effort to mold a new ideology of egalitarianism and, as Ith Sarin put it, "to strengthen 'political sentiment' more than just 'sentiment' by itself."[86]

However, kin terms evidently continued to be used, at least to some extent, among the populace.[87] Moreover, as is the case with many polities, Democratic Kampuchea selectively used an "idiom of kinship" to forge political bonds by taking kinship ties as a model or metaphor for organization. This might explain a seeming non sequitur in Ponchaud where on one page he suggests that kin terms were replaced by "*mit*," and on the following page he reports that all adults were called "Mom" and "Dad"; age peers were addressed as "senior comrade" and "junior comrade" (just as siblings and cousins were distinguished as older or younger than oneself in kin terminology); and children were "comrade child" because "all children are regarded as the infants of the *Angkar*."[88] If this was indeed the case, the use of appellatives drawn from the family and extended to (all?) others is intriguing because it simultaneously negates the distinctiveness of individual families while evoking sentiments of cohesion associated with familial ties. As one revolutionary song put it: "O solidarity group, you are a new kind of family, special, beautiful, and unique."[89] Some local leaders were designated as Grandfather (*Ta*) or Mother (*Me*). Going further, the commonly used conception of nation as family writ large was invoked in some references to Pol Pot as "*bóng* number one."[90] This designation is interesting not only for its unintended connotations of Orwell's "Big Brother,"[91] but because the term chosen was "elder sibling" rather than "father" or "grandfather." Use of the sibling term evokes an egalitarian sense of one's peers, but in the old society an elder sibling was also a peer who had some authority over and responsibility for the care of younger siblings.

Community Sociopolitical Organization

Where villages traditionally had been aggregates of kinsmen and friends who were well known to one another, the influx of urban evacuees into the countryside obviously altered the composition of numerous communities. The resulting mix of people, coupled with the new social distinctions discussed earlier, gave rise to a restructuring of traditional social relationships and, in various instances, to social and political tensions.[92] Vickery (1981: 12) suggests that the situation in communities was "best where a small number of new people were thoroughly mixed with the old peasantry, and worst where a large community of new people were segregated from the base population." Despite the egalitarianism espoused by DK political ideology, new social rankings emerged. High status was given to party cadres, base people, lower and lower-middle peasants, and "full rights" people, although, as noted earlier, these variables could combine in different ways in specific individuals. Conversely, low status was generally conferred on new people, urbanites of middle and upper class backgrounds, and depositees. Political reliability thus replaced age, good character, and other traditional criteria for respect.

Local political officials operated within a hierarchy that ascended from the level of the *krom* work team to the various higher-order groupings of units (e.g., company, battalion). A committee of three or more people, including a chair, deputy chair, and other members for particular responsibilities, headed each *krom* and higher-level group. Similar three-person leadership is reported for various other units: mobile work teams, villages or cooperatives, territorial administrative divisions (subdistrict, district, and up), and associations.[93] In early Democratic Kampuchea, according to Carney, lower- and middle-level positions were held largely by demobilized soldiers and sometimes by nonmilitary party cadres or, presumably if cadre were in short supply, even new people.[94] Later, in 1976–77, the situation became more complicated as the national administration placed party cadres in administrative positions held by local base people and as cadres loyal to Pol Pot replaced those associated with other factions. This was often accompanied by large-scale purges and executions in 1977–78 at both local and higher levels.[95]

Those in the low-valued social categories were handicapped in their access not only to political office but, in some places and periods, to decent living conditions as well. New people, urban evacuees, and depositees fared reasonably in some communities, especially prior to 1977–78.[96] But in other areas they ranked lowest in priority with regard to distribution of food and goods and sometimes were targets for execution.[97] It should be noted, however, that generally harsher conditions, such as cuts in food rations, affected virtually everyone in the country as a whole after 1977–78.[98]

While Democratic Kampuchea diminished the importance of the family, they stimulated the development of another kind of social unit that had not previously existed in rural communities—associations. There were associations for men, women, school children, and young people. Of special importance was the Youth League, which was a means of indoctrinating and organizing young people who were considered to be critical figures in carrying forth the revolution.[99] Except for the Youth League, it is not clear how active such associations actually were. In terms of social structure, however, they offered a means of establishing ties that crosscut those of labor teams and hence were another way of organizing people in groups supervised by the regime.

Revolutionary Ideology and Behavior

The final point to be discussed regarding major sociocultural changes in Democratic Kampuchea is the suppression of the Buddhist religion and the attempt to instill new political ideology and behavior. While Buddhism had been the state religion of the old society, its practice was discouraged in pre-1975 DK cooperatives and eventually forbidden by the end of 1975.[100] Although the 1976 Constitution explicitly granted people "the right to worship according to any religion," it also immediately added a strict prohibition against "all reactionary religions that are detrimental to Democratic Kampuchea and the Kampuchean people."[101] While the Constitution does not explicitly name Buddhism, it is clear from DK rhetoric and actions that Buddhism (as well as Islam and Christianity practiced by minority populations) was considered reactionary, feudalistic, and exploitative.[102] Buddhism, of course, would have competed with the state for manpower, resources, and loyalties. DK class analysis categorized monks as belonging to a "special class" and comparable to "subcapitalists" or petty bourgeoisie;[103] contributions to and maintenance of temples were seen as drains on people's resources; and the other-worldly orientation of Buddhist teachings was viewed as detrimental to Democratic Kampuchea's desire for active transformation of this world.[104] While the national emblem of the Khmer Republic had Buddhist motifs, that of Democratic Kampuchea showed a "network of dikes and canals" and a factory.[105]

Democratic Kampuchea's attempts to eradicate religion involved not simply rhetorical attacks but the symbolic and literal destruction or desecration of two critical elements of Buddhism: the *sangha* and the temples that had served as moral and ceremonial centers for the populace. Following the DK victory in 1975, several leading monks were killed, while treatment of more ordinary monks varied: some were put in special, separate labor teams; some were forced to leave their temples and resume secular life; and still others (few and only in the initial days of the regime) were left in place.[106] Buddhist temples were either physically demolished or profaned by being used as storage or housing facilities for soldiers.[107]

Buddhism was now replaced by a revolutionary political ideology upholding new and different values and codes of behavior. With the need to socialize the young and resocialize their elders into what was in some respects a new culture, Democratic Kampuchea undertook an active program of indoctrination into their new ideology. In base areas before 1975 and widespread thereafter was the institution of another phenomenon new to rural life: the meeting. Such meetings, held on a regular basis for cooperatives or work teams, might discuss some practical problems of work organization or productivity; but, probably more importantly, they attempted to instill revolutionary and socialist consciousness through self- and mutual-criticism sessions, speeches, songs, and the like.[108] Associations, political study schools, radio broadcasts, magazines, newspapers, and theatrical performances with song and mime were further vehicles for instruction.

Some of the DK codes for behavior were not new. For example, "women, liquor, and gambling," viewed as "aspects of delinquency" by the Khmer Rouge, were the same three things constantly cited as attributes of "bad character" by the villagers I knew. And various rules of conduct for early Khmer Rouge cadres accorded with traditional notions of common courtesy and decency.[109] But what was new were not only revolutionary-socialist values but also the high degree of regimentation and discipline imposed on individual behavior. In early Khmer Rouge base camps there were "life style meetings" that included criticism sessions on proper modes of "drinking, sleeping, walking, talking" as part of molding individuals to become "robot(s)" in the service of the revolution.[110] There were also rules dictating other details of life, such as short haircuts for women and pants of a certain length. Certainly there were norms of behavior in old-society village life, some of which were strictly observed and rarely transgressed. But there was also considerable latitude for individual variations of behavior within certain broad limits; mistakes were tolerated if an individual showed contrition, and the breaking of some rules often brought only a shrug of the shoulders or behind-the-back gossip. Democratic Kampuchea, however, regarded individualistic behavior as "reactionary traits" that had to be suppressed because they were inimical to collective interests.[111] Deviation from or infraction of rules in Democratic Kampuchea was dealt with at the simplest level by criticism sessions* and, more severely, by stringent punishments that could include hard labor, beatings, imprisonment, and execution. "An individual who tries hard but fails despite following instruction still takes the blame because the party does not make mistakes."[112] "Laziness, resistance, even verbal, to policy or instructions, and boasting or pretension" were considered "crimes" punishable by death.[113]

In the sense that Democratic Kampuchea was a revolutionary regime attempting to change long-established institutions and to construct a new social order, it was not unlike new states in other times and places which faced the problem of achieving firm, centralized control over its polity. As Cohen noted in a historical and cross-cultural discussion of the development of state systems, an "inchoate" state may use a variety of means to consolidate its power. These include (a) legitimization of authority and change from one mode of organization to another by the promulgation of new ideologies and selective use of familiar "idioms" (while often subverting them); (b) imposition of strict controls to ensure conformity to behavior until the new ideology has been firmly instilled and there is "assurance that the polity will obey and follow"; and (c) elimination of possible sources of subversion and competition for loyalty, along with "transfer of loyalties from local nexuses to the state" by, for example, dispersion of the

*Criticism sessions were another significant innovation, in view of the fact that face-to-face censure was carefully avoided in traditional village life except within the family.

population and forced relocation that breaks ties to kin and friends and minimizes possibilities of conspiracy because strangers are "unlikely to trust each other enough to engage in conspiracy." Thus, while Democratic Kampuchea was unusual in many respects, in other ways it exhibited processes that have been manifest in other "early" states and revolutionary regimes.[114]

People's Republic of Kampuchea

Democratic Kampuchea's effort to reconstruct and restructure the country into a new social formation was called abruptly to a halt with the takeover by the Vietnamese in January 1979. DK forces had too brief a period of rule—only about four years for the entire nation and a decade or less for those regions in which DK bases were established before 1975—for their radical reformulations to have taken hold firmly and be perpetuated, except in the DK resistance camps that are still operating. There is, in any event, a question as to how wholeheartedly much of the population had adopted DK doctrine. Cadre terrorism in various places engendered distrust and hatred of the Pol Pot regime. DK leaders had largely ignored (except for select purposes) or underestimated the shared and deep-seated commitment to various institutions and norms that constitutes a people's culture; they evidently assumed further that a society could be remolded by fiat and force. Once DK hegemony was ended, many basic elements of traditional (pre-1975) culture were resurrected under the more lenient policies of the People's Republic of Kampuchea.

If one were to use a dialectical model of development with pre-1975 Cambodian culture as the thesis, then in broad terms Democratic Kampuchea was an attempted antithesis of the old society in many ways, and community organization in PRK can be seen as a synthesis—or what might be called a syncretic combination—of "old" (pre-1975) and "new" (socialist) elements. While the Vietnamese have imposed a certain political framework and are evidently trying to construct a semisocialist economic system both similar to and different from that established by Democratic Kampuchea, they have also allowed many aspects of traditional Khmer culture to be resumed. Accounts of local existence in the People's Republic of Kampuchea describe many things similar to the village life I knew. Such reports are also very revealing of peasant adaptive strategies and how people manipulate and maneuver within given frameworks to survive under various ecological, economic, social, and political pressures.[115]

Reports on the People's Republic of Kampuchea speak once again of "villages" as the local territorial and sociopolitical unit in which people reside.[116] After the DK communes and mobile work teams were disbanded, people were permitted to go where they wished. In the initial months of the new regime there were once again population movements, but this time of a voluntary nature as people searched for lost family and relatives or returned to their former homes in both rural and urban areas.[117] Villages that had been abandoned or destroyed were repopulated and rebuilt. Other communities that had remained physically intact, but which had experienced considerable disruption (e.g., an influx of new people, relocation or killing of parts of their population, reorganization into large-scale cooperative units) returned to more stable conditions.[118] Also, the composition of various villages obviously changed with the exodus of DK cadres, urbanites, and those relocated from other rural areas. Heder suggests that villages, like the country as a whole, became numerically more dominated by the peasantry.

Economy

At the beginning of its administration, the People's Republic of Kampuchea faced serious economic problems due to various dislocations of the economic infrastructure that had occurred in Democratic Kampuchea and during the subsequent turmoil of the Vietnamese invasion. There was a grave shortage of food (alarming famine conditions in 1979 drew worldwide attention) as well as of draft animals, seed, and manpower; cultivation was disrupted by population movements and adjustments to new policies. Aid from foreign countries and international relief agencies was necessary to sustain the economy.[119] While the period of starvation passed and much of the country returned to relatively stable conditions, Kampuchea still has problems regarding food production and distribution, shortages of animal and human labor power, reestablishment of industry, repair of irrigation works, reclamation of lands, and need for foreign aid to make up food deficits and reconstruct various parts of the economy.[120]

A semisocialist economy has developed in Kampuchea, partly through design and partly through necessity because of the economic problems just noted. Almost all peasants are organized into so-called Increasing Production Solidarity Teams (*krom samaki*), composed of five to twenty families (generally ten to fifteen) who communally cultivate certain lands held by the *krom* as a whole.[121] Given shortages of draft animals, manpower, and tools, the government has appealed once again to the old tradition of cooperative labor as necessary for economic reconstruction. *Krom* are also thought to be useful for surveillance and protection against enemy guerillas, since various regions still harbor DK forces.[122]

The *krom*, however, is no longer *the* basic organizational unit of life that it was in Democratic Kampuchea because, in addition to this communal system, there has also been a significant reemergence of "private" production, consumption, and distribution. The PRK administration, recognizing the popular reaction to the regimented communalism of Democratic Kampuchea and faced with very serious practical problems of production and distribution of food and goods, has allowed part of the economy to revert to private enterprise. Thus, while some land is given over to *krom* ownership, production, and distribution, other lands are allocated to individual households for private consumption and marketing.[123]

In the early days of the People's Repubic of Kampuchea there was evidently great variability among villages as to how government directives were carried out regarding land allocations to *krom* and households and work arrangements. This was due sometimes to genuine confusion and sometimes to "interpretations" of policy according to the political sentiments of local officials and residents. Thus, in 1980 some villages continued more or less collective farming, others allocated all paddies to households, and still others combined communal and household plots.[124] It is not clear whether such local variation persists today; reports from 1981–82 indicate some combination of communal land and household plots.[125] According to Quinn-Judge, land amounting to one hectare per family is placed under *krom* control (e.g., if a *krom* were made up of ten families, it would collectively work ten hectares). Each family also has a private plot of 1,500 to 2,000 square meters and may, in addition, "'borrow' more land from the *krom samaki* if it is not being used."

Through time there was also confusion, debate, and variable practice concerning tools of production—draft animals, plows, carts, and other implements. There is now de facto ownership, although the owners are supposed to make such items available for use in communal production.[126] Given shortages of tools, as well as of male labor, efforts are presumably made to see that a *krom* has sufficient means to carry out adequate production.[127] According to one official, local labor may be supplemented on weekends or during busy periods of cultivation by teams of government workers and soldiers.[128] One recent report notes a government-sponsored "food for work" program in several provinces whereby voluntary agencies provide food to young people selected and sent by their villages to work for two weeks on projects such as irrigation systems. These are not, however, the permanent mobile work teams of the DK period, and such work stints are even welcomed by young women as a chance to get out of their villages.[129]

Krom team produce is distributed to its able-bodied members according to the number of days of labor per year expended in the "main work force." Provision is also made for giving lesser amounts of produce to those in the "auxiliary work force" contributing light labor, to dependents such as the elderly and very young, and to noncultivators such as teachers and nurses. In a number of areas, additional compensation is given to the owners of draft animals used in team production. Thus, while all members of a *krom* are ultimately provided for, those families with draft animals and a number of able-bodied workers benefit in the distribution of *krom* produce. These individuals may also be more productive in the use of their private plots and pursuit of other economic activities.[130]

The People's Republic of Kampuchea has not only permitted but actively supported private household production, called "the family economy," as a necessary supplement to communal organization.[131] The *krom* evidently function primarily for rice cultivation; during periods when the rice does not need tending, families engage in varied pursuits of their own, just as they did in old-society villages. Food and income are supplemented by growing vegetables and other crops on household plots, fishing, tending poultry, making palm sugar, handicrafts (weaving, pottery-making, etc.), working at odd jobs in urban centers, and marketing produce or goods.[132] In a number of cases there is what Boua calls a "mixed domestic economy" in which different members of the family pursue varied economic endeavors. Communal *krom* production also frees some family members to engage in private economic pursuits such as trading, with the result that the "collective system has provided a basis for the development of peasant enterprise."[133] According to the government, a family may dispose of its private produce as it wishes; such produce cannot be taxed or claimed by the *krom*, village, or state; and it may be sold either to state stores or on the open market.[134] There has reemerged, then, a partial domestic mode of production, quite similar to what existed in the old society, as a way to make ends meet. But, in the PRK synthesis, it coexists with collectivization of the major activity of rice cultivation.

As implied by the preceding, there has also been a renascence of a free-market system and reintroduction of money.[135] Although there are state stores that buy paddy from peasants and sell some basic items (salt, cloth, household items) at low cost, industrial production is still limited and inadequate transportation facilities make distribution difficult. Thus, various kinds of trade have been permitted to flourish on both small and large scales: individual barter, street vending of produce and other items, and markets with an array of foods and goods.[136] In connection with the latter, a major development has been the establishment of extensive trade and smuggling between Kampuchea and Thailand. Vast numbers of people, both full-time professional traders (or smugglers) and villagers engaged in occasional trade, go to and from the Thai border to obtain various items which then find their way to regional markets and homes. While the government does not formally approve of this, it lacks means to control this border traffic and, in the words of one official, "We winked at it because our people were so needy."[137]

The Family

With the dissolution of the DK communes, another major development in the People's Republic of Kampuchea was the reconstitution of the family as a social unit. Comments in various accounts indicate feelings of homesickness or strong yearnings to find and be with family and kin. Thus, when DK control was lifted, many people went in search of family and relatives who had been dispersed in the preceding years, returned to their home villages, or settled in communities where they had kin or friends. Families and kin networks were reestablished as best possible, given the losses suffered during the ravages both before and after 1975.[138] Not infrequently, households take the form of extended families composed of various relatives, following the old-society pattern of taking in and caring for orphaned or needy kinsmen and sometimes even nonrelatives who are "adopted" into the family.[139] In some instances, however, there were divorces rather than reunions of spouses, specifically on the part of women who had been forced into distasteful marriages under the DK regime.[140]

Family composition and labor are affected by a current demographic imbalance between the number of men and women in the population. Various sources report that females at present constitute 54–60 percent of the overall Kampuchean population, with the proportion rising even higher (estimates are two-thirds to three-quarters of some communities) if one considers only adults. Boua suggests that this is due to men having been more frequently the victims of killings during the DK period and having been less able to survive food shortages.[141] Numerous widows and unmarried women thus bear the burden—partly relieved by the *krom* system—of supporting families; indeed, some families may be composed entirely of females.[142] Women are pursuing a variety of endeavors (e.g., raising secondary crops, tending pigs, vending produce, making craft items), including tasks such as plowing that are customarily performed by men. One should recall, however, that even in the old society the sexual division of labor was never absolutely rigid, especially in times of necessity, and that rural women in particular were notable for their independence, authority, and resourcefulness in maintaining their households and handling economic matters.[143] Thus, it is not a new pattern for village women to display considerable strength and energy in undertaking a host of endeavors to make ends meet.[144]

This is not meant, however, to minimize the fact that contemporary Kampuchean women bear a particularly heavy burden because of the ravages of the preceding decade, not only economically but psychologically because of the traumas endured. There is also a social problem: the shortage of men means that many women will not be able to marry or remarry, although Boua mentions that one group of peasant women was against the idea of reviving polygyny. She also suggests that premarital and extramarital sexual activity is not uncommon, following the rigid segregation of the sexes in Democratic Kampuchea and the fact that many persons have been left widowed or unwed. She notes that "such license is greater in the city," implying that some of this sexual behavior occurs in rural areas as well (as it always did, though to a very limited extent because of the difficulties of keeping anything secret in a small community). However, pregnancy out of wedlock is not condoned and sometimes terminated by abortion, although the latter is rarely practiced by married women.[145] Children are desired and treasured in traditional Khmer culture,[146] and there is evidently a "baby boom" in Kampuchea as new marriages occur and as couples, freed from the restraints and separations imposed by the DK regime, make up for lost time.[147] The current increase in birth rate is probably due also to amelioration of the severe conditions in Democratic Kampuchea (such as malnutrition, hard labor, stress, and inadequate medical facilities) that suppressed female fertility and heightened the likelihood of infant and maternal mortality.

Political Organization and Stratification

The political framework instituted by the People's Republic calls, at least formally, for village administration to be headed by a three-person committee (chair, vice-chair, and a third member) to handle general village affairs, the organization of *krom* teams, the division of communal and private lands, etc.[148] Each *krom* has its own chief who oversees team activities (e.g., recording days of labor for team members), acts as liaison between the *krom* and village officials, and forms part of the overall local administration.[149] There is also a village militia charged, as in previous regimes, with guard duty at night and keeping watch for thieves and subversives. The village is, of course, part of the larger system of territorial-political divisions, with the familiar subdistricts, districts, and provinces each having its own administration. While the PRK state exercises much less rigid and autocratic control than did Democratic Kampuchea, through time there has been a "gradual tightening and regularization of certain kinds of *khum* (subdistrict) and *srok* (district) supervision of village affairs."[150] There is some question about the general nature of the People's Republic. For example, while Heder has suggested that "the Vietnamese feel obliged to play a classic direct colonial role of military occupation and comprehensive supervision of the social structure down to the district level and sometimes beyond," Kiernan implies that the Vietnamese are acting primarily in an advisory capacity and have been gradually withdrawing government advisers and troops.[151]

At the local level, the "selection of village leaders (has) tended to revert to traditional patterns." The People's Republic initially issued general directives that encouraged selection of persons with "clean backgrounds" not connected to former regimes and from poor or lower-middle classes. While such instructions were followed to some extent, Heder suggests a general tendency, coincident with a move toward decooperativization, for village officials to be chosen on the basis of "traditional leadership characteristics as they had been defined in the old society," i.e., "persons of 'good character,' who had some economic and/or educational status in the old society and who displayed intelligence, charisma, and/or nonantagonistic forcefulness."[152] And, as in the past, these village administrators serve as "hinges" or "brokers"[153] between the local community and the larger political system, between "us" and "them." Whereas in old-society Sobay, however, village and subdistrict officials were considered basically one of "us," in the People's Republic of Kampuchea they may very well be identified as part of "them."[154]

Within villages there was evidently gradual muting of the social distinctions, such as that between new people and base people, that had existed in Democratic Kampuchea. Within the society as a whole, however, the change in government and reestablishment of a market economy has led to reemergence of socioeconomic strata. According to Heder's analysis, government officials, professional traders, urbanites, and the Vietnamese have come to be viewed by the peasants as constituting a higher political-economic stratum of ''privileged individuals.'' Furthermore, within the peasantry itself, economic differentiation has once again become possible with restoration of some private ownership of means of production. For example, those villagers with productive or large household plots or with draft animals and carts that can be used for transportation and trade have opportunities for economic gain despite—or in addition to—the existence of communal production and distribution. Thus, Heder sees the development of an ''upper stratum of peasantry'' throughout the country, many of whom had been middle or prosperous peasants in the old society. He also suggests, however, a contradictory trend toward a sense of solidarity among the peasantry as a whole in opposition to the nonpeasant ''elite'' noted earlier, along with the decrease in the hostilities (e.g., between new and base people) that had divided peasants in Democratic Kampuchea.[155]

Another aspect of political organization in the People's Republic, which continues an innovation begun in Democratic Kampuchea, is the attempt to create local associations of women and youth to serve as vehicles for instruction in political doctrine. These associations may also provide various services. It is not certain how widespread or active such groups may be at present; one report states that the Youth Association seems to have ''more drive and impact'' than the Women's Association. Apart from associations, there is also the Communist Party, but its membership is very limited. Most villages have no party members, and there have been few efforts to recruit people at the local level.[156]

In addition to the associations, there are two other vehicles for disseminating and instilling political sentiments and ideology. The first is the educational system, which has been restored with primary, secondary, and advanced technical/professional schools.[157] The curricula at all levels include not only standard subjects but instruction in ''politics'' and the Vietnamese language. Second, there is government support for the revival of ''traditional'' dance, music, and popular theater, including the establishment of district orchestras or theatrical troupes that perform in rural areas.[158] There is a syncretic combination of the traditional and new, in that the messages communicated in song, drama, or comedy are often political ones, expressing in particular strong anti-Pol Pot feelings.

Considering the centuries-old enmity that existed between the Khmer and Vietnamese, a major question is, of course, popular reaction to the Vietnamese-supported PRK regime. Assessments of this matter vary, correlating in large part, if not totally, with judgments regarding Democratic Kampuchea.[159] Some observers report that the Kampuchean people are generally grateful to the Vietnamese for delivering them from Pol Pot; that initial fears of and traditional hostility toward the Vietnamese[160] have largely subsided in the face of a moderate administration, improvement of living conditions, and the good behavior of Vietnamese occupation troops; that people prefer the Vietnamese to the return of Democratic Kampuchea, although there is some concern regarding a long-term Vietnamese presence.[161]

Heder, on the other hand, is less sanguine about Kampucheans' feelings toward the Vietnamese. He believes that although the people were indeed relieved to be liberated from Democratic Kampuchea and to find the Vietnamese more benign than expected, "expectation of the worst did not translate into support for the bad." He suggests that by August 1980, among perhaps three-quarters of the peasantry there was a "tendency toward the emergence of a more or less insistent and consistent set of complaints and dissatisfactions about the superior political and economic status of government functionaries, political traders, and Vietnamese, who tended to be seen as an interrelated and interdependent distant whole, defining, as a collection of privileged individuals, an uncaring, self-serving, and foreign-dominated regime."[162]

Heder sees both lower and more prosperous strata of peasants as having grievances of various sorts toward the Vietnamese-sponsored regime, and both also favoring "the dismantling (of) semi-socialist and socialist structures in the countryside" (1980c: 110). Boua, however, believes that "after the relative success of the 1980 harvest, most peasants seem to accept the new system (of *krom samaki*)."[163] Peasant attitudes toward the People's Republic thus remain a moot point. There is undoubtedly varying opinion on the matter, as well as ambivalence and potential for change through time.

Religion

Another critical aspect of the old society that has been restored in the People's Republic of Kampuchea is the practice of Buddhism. Temples are evidently being rebuilt and, although there are relatively few monks compared to pre-1975 figures, they are scattered throughout the country. (David Hawk has estimated that there were 5,000 monks in 1984 as compared to 40–80,000 in the old society.) In 1980, Kiernan saw celebrations of *Prachum* and *Katún* festivals which, along with the New Year, were the major Buddhist celebrations in prerevolutionary Cambodia. His observations of large crowds of people going to *wat*s for Holy Day is also very significant because, in Sobay at least, these were commonly observed only by the elderly and did not draw great numbers to the temple.[164] Such activity indicates restoration of the religious ceremonies that were important for reinforcing Buddhist doctrine and social cohesion. Religion is especially important now in offering social and psychological support to a people who have undergone years of extreme stress.[165] In the light of occupation by a foreign power, the practice of Theravada may also be a means of asserting Khmer identity and distinctiveness from the Vietnamese, who have different religious traditions.

But while the People's Republic of Kampuchea has permitted Buddhism to be revived, it is likely to be kept under careful government surveillance and control.[166] Kiernan (pp. 8, 368) suggests that the number of monks will probably be limited (only older men are being allowed to enter the *sangha*); and a number of points in a government policy statement on religion[167] indicate concern that monks might be potential agents of subversion or threats to the administration. For example, while "festivals according to national customs" may be observed, they must not "disturb people's production and cause insecurity or social disorder"; monks no longer seem to enjoy a separate and privileged social-legal status but have "rights and duties . . . the same as the rights and duties of citizens"; and monks in rural areas must be issued "permits" by local authorities, doubtless as a means to keep track of their numbers and whereabouts. It seems, therefore, likely that Buddhism will be kept from becoming too large and powerful a force vis-à-vis the state.

The Future and Future Research

While events of the past decade in Kampuchea seem, and are, in some respects extraordinary and unique, they also manifest general sociocultural processes. The revolutionary overthrow of an "old regime" is certainly not unusual in world history, and it was previously suggested that the policies of Democratic Kampuchea were similar to means used by other "early states" to consolidate control over a new polity. While DK leaders and cadres were ethnically Khmer, their ideology was so different from much of the Cambodian populace that their effort to create a new social formation was not unlike the situation of a foreign power seeking to impose a new culture on a subject population. But various anthropological studies have indicated that people will not easily accept transformations of fundamental features of economic, social, and ideological systems unless they are convinced that there are good reasons or demonstrable advantages for doing so.[168] Democratic Kampuchea instituted radically new forms of organization, behavior, and norms primarily by force and was not able, in the few years of the regime's rule, to convince most of the populace that there were "good reasons" for these changes. Indeed, the severity of their measures had the reverse effect of engendering considerable hostility to their regime.

Thus, when DK controls were ended, elements of pre-1975 culture were revived under the more lenient policies of the People's Republic of Kampuchea. This is not to suggest the old and by now superseded notion that peasants are inherently tradition-bound, conservative people who will revert to "old ways" whenever given a chance or unless forced to change. It does, however, suggest the tenacity of culture—especially, in this case, deep-rooted aspects of Khmer peasant existence such as the importance of family and kin, household production and consumption, and Buddhism, that have reemerged despite the DK attempt to remold or suppress them. These and other cultural elements persist not simply out of force of habit or as meaningless remnants from the past, but because culture is a critical and powerful means of providing a framework for people's behavior, social organization, group cohesion, and ethnic identity. This is particularly important for a people struggling to reorder their lives after years of enormous upheaval and dislocation. Maintenance of certain cultural elements can also be a means, largely covert, of resistance to domination by others, such that the Kampucheans' return to certain cultural forms such as the practice of Theravada Buddhism might also be viewed as reassertion of "Khmerness" in the face of political occupation by the Vietnamese. All cultures are basically in process rather than static entities, and Kampuchean culture is at present in a state of particular flux, with its syncretic mixture of "old" and "new," and with confusions and contradictions at both the local level and in the society at large. Such processes will continue, and it is difficult to surmise what further reformulations may occur in Khmer society and culture, given Kampuchea's uncertain political future. At this moment, several critical factors are unpredictable: whether Kampuchea will continue to be more or less under the aegis of Vietnam, or if it can reestablish itself as a viable, independent polity; what sort of economic and social system the government (whatever it may be) will foster; and what will be the consequences, both intended and unintended, of state policies on the society at large.

Although the future is unclear, there is much to be gained from closer examination of Kampuchea's recent history and contemporary life. At present, various scholars cited in this chapter are accumulating essential information on political history and the broad outlines of socioeconomic organization in prerevolutionary Cambodia, Democratic Kampuchea, and the People's Republic of Kampuchea.[169] However, the radical changes that occurred in Kampuchea were unusual and complex phenomena that merit further inquiry and analysis. We need additional data on a number of topics, as well as consideration of how the Kampuchean material has relevance for general theoretical problems such as processes of change.

A number of different questions could be raised as topics for further study, depending on one's research interests. As an anthropologist whose primary focus has been on peasant villagers, I believe that it is especially important to learn more about local-level existence and the ordinary people who comprise the majority of the population. This is not an easy task, but the existing literature shows that it is possible to retrieve considerable information on the recent past from documentary materials and interviews. There is also some opportunity, though still limited, for firsthand observation of conditions in the People's Republic of Kampuchea.

I suggest below some general areas that warrant further inquiry.[170]

1) The condition(s) of Cambodian villagers in the late 1960s and early 1970s. While several works provide some detail on Cambodian peasantry in the 1950s and early 1960s, we need more studies (such as Kiernan's on the Samlaut rebellion) that explore local life and the forces that impinged upon it in the decade preceding the revolution. This is a difficult topic because of regional and other variations within the peasantry, but more attention to the micro-level of village

life is essential as background for the following topic.

2) The spread of Khmer Rouge/Democratic Kampuchean forces throughout the countryside prior to 1975. A paramount question is how and why the Khmer Rouge managed to expand so widely in the course of just a few years. The subject has been discussed to some extent[171] but warrants further analysis. A dimension that particularly demands more scrutiny is the nature of grass-roots interaction between DK cadre and country folk: the modes of persuasion and coercion used by the Khmer Rouge to extend control over rural areas; villagers' reactions to events of the early 1970s; and why there was acceptance or rejection of revolutionary ideology and organization. This and the first topic are critical for understanding both the Kampuchean revolution in particular and revolutionary movements in general. Historical and cross-cultural comparisons should be made between the Kampuchean material and revolutions elsewhere, perhaps especially Vietnam and China. Such a perspective would elucidate the ways in which the Kampuchean experience was similar to and different from other cases and would also make an important contribution to discussions concerning different models of peasant participation in revolutionary movements.[172]

3) Aspects of local society and culture in Democratic Kampuchea. The past few years have seen substantial additions to our knowledge of conditions in Democratic Kampuchea, as evidenced by materials in this chapter. Various aspects of life, however, remain unclear or only sketchily known, some of which are noted below.

We know certain basic points regarding local social organization, but it would be desirable to have more detail on topics such as the impact of DK policies and stringent circumstances on customary norms, values, and role relationships within the family and kindred; the organization and functioning of associations, including those for children, men, and women as well as the Youth League; changes that may have occurred in the statuses and roles of women, given the presumed equality of sexes and new forms of social organization; and the extent to which traditional patterns of interaction, such as avoidance of confrontations, were modified by exposure to harsh circumstances and the behavior of DK cadres.

While a good deal has been learned about the organization of cooperatives, some points remain unclear. For example, were there cooperatives whose subsistence base was something other than rice cultivation (e.g., fishing, animal husbandry), and how did these operate? What effect did the consolidation of cooperatives under *khum* subdistrict administration have on settlement patterns and supervision of labor? How were essential goods, such as clothing, distributed? Were any private plots and produce permitted in the later years of Democratic Kampuchea? Finally, although the regime placed strict controls over behavior, were there forms of covert (or overt) resistance to DK policies? For instance, although religion was firmly suppressed, was there hidden adherence to Buddhism or the folk religion?

4) Contemporary People's Republic of Kampuchea. As with Democratic Kampuchea, some things about the People's Republic are known in broad terms but would benefit from more data, while other areas of life are relatively unknown. The semisocialist economy raises questions as to precisely how the combination of collectivized and "private" modes of production-distribution operates locally. For example, it would be instructive to have more detail on how access to resources is negotiated, allocated, and controlled by *krom* teams and individual households (including possible manipulations and subversions of official policies); how local production and distribution intersect with the national economy; and positive and negative reactions on the part of different kinds of people to the present economic system, now that it has weathered the confusion and extreme food shortages that marked the beginning of the PRK regime.

While various aspects of old-society culture have presumably been revived, they obviously bear the effects of what happened in Democratic Kampuchea and present-day conditions in the People's Republic. Thus, another major area for study is how and why certain "traditional" cultural elements have been altered. Of particular interest is change in facets of family life such as courtship and marriage patterns; the composition and functioning of households, given the current imbalance in the sex ratio; interpersonal relationships (e.g., do children socialized during the DK period continue to be estranged from parents? Have elders permanently lost their authority?); and whether old-society norms of family-kin obligations and solidarity have been weakened by experiences during Democratic Kampuchea or, on the contrary, strengthened because of the need for both emotional and material support after a time of great hardship.

Other facets of social organization also deserve study with respect to both current conditions and future developments. For example, will the position of women in the People's Republic of Kampuchea undergo changes because women are assuming new responsibilities and activities due to the skewed sex ratio? Will associations become important forms of local organization? Will some sort of class structure develop despite a socialist-egalitarian ideology?

Another important topic, about which little is known, is the condition of ethnic minorities such as the Chams, Chinese, Vietnamese, and tribal peoples, many of whom were killed or expelled in Democratic Kampuchea. Their current situation and how they are being socially and politically integrated into the People's Republic are unclear. In the political sphere, we must ask about the relation between the village and higher administration, including the role of village officials as "brokers" and local stratagems for dealing with state policies; the extent to which villagers' political consciousness may have been raised by experiences of the past decade (e.g., do they now pay greater attention to national and international politics?); and positive and negative attitudes toward socialist-communist ideology and the Vietnamese presence.

Another realm of considerable significance is religion. While Buddhism is once again practiced, one might question whether some of the populace may have become disaffected or disillusioned with religion because of experiences in

Democratic Kampuchea, especially young people raised during that time but others as well.[173] Other queries involve the role of the *sangha* and *wat* in village life, given the substantial reduction in the number of monks and temples, and whether the state will try to limit or use religion in pursuit of its goals. Additionally, I have seen no information for either Democratic Kampuchea or the People's Republic on what happened to the beliefs, practices, and specialists associated with the folk religion, which was very important among the villagers. Finally, it would be interesting to know current practices with regard to life-cycle observances such as birth rituals, weddings, and funerals.

For both Democratic Kampuchea and the People's Republic it would be extremely valuable to have case studies of particular cooperatives and communities to supplement existing works dealing with macro-level organization or offering general overviews of topics. Such specific case studies would provide greater sense of how people, groups, and institutions are articulated within a local unit, their day-to-day functioning, and the ways in which the local level is connected to the larger society. Another useful avenue of inquiry would be comparison of forms of organization in Democratic Kampuchea and the People's Republic to other communist/socialist systems to see what was and was not unique in the Kampuchean experience.

I have proposed only a limited sample of subjects for further investigation; much remains to be studied. Events of the past decade in Cambodia/Kampuchea constitute an exceptionally rich and important domain for research, not only because the data are fascinating in and of themselves, but because they can also make significant contributions to larger theoretical questions such as the nature of revolutionary movements, the development of new social forms, and how sociocultural change is effected. Kampuchea specialists face avenues of inquiry that are very challenging, but that promise to yield material of enormous interest and significance.

Notes

1. Cf. May Ebihara, "Perspectives on Sociopolitical Transformations in Cambodia/Kampuchea," *Journal of Asian Studies* 41 (1981):63–71.

2. Research was supported by a Ford Foundation Foreign Area Training Fellowship. I would also like to thank Margo Matwychuk and Gerald Creed for helpful discussions of parts of this paper.

3. See Jean Delvert, *Le Paysan cambodgien,* Le Monde d'Outre-Mer, Passé et Présent, Première Série, Études no. 10 (Paris: Mouton, 1961), for a general survey of Cambodian peasantry; also Gabrielle Martel, *Lovea, Village des environs d'Angkor* (Paris: Publications d'École Française d'Extrême-Orient, 1975), and Milada Kalab, "Study of a Cambodian Village," *Geographical Journal* 134, 4 (1968):521–36, on villages elsewhere.

4. In West Hamlet, four households had no paddies but rented them from others and/or had nonagricultural jobs.

5. A husband and wife together could purchase land, which was then considered to be

held in joint ownership, but this was relatively rare. For the most part, paddy land was inherited by individuals.

6. While land is owned individually, average paddy holdings are stated in terms of households because spouses share resources and have usufruct of one another's lands during marriage. In some cases, both husband and wife had inherited land; in other instances, one or the other owned the paddies used to support the entire family. In addition to rice paddies, individuals could also own land within the village for house sites, uncleared land, and palm trees.

The one household with four hectares of paddies was distinctly unusual. Four households owned two hectares of fields. For the country as a whole, 80 percent of owner-cultivators owned less than two hectares of land; and in the subdistrict (*khum*) in which Sobay was located, 83 percent of households had less than one hectare (Delvert, *Le Paysan cambodgien,* p. 496). In Martel's village near Angkor, 50 percent of families had less than one hectare of paddies, while about 28 percent had one to two hectares (Martel, *Lovea*, p. 127). Thus, the small holdings in Sobay were not unusual, although there were regional variations in holdings per household nationwide (see Delvert, *Le Paysan cambodgien,* pp. 470–74). Kiernan's discussion of land ownership in my village, in Ben Kiernan and Chanthou Boua, eds., *Peasants and Politics in Kampuchea, 1942–1981* (Armonk, NY: M.E. Sharpe, 1982), p. 5, tends to overemphasize the degree of poverty and indebtedness, although the latter certainly did occur. The percentage of land owned by the richest families is somewhat skewed because of the one atypical household with four hectares. There are also some problems with the discussions of peasantry by Khieu Samphan, *Cambodia's Economy and Industrial Development* (Cornell University Southeast Asia Program, Data paper no. 111, 1979), and Hu Nim, "Land Tenure and Social Structure in Kampuchea," in Kiernan and Boua, eds., *Peasants and Politics in Kampuchea*; cf. Timothy Carney, *Communist Party Power in Kampuchea (Cambodia): Documents and Discussion* (Cornell University Southeast Asia Program, Data Paper no. 106, 1977), pp. 14–17.

7. The distinction between "family" and "household" has been debated within anthropology and is a problem in the Khmer case. By household I mean a group of individuals living under one roof, while family refers to kin units commonly designated nuclear, stem, or extended families. They were largely, although not completely, isomorphic in Sobay, such that the two terms will be used synonymously in this paper. However, in some cases conjugal units within a household acted separately (see May Ebihara, *A Khmer Village in Cambodia* [Ann Arbor: University Microfilms, 1968; Ph.D. diss., Columbia University], pp. 106ff.).

8. It was also possible to hire labor for such tasks as transplanting, but this was rarely done because most villagers could not afford it. Such hired workers came from other villages, not Sobay itself.

9. In Sobay, housewives estimated annual rice consumption at approximately 180–270 kilograms per person, or about 540–1,000 kilograms per family depending on family size and age/sex composition. Some of the annual crop was also used for other purposes, such as barter and seed (see Ebihara, *Khmer Village*, pp. 268ff., on rice yields). Kiernan states that one hectare of paddies will sustain a family of three or four members (Kiernan and Boua, eds., *Peasants and Politics in Kampuchea*, p. 4), while Khieu Samphan, *Cambodia's Economy,* p. 39, says that two hectares are needed to support a family of five.

10. The extent of indebtedness among old-society Cambodian peasants has been de-

bated. Kiernan, Khieu Samphan, and Hu Nim have argued that there was a high degree of indebtedness; while Delvert, Carney (*Communist Party Power*), and my data suggest that although indebtedness certainly did exist, it was not so severe or widespread. Taxes were relatively small in 1959–1960.

11. The lontar palm (*Borassus flabillefera*), which is ubiquitous in the Cambodian landscape, produces a liquid that can be boiled down into sugar.

12. Marshall Sahlins, *Stone Age Economics* (Chicago: Aldine, 1972), p. 68.

13. Cf. Eric Wolf, "Aspects of Group Relations in a Complex Society: Mexico," *American Anthropologist* 58 (1956):1065–78.

14. In 1959–1960, Khmer communists were very few in number, but villagers had heard of communism in connection with Russia, toward whom Sihanouk maintained alternately positive and negative stances in his famous balancing act between the Soviet Union and the United States.

It is curious that a decade or so before my stay, some Sobay men had been involved in the Khmer Issarak movement agitating for independence from French colonial rule. The Issarak, however, were splintered into a number of disparate groups, some genuinely political and others semibandit bands. I suspect that the Issarak in the Sobay area were more the latter, and most Sobay residents evidently were neutral or even hostile to the Issarak. See Ebihara, *Khmer Village*, pp. 540–43.

15. Some Khmer words in the text are my transcriptions in the International Phonetic Alphabet of spoken rather than written Khmer. *I* designates *i* as in "h-i-t"; *ĕ* designates the linguist's shwa; ó represents *o* as in "h-o-t"; and ú as in "b-u-t."

16. The villagers' categorization corresponds in part to different levels of peasantry outlined in "class analysis" discussions by Khieu Samphan, Hu Nim, and DK documents. What Sobay villagers called *neak miin* and *neak kuĕsóm* are comparable to "middle peasants" owning two to five/seven hectares of paddies, while the remainder fall into Hu Nim's category of "poor peasants" with less than two hectares of fields (although Sobay villagers do not support his implication that such peasants lack plows and oxen to work the land).

17. See, for example, Kiernan and Boua, eds., *Peasants and Politics in Kampuchea*; Ben Kiernan, *The Samlaut Rebellion and Its Aftermath, 1967–1970: The Origins of Cambodia's Liberation Movement* (Melbourne: Monash University Centre of Southeast Asian Studies, working papers nos. 4–5, n.d.; reprinted in Kiernan and Boua, eds., *Peasants and Politics in Kampuchea*); Carney, *Communist Party Power* and "Cambodia: The Unexpected Victory," ms., 1981; and Michael Vickery, "Democratic Kampuchea—Theme and Variations," paper presented at Social Science Research Council Conference on Kampuchea, Chiang Mai, Thailand, 1981.

18. See William Shawcross, *Sideshow: Kissinger, Nixon, and the Destruction of Cambodia* (New York: Simon and Schuster, 1979).

19. The extent and depth of peasant support for the Khmer Rouge insurgency is an extremely complex question that would require another essay to discuss. While some sources (Kiernan, *Samlaut Rebellion*) imply that the peasants were ripe for revolution, comments in various works suggest that the communists used both persuasion and coercion to extend their control over the countryside.

20. See news reports by Henry Kamm, Sydney Schanberg, and Malcolm Browne in the *New York Times* of June 14, 15, 26; July 3, 12; August 1, 18; September 6, 1973; also reports by Barry Hillenbrand in *Time*, July 23, 1973, and Dennis Neeld in the *New York*

Post, August 16, 1973.

21. I have heard two subsequent reports from visitors to Kampuchea that villages in the vicinity of Sobay are being rebuilt, although I have no information on Sobay itself.

22. Francois Ponchaud, *Cambodia Year Zero* (New York: Holt Rinehart and Winston, 1977), p. 121; Carney, *Communist Party Power*; see also Vickery, "Democratic Kampuchea," who argues that various aspects of life presumed to be new developments under DK rule were actually present in the old society. Thion's essay in this volume makes a similar point regarding political dynamics. (See Ebihara 1981 for a review of some of the literature on Democratic Kampuchea.)

23. Paul Mus, *Vietnam, Sociologie d'une guerre* (Paris: Editions du Seuil, 1952); Eric Wolf, *Peasant Wars of the Twentieth Century* (New York: Harper and Row, 1968), p. 189.

24. Article 16 of the DK Constitution, in Ponchaud, *Cambodia*, p. 204; see also David Chandler, "The Constitution of Democratic Kampuchea (Cambodia): The Semantics of Revolutionary Change," *Pacific Affairs* 49, 3 (1976):511–12; Carney, *Communist Party Power*, p. 3. This flag had other symbolic echoes: first, of the Khmer Republic flag which had Angkor in one corner; and second, of the Khmer Viet Minh flag which was red with five towers. The very name given to the new regime also combined old and new: the Westernized designation "Cambodia/Cambodge" was replaced by "Kampuchea," which presumably approximated more closely the ancient term for the Khmer kingdom, but was coupled with "Democratic" to represent the new egalitarian order.

25. Black clothing was not worn constantly in village life, but usually for doing arduous labor in which clothing would get soiled. Ith Sarin in Carney, *Communist Party Power*, p. 46, notes that "the Khmer Rouge require black dress with the *krama* around the neck" in order to be "modest and simple." See also Vickery, "Democratic Kampuchea," p. 5, and Laura Summers, "Cooperatives in Democratic Kampuchea," paper presented at the Social Science Research Council Conference on Kampuchea, Chiang Mai, Thailand, 1981, p. 19, on this clothing as a symbol of social equality. Peang Sophi's account (David Chandler, *The Early Phases of Liberation in Northwest Cambodia: Conversations with Peang Sophi* [Melbourne: Monash University Centre of Southeast Asian Studies, working papers no. 10, n.d.], p. 4) notes that even after he had dyed a pair of "old society" trousers black, he was told to cut the pants to a certain length and sew up the pockets.

26. See, for example, Summers, "Cooperatives," pp. 1–5, on the objectives of collectivization in Marxist-Leninist thought; Carney, *Communist Party Power,* "Cambodia," and "The Organization of Power in Democratic Kampuchea," ms., 1982; and others.

27. Ben Kiernan, "Conflict in the Kampuchean Communist Movement," *Journal of Contemporary Asia* 10 (1980):8, 58–60.

28. On details of general political structure, see Kiernan, "Conflict"; Carney, *Communist Party Power*; Vickery, "Democratic Kampuchea"; Ponchaud, *Cambodia.*

29. Subsequently, however, Phnom Penh was partially repopulated when limited industrial redevelopment was begun. See George Hildebrand and Gareth Porter, *Cambodia: Starvation and Revolution* (New York: Monthly Review Press, 1976); Ponchaud, *Cambodia*; Timothy Carney, "The Organization of Power in Democratic Kampuchea," ms., 1982.

30. Having lived in an agricultural village, I was particularly struck that the evacuation of Phnom Penh occurred in April just before the beginning of the rainy season, when preparations for rice cultivation must start in order to have a crop for that year. While

there has been debate as to whether food supplies were or were not stored in the countryside, there was an enormous population in Phnom Penh that had to be fed without the foreign aid food-lifts which had sustained the urban populace for some time. I do not mean to suggest that the necessity for food was the only or major reason for the evacuation of cities, but the seasonal cycle of cultivation must have been an important pragmatic consideration.

31. Summers, "Cooperatives," cited in Carney, "Organization of Power," p. 13, also 9, 11–16; see also Carney, *Communist Party Power,* pp. 12–14 30-31. In a chart presented in Carney, "Cambodia," p. 13, "middle" and "poor" peasants are even further subdivided into upper, middle, and lower depending on several criteria.

32. Carney, "Organization of Power," p. 15.

33. Vickery, "Democratic Kampuchea," p. 10.

34. Stephen Heder, *Kampuchean Occupation and Resistance* (Bangkok: Chulalongkorn University, Institute of Asian Studies, Asian Studies Monograph no. 027, 1980), p. 5.

35. Ibid., pp. 6–7; see also Carney, "Organization of Power," pp. 17–18; Summers, "Cooperatives," pp. 14–15.

36. Heder, *Kampuchean Occupation and Resistance*, p. 6; see also Summers, "Cooperatives," pp. 20-21.

37. Heder, *Kampuchean Occupation and Resistance*, pp. 6–7; Vickery, "Democratic Kampuchea," pp. 10–12; Chandler, *Early Phases of Liberation*.

38. Summers, "Cooperatives," p. 25 and n. 35.

39. For example, in one refugee account, Peang Sophi, an urban, new person, factory worker became one of the foremen of an 800-man work team (Chandler, *Early Phases of Liberation*, pp. i, 4).

40. See, for example, Vickery, "Democratic Kampuchea," p. 12; Kiernan, "Conflict"; refugee accounts in Kiernan and Boua, eds., *Peasants and Politics in Kampuchea*, and Chandler, *Early Phases of Liberation*.

41. Carney, "Organization of Power," p. 18; see also Summers, "Cooperatives," p. 27.

42. See Ponchaud, *Cambodia*, pp. 56–58, 93; Kiernan, "Conflict," pp. 44-45; Summers, "Cooperatives," pp. 20, 23; Carney, *Communist Party Power*, pp. 18–19, and "Organization of Power," pp. 18–19.

43. For base areas prior to 1975, several different forms of collectivization are reported: 1) cooperative labor only; 2) families cultivated their own paddies but there was group harvesting and a communal granary from which daily rations were distributed; and 3) cooperative labor, communal ownership of land, oxen, and tools, with a common granary (Ponchaud, *Cambodia*, p. 93; Carney, *Communist Party Power*, pp. 18–19; Summers, "Cooperatives"). Which system was used evidently depended on time and place; e.g., types 1 and 2 were earlier forms of limited collectivization.

44. Ponchaud, *Cambodia*, p. 200.

45. Stephen Heder, "From Pol Pot to Pen Sovan in the Villages," paper prepared for the International Conference on Indochina and Problems of Security and Stability in Southeast Asia, Chulalongkorn University, Bangkok, 1980; Summers, "Cooperatives," p. 10 and n. 34.

46. Such irrigation works (as well as cooperative organization) were advocated in Khieu Samphan's doctoral thesis (1979). But there is some question as to whether this

thesis did (e.g., William E. Willmott, ''Analytic Errors of the Kampuchean Communist Party,'' *Pacific Affairs* 54 [1981]:212; Summers' introduction to Khieu Samphan, *Cambodia's Economy*) or did not (Ben Kiernan, personal communication, November 12, 1982; Khieu Samphan, personal communication, 1982) actually influence DK agricultural policy.

47. See Ponchaud, *Cambodia*, pp. 75–78; Hildebrand and Porter, *Cambodia*, pp. 73ff., 89.

48. DK Constitution, article 17, in Ponchaud, *Cambodia*, p. 204.

49. See Ponchaud, *Cambodia*, pp. 55, 79–80; Hildebrand and Porter, *Cambodia*, pp. 82–84.

50. See also Ponchaud, *Cambodia*, pp. 102–103; Hildebrand and Porter, *Cambodia*, p. 69; Kiernan, ''Conflict,'' p. 14.

51. Ponchaud, *Cambodia*, p. 89; Summers, ''Cooperatives,'' p. 8, with regard to pre-1975 cooperatives; Carney, *Communist Party Power*, pp. 18–19, and ''Organization of Power,'' p. 19; Chandler, *Early Phases of Liberation*, p. 5.

52. Summers, ''Cooperatives,'' n. 22.

53. Ibid., p. 8; cf. Ponchaud, *Cambodia*, p. 94; Carney, *Communist Party Power*, p. 18.

54. Carney speaks of regiments as numbering up to 1,000 families (''Organization of Power,'' p. 19); Peang Sophi's account of Battambang notes work teams of 30–1,000 or more persons (Chandler, *Early Phases of Liberation*, pp. 5–6). See also Summers, ''Cooperatives,'' p. 3; Ponchaud, *Cambodia*, p. 97.

55. See Ponchaud, *Cambodia*, pp. 91–92; Summers, ''Cooperatives,'' p. 13; Chandler, *Early Phases of Liberation*, pp. 5–6; Kiernan and Boua, eds., *Peasants and Politics in Kampuchea*, pp. 347, 355.

56. Ponchaud, *Cambodia*, pp. 92, 123–24; Kiernan and Boua, eds., *Peasants and Politics in Kampuchea*, pp. 333, 335, 353–54. While it was common for the elderly in old-society villages to pursue nonarduous but useful tasks such as child tending and making baskets, children (at least in Sobay) generally had few or no chores until adolescence. Their early incorporation into the DK labor force can be viewed as not only pragmatically useful but as another means of socializing the young into ''revolutionary'' organization and ideology.

57. Ponchaud, *Cambodia*, pp. 55, 96; Chandler, *Early Phases of Liberation*, p. 7; Kiernan, ''Conflict,'' pp. 58–59; Kiernan and Boua, eds., *Peasants and Politics in Kampuchea*, pp. 332, 337, 338, 340, 346, 355; Boua, ''Women in Today's Cambodia,'' *New Left Review* 131 (1982):45–61.

58. Elizabeth Becker in the *Washington Post*, December 30, 1978, reported a cycle of ten days of work and one day of rest. This was a much more strenuous schedule than labor in old-society villages, except during the busiest seasons of rice cultivation.

59. Currency was abolished in Democratic Kampuchea. Various accounts indicate the surreptitious use of gold as a medium of exchange for goods or favors. In the old society, villagers (and presumably urbanites as well) often used gold jewelry as a sort of ''savings account'' for surplus money because gold was portable and easily convertible.

60. Ponchaud, *Cambodia*, p. 56; Kiernan, ''Conflict,'' p. 18.

61. Ben Kiernan, ''Kampuchea's Choices for Survival,'' *Southeast Asia Chronicle* 77 (1981):58, relates this to the ascendancy of the Pol Pot faction within the Kampuchean Communist Party. It is not entirely clear to me exactly what groups such communal dining

halls served—a village, an entire cooperative, some level of work team, or what. It was probably variable (cf. Chandler, *Early Phases of Liberation*; Kiernan and Boua, eds., *Peasants and Politics in Kampuchea*, pp. 322, 354). Summers notes a passage in a work by Ponchaud stating that in 1978 in some areas, communal dining was practiced by "smaller groups," which Summers interprets as referring apparently to work teams of thirty persons ("Cooperatives," n. 34). See also Ponchaud, *Cambodia*, pp. 61, 93; Kiernan, "Conflict," p. 58; Summers, "Cooperatives," p. 24; Chandler, *Early Phases of Liberation*, p. 7; Boua, "Women," p. 60; Kiernan and Boua, eds., *Peasants and Politics in Kampuchea*, pp. 333, 336, 338, 342, 360.

62. Items from kitchen gardens (if indeed such plots existed) were presumably exempt, but Boua ("Women," p. 60, also Kiernan and Boua, eds., *Peasants and Politics in Kampuchea*, p. 360) states that produce from family gardens went to the communal dining hall. Summers ("Cooperatives," p. 10) notes that in pre-1975 cooperatives, various foods such as fish, wild roots, and fruits were distributed equally. See also Kiernan, "Conflict," pp. 19, 58, and Princeton Conference, November 13, 1982.

63. Summers ("Cooperatives," pp. 8–10) has some points regarding this for pre-1975 cooperatives, but I do not know whether they hold true for a later period. Democratic Kampuchea attempted to resume light industrial production of items such as textiles, cigarettes, tires, and jute sacks (Hildebrand and Porter, *Cambodia*, p. 90; Ponchaud, *Cambodia*, p. 81; Chandler, *Early Phases of Liberation*).

64. Ponchaud, *Cambodia*, pp. 99–102, 123; Summers, "Cooperatives," p. 8. In the old society, the school system expanded greatly after World War II, and virtually all Sobay children attended primary school, although relatively few went on to secondary schooling. Sobay villagers went primarily to local curing specialists for various ailments. (See Ebihara, *Khmer Village*, on both topics.) Folk medicines were evidently widely used in Democratic Kampuchea because of the lack of medical personnel and manufactured medicines.

65. I do not know if this was ever explicitly articulated as official policy in any written documents, but certainly various measures imposed by the government had this effect. The DK Constitution nowhere mentions anything pertaining to the family except for a provision outlawing polygamy and a metaphorical reference to the "family of nonaligned nations."

66. Ponchaud, *Cambodia*, p. 92; Boua, "Women," p. 46.

67. While Summers speaks of "nuclear households" as "accounting units" ("Cooperatives," n. 22), such households would have been composed, then, of only parents and very young children. Kiernan has noted an intriguing semantic change in the meaning of the term *kruèsóó*: where formerly this meant "family," it evidently now refers only to "husband and wife" (personal communication November 13, 1982). When this semantic shift took place is uncertain, but it may have occurred during Democratic Kampuchea because of the various measures I am discussing. This raises a further question as to whether extended families, which were widespread in both rural and urban segments of the old society, were usually broken up into nuclear units in Democratic Kampuchea. Summers (p. 19, n. 22) suggests that urban evacuees with large extended families would have been at a "disadvantage in finding rural communities which would accept them" because the number of young and elderly might not be offset by the number of working adults in the family, and hence would constitute a burden on the cooperative as a whole. However, given high mortality both before and after 1975, especially of men, there must have been many

"denuded" nuclear units with only one parent and, following customary practice, perhaps nuclear units with other relatives attached (e.g., an elderly parent).

68. See Ponchaud, *Cambodia*, p. 122; Kiernan, "Conflict," p. 58; Kiernan and Boua, eds., *Peasants and Politics in Kampuchea*, pp. 333, 335, 338, 342, 354–54.

69. Carney, *Communist Party Power*, pp. 10, 27–33, and "Cambodia," p. 22; Kiernan, "Conflict," pp. 22, 25; Chandler, *Early Phases of Liberation*, p. 9; Ponchaud, *Cambodia*, pp. 122–24; Summers, "Cooperatives," p. 13.

70. See also Ponchaud, *Cambodia*, p. 23; Summers, "Cooperatives," p. 13.

71. See also Ponchaud, *Cambodia*, p. 123; David Hunt, "Village Culture and the Vietnamese Revolution," *Past and Present* 94 (1982):131–57.

72. See May Ebihara, "Khmer Village Women in Cambodia," in Carolyn Matthiasson, ed., *Many Sisters: Women in Cross-Cultural Perspective* (New York: Free Press, 1974), pp. 315–17.

73. Boua notes that "parents felt deprived because the marriage system in the Pol Pot period prevented them from having their traditional say in the choice of their offspring's partner" ("Women," p. 47). See also Ponchaud, *Cambodia*, p. 125.

74. It should also be noted that all marriages now had to be monogamous. Polygyny was legally permitted in the old society but it occurred primarily only among the relatively wealthy, although village men sometimes entertained thoughts of second wives (see Ebihara, "Khmer Village Women"). While the rationale for the DK prohibition of polygamy is not stated, it is possible that polygyny was viewed as inimical to the proclaimed equality of the sexes, as a manifestation of old regimes in which multiple wives and concubines were marks of high status, and as offensive to Democratic Kampuchea's rather puritanical morality. While Elizabeth Becker reported that people were encouraged to marry before the age of twenty-one (*Washington Post*, December 30, 1978), Ponchaud suggests that age of marriage may have been delayed for military personnel (twenty to twenty-five for women, twenty-five to thirty for men) (*Cambodia*, p. 125). Such postponement of marriage would have been useful for the state in terms of keeping soldiers' energies and sentiments focused on military rather than domestic responsibilities.

75. By contrast, in the old society theoretically only males did so, although it was not unheard of for the parents of a young woman to scout around for potential husbands if she had not attracted any proposals (see Ebihara, "Khmer Village Women").

76. Ponchaud, *Cambodia*, pp. 125–26. It is not clear whether the "group" noted by Ponchaud refers to a work team or youth association.

77. Ibid., p. 125; Boua, "Women," p. 46; see also Kiernan and Boua, eds., *Peasants and Politics in Kampuchea*, pp. 333, 351. In old-society village life, weddings were festive observances lasting a day and a half and overseen by ritual specialists, parents, and honored elders (Ebihara, *Khmer Village*, pp. 474ff., and "Khmer Village Women"). In Democratic Kampuchea they became hasty group affairs, presided over by local officials, in which a number of couples were married off at once (Ponchaud, *Cambodia*, p. 126; Boua, "Women," p. 46; Kiernan and Boua, eds., *Peasants and Politics in Kampuchea*, pp. 333, 351). Cf. Hunt, "Village Culture," p. 154, on NLF marriage ceremonies in Vietnam for a similar situation in which political officials replaced parents and elders, thus symbolizing a similar decline in the authority of the latter.

78. Boua, "Women"; Kiernan and Boua, eds., *Peasants and Politics in Kampuchea*, pp. 351, 356. In an inversion of the "separate but equal" notion, the DK Constitution proclaimed that men and women are equal, but DK policies certainly kept them segre-

gated. In old-society village life, while there was some effort made to "chaperone" unmarried young women to prevent premarital sex, there was generally no strict separation of the sexes.

81. Boua, "Women," p. 60.

82. See Ebihara, "Khmer Village Women."

83. Kiernan and Boua, eds., *Peasants and Politics in Kampuchea*, pp. 345, 350.

84. Summer notes that in the pre-1975 base area cooperatives, the labor teams and even whole cooperatives "coincided with neighborhood, kin, and village friendship groups" ("Cooperatives," p. 13). After 1975, however, the influx of new people into various communities required formal means of organization rather than cooperation based, as it were, on "sentiment."

85. Ponchaud, *Cambodia*, p. 120; Carney, *Communist Party Power*, p. 47; Kiernan and Boua, eds., *Peasants and Politics in Kampuchea*, p. 342.

86. Carney, *Communist Party Power*, p. 47; see also Ponchaud, *Cambodia*, pp. 120–21. For discussion of DK linguistic usages as reflections of political ideology, see in particular Chandler, "Constitution," also Ponchaud, *Cambodia*, pp. 120–21, 124–25.

87. Carney, *Communist Party Power*, p. 47; Ponchaud, *Cambodia*, p. 121; Kiernan and Boua, eds., *Peasants and Politics in Kampuchea*, p. 342; Kiernan, personal communication, November 13, 1982.

88. Ponchaud, *Cambodia*, pp. 120–21. What Ponchaud translates as "senior" and "junior" are, I would assume, the terms *bóng* and *p'on*, which mean "older" and "younger." In Khmer kinship terminology, they specify siblings, cousins, and siblings-in-law older or younger than oneself. The phrase "*bóng-p'on*" can also be used to refer to relatives in general (see Ebihara, *Khmer Village*, pp. 656–77, on Khmer kin terms).

It may also be the case, of course, that usage was variable in time and place, or possibly even according to individual or situation (see, e.g., Kiernan and Boua, eds., *Peasants and Politics in Kampuchea*, p. 342).

89. Chandler, *Early Phases of Liberation*, p. 15; cf. Hunt ("Village Culture," p. 149), who notes that former NLF guerillas in Vietnam often spoke of the movement as "a kind of family," including one young woman who had rejected some traditional kinship roles but spoke of herself as "a daughter of the revolution."

90. Kiernan and Boua, eds., *Peasants and Politics in Kampuchea*, pp. 346, 350; Carney, "Organization of Power," pp. 40, 43; Kiernan, personal communication, November 12, 1982; Chanthou Boua, Ben Kiernan, and Anthony Barnett, "Bureaucracy of Death: Documents from Inside Pol Pot's Torture Machine," *New Statesman* 2 (May 1980):669–76. In the old society, Sihanouk sometimes spoke of how "all we Khmer are *bóng-p'on*." Cf. David Schneider, "Kinship, Nationality, and Religion in American Culture: Toward a Definition of Kinship," in Robert Spencer, ed., *Forms of Symbolic Action*, Proceedings of the American Ethnological Society (Seattle: University of Washington Press, 1969), on the ways in which conceptions of kinship and nationality share identical symbols.

91. Kiernan, personal communication, November 13, 1982.

92. Vickery, "Democratic Kampuchea," pp. 10–12; refugee accounts in Kiernan and Boua, eds., *Peasants and Politics in Kampuchea*, pp. 340, 342, 345, 346, 356, 359, 360.

93. See Carney, *Communist Party Power*, pp. 8, 53–54, and "Organization of Power," p. 19; Summers, "Cooperatives," pp. 8–9, 23; Vickery, "Democratic Kampuchea," p. 4; Ponchaud, *Cambodia*, pp. 89–91.

94. Carney, ''Cambodia,'' p. 21; cf. Ponchaud, *Cambodia*, pp. 89, 94. Cf. Summers, ''Cooperatives,'' p. 9, who states that in pre-1975 base areas, the three-person committees heading cooperatives were existing village administrators or local party cadres, all poor or lower-middle peasants. Ponchaud (pp. 89–90) also implies that local people from poor backgrounds were *krom*, ''village,'' and cooperative chairs, though district (*srok*) authorities were said to be former Khmer Rouge army officers.

95. Summers, ''Cooperatives,'' p. 20; Kiernan, ''Conflict,'' pp. 58ff.; Vickery, ''Democratic Kampuchea''; Carney, ''Organization of Power.''

96. See, for example, Vickery, ''Democratic Kampuchea,'' p. 17; Chandler, *Early Phases of Liberation*; Kiernan and Boua, eds., *Peasants and Politics in Kampuchea*, p. 356.

97. Heder, *Occupation and Resistance*, p. 8; Summers, ''Cooperatives,'' pp. 20–21; Vickery, ''Democratic Kampuchea''; Carney, ''Organization of Power,'' pp. 17–18.

98. Kiernan, ''Conflict,'' pp. 58ff.

99. Carney, *Communist Party Power*, pp. 10, 27–33, 50, and ''Cambodia,'' pp. 27ff; Summers, ''Cooperatives,'' n. 14; Ponchaud, *Cambodia*.

100. Summers, ''Cooperatives,'' pp. 10, 12, 22; Kiernan, ''Conflict,'' p. 45. Prior to 1975 there was occasional use of Buddhism to support the revolutionary cause: Buddhist teachings were sometimes invoked to support DK ideology in propaganda aimed at Khmer Republican areas (Ponchaud, *Cambodia*, p. 127), and one 1973 DK radio broadcast reported an association called Patriotic Monks of Kampuchea (Carney, ''Cambodia,'' p. 28). More recently, after DK leaders disavowed the ''former excesses'' of their regime and tried to adopt a more moderate posture, a Buddhist monk was taken in 1982 under DK auspices to visit their resistance bases in Kampuchea (see press release of April 1982, Permanent Mission of Democratic Kampuchea to the United Nations).

101. Article 20, in Ponchaud, *Cambodia*, p. 205.

102. Carney, *Communist Party Power*, p. 26, and ''Cambodia,'' n. 14; Ponchaud, *Cambodia*, pp. 130–32; Summers, ''Cooperatives,'' p. 22.

103. Summers, ''Cooperatives,'' p. 17; Carney, ''Organization of Power,'' p. 13.

104. See also Ponchaud, *Cambodia*, p. 127; Summers, ''Cooperatives,'' p. 13.

105. Article 17 of the DK Constitution, in Ponchaud, *Cambodia*, p. 204; Chandler, ''Constitution,'' pp. 511–12.

106. Ponchaud, *Cambodia*, pp. 127–29; Chandler, *Early Phases of Liberation*, p. 12; Kiernan, ''Conflict,'' p. 45; Kiernan and Boua, eds., *Peasants and Politics in Kampuchea*, pp. 330, 341, 356.

107. Ponchaud, *Cambodia*, p. 131; Kiernan and Boua, eds., *Peasants and Politics in Kampuchea*, p. 360. As Kiernan notes (''Conflict,'' p. 19), certain acts of physical destruction also carry dramatic symbolic weight, e.g., the dynamiting of the National Bank and destruction of the Catholic cathedral in Phnom Penh.

108. Ith Sarin in Carney, *Communist Party Power*, pp. 40, 47, 52–53, also pp. 27–33; Chandler, *Early Phases of Liberation*, pp. 10, 11, 14–16; Ponchaud, *Cambodia*, p. 87; Carney, ''Organization of Power,'' p. 22; Kiernan and Boua, eds., *Peasants and Politics in Kampuchea*, pp. 336, 356. The selective conjunction of the old and new occurred here also; Chandler (p. 14) notes that some of these songs used ''existing folk tunes and traditional rhythms''; but their lyrics clearly promoted revolutionary and collective ideals, and the songs were sung by the group rather than by individuals at political meetings.

109. See Carney, *Communist Party Power*, pp. 45–47, 50–51.

110. Ith Sarin in ibid., p. 47; see also pp. 51–52.

111. Ibid., p. 47; see also Chandler, *Early Phases of Liberation*, p. 7. Ith Sarin and others have characterized traditional Khmer culture as being individualistic, although this is a moot point (see Ebihara, *Khmer Village*, pp. 203ff., and ''Interrelations Between Buddhism and Social Systems in Cambodian Peasant Culture,'' in Manning Nash et al., *Anthropological Studies in Theravada Buddhism*, Yale University Southeast Asia Studies, Cultural Report Series no. 13, 1966). Certainly, however, collective action became paramount in Democratic Kampuchea to a degree unknown in the old society. But even here, Kiernan, ''Conflict,'' pp. 18–19, notes that the focus was not exclusively collective: at some point daily work requirements were set on an individual rather than team basis.

112. Carney, *Communist Party Power*, p. 11.

113. Vickery, ''Democratic Kampuchea,'' p. 10.

114. Yeheudi Cohen, ''Ends and Means in Political Control: State Organization and the Punishment of Adultery, Incest, and Violation of Celibacy,'' *American Anthropologist* 71 (1969):658–87. While the case of Democratic Kampuchea does not precisely fit Cohen's models of ''incorporative'' and ''expropriative'' states but partakes of features of both, a number of the processes of state consolidation that he discusses are certainly applicable.

115. Cf. Eric Wolf, *Peasants* (Englewood Cliffs, NJ: Prentice-Hall, 1966), p. 77.

116. Description and discussion of the People's Republic of Kampuchea will rely primarily on the work of Heder, Kiernan, and Boua. Heder's materials derive from interviews with Khmer in Thai refugee camps and along the border region in 1980; Boua spent eight months in 1980 and one month in 1981 in Kampuchea; Kiernan spent four months there in 1980 and visited in 1981. Gough's article is based on interviews with several PRK officials on a three-week visit in February 1982. Information on present-day conditions in Kampuchea is still relatively limited.

117. In some cases people chose to stay where they had ended up living during the DK period, while others moved to more prosperous (or, in the case of traders, more profitable) regions or settled (perhaps temporarily) around urban centers (see Stephen Heder, ''Kampuchea, October 1979 to August 1980, the Democratic Kampuchea Resistance, the Kampuchean Countryside, and the Sereikar,'' ms., 1980, pp. 86–88, 90, and *Occupation and Resistance*, pp. 39f.).

118. Boua, ''Observations of the Heng Samrin Government, 1980–1982,'' paper presented at the Social Science Research Council Conference on Kampuchea, Chiang Mai, Thailand, 1981, p. 1; Heder, ''Kampuchea,'' p. 86.

119. For an excellent, detailed discussion of unsettled conditions in the early People's Republic of Kampuchea, see especially Heder, *Occupation and Resistance*, pp. 27ff., also ''Pol Pot,'' and ''Kampuchea.''

120. Kathleen Gough, ''Interviews in Kampuchea,'' *Bulletin of Concerned Asian Scholars* 14 (1982):55–65; Boua, ''Women''; Sophie Quinn-Judge, ''Working for the Basics in Kampuchea,'' *Southeast Asia Chronicle* 87 (1982):17–25.

121. Boua, ''Observations,'' pp. 4–6, and ''Women,'' p. 48; Heder, ''Pol Pot,'' p. 19, ''Kampuchea,'' pp. 91–92, and ''Kampuchea 1980: Anatomy of a Crisis,'' *Southeast Asia Chronicle* 77 (1981):8; Gough, ''Interviews''; Quinn-Judge, ''Working for the Basics.'' Heder notes that in the early days of the People's Republic of Kampuchea there was confusion as to the nature and designation of these teams (''Pol Pot,'' p. 28, ''Kampuchea,'' p. 91). This confusion, as well as change through time or local variation, may

account for differing statements as to the number of families within a *krom samaki*. Heder also suggests that in the early People's Republic of Kampuchea there was a slight tendency to move from larger (thirty-household) to smaller (ten-household) teams ("Kampuchea," p. 95). According to the minister of agriculture quoted in Boua, some 95 percent of the population is organized into teams ("Observations," p. 7). Boua also notes that the Kampuchean work teams appear to have no parallel in Vietnamese rural organization (p. 9).

122. Boua, "Observations," pp. 5, 8; Gough, "Interviews."

123. See Heder, "Pol Pot" and "Kampuchea"; Boua, "Observations," pp. 11–13, "Women," p. 47; Gough, "Interviews," p. 61; Quinn-Judge, "Working for the Basics," p. 20.

124. Heder, "Kampuchea," pp. 92–94, "Pol Pot," pp. 28ff. Boua notes government efforts "to work out clear administrative details so that lower-level cadres who have to implement the orders cannot misinterpret them" ("Observations," p. 27).

125. Boua, "Observations" and "Women"; Gough, "Interviews"; Quinn-Judge, "Working for the Basics."

126. Heder, "Kampuchea," p. 95; Boua, "Observations," p. 4; Quinn-Judge, "Working for the Basics," pp. 21–22.

127. Boua, "Women," p. 48; Gough, "Interviews"; Quinn-Judge, "Working for the Basics," p. 21.

128. Gough, "Interviews," p. 61.

129. Quinn-Judge, "Working for the Basics," p. 23. The irrigation works constructed in Democratic Kampuchea were evidently laid out incorrectly in some areas and aggravated rather than controlled flooding; in other regions they were rebuilt to suit local needs at the expense of the larger system; and still elsewhere the channels have deteriorated for lack of maintenance (Heder, *Occupation and Resistance*, pp. 41–42, "Kampuchea," p. 94; Gough, "Interviews," p. 62, Quinn-Judge, "Working for the Basics," p. 23).

130. On distribution of produce, see Boua, "Observations," pp. 9–10, "Women," p. 47; Quinn-Judge, "Working for the Basics," p. 20. On draft animals, see Heder, "Kampuchea," p. 94; Quinn-Judge, "Working for the Basics," p. 21. Those families owning draft animals and carts can rent them out for, or themselves engage in, trade (Heder, "Pol Pot" and "Kampuchea").

131. Heder, "Kampuchea," p. 95; Boua, "Observations," pp. 12–13.

132. Heder, *Occupation and Resistance*, p. 47, "Pol Pot," pp. 47f.; "Kampuchea," p. 95; Boua, "Observations," pp. 12–13, "Women," pp. 47–49, 53; Quinn-Judge, "Working for the Basics," p. 22.

133. Boua, "Observations," p. 20.

134. Ibid., p. 13; see also Gough, "Interviews," p. 62.

135. While currency was reinstated, much of the border trade is evidently transacted in gold. Thus, another government concern is the outflow of gold from Kampuchea (Gough, "Interviews," p. 64; Boua, "Women," p. 50).

136. Boua, "Observations," pp. 11, 13, 14ff., 18–24, and "Women," pp. 49–50, 53; Gough, "Interviews," pp. 62, 64–65; Heder, *Occupation and Resistance*, "Pol Pot," "Kampuchea."

137. Gough, "Interviews," p. 64; see also Heder, "Pol Pot," pp. 48f.; Boua, "Observations," pp. 19f.

138. Heder, "Kampuchea," mentions the use of kin and friendship bonds with regard

to choice of residence, trade connections, etc.

139. See, for example, various families mentioned in Boua, ''Women,'' pp. 50, 52, 53, 57.

140. Ibid., pp. 46–47; Gough, ''Interviews,'' p. 57.

141. Boua, ''Observations,'' pp. 4–5, ''Women,'' p. 45; Kiernan and Boua, eds., *Peasants and Politics in Kampuchea*, p. 380; Gough, ''Interviews,'' pp. 57, 61.

142. Gough, ''Interviews,'' p. 61.

143. Ebihara, *Khmer Village*, p. 682, and ''Khmer Village Women.''

144. It would seem that men still dominate the higher political echelons, just as they did in the old society. In Democratic Kampuchea relatively few women occupied top government posts and those that did were wives or kin of high officials. Similarly, few women hold top posts in the People's Republic. See Carney, *Communist Party Power*, pp. 62–63; Boua, ''Women,'' pp. 52, 61. While equality of the sexes is formally supported by the PRK state, Boua believes that there is still prejudice against women in higher administration (p. 61).

145. Boua, ''Women,'' pp. 46, 47, 58. It is unclear whether such abortions, said to be very expensive, are resorted to primarily by urban women or if they occur also in rural areas. I find this practice quite startling because in old-society Sobay, abortion was considered tantamount to killing and hence a grievous sin according to Buddhist doctrine. Furthermore, while premarital pregnancy was certainly not condoned in Sobay, it was generally shrugged off if the woman eventually married (see Ebihara, *Khmer Village*, pp. 445ff., ''Khmer Village Women''). With the current shortage of men in the People's Republic of Kampuchea, however, such marriages may not be possible.

146. See Ebihara, *Khmer Village*, p. 444.

147. Boua, ''Women,'' p. 60; Kiernan and Boua, *Peasants and Politics in Kampuchea*, p. 380.

148. Heder, ''Kampuchea,'' pp. 22–23, 26, 95. The terms for these local officials as given by Heder are identical to those used in old-society village administration. The three-person committees also, however, recall DK local administration.

149. While one government official claimed that village and higher level officials cannot influence decisions of the *krom* (Boua, ''Observations,'' p. 6), this seems questionable. The same official is elsewhere quoted as saying that the *krom* system makes it ''easier for higher authorities to direct the local levels'' (ibid., p. 10).

150. Heder, ''Kampuchea,'' pp. 97–99, ''Kampuchea 1980,'' p. 8; see also Boua, ''Observations,'' p. 27.

151. Kiernan, ''Kampuchea's Choices,'' p. 7; Kiernan and Boua, *Peasants and Politics in Kampuchea*, p. 382f.

152. Heder, ''Kampuchea,'' pp. 96, 97; see also Boua, ''Observations,'' p. 6.

153. Cf. Wolf, ''Group Relations.''

154. Heder, ''Kampuchea,'' pp. 109, 107.

155. Ibid., pp. 104, 107, 108–109, though cf. Boua, ''Women,'' p. 61. 150; Heder, ''Kampuchea 1980,'' p. 7; Kiernan, ''Kampuchea's Choices,'' p. 18; Kiernan and Boua, eds., *Peasants and Politics in Kampuchea*, pp. 382f. Heder also suggests that another factor contributing to a sense of peasant solidarity was the mixing of peasants from different communities that occurred during the DK period and continued to some extent in the People's Republic; this meant that ''villages were no longer, as they had been traditionally, relatively encapsulated and closed entities containing peasants mostly related to each

other through interlocking kin clusters but relatively cut off from other villages by the lack of such ties'' (''Kampuchea,'' p. 90, see also 91). His implication that peasants did not often mingle with people from other locales and his characterization of ''closed'' villages in the old society are incorrect. The practice of village exogamy, which occurred frequently (Ebihara, *Khmer Village*, ''Khmer Village Women,'' and ''Residence Patterns in a Khmer Peasant Village,'' in Stanley Freed, ed., *Anthropology and the Climate of Opinion*, New York Academy of Sciences Annals, vol 293, 1977; Martel, *Louvea*), automatically created kin links to other communities. And, at least in the case of Sobay residents, there was a good deal of interaction in various contexts with people outside the village (see Ebihara ''Intervillage, Village-Town, and Village-City Relations in Cambodia,'' in Anthony LaRuffa et al., eds., *City and Peasants: A Study in Sociocultural Dynamics*, New York Academy of Social Sciences Annals, vol. 220, 1974). Furthermore, I would suggest that some sense of peasant identity (if not solidarity) existed in the old society insofar as villagers conceived of themselves as ''poor country folk'' in contrast to urbanites, wealthier strata, etc. It is, however, true that villagers tended to be wary of persons from other communities who were not kin or friends.

156. Heder, ''Kampuchea,'' pp. 97, 98; Boua, ''Women,'' pp. 58–59; Gough, ''Interviews,'' pp. 56–57.

157. Kiernan and Boua, eds., *Peasants and Politics in Kampuchea*, p. 370; Colin Campbell, *New York Times*, April 3, 1983.

158. Kiernan and Boua, eds., *Peasants and Politics in Kampuchea*, pp. 365–68.

159. See Ebihara, ''Sociopolitical Transformations.''

160. Kiernan's ''impression . . . that, among most Khmer peasants, anti-Vietnamese feeling has seldom been as strong as among the urban and educated classes'' (Kiernan and Boua, eds., *Peasants and Politics in Kampuchea*, p. 374) is one that I would question on the basis of Sobay villagers whom I knew. Of the various ethnic groups with whom villagers were familiar, the Vietnamese were the most disliked (Ebihara, *Khmer Village*, pp. 580f.), and such sentiments were reinforced by government propaganda at the time.

161. See, for example, Kiernan and Boua, eds., *Peasants and Politics in Kampuchea*, pp. 370–77; Boua, ''Observations.''

162. Heder, ''Kampuchea,'' pp. 107–108.

163. Boua, ''Women,'' p. 48. It might be noted here that Heder's discussions deal with Kampuchea up to August 1980 and are based on interviews with some 800 people along the Thai border region, many of whom were probably dissatisfied with life in the People's Republic of Kampuchea. Kiernan and Boua had stays of varying lengths in the People's Republic in 1980–81 (see note 60), speaking to a variety of people who had stayed in Kampuchea for better or worse. The samples of informants are therefore somewhat dissimilar. Moreover, while Heder, Kiernan, and Boua have all had negative reactions to Democratic Kampuchea, Kiernan and Boua are more pro-People's Republic of Kampuchea than is Heder (see also Ebihara, ''Sociopolitical Transformations'').

164. Cf. Ebihara, ''Interrelations.''

165. Cf. Hunt, ''Village Culture,'' p. 145, on the importance of ritual as a way of establishing social solidarity and strength in rebuilding war-torn Vietnam.

166. Boua, ''Observations,'' p. 29; Kiernan and Boua, eds., *Peasants and Politics in Kampuchea*, pp. 8, 368.

167. Quoted in Boua, ''Observations,'' p. 29.

168. See, for example, Felix Keesing, *Culture Change*, Stanford Anthropological

Series, no. 1 (Stanford: Stanford University Press, 1953).

169. See also Ebihara, "Sociopolitical Transformations."

170. At the time of writing, there were a number of works on Kampuchea in preparation or in press. It may well be that some of the questions I raise will be addressed in these studies, which include doctoral dissertations by Ben Kiernan (Monash University, 1983), Stephen Heder (Cornell University, 1983), and Laura Summers (Cornell University); David Chandler's *A History of Cambodia* (Boulder, Colorado: Westview Press, 1983); David Chandler and Ben Kiernan, eds., *Revolution and Its Aftermath in Kampuchea: Eight Essays* (Yale University Southeast Asia Studies, Monograph Series no. 25, 1983); a collection of articles edited by Karl Jackson, and books by Michael Vickery and Anthony Barnett.

171. For example, Carney, "Organization of Power."

172. For example, Wolf, *Peasant Wars*; James C. Scott, *The Moral Economy of the Peasant: Rebellion and Subsistence in Southeast Asia* (New Haven: Yale University Press, 1976); Samuel Popkin, *The Rational Peasant: The Political Economy of Rural Society in Vietnam* (Berkeley: University of California Press, 1979).

173. I have heard reports of conversions to Christianity among Khmer in refugee camps, but it is not clear whether these are due to disillusionment with Buddhism or to hopes that espousal of Christianity might improve chances of emigration to Western countries.

INTERNATIONAL RELATIONS

David W. P. Elliott

DEADLOCK DIPLOMACY: THAI AND VIETNAMESE INTERESTS IN KAMPUCHEA

There is no single key to unlock the impasse in Kampuchea. A solution to this complex problem will necessarily involve at least three distinct sets of actors. The ultimate solution, of course, will have to be worked out within Kampuchea by Kampucheans. But the outcome of this process will be largely dependent on the interests and perceptions of external powers, for whom Kampuchea is one part of a larger interest. The two external powers most directly involved are Vietnam and Thailand, the focus of this paper. The third dimension of the Kampuchean problem is the larger regional and global context, involving Thailand's Association of South East Asian Nations (ASEAN) partners, China, the United States, and the Soviet Union. It is conceivable, though unlikely, that the Kampuchean problem could be resolved independent of the larger regional and global context, but it is difficult to envision a settlement that does not involve Vietnam and Thailand.

An obvious first step in locating the common ground for a settlement would be a close examination of the national interests of Vietnam and Thailand in order to identify the boundaries of conceivable compromise. This is fairly easily done, but it may not take us where we want to go. Interests do not necessarily have predictive or prescriptive value, and the policy implications of any interest or set of interests may be ambiguous. Also, for Vietnam and Thailand, Kampuchea is not the only issue that divides them, but is a component of larger problems that make it difficult to posit an internal Kampuchean settlement as the sole basis of a broader rapprochement between Vietnam and Thailand. It is more probable that such a rapprochement would precede a Kampuchean settlement rather than follow it. Finally, as this chapter illustrates, it is not clear that the interests of either Vietnam or Thailand are driving them toward a settlement in Kampuchea. There are a number of indications that a perpetuation of the status quo may be preferable to the presumptive costs of doing what is necessary to achieve a solution in Kampuchea, and that the current situation of "neither war nor peace" may indeed be the solution that both countries have settled on, for the short and medium term at least.

The prolonged standoff between ASEAN and Vietnam over the Kampuchean issue has been punctuated by periodic signs of diplomatic movement and corresponding upswings of optimism. ASEAN's diplomatic victories at the

United Nations, the formation of a coalition government by a tenuous marriage of convenience between Pol Pot's Democratic Kampuchean (DK) forces and the noncommunist followers of Sihanouk and Son Sann, a series of Vietnamese initiatives accompanied by a new, softer tone emanating from Hanoi in 1982, and a new Thai emphasis on "flexibility" accompanying Prime Minister Prem Tinsulanond's reappointment in 1983 are all part of this pattern. To date, none of these initiatives has created any sustained momentum toward the resolution of the basic issues in question. Indeed, despite a flurry of proposals and counter-proposals between Vietnam and Thailand in the spring of 1983, Dr. Sarasin Wiraphon, director of policy and planning in the Thai Foreign Ministry, stated that the situation remained "essentially a stalemate." Dr. Sarasin added that "We are not even at the stage of negotiation; we are just trying to read each other's position."

From the outside it appears puzzling that the obvious benefits to all parties of ending the Kampuchean conflict have not led the concerned parties farther along the road to a settlement. This is especially true in the case of Thailand, which has been unwillingly pulled into a conflict instigated by others, and for Vietnam, for whom the heavy economic and diplomatic costs of military involvement in Kampuchea are an undesirable addition to its formidable burdens. Why, then, has it proven so difficult to find a settlement based on the apparent shared interests of Kampuchea's neighbors in terminating an unwanted conflict? One answer may be that each state has mixed interests, some favoring a settlement and some impeding it. A related problem is that neither side agrees about what the fundamental issues are, which is one reason that difficulties occur in trying to "read each other's position." Failure to agree on what the essence of the problem is reflects the diversity of interests involved, not only between states but within them. For some purposes it might be expedient for a state to pose the issues in one way, while other interests might dictate a different formulation of the problem.

One of the most difficult tasks in diplomacy is to find a common definition of the problem, and to agree on a unified framework to place it in. For Vietnam, Kampuchea is an issue linked to a hostile regional system organized by Vietnam's main antagonist, China. For Thailand, Kampuchea is a key element in a potentially destabilizing bloc, the "Indochinese Confederation," which places both Laos and Kampuchea under the domination of Vietnam and is the conduit through which great power influences (from the Soviet Union) filter into the region. A Kampuchean settlement will involve either the untangling of specific Kampuchean problems from these larger concerns, or a more comprehensive settlement that tries to resolve the broader issues as well. In general, Thailand sees its interests leading to a broadening of the issues, while Vietnam's perceived interests led toward narrowing them—though Vietnam has also suggested that if issues that it considers extraneous to the central problem of Kampuchea are raised, it will counter by throwing in larger issues of its own.

The primary Thai interests that can be inferred from Bangkok's comments on the Kampuchean situation may be summarized as follows: 1) Upholding a principle of acceptable international behavior—aggression should not be rewarded. This is perceived to be the political basis for undermining the Vietnamese position in Kampuchea and rallying international opposition against Vietnam. 2) A neutral and independent Kampuchea to serve as a buffer between Thailand and Vietnam, and to prevent a Vietnamese-dominated Kampuchea from swinging the regional power balance against Thailand. 3) A regional order in Southeast Asia that would insulate Thailand from being caught up in conflicts outside the region. 4) Isolating the internal challenge to Thai authority from external support.

In the view of a leading Thai academic specialist on international relations, Dr. Khien Theeravit, the Vietnamese invasion of Kampuchea has damaged Thai interests in four ways: it has compromised Kampuchean independence and prevented Kampuchea from playing its traditional role as a buffer state, destabilized the area, created a massive refugee problem for Thailand, and stimulated the growth of military influence in Thai politics.[1]

Vietnam's stated objectives are 1) ending the perceived threat posed by continued Chinese support for Pol Pot, and Chinese attempts to become a major power factor in Southeast Asia; 2) consolidation of power in Kampuchea by a reliably friendly regime; 3) consolidation of the bloc of Indochinese countries under Vietnamese leadership, at least in the foreign policy sphere; 4) neutralization of all Southeast Asia; 5) avoiding total dependence on a single power and ending its diplomatic isolation.

Were Kampuchea the only problem for Vietnam and Thailand, an obvious way to harmonize their interests would be a Thai derecognition of Pol Pot (which would entail the elimination of sanctuaries and supply lines) in return for a Vietnamese troop withdrawal from Kampuchea. This would not, however, necessarily lead to a neutral Kampuchea. As can be inferred from the broad range of interests involved, the question is not only how Thai and Vietnamese interests in Kampuchea can be resolved, but also the significantly different one of how a solution in Kampuchea will affect matters that go beyond Kampuchea itself.

One reason why it has proven so difficult to discuss the outlines of a solution is that the parties do not agree on a common framework for settlement. Each party has added layers of external issues onto the already intractable problem of Kampuchea. The Thai focus on the issue of neutralization and Vietnamese withdrawal is complicated by its facilitation of Chinese aid to Pol Pot and the members of the DK coalition, the linkage of the Kampuchean issue with broader ASEAN concerns, and insistence on the United Nations, in the form of the International Conference on Kampuchea (ICK), as the proper forum for resolving the issue. The Vietnamese insist that the real issue is the link between China and Pol Pot, counter with a demand for the neutralization of all Southeast Asia, and propose a conference outside the UN framework.

If a settlement is to be reached, it will involve either each side stripping away the extraneous issues and moving to the irreducible core of Kampuchean problems, or an overall package deal that will try to piece together some solution out of the tangle of interwoven issues linking the local parties and the external powers. The Middle East offers an instructive case of the advantages and pitfalls of each approach, which suggests that both may have their place. The Geneva negotiating format clusters all relevant issues together, in the hopes that putting everything on the table at once will create the possibility for the trade-offs and compromises that would not emerge from a consideration of a few core issues by the parties directly involved. Regional issues are linked to superpower concerns in an attempt to reach a broad settlement. The alternative approach is a more modest, step-by-step effort, focusing on narrower questions of immediate interest to the parties directly involved in the conflict. Often such an approach must try to uncouple issues rather than link them together, and concentrate on peripheral or technical questions rather than core issues. These approaches are not mutually exclusive, though it is difficult to engage in both simultaneously. Each can be tried to break a stalemate, but the sequence and timing of each approach is vital, especially the decision on whether to focus on the broader or narrower aspects of the problem as a means of generating momentum toward a resolution of the conflict.

Vietnam has implicitly proposed two modes of settlement: temporary stabilization or long-term resolution. It has said that if the main Thai concern is the danger of a spillover of conflict from Kampuchea, this could be avoided by the creation of a safety zone—which would resolve the problem of Pol Pot as well, and which would presumably reduce the Chinese involvement in the Kampuchean situation and thus respond to the primary Vietnamese concern about the "Chinese threat"—though Vietnamese Foreign Minister Nguyen Co Thach in late 1983 stated that even if Thailand stops giving aid to the Khmer Rouge, "China can still use the sea to supply Pol Pot."[2] If, however, Thailand feels that this does not address the larger and more fundamental issues, Vietnam has suggested that it is willing to deal with those as well. The catch is that Vietnam's price for including broader diplomatic concerns on the negotiating agenda is a Thai commitment to discuss the neutralization of all Southeast Asia, and sensitive issues involving Thai friends and allies, such as American bases in the Philippines and the Indonesian occupation and absorption of East Timor. Neutralization would pose an interesting dilemma for Vietnam if, as the policy implies, it substantially reduced Soviet military access to Vietnam, which is one of the few things Vietnam can offer in repayment for Soviet aid. Yet Vietnam also has shown a concern for the U.S. reinvolvement in Southeast Asia and, therefore, some incentive to reduce the involvement of both superpowers.

There have been signs of change in both Vietnamese and Thai positions, but the difficulty in proceeding from an elaborate diplomatic minuet of "signalling" to substantive bargaining indicates that the impasse stemming from the fundamentally different formulations of the problem proferred by each side will be difficult to break through.

As the context of the Kampuchean problem expands or contracts, so too do the calculations of costs and benefits and the combinations and rank ordering of the interests involved. The safety zone issue raises the related problem of Thai support for the DK government and facilitation of Chinese aid to Pol Pot and the other components of the coalition. The Thais understandably insist that an entirely unsatisfactory situation has been thrust on them. It was, after all, Hanoi, not Bangkok, that invaded Kampuchea. But to insist that Thailand is not a participant, however reluctant, in the conflict is to deny a basic reality at the center of the issues that need to be resolved if peace is to come to Kampuchea. Because of the unalterable geography of the situation, Thailand holds the key to determining whether the issue will be placed in a local, regional, or global framework simply by the degree of cooperation it affords external supporters of the Democratic Kampuchea coalition.

Perhaps the most important point about the interests involved in the Kampuchean question is that there is a distinction to be made between balance-of-power interests and conflict-termination interests. The policies that stem from attempts to create a favorable balance are not necessarily those that lead to peace—indeed they may head in the opposite direction. Vietnam hopes to secure its interests by consolidating the Indochinese states under its dominant influence, and it relies on Soviet support as a counterweight to an unfavorable regional balance. Its invasion of Kampuchea was largely an attempt to prevent China from staking out a position of influence in an area that the Socialist Republic of Vietnam (SRV) regards as vital to its security. Thailand supports a policy of continuing conflict in preference to accepting a Vietnamese-dominated Kampuchea, and it relies on a Chinese deterrent along with ASEAN support and UN influence to counter Vietnam's position of influence in Indochina.

There is a shared interest in ending the conflict, but it is not as high a priority for either Vietnam or Thailand as striving to maintain a favorable balance in the region. This raises a final point about interests. There are expressed interests and objective interests. No one in either Vietnam or Thailand has spoken of an advantage in pursuing a confrontation; the costs and dangers are well noted by both sides. But a *lack* of a compelling interest to change a policy can be just as much of a factor in policy determination as striving to advance a perceived interest. The various interests ascribed to both the Thais and the Vietnamese in keeping Pol Pot afloat are a case in point. The status quo strengthens Thailand's position in ASEAN, fragments Kampuchea and creates an opening for the expansion of Thai influence there, attracts support from external powers such as the United States and China, and weakens Vietnam. In turn, the continuing threat of a return of Pol Pot further strengthens the Vietnamese position in Kampuchea.

Any negotiated solution will require changing the incentive structure in ways that lead the key parties to alter their policies in the direction of settlement. Lack of incentives for change has led to a prolongation of the conflict. This, in itself, is not necessarily a bad policy for the side that feels time is working to its advantage. For Thailand and ASEAN, the short-term costs of supporting the coalition are not onerous even if, in the end, it fails. An unsatisfactory but tolerable status quo seems preferable to acquiescence to a Vietnamese-dominated Kampuchea, and the strains imposed on Vietnam by a continuing drain on its resources weaken a regional rival. For the Vietnamese, the costs of maintaining the status quo are considerably greater. Whether the direct costs will increase or decrease over time remains to be seen. The indirect costs in terms of diplomatic isolation and lost development opportunities seem much more troubling in the long term.

But given the present uncertainties about how a change in the current status quo could accommodate its interests, Vietnam has a strong inducement to hold to its present course. A number of actions could be taken by either side that would change the existing incentive structure, but each one is counterbalanced by a persuasive opposing rationale. For example, Vietnam's objections to the UN format for negotiations could be met by devising an alternative forum involving more or less the same parties, but that would weaken the ASEAN advantage.

The Pol Pot card is used by Thailand and ASEAN to turn a position of weakness into one that can deflect Vietnam from its chosen course. But it is precisely this card that is the greatest stumbling block to a settlement. It also tends to forestall some of the problems an unpopular occupation would normally encounter, thus relieving an important potential source of pressure on the Vietnamese to compromise. Similarly, the strategy of isolating Vietnam to raise the costs of its occupation merely confirms Hanoi's view that in a hostile environment a reliably controlled buffer is essential, and that no risks that might result in an unreliable neighbor can be taken. It is this, as well as the failure of ASEAN to detail the precise mechanisms within its proposed solution that would safeguard Vietnamese interests against the return of a vengeful Pol Pot aligned with China, that leads the Vietnamese to adhere to their present course in preference to any of the alternatives that have so far been presented. The Thais are frustrated with having to cope with a threatening situation not of their own making, but they calculate that the dangers of a direct security threat to Thailand by a spillover of fighting and the indirect dangers of a further entrenchment of Soviet influence in the region are outweighed by the potential advantages of having a Kampuchea that lies more in the Thai than in the Vietnamese orbit by virtue of both what the opponents of the current Phnom Penh government would owe Bangkok for its support and the powerful pull on any Kampuchean government that economic advantages from close links to Thailand would exert. Even if the DK coalition fails, the advantages of keeping Vietnam embroiled in a protracted and costly protectorate might in itself serve Thai interests.

As usual, it is the weakest factor in this equation, the Kampuchean people, who will pay the price for safeguarding the interests of their more powerful neighbors. But to point out that the structure of competing interests does not favor a solution does not mean that a solution is impossible. This will depend, however, on the willingness of both Vietnam and Thailand to find alternate ways to address their bilateral problems that go beyond Kampuchea. The current impasse in the Kampuchean problem is not so much an unresolvable clash of interests as it is a failure of the imagination in devising ways to accommodate those interests. Until the hidden agenda of negative incentives that perpetuate the status quo is transformed, the impasse will probably persist.

Kampuchea in Thai-Vietnamese Relations

At the center of the disagreement between Thailand and Vietnam over the appropriate way to resolve the Kampuchean question is a fundamental conflict on the diplomatic framework within which the issues can be addressed. Both countries agree that the internal problems of Kampuchea reverberate well beyond its boundaries, and that Kampuchean problems are inevitably linked to other issues. But the precise nature of these linkages, and the extent to which the issues specific to Kampuchea may be dealt with in isolation from the larger context, are basic points of contention. The diplomatic foundations of the opposing positions are obvious. It is in Vietnam's interest to represent the Kampuchean question as a regional issue involving only Vietnam and ASEAN since this would help neutralize the strong international support given Thailand by its friends and allies. Similarly, Thailand would like to compensate for its limited ability to alter local and regional realities by broadening the forum of negotiations to provide more diplomatic leverage, while narrowing the range of issues discussed to block Vietnam from raising the question of the "China threat."

Although this diplomatic strategy has much to recommend it in terms of Thailand's relatively limited options in dealing with the Kampuchean problem, and though it has been quite skillfully managed by Bangkok, there are costs as well as benefits in pursuing this approach. Unless there is a major and unlikely shift in the Vietnamese position, this definition of the problem will lead to a continuation of the stalemate because it neither addresses the basic concerns of the Vietnamese (and far less Hanoi's allies in Phnom Penh) nor provides strong enough leverage to move the situation toward a resolution favorable to Thailand. It also rests on a diplomatic fiction comparable to Hanoi's insistence during the Vietnam conflict that there were no troops from the north fighting on southern battlefields. In the case of Kampuchea, Bangkok resolutely denies having any involvement whatever in the conflict. A representative quasi-official formulation of this position is that "Thailand has never been a party to conflict in Kampuchea and has no desire whatsoever to become involved in the fighting in Kampuchea. Thailand neither harbors any sanctuary for armed Kampuchean factions nor maintains any armed presence in foreign countries, unlike the presence of the Vietnamese troops in Kampuchea, which is the root cause of serious consequence for regional and international peace and security."[3]

It is certainly true that Thailand is an extremely reluctant and unwilling participant in the Kampuchean conflict. It is also true that there is a persuasive diplomatic rationale for denying that Thailand offers sanctuary or facilitates the arming of any of the Kampuchean factions. And this diplomatic evasion may pay off at some future date in much the same way Vietnam's dogged refusal to acknowledge northern troops in the south gave it a major advantage in the Paris accords, which skirted the issue and implicitly condoned the retention of northern forces on southern soil. The effectiveness of this ploy depends on what the realistic options are for Thailand, and whether or not the benefits of this subterfuge compensate for the costs of complicating a settlement.

Hanoi's response is to press Thailand to acknowledge that it is a party to the conflict as the necessary first step toward a negotiated solution. "If the situation along the Kampuchean-Thai border is settled," states a typical commentary in the party newspaper *Nhan Dan*, "then the relations between the People's Republic of Kampuchea, Laos, and Vietnam on the one side and the ASEAN countries on the other will change for the good of all nations in the region, to the disappointment of Beijing."[4] Vietnam also notes that it is in Thailand's interest to reduce the potential for conflict involving Vietnam and Thailand along the Thai-Khmer border. "Why has the situation on the Kampuchean-Thai border become tense over the past few years?" asks *Nhan Dan*, "And who is responsible for this situation? . . . The Thai authorities have often argued that they do not take sides in the conflict along the Thai-Kampuchean border, but their deeds have contradicted their words."[5] Hanoi feels that it is logically inconsistent to argue that, on the one hand, no sanctuaries for any Khmer factions exist on Thai soil but that, on the other hand, Vietnamese and Phnom Penh troops attacking these sanctuaries have violated Thai territory. The Vietnamese attempt to balance appeals to Bangkok to act in its own interests (as seen from Hanoi) with veiled warnings of the consequences of ignoring this advice. "The policy of the Thai authorities on the Thai-Kampuchean border is fraught with disastrous consequences for the peoples in the region and runs counter to the legitimate interests and sovereignty of the Kingdom of Thailand," says *Nhan Dan*.[6] The immediate hazard is the continuing threat of direct clashes between Vietnamese and Thai troops along the border. Prior to the spring 1983 Vietnamese offensive along the border a "senior Thai intelligence official" noted the dangers that this could pose to Thai forces and called it "frankly, alarming."[7] Bangkok also reacted strongly to reports from Tokyo that a ranking Vietnamese diplomat had implied that Vietnam would exercise the right of hot pursuit to track down DK troops fleeing into Thai territory. The longer term danger to Thailand, which Hanoi hopes to impress on the Bangkok government, is that involvement in supporting Khmer rebels may open the way for internal problems. "The involvement in the so-called Kampuchea problem is a pernicious mistake," said a Radio Hanoi broadcast. "In other words, it has started a fire near its own house and the house is filled with fuel. Moreover, tailing after Beijing has created more opportunities [for China] to interfere deeper in Thailand's internal affairs."[8] Thailand, as will be seen below, does not regard this as a serious problem.

Both Vietnam and Thailand agree that the main sources of tension in their relations do not stem from bilateral issues. During talks with Soviet Deputy Foreign Minister Kapitsa in early 1983, Thailand informed the Soviet Union that the Kampuchean problem would "remain unsolved even if Thailand and Vietnam hold negotiations. This is because the Kampuchean problem involves many countries."[9] Vietnamese Foreign Minister Nguyen Co Thach has repeatedly asserted that there are no major bilateral problems between Vietnam and Thailand. In a meeting with Thai Foreign Minister Sitti following a flurry of controversy about a reported Thach statement in Singapore criticizing the Thai government for "indulging" the Khmer rebels and threatening hot pursuit into Thai territory,[10] Thach said Vietnam did not want confrontation with Thailand or ASEAN, and that this was a shared interest.[11] Despite the border tension and the Vietnamese criticism of the Thai role in supporting Democratic Kampuchea, Thach has observed that "We have always said that there is not a problem between Cambodia and Thailand, but between Indochina and China."[12]

Thailand has refused to attempt a resolution of the border problem by focusing on the narrow and immediate issues involved, because in doing so it might prejudice the outcome of the larger issues, resulting in an acceptance of a Vietnamese-dominated Kampuchea. A compromise by Thailand that would achieve border stability but would leave a Vietnamese-dominated Kampuchean regime in power might fritter away the international support Thailand has enjoyed as a consequence of a successful effort to portray itself as the indirect victim of aggression. Also, compromise with Vietnam would certainly strain relations with China. Thai National Security Council Secretary General Squadron Leader Prasong Sunsiri, an official with a direct and immediate concern in coping with the threat of border conflict, told Thach during his July 1982 visit to Bangkok that all existing problems, including border security, refugees, and other security issues, have to be dealt with as a package, and not individually.[13] Another key Thai official with security responsibilities characterized the Thai-Kampuchea border problem as a "conflict among superpowers."[14]

The extent to which Thai officials regard the border problem as a direct and immediate threat to Thai security has fluctuated over time, and there have been significant differences on this question among Thai decisionmakers. In addition, the border question is a prime example of the tension between policies that aim mainly at creating or maintaining a favorable balance of power and policies devised to hasten the termination of the conflict. Several factors must be weighed in assessing how to handle this issue. The greater the immediate threat of a spillover of fighting into Thai territory and, therefore, of a direct confrontation between Thai and Vietnamese troops, the more urgent a stabilization of the local situation would appear, even if it did not resolve the larger issues. This consideration, along with the apparent lack of progress in moving toward an overall solution, may have prompted Foreign Minister Sitti to make a proposal to Vietnam that seemed to represent a slight shift in position.

During the Thai electoral campaign in spring 1983, Sitti proposed that all Vietnamese military forces be withdrawn thirty kilometers from the border as a condition for talks leading to a reduction of tension. This proposal resulted in extensive diplomatic speculation and analysis, and it highlights the interlocking structure of the diplomatic elements of the Kampuchean problem. This apparently straightforward proposal raised a host of questions, which have not been completely answered. The reason the distance of thirty kilometers was chosen was that it moves Vietnamese artillery out of range of the Thai border. But, of course, this step would not in itself address the question of supplies and sanctuary to the Khmer rebels. Why, one might ask, would Vietnam be interested in such a proposal? The incentive for Vietnam is that any move toward direct talks between Thailand and Vietnam on any aspect of the Kampuchean problem would represent a departure from the Thai (and ASEAN) insistence that the International Conference on Kampuchea is the only valid framework for discussing the Kampuchean question. Further, it might be interpreted as evidence that some elements of the Thai strategy for dealing with the Kampuchean issue are being reevaluated. These include the efficacy of the DK coalition as a military and diplomatic instrument, the utility of the Chinese (and U.S.) deterrent, the success of the international pressures put on Vietnam, and the likelihood of Vietnam being forced to compromise by the pressures of economic realities.

Shortly after Sitti's proposal Nayan Chanda, the authoritative diplomatic analyst of the *Far Eastern Economic Review* (*FEER*), noted that there appeared to be a Thai deemphasis on the ICK as the sole diplomatic framework for negotiating with Vietnam.

> After an abortive proposal by Malaysian Foreign Minister Tan Sri Ghazali Shafie for talks on Cambodia between ASEAN on the one hand and Vietnam and Laos on the other, ASEAN foreign ministers issued a statement in March in which the ICK was invoked twice to reassure the world that the organization was not deviating from the UN framework for settling the Cambodia question. At the time, at least one senior ASEAN leader privately expressed his frustration over the ritualistic reference to the ICK, which he saw as a red flag waved at the Vietnamese bull. Even hardline Singapore Foreign Minister S. Dhanabalan told the *Review* then that the ICK was not the last word on the subject. "It is a starting point for negotiations." . . . More than a month after the uproar over Ghazali's initiative, Thailand now appears ready to seek a Cambodia solution similar to the one embodied in the UN resolution, but without the near-religious significance the ICK has assumed.[15]

Despite subsequent reaffirmations that the essential framework of Thai and ASEAN diplomacy had remained unchanged, and a successful attempt by Sitti to secure the endorsement of other ASEAN ambassadors for this proposal, it seemed to be an indication of a rethinking of the relationship between elements of Thai/ASEAN diplomatic strategy, while adhering to the broad outlines of the established approach.

Perhaps the most striking change of emphasis was the close linkage by Thailand of the Kampuchean problem with Thai-Vietnamese relations. Just prior to the June 1983 meeting between Sitti and Thach, the Thai Foreign Ministry issued a statement: "The Thai and Vietnamese foreign ministers have already met three times without much progress in the search for a solution to the Kampuchean problem, which has obstructed the development of Thai-Vietnamese relations. . . . Thailand and Vietnam can contribute to the international efforts for a permanent and just settlement of the Kampuchean problem. Thailand and Vietnam are neighbors."[16] This represents somewhat of a departure from the Thai strategy of refusing to deal bilaterally with Vietnam on Kampuchean issues in order to underline Bangkok's insistence that the United Nations is the only proper framework for handling this issue. The problem with this approach was that it did not seem to bring the Kampuchean problem any closer to settlement and, at the same time, made it impossible to make progress on any area of Thai-Vietnamese relations. The Thai statement noted that Thailand is "determined to see peaceful coexistence materialize in the region" and that it has "attached special importance to the neighboring countries." It was, therefore, "regrettable that for the past four to five years Thai-Vietnamese relations have been negatively affected by the Vietnamese invasion and occupation of Kampuchea." The Foreign Ministry reaffirmed its adherence to the UN framework, but acknowledged that "there are several paths toward permanent peace."[17]

Although Thai objectives remain unchanged, there is increasing debate in Bangkok about how to achieve them. Within the ruling coalition some prominent voices, such as that of Deputy Prime Minister (and former Foreign Minister) Phichai Rattakun, have argued for a resumption of official commercial ties with Vietnam "as a path leading to a political solution of the Kampuchean problem within the framework of the UN resolutions."[18] Phichai denied that this was a departure from the Foreign Ministry's policy, but added that "it's the means toward the same goal that should be scrutinized." The proposal to expand trade with Vietnam was endorsed by former Prime Minister Kriangsak Chamanan, who also sought to travel to Hanoi in his capacity as chairman of the Standing Committee on Foreign Affairs of the Thai House of Representatives to sound the Vietnamese out on this subject.[19]

China and ASEAN in Thai Kampuchea Policy

Restoration of official commercial ties with Vietnam would represent a significant distancing of Bangkok from the Chinese strategy of "bleeding Vietnam white." Such an approach would be a modification of the official view expressed at the time of Prime Minister Prem's China visit in November 1982 that "Vietnamese adventurism in Kampuchea was bleeding Vietnam economically and served it a costly lesson."[20] The logical corollary of this view is that economic isolation of Vietnam is an effective lever to pry its troops out of Kampuchea. The implications of a policy shift in this area both for Thai-Chinese relations and for Bangkok's ASEAN partners are numerous and complex. Moreover, there are a variety of opinions among key decisionmakers in Thailand on this subject and, according to Deputy Prime Minister Phichai, debate on this question within the government has "started to heat up."[21]

For these reasons, Bangkok is sure to move very cautiously in this area. China has played an important role in Thai policy toward Vietnam, and any policy change that might complicate Thai-Chinese relations would have to be supported by a very persuasive rationale. Bangkok's sensitivity toward Chinese reactions to Thai policy initiatives involving Vietnam and Kampuchea was illustrated by Foreign Minister Sitti's careful mention of the fact that the Chinese ambassador had called on him to support his thirty-kilometer-pullback proposal, following rumors of Chinese displeasure with this initiative.[22] Sihanouk stated in early 1983 that Chinese leaders had told him they "want Vietnam to withdraw from Kampuchea totally and unconditionally, not step by step" and that a partial Vietnamese withdrawal would not fulfill a key Chinese condition for normalized relations with the Soviet Union that calls for a cessation of Soviet assistance to the Vietnamese occupation and a total Vietnamese withdrawal.[23]

China's role in Thailand's Indochina policy has been both direct and indirect. There has been an implicit assumption that common interests binding the two countries together would ultimately induce China to end its support for the revolutionary movement in Thailand, and indeed, at the time of Prime Minister Prem's fall 1982 visit to China, Bangkok was reported to have concluded that Beijing was no longer giving material support to the Communist Party of Thailand (CPT).[24]

China has taken a direct role by threatening military action against Vietnam in the event of a border crossing into Thai territory. Hanoi indignantly reported that during Prem's visit he was given a "sensational present," namely, in the event of a Vietnamese attack against Thailand, China would "help Thailand counter this aggression."[25] In addition, Chinese Premier Zhao Ziyang was reported to have told Prem that "China's confidence in and support for Thailand is important in the so-called control of Vietnam's regional hegemony in order to protect peace and stability in Southeast Asia."[26] In fact, China did intensify military activity on its border with Vietnam at the time of Hanoi's spring 1983 offensive along the Thai border. This appears, however, to have been more a symbolic gesture than a serious deterrent threat. "China knows very well that short of another invasion of Vietnam, no action would make any difference to

Vietnam's military position in Cambodia,'' said a source close to official thinking in Beijing. He added, ''It is just a symbolic act.''[27]

In view of the limited effectiveness of the Chinese deterrent, and the diplomatic costs of too heavy reliance on Beijing, Thailand has been diversifying its options. Evidence of Bangkok's desire to distance itself somewhat from China surfaced during the visit of People's Liberation Army Chief of Staff Yang Dezhi to Thailand in February 1983. General Saiyut Koetphon pointedly observed that ''he had explained to Mr. Yang that Thailand supported regional independence and peace because any instability could have great repercussions internationally,'' and that Thailand was ''a key to regional stability, but that it was important that the regional leaders strive to create better understanding among themselves.'' The general added that while he was concerned about the fighting near the Thai border, he was confident that Thailand could handle the situation.[28] During the intensification of the border fighting in April 1983, the Foreign Ministry thanked the Chinese ambassador for Beijing's ''sincerity'' in opposing the alleged Vietnamese intrusion into Thailand, but called on a host of other friendly countries to put pressure on Vietnam.[29] A few days earlier it had been announced that the Chinese foreign minister was postponing his trip to Bangkok for ''undisclosed reasons''—possibly indicating Beijing's displeasure or uncertainty about the direction of Thai policy toward Indochina.[30] This also coincided with the visit of the French foreign minister to Bangkok, and the evident Thai interest in French proposals for a diplomatic solution that appeared to rule out both the economic sanctions approach and primary reliance on the United Nations as the framework of negotiations.[31] General Saiyut travelled to Beijing in August 1983 to ''study the Chinese military experiences'' and to discuss the problem of Kampuchea with Chinese military leaders. But although he expressed the view that both countries should cooperate to ''defend the region,'' he also said he would tell the Chinese that ''we will stand on our own feet'' in a posture of self-reliance.[32]

Signs of a more assertive Thai foreign policy were already evident by late 1982, and the evolving diplomatic context in Southeast Asia encouraged this trend. A hard-line, confrontationist policy based on isolating Vietnam seemed to be an attractive alternative for Thailand as long as both the costs and the risks were minimal. It kept open the possibility of a neutral or pro-Thai Kampuchea while bogging down Thailand's main regional rival in a costly and protracted occupation. But 1982–83 saw major changes in other important elements of the equation. Despite the fact that Thailand had been quite successful in solidifying support from its ASEAN partners, there was a lingering suspicion among some ASEAN members that Thailand had become too heavily dependent on Chinese support in its search for ways to reduce costs and risks in the continuing confrontation over Kampuchea. Moreover, China's own negotiations with the Soviet Union created an additional element of uncertainty, evidenced in Premier Prem's reaction to the initial rounds of talks between Moscow and Peking. Asked what he thought the implications of Sino-Soviet normalization might be for the situation along the Thai border, Prem responded cautiously that it could be "either good or bad for Thailand."[33] Thailand also attempted to deal directly with the Soviet Union to undercut Hanoi but, apparently, with little success. Other important factors were the shifts in Kampuchean policy by important allies such as France and Australia. It was French opposition to the strong confrontationist language proposed by ASEAN in its annual meeting with European Community countries that led to substantial compromises in the final joint statement.[34]

Foreign Minister Sitti's fall 1982 speech at the United Nations on international matters went beyond regional concerns to present the Thai position on broader global issues, and it attempted rhetorically to distance Thailand from the close ties with large powers that had been the mainstay of its previous security policy. Sitti observed that in Kampuchea "the present situation has interrupted the process of constructive cooperation among countries of Southeast Asia and brought about an intensification of big-power rivalry in the region," a statement which could be interpreted as a precursor of the subsequent Thai emphasis on the importance of regional actors (e.g., Thailand and Vietnam) in stabilizing the region.[35] Noting the foreign minister's stress on identification with the third world, the *Bangkok Post*, perhaps overenthusiastically, hailed the speech as a "historic watershed in Thai diplomacy."[36]

A combination of internal and external pressures may have contributed to this apparent shift in some foreign policy areas. At a seminar at Chulalongkorn University, several leading academic foreign policy experts offered a critical evaluation of key areas of Thai foreign policy. Dr. Kusuma Santiwong asserted that some recent policy approaches have been "incoherent" and chided the Foreign Ministry for monopolizing foreign policy among a narrow group of experts. Dr. Chaianan Samutwanit urged the Foreign Ministry to be more "innovative" in dealing with the Kampuchean issue, and he warned that the military was beginning to develop its own foreign policy.[37] As noted earlier, Deputy Prime Minister Phichai, in calling for a study of the resumption of official commercial ties with Vietnam, said that "the debate on the policy against trading with Vietnam had started to heat up within the government."[38]

Concurrently, some important diplomatic partners began to develop new approaches to the Kampuchean issue. Although Australia later modified the more sweeping aspects of Labor's campaign positions on Vietnam and Kampuchea, the new government still effected a major departure from the Thai/ASEAN line. France, which also modified its initial position somewhat, created an important diplomatic impact with Foreign Minister Cheysson's trip to Southeast Asia. While in Bangkok, Cheysson bluntly noted that supply lines to the DK coalition "necessarily run through Thailand"—a "reality" of the Kampuchean conflict. More specifically, "this reality is that Chinese arms do go through here [Thailand]. When they suffer a defeat in Cambodia, anti-Vietnamese Cambodians move over here, then take off again." It is the Thais, observed Cheysson, who must live with this reality. An Agence France Press correspondent wrote that the

French foreign minister "also hinted that Thailand had reasons to mistrust China," in part because of its geographic location. "Pointing to China's proximity, he noted that in the past Thailand had 'often' had to face 'dangers' from China similar to those posed by Vietnam's military control of Cambodia." Cheysson pointed out that the Thai leaders are forced to speak "concretely" in discussing Kampuchea as "they have to live with the Vietnamese, the Chinese, and the Cambodians." While other members of ASEAN, more removed from the problem, "can stick to declarations of principles . . . [the Thais] cannot do so. . . . When you talk with Thai Foreign Minister Marshal Sitti Sawetsila," said Cheysson, "you discuss concrete issues, not in terms of United Nations resolutions."[39]

Within ASEAN itself, there are significant differences over the proper way to handle the Indochina problem. ASEAN has been remarkably successful in maintaining a reasonably unified public stance on this issue, but the underlying reasons for the divergent assessments of the problem have to be taken into account in Bangkok. Indonesia has long been critical of Thailand's reliance on the Chinese deterrent and, more recently, has said that Thailand's other great-power ally, the United States, has overstated the Soviet threat. Hanoi's nationalist heritage, Indonesia asserts, will keep it from becoming a Soviet puppet. Jakarta further hopes tht ASEAN can offer Vietnam an alternative to its growing dependence on Moscow, which is an implicit rejection of the policy of isolation and confrontation. Indonesia is also sensitive to the dangers of continuing polarization, which affords the great powers an opportunity to intervene in the region. The Indonesian foreign minister "expressed confidence that the current polarization between ASEAN, on the one hand, and the Indochinese states on the other, could be overcome in perhaps five years. 'It's not such a far-fetched dream,' he said, adding that regional states were already 'nearer to keeping the big boys out.'"[40]

Malaysia, of course, was the originator of the concept of a neutralized Southeast Asia that would exclude great-power influence from the region. Like Indonesia, Malaysia has been concerned about the continuing polarization in the region affording an opportunity for external intervention. But Malaysia quite dramatically shifted its approach to this problem from an attempt to end the polarization by encouraging diplomatic accommodation with Hanoi (the "Kuantan formula") to sponsorship of the idea of expanding Democratic Kampuchea even while creating a coalition that would contain forces that could become an alternative to the Beijing-oriented Khmer Rouge. The essential element of this formula is creating a credible, noncommunist military force as the nucleus for a political settlement that would either exclude or neutralize the Khmer Rouge, and the weaning of Prince Sihanouk from dependence on the Chinese. Despite their role in invigorating the anti-Vietnamese forces and thus escalating the conflict, it was Malaysia that apparently suggested the possibility of direct talks between ASEAN and the Indochinese bloc in the form of Vietnam and Laos, excluding the Phnom Penh regime.

The obvious clash between the competing objectives of stabilizing the region on the one hand and opposing Vietnamese control of Kampuchea by pursuing a policy of supporting armed resistance on the other is a strong undercurrent beneath the public ASEAN discussion of the Kampuchean problem, and it is getting increasing attention from some of Thailand's partners. After Vietnamese Foreign Minister Nguyen Co Thach had visited the Philippines, President Marcos told the press that he had been informed that "Vietnam was willing to sit down anytime on matters relating to the general stability of the region and is willing to disregard the Kampuchean question for a while," and that Vietnam would not allow the situation in Kampuchea "to contaminate relations with other countries in Southeast Asia." Thach proposed that Vietnam and the ASEAN countries "disregard the Kampuchean question for the moment and determine a long-range plan for the region."

Underlining the physical and, perhaps, political distance from Kampuchea, Marcos also noted that "he was given assurances that Vietnam has no hostile intention against the Filipino people, due to existing friendly relations between both countries." In accord with the Vietnamese position that it was willing to focus on relations among the states in the region, the question of U.S. bases in the Philippines was, apparently, not raised. As an Agence France Press account of Thach's visit remarked, Marcos "said that he believes that Vietnam has no intention of attacking any country, underscoring once more his government's two-track approach to Hanoi and the Cambodian problem—and confusing diplomats here anew. Diplomatic observers here said that while the Philippine government has officially maintained a hardline stance on the presence of 150–180,000 Vietnamese troops in Cambodia, the president himself has been much more conciliatory toward Hanoi."[41]

These external developments, along with the internal debate over the direction of Thai foreign policy, have led to an increasing emphasis on the importance of serving Thai interests, and an implied recognition that those interests may not always parallel those of Thailand's allies. Foreign Minister Sitti characterized this new approach as involving more "flexibility" and aiming at a "balance of interests" with other countries.[42] At a Bangkok news conference in mid-May 1983 Sitti said that

> We will stick to previous policies but will be more flexible. We will expand our policies regarding trade—will put more emphasis on trade. ASEAN is the center of our policy . . . we do not regard Indochina as our enemy. It will be stipulated in our foreign policy that we—like the other ASEAN countries—want a zone of peace, freedom, and neutrality. The Kampuchean conflict is one of the problems that must be resolved in order for the zone of peace, freedom, and neutrality to materialize. As for bigger countries, we will see if their presence is constructive. We do not want a

> presence that will result in disputes or armed conflicts. More than anything else, we will attach importance to the benefits their presence brings to our region. We do not want our region to be the cause of conflict between the superpowers. As a small country, we will do the best we can. We will not consider ourself the enemy of any country. We will preserve our interests. We will trade with all countries. We will separate trade and politics.[43]

Insofar as this new emphasis has specific application to Indochina, it would appear that a higher priority is being put on measures aimed at regional stabilization, and a gradual distancing from the Chinese policy of unbending confrontation is taking place.

This does not necessarily mean that there is a strong compulsion for Thailand to make any immediate conciliatory moves toward Hanoi or Phnom Penh. Indeed, while the Indochinese conflict has brought about a heightened Soviet military presence in Southeast Asia this, in turn, has led to a resurgence of American interest in Southeast Asia as part of an attempt to counter the Soviet Union on a global scale. Sitti's October 1983 speech in New York at the Asia Society sounded a very different note from his UN address of the preceding year. He called on the United States and other ASEAN allies to maintain the "larger balance of power" in the region stretching from the Persian Gulf to East Asia. "In spite of the bitter legacy of the Vietnam war," said Sitti, "the security of the Southeast Asian region should still be of great concern to the countries of the free world, particularly the United States."[44] From this Thai perspective, the tension in Indochina may have the positive impact of revitalizing the U.S. military presence in the region. This is another example of balance-of-power interests leading in a different direction from conflict-termination interests.

There has been some evidence of diplomatic movement from the Thai side. In addition to the thirty-kilometer-pullback proposal, and the hints that some other forum than the ICK may be acceptable, Thailand and ASEAN have elaborated the pullback proposal into a zone-by-zone removal of Vietnamese troops that would be easier to verify.[45] However, Vietnam has rejected this on the grounds that it would only provide an opportunity for the Pol Pot forces to fill the vacuum left by a Vietnamese withdrawal. In his talks with Vietnamese Foreign Minister Nguyen Co Thach at the fall 1983 General Assembly meeting, Sitti said that it was unlikely that there would be any further Thai initiatives, since Vietnam had not responded positively to these proposals.[46]

Whether or not there will be any break in the Kampuchean deadlock depends on a number of factors: the Vietnamese response to the Thai/ASEAN proposals, shifts in the broader diplomatic context, and the evolution of the situation within Kampuchea itself (including the ability of the DK coalition to expand the scope of its operations and the capacity of the Phnom Penh government to consolidate its administrative grip, political impact, and military/security power). The interplay of these evolving internal and external factors raises the ultimate question: each side has maintained that time is on its side (indeed, all sides' fundamental strategies are based on this assumption), but which side has the best reason to make this claim?

Vietnamese Foreign Policy and Kampuchea

Vietnam's objectives in Kampuchea are clear: to ensure that its neighbor is a reliable buffer against hostile third-party influences, and that it does not itself turn against Vietnam as during the Pol Pot period. The only questions that provide a spectrum of uncertainty within which diplomacy is possible are Hanoi's capability to ensure these objectives, and the price it is willing to pay. The incentive structure of the diplomatic aspects of the Kampuchean problem must, therefore, be analyzed from Hanoi's perspective to provide some basis for assessing the costs Vietnam is willing to pay to maintain hegemony in Kampuchea, and to determine whose side time is on. Vietnamese-Thai relations are an essential ingredient of this question, because it is ultimately Thailand that will determine whether the situation will be resolved in a local, regional, or global framework.

It is difficult to compare Vietnam's Kampuchean policy with that of Thailand because of fundamental asymmetries in their respective positions. To some extent, Vietnam already has what it wants—control of Kampuchea. Thailand does not yet have what it wants—a Kampuchea not under the control of Vietnam. Stated in these simple terms, this leaves Thailand with only two options: to find a way of forcing Vietnam to get out with a combination of military and diplomatic pressures (since the Thais have concluded that neither of these elements will succeed alone) or to hope that Vietnam will find its commitment too burdensome and get out on its own.

Just as Thailand initially refused to deal directly with Vietnam on questions relating to Kampuchea, on the grounds that it was not involved as a belligerent party and because it hoped to deal with the issue in an international arena from greater strength, Vietnam insisted that there were no inherent bilateral problems between Vietnam and Thailand, and that only Bangkok's complicity in aiding Chinese intervention in the region clouded their relationship. Moreover, Vietnam insisted that it would not negotiate anything involving Kampuchea's political future without the presence of Phnom Penh authorities at the table. The apparent receptivity of Vietnam to the abortive Malaysian initiative at the nonaligned conference suggests a slight departure from this position. Nonetheless, Vietnam continues to insist that the essential problem in Kampuchea is Chinese interference. This would clearly be a major item of Vietnamese concern in any bilateral discussion with Bangkok on security issues related to the Kampuchean conflict, and it explains Thai reluctance to discuss Kampuchea in this framework. Foreign Minister Sitti's proposal for direct talks with Hanoi on the subject of Kampuchean-related security issues is, therefore, a significant development in the diplomacy.

It may well be, however, that this new direction comes too late to offer any real prospect for constructive dialogue. When Vietnam was seriously concerned about the military potential of Pol Pot and the opening wedge for Chinese influence in Indochina that a successful DK insurgency might afford, and while it was still struggling to hold together an economic system that was in a state of near collapse, there was a strong incentive for Hanoi to make a diplomatic compromise in Kampuchea. While Vietnam is still in serious economic trouble, it has edged away from the brink of disaster.[47] And though there is mixed evidence as to how well the Phnom Penh regime is doing, and the extent of the problems that Hanoi is experiencing with its client, the immediate threat to Vietnamese control in Kampuchea seems to be troublesome but manageable, for the moment at least. Indeed, the gamble taken by ASEAN in forming the enlarged DK coalition was that it would at least keep the future of Kampuchea open while time and changes in the situation worked against Vietnam. In the short run, it has not worked out this way. The coalition has been unable effectively to challenge Hanoi's control.

At the same time, some of the costs inherent in this strategy, which ASEAN did not feel to be major at the time, are now becoming more obvious. By tying the noncommunist forces to the Khmer Rouge, the coalition formula has significantly reduced the diplomatic margin for maneuver. In the precoalition period, it was conceivable that Hanoi (and the Phnom Penh government, whose needs must be taken into some account) would accept a cosmetic broadening of the political base. But with most of the political opposition to Phnom Penh linked to Pol Pot, this is now much less likely, although some observers have detected a slight softening of Phnom Penh's earlier categorical rejection of any future political role for the noncommunist members of the coalition.[48]

From Hanoi's perspective, the essential elements of a regional diplomatic settlement in Kampuchea remain a stabilization of the Thai border area and reduction of tension in Southeast Asia in return for an end to Thai, ASEAN, and Chinese support for Pol Pot. Such a settlement would clearly be a major success for Vietnam, since it would give up nothing substantial and gain all of its major objectives. In Hanoi's view, Thailand and ASEAN have not put forward any realistic diplomatic alternatives, since their strategy necessarily cannot exclude the possibility of a return of Pol Pot, whose exclusion from any future role in Kampuchea is, in Vietnam's eyes, an "irreversible" commitment. The electoral formulas that have been proposed do not guarantee that Pol Pot will not return; indeed the insistence on supervision by the United Nations, which recognized Pol Pot as the legitimate government of Kampuchea, is certainly not acceptable to Vietnam. As long as the alternatives are framed in those terms, the only possibility of creating any movement on the part of Vietnam is through the leverage of compulsion, which would require a much stronger hand than any of its opponents holds at present.

Because Vietnam was able to ride out its period of greatest vulnerability, while its oppositon narrowed the options, it is not surprising that it is now playing a waiting game. After an intense period of diplomatic probing throughout much of 1982, Hanoi paused to await a reaction. In late October 1982 Foreign Minister Thach was asked in Bangkok whether he had any new proposals to make to ASEAN. He replied, "How can I come forward with a new proposal when ASEAN has not as yet responded to our latest proposal made last July [on border security]?" Thach added, "The ball is now in ASEAN's court. . . . We have made too many proposals and we are also not ambitious."[49] Scarcely ten days prior to Thach's indication of a more reserved diplomatic posture, Deputy Foreign Minister Vo Dong Giang noted in passing that "Vietnam is still facing many difficulties, but the worst period is over," a reflection of a new sense of confidence in Hanoi that Vietnam can hold to its fundamental position.[50]

Of course, this more aloof stance may itself be a diplomatic ploy. If so, it has, at least, been consistent. The Vietnamese ambassador in Bangkok gave a detailed outline of his country's policy toward Thailand in February 1983. Noting that Thailand had not, in his view, responded either to the July 1982 offer to stabilize the border or to the ostensible partial troop withdrawal, the ambassador stated that, "We are waiting for a reciprocal response to our good will so as to gradually restore peace along the Thai-Kampuchean border." Asked if Vietnam would withdraw more troops if the Thais responded positively, and whether or not this reciprocation was a condition for further withdrawals, the ambassador replied, "Yes, it is a condition in accordance with the [Indochinese summit] joint communiqué. We are waiting for reciprocation from the Thai side, which has not . . . reciprocated our good will, but has even rejected it."[51]

This stress on reciprocal action was echoed in an early 1983 article in *Le Monde,* which noted "the very recent signing of a 'Document on Increasing Cooperation Between the Vietnamese People's Army and the Kampuchean People's Army,'" which "illustrates the Hanoi leaders' determination to preserve the fragile birth of their Indochinese ally which is perhaps more inclined than before to mark its distance to some extent—at least in words—from its powerful protector. 'No unilateral withdrawal by our troops,' Mr. Thach stated. 'That is unacceptable. For us it is a question of principle and security, a question of life or death.'"[52] Yet Hanoi did subsequently withdraw a division (whether it was replaced or not is another question), signalling that it would take some initiative without reciprocation, but again insisting that if the process were to continue some response was needed. Vietnam has, in turn, expressed irritation with the Thai border pullback proposal, which Hanoi interprets as constituting a Thai precondition for substantive talks on Kampuchea, and Bangkok insists is a confidence-building measure and a test of Vietnam's sincerity.[53] As each side waits for the other, the impasse continues.

Evidently the political gesture that the Vietnamese troop withdrawal represented was aimed at China as well as Thailand and ASEAN. Foreign Minister Thach in announcing the decision said that it was based on the strengthening of the Kampuchean army. "But it is also a goodwill gesture toward many countries," he added. According to Agence France Press, "observers" of the Hanoi diplomatic scene maintained that this applied to China in particular. "China demands a unilateral timetable for the Vietnamese withdrawal," said Thach, "But two timetables are required." This must "include a timetable for an end to the 'Chinese threat'—an end to Beijing's aid to the ousted Khmer Rouge—and the end of the 'use of Thai territory as a sanctuary.' " Thach reiterated the theme that "We have always said that there is not a problem between Kampuchea and Thailand, but between Indochina and China."[54] Still, Hanoi insists on reciprocity from Thailand as well, and it seems to feel that it will be difficult to separate the issues of China and Thailand. The Vietnamese party newspaper, *Nhan Dan*, commenting on the Sitti disengagement proposal, said that "certain rightist forces in the ASEAN countries, acting upon Beijing's orders, are obstructing dialogue to maintain tension." Thus, the paper noted, "in such a situation Thai Foreign Minister Sitti Sawetsila's April 16 statement is worth noticing. However, this proposal has failed to mention necessary action on the part of Thailand. Nor did it say anything about the three Indochinese countries' proposal for a security zone along the Kampuchean-Thai border to ensure security for both sides." The paper conceded that "Thailand is right to be concerned about its security," while insisting that Thailand should acknowledge that "the People's Republic of Kampuchea has the same problem." So only those proposals "which aim at ensuring security for both sides can provide the real groundwork for dialogue." *Nhan Dan* concluded that "The best way is for the Indochinese and ASEAN countries to sit down and, together, to tackle all problems of mutual concern in the spirit of negotiations and reconciliation."[55]

Thai-Vietnamese Relations and the Future of Kampuchea

There are many questions left by the developments of 1982–83: Is there any way of disengaging the Chinese dimension from Vietnamese-Thai relations? Can any progress be made in other areas as long as the impasse on Kampuchea persists? Will the passage of time render the position of one side or the other untenable? Will the deadlock persist indefinitely, or will other issues emerge that will provide an external impetus for settlement? Foreign Minister Thach has offered several possible scenarios involving the future of Kampuchea. In a comprehensive, five-point scenario Thach saw the possibilities as 1) continuing deadlock between Vietnam and Indochina on the one hand and China (in collusion with the United States) and ASEAN on the other, resulting in continuing confrontation; 2) ASEAN, Vietnam, and China all improve their relations; 3) China (with U.S. support) along with ASEAN launch a large-scale war against Indochina; 4) ASEAN and the Indochinese states improve relations while China continues its opposition; and 5) China and the Indochinese states improve relations while

ASEAN continues a hostile policy toward Vietnam. With respect to Vietnamese-Thai relations, Thach saw the following possibilities: 1) the situation of confrontation between Thailand and Indochina will continue; 2) this situation will escalate beyond the control of the concerned sides and will lead to unforeseen serious consequences; 3) there is a gradual improvement of relations, during which "the concerned parties will agree to adopt severe [sic] measures aimed at ensuring peace and stability equally for both sides in the Kampuchean-Thai border areas"; 4) pending the settlement of differences, confidence-building measures are taken to "prevent an aggravated situation and to control the situation."[56]

If this is what Vietnam wants, what incentive do Thailand and ASEAN have to normalize relations with Vietnam in a way that would acknowledge and perpetuate Vietnamese dominance in Kampuchea? Thach's response is that since there is no way to alter the outcome in Kampuchea, and since continuing confrontation may be dangerous for Bangkok (the implied threat of escalation mentioned in his second hypothesis), it is in the Thai interest to normalize relations even without a Kampuchean settlement. Passing through Bangkok in June 1983, "Thach suggested that ASEAN put aside the question of Cambodia and search for a 'general framework' for peace and security in the region. The ASEAN approach of seeking a solution to Cambodia first had 'in our opinion' failed, he said. Continuing along this line, he added, would only lead to further escalation of the confrontation and mistrust that have bedevilled Indochina and Southeast Asia for the past 40 years."[57] Nayan Chanda points out that this position, advocated by Hanoi for several years, has consistently been rejected and that "there is no sign that ASEAN is about to change its mind on the subject."[58]

The complex character of Thai-Vietnamese relations was summed up by Vietnam's deputy foreign minister in late 1982: "On the one hand dialogue and reconciliation between the two countries tend to make headway. On the other hand there is a state of stagnation, as a result of Thailand's vain hopes in its miscalculations." But, he added, "the time will come when Thailand can see that such a stagnation is detrimental to itself—primarily to itself—and the only people to benefit from it are the forces of expansionism and hegemonism which are seeking to undermine peace and stability in the region."[59] For the time being, however, Hanoi maintains that in building up the "trump" of the DK coalition as a solution for Kampuchea "they are making efforts detrimental to dialogue and détente in Southeast Asia. They hope that time will be in their favour. But time will go by and will teach them the necessary lesson that they have harboured vain hopes and have miscalculated."[60]

Hanoi has significantly reevaluated its view of the larger diplomatic context, and it now sees a split between the purposes of the United States and China, which it feels is also working in its favor. The deterioration of relations between the United States and China, in Hanoi's view, is both a cause and an effect of a more assertive U.S. role in Southeast Asia. Washington is now, in Vietnam's view, attempting to "use ASEAN countries to restrain Beijing's growing influence on Maoist groups in the region."[61] And, Hanoi feels, "The time will come when the United States realizes that it can make no condition(s) to Vietnam . . . [and] that there is a price to pay for its collusion with China against the three Indochinese countries. When such a time comes, relations between Vietnam and the United States will be normalized."[62] Vietnam has been more nervous about the implications of the Sino-Soviet negotiations, but has concluded that it cannot fundamentally alter the regional situation in the short run and, were normalization to take place, it might work to its advantage in the long run by clearing the way for an improvement in Sino-Vietnamese relations and diminishing the "China threat."[63]

Whether or not time is on their side, the Hanoi leaders almost certainly believe that it is, and this belief will guide their actions. A representative formulation of this attitude can be found in the following *Le Monde* interview with the editor of the army newspaper *Quan Doi Nhan Dan*:

> The equipment, fuel, food, and the soldiers who would otherwise be students, workers, or farmers "cost us dear," Colonel Tin admitted. The burden is also increased by the desertions and the bought exemptions from military service and the population's weariness with a war to which no end can be seen. No matter, "the job must be completed," the [*Quan Doi Nhan Dan*] editor said. "In the past our struggles have always lasted tens of years; time is on our side."[64]

That this view is not entirely without foundation is evidenced by the report of former U.S. Ambassador to Cambodia Emory Swank, following his early 1983 visit to Vietnam and Kampuchea:

> Because I admire the noncommunists in the coalition, I deeply regret that I must conclude, following a recent study mission to Vietnam and Cambodia, that their aspirations to achieve a broadened government in Phnom Penh and a timetable for withdrawal of the Vietnamese forces are probably illusory.

He also concluded from his talks with the Vietnamese and Khmer leaders that "there is little or no room for bargaining, since Vietnam and the PRK see their fundamental interests served by further consolidating the power of an unaltered PRK." Vietnam's leaders, he observed, have concluded that "the economic deprivation and diplomatic isolation they are experiencing are out-balanced by retaining the dominant position in Indochina for which they fought so long and sacrificed so much to achieve."[65]

Despite the ebbs and flows of diplomatic initiatives and atmospheric changes, it seems likely that both Vietnam and Thailand prefer a continuation of an unsatisfactory status quo to what they calculate are the even more unsatisfactory risks of taking the steps necessary to move toward a resolution of the Kampuchean problems. Vietnam, despite its constant reiteration of the dangers of the "China threat," has concluded that its overweening neighbor can be managed and cannot inflict unacceptable damage on Vietnam. "They look on our (Indochinese) countries as vassal states," said Vietnam's foreign minister, "But wanting something is one thing, achieving it another. An old man cannot do what he wants with a young girl."[66] Moreover, there have been some signs of a slight improvement in Chinese-Vietnamese relations.[67] Just as Thailand prefers to accept a continuation and possible escalation of the Kampuchean conflict because the increase in support from friends and the entanglement, isolation, and debilitation of its adversaries more than compensate for the costs of this policy, Vietnam also feels that the pressures directed against it by China can be handled and will not force it to seek a speedy termination of the conflict that might diminish its influence in Kampuchea with adverse consequences for the regional balance of power. By the summer of 1983, some observers felt they detected a more confident and inflexible stance by Hanoi.

> Indeed, a top Vietnamese official is reported to have told an ASEAN ambassador recently that the Vietnamese withdrawal from Cambodia can come about in two ways. Coinciding with measures taken by Thailand to improve security along the border, Vietnam would withdraw the bulk of its troops. If this does not happen, the Vietnamese would continue their partial annual withdrawal and would complete it "in five to 10 years, when there will be nothing to negotiate." ASEAN diplomats believe Hanoi may actually have that time-frame in mind in view of other considerations, chiefly a realignment of Sino-Soviet relations and the creation of a dependable Khmer administration in Phnom Penh. Whatever the calculations, there is no mistaking the hardening of the Indochinese position.[68]

By late 1983, Foreign Minister Thach had taken up the theme that Vietnam would not fall into the dual traps set for it by the Chinese: "either a unilateral Vietnamese withdrawal or a long war that would bleed us white."[69]

Although Bangkok is concerned about the border problem and the risks of escalation, and although there is evidently some internal disagreement on how urgent a diplomatic attempt to neutralize this threat is, it seems unlikely that Thailand will, in the short run, accede to any solution that confirms Vietnam's control over Kampuchea. The director of the Thai Supreme Command Information Office noted at a press conference that "a major factor for the problem at the Thai-Kampuchean border is the lack of definite border demarcation," and that "a way to solve the problem is to negotiate with Kampuchea and set up border markers, similar to what has been done on the Thai-Malaysian border." If the main problem was seen to be the danger of a spillover of fighting from Kampuchea into Thailand, this would indeed be a logical solution. However, the director also noted that while "efforts had been made . . . results depend on the settlement of the international problem."[70] This suggests that the threat of border flare-ups and possible escalation are not the primary determinants of Thai policy toward Kampuchea.

Many of the public statements by Thailand's top military officials have tended to discount the capacity (and, by implication, the intention) of Vietnam posing a direct military threat to Thailand. Indeed some prominent Thai military leaders, such as the commander of the navy, have said that they do not feel that Vietnam would violate Thai territory except "in unintentional cases as in hot pursuit of the Khmer resistance forces."[71]

Not all Thai military leaders are that relaxed about the presence of Vietnamese troops near their border. Whatever the differences among them about the exact nature and urgency of the threat posed by this presence, the basic strategic rationale seems to point to a willingness to adjust to the status quo. The assistant army chief of staff made a distinction between the threat posed by the very presence of Vietnamese troops in Kampuchea and the threat (which he discounted) of a deliberate, planned invasion. "From the security point of view," he said, ". . . we have to keep Indochina [as a buffer] to ward off a possible outside invasion—frankly speaking, from Vietnam. And we also have to keep Vietnam as a buffer to prevent [a] potential threat from some country,"—presumably China. Speaking of the pattern of Vietnamese troop deployment in Indochina, Lt. General Chawalit Yongchaiyut said, "no country which intends to invade another will deploy its forces in that way. So I consider this a threat against Thailand, but Vietnam's intention to threaten us is not very clear."[72] This analysis emphasizes the importance of balance-of-power concerns; that the mere presence of a force capable of posing a threat is a danger, though only in a hypothetical sense since there appears to be no likely intention of using that force aggressively against Thailand. On the other hand, Vietnam is apparently viewed as a counterweight to China. In combination, these considerations suggest that though the current balance of power is not ideal for Thailand, it is something that Bangkok can live with while waiting for the emergence of conditions that would attain its preferred goal—the elimination of Vietnamese influence from Kampuchea and its reconstitution as a buffer between Thailand and its major regional rival.

As noted earlier, it is not disadvantageous to Thailand to see its most formidable regional rival enmeshed in a costly and debilitating occupation, not just in the near future, but in the long term as well. Not only will Vietnam pay heavy current costs for its occupation, but the opportunity costs of forgone developmental gains and external assistance will be a negative legacy for the future as well. And, even if Vietnam consolidates its hold over Kampuchea, the powerful pull of Thailand's more dynamic economy will inevitably pose a formidable challenge to Vietnam's ability to tie Kampuchea to the Indochinese bloc.

For Kampuchea and its people, the implications of this stalemate are not promising. Despite considerable incentives for both Vietnam and Thailand to move toward an end of armed hostilities, there are even more powerful factors that lock them into diplomatic immobility. Ironically, both sides have a shared interest in the sustenance of Pol Pot, whose forces represent the most sinister aspects of both past and future of the Kampuchean conflict and pose the greatest obstacle to a peaceful settlement of the issue. Thailand needs Pol Pot to keep the issue of Kampuchea's future government open and with it the chance of a neutral or pro-Thai neighbor. Pol Pot also helps to justify the Vietnamese military presence and neutralize much of the adverse reaction to an otherwise intolerable foreign occupation among the fiercely nationalistic Kampucheans. For the moment, at least, Vietnam holds most of the cards. Only the lack of any pressing reason to move toward an effective and, in the French foreign minister's term, "concrete" solution to the problem can explain the single-minded Thai-ASEAN commitment to the coalition formula, with its contradictory assumptions and its remote prospects for success. The coalition, it is conceded by its backers, is only a political card and cannot be a decisive military influence. But it is not likely that the noncommunist elements can ever dispense with the need for Pol Pot's military forces, although even with them the prospects of mounting an effective military challenge to Vietnam seem remote. Pressuring Vietnam through confrontation and isolation has not worked so far, and probably will not work short of an unanticipated total collapse of Vietnam's economy and institutions. Moreover, the support for this policy by important Thai allies is beginning to erode.

Yet, in a perverse way, the current stalemate does have a certain logic from both Thailand's and Vietnam's perspectives. Vietnam has what it wants (even though this was neither its preferred nor its anticipated means of achieving its objectives in Kampuchea) and, having paid a heavy price already, seems content to wait until the options of its opposition are exhausted. As noted above, while ostensibly sitting on a powder keg, Thailand has actually neutralized its most formidable neighbor and has attracted political and diplomatic support from a broad spectrum of forces, while incurring minimal costs for its indirect involvement in Kampuchea. It may be that dramatic reversals either on the international scene or within Kampuchea itself will break the stalemate. But the more likely pattern seems to be a continuation of more of the same, in which case the Kampuchean people will continue to bear the heaviest costs of the political and diplomatic convenience of others.

Notes

1. Khien Theeravit, "Thai-Kampuchean Relations: Problems and Prospects," *Asian Survey* 6 (June 1982)L569–92. Another authoritative collection of Thai view on security interests can be found in the ISIS Bulletin of the Institute of Security and International Studies, Faculty of Political Science, Chulalongkorn University, 1 (July 1982).
2. *Far Eastern Economic Review* (hereafter *FEER*), December 15, 1983, p. 18.
3. *Foreign Broadcast Information Service Daily Report: Asia and Pacific* (hereafter *FBIS*), February 9, 1983.
4. Ibid., May 10, 1983.
5. Ibid.
6. Ibid.
7. Ibid., November 19, 1982.
8. Ibid., January 10, 1983.
9. Ibid., February 9, 1983, p. J1.
10. Ibid., July 22, 1982.
11. Ibid., July 30, 1982.
12. Ibid., February 28, 1983.
13. Ibid., July 30, 1982.
14. Ibid., February 7, 1983.
15. Nayan Chanda, "Subtracting the ICK Factor," *FEER*, May 26, 1983.
16. *FBIS*, June 8, 1983.
17. Ibid.
18. Ibid., June 1, 1983.
19. Ibid., June 13, 1983.
20. Ibid., November 19, 1983.
21. Ibid., June 1, 1983.
22. Ibid., November 19, 1982.
23. Ibid., January 31, 1983.
24. Ibid., November 17, 1982.
25. Ibid., November 26, 1983.
26. Ibid.
27. Nayan Chanda, "A Symbolic Offensive," *FEER*, May 5, 1983, p. 43.
28. *FBIS*, February 8, 1983.
29. Ibid., April 6, 1983.
30. Ibid., March 31, 1983.
31. Ibid., March 31, 1983.
32. Ibid., August 16, 1983.
33. Ibid., December 23, 1983.
34. Nayan Chanda, "Vive la Différence," *FEER*, April 7, 1983, p. 12.
35. *FBIS*, October 1, 1982.
36. Ibid.
37. Ibid., February 15, 1983.
38. Ibid., June 1, 1983.
39. Ibid., March 31, 1983.
40. Ibid.

41. Ibid., June 13, 1983.
42. Ibid., May 11, 1983.
43. Ibid., May 17, 1983.
44. Ibid., October 12, 1983.
45. Michael Richardson, "A Force for Peace," *FEER*, December 1, 1983, pp. 26–29.
46. *FBIS*, October 17, 1983.
47. Cf. Nayan Chanda, "Vietnam's Economy: 'Bad, but Not Worse,'" Center for International Policy, Indochina Project, *Indochina Issues* 41 (October 1983).
48. Cf. Willy van Damme, "Closer to Compromise?" *FEER*, December 15, 1983, p. 18.
49. *FBIS*, October 28, 1982.
50. Ibid., October 18, 1982.
51. Ibid., February 10, 1983.
52. Ibid., January 11, 1983.
53. Ibid., July 26, 1983.
54. Ibid., February 25, 1983.
55. Ibid., April 21, 1983.
56. Ibid., July 23, 1982.
57. Ibid., June 13, 1983.
58. Nayan Chanda, "Thach's Try in Thailand," *FEER*, June 23, 1983.
59. *FBIS*, October 18, 1982.
60. Ibid.
61. Ibid., October 11, 1982.
62. Ibid., October 18, 1982.
63. Cf. David W. P. Elliott, "Vietnam in Asia: Strategy and Diplomacy in a New Context," *International Journal* (Toronto) 2 (Spring 1983):303–304.
64. Jacques de Barrin, "Time for Pragmatism in Vietnam," *Le Monde*, January 7, 1983, in *FBIS*, January 11, 1983.
65. "Cambodia: The Rebirth of a Nation," *FEER*, March 17, 1983.
66. *FBIS*, December 10, 1982.
67. Cf. Nayan Chanda, "Romanian Rendezvous," *FEER*, March 17, 1983, and "Thach's Try in Thailand."
68. Nayan Chanda, "Candour and Confidence," *FEER*, August 18, 1983.
69. *FBIS*, October 17, 1983.
70. Ibid., October 24, 1983.
71. Ibid., February 10, 1983.
72. Ibid., July 14, 1983.

Anthony Barnett

CAMBODIAN POSSIBILITIES

How will the fate of Cambodia* be decided? This question is often discussed in a way that may prove to be unrealistic. Statements—usually pious—about the democratic and national rights of the Cambodian people are coupled with references to the balance of forces, tactically in the frontier region with Thailand and strategically vis-à-vis Vietnam. In this way much of the debate about Cambodia implies that its destiny will be determined by the outcome of a contest between Hanoi and a combination of ASEAN and Washington; one in which the latter exploit Beijing's hostility toward Hanoi to obtain a result favorable to the West.

A striking example of such an attitude was Son Sann's claim (he was speaking before the U.S. invasion of Grenada) that Cambodia represents an opportunity to achieve "the first liberation" of any country from communism.[1] There is something rather odd about this assertion, given that the major armed force in the Son Sann/Sihanouk/Pol Pot coalition of Democratic Kampuchea is the Khmer Rouge. True, some of the spokesmen from the exiled Pol Pot government now project themselves as convinced believers in all manner of capitalist freedoms. But to put their sincerity to the test, they will have to be returned to supreme power in Cambodia. If you listen to Thai representatives in private, however, or American officials in public, they insist with every sign of sincerity that Pol Pot will not be returned to power under any circumstances, so far as they are concerned. Western policy is to use Pol Pot against the Vietnamese, it seems, much as the Pentagon once used Agent Orange: we can rest assured that both come with the similar label, which says that there will be no adverse aftereffects.

This chapter presents an alternative scenario to that which is usually presumed. Namely, the possibility of a "solution" to the Cambodian problem that stems from the North rather than the West, from a détente between Hanoi and

*I call the country "Cambodia" rather than "Kampuchea" because the latter is a Pol-Potism: countries have the right to change their name, but not to adjudicate how the same word is spelled and pronounced in other languages. Such nationalist ultrasensitivity is, of course, quite undemocratic in its actual effect, which was witnessed inside Cambodia on the most terrible scale after 1975. Many more English-speaking people know where Cambodia is and what happened to it than comprehend the slightly arcane and fuzzy place called—so oddly—Kampuchea. The same thing would happen to Spain if a regime there insisted that Anglo-Saxons were obliged to call it Espagna.

Beijing rather than the current *de facto* collaboration in Thailand of Beijing and Washington. To discuss such a possibility in cold print is more difficult than to debate it in a verbal session. What follows is necessarily tentative and hypothetical.

In part this is for structural reasons inherent in any attempt to decipher the diplomacy that surrounds Indochina. In most cases, the posture a country adopts toward Cambodia is a function of its policy toward Vietnam. Its policy toward Vietnam, however, will be a function of its policy toward China (Cambodia, it should be noted, comes at the bottom of the pile). Thus both the American and the Soviet approaches to the Cambodian conflict are primarily determined by their respective attitudes toward Vietnam, which are governed in turn by their respective China policies. The same rule applies to ASEAN countries and to Europe and Japan. But it cannot hold, at least not in the same way, for Hanoi and Beijing themselves. Certainly for both, their policy toward Cambodia is a direct function of their policy toward each other. But their policy toward each other is central (although in different ways) to the entirety of their conduct of foreign affairs.

One consequence has been that in Cambodia itself the liberation struggle has swung back and forth under the magnetic influences of the two poles of Beijing and Hanoi. Yet while they are opposites, they too are held together, in a single field. The dispute between them over Cambodia may prove to be irreconcilable for another generation or more. Nonetheless, the starting point for any analysis of the diplomatic prospects that lie in store for Cambodia, and that may determine its position amongst the blocs, is the base line of *agreement* about Cambodia that exists between China and Vietnam. The emphatic nature of this unity of view—yet again a paradox—may be seen by looking at the repeated commitment made by Hanoi to withdraw its armed forces from Cambodia when it reaches a settlement with China—one that lifts what it sees as the Chinese threat to the integrity, separate and joint, of the three countries of Indochina. What is the meaning of this commitment, so overtly antagonistic to China, that could possibly be described as drawing Hanoi into fundamental agreement with Beijing? It is simply the fact that the one thing both Hanoi and Beijing agree upon is that they, and really they alone, should be the ones who decide the final shape of any settlement. They may not agree about anything else, but they hold in common the belief that they should be the decisive arbiters. Insofar as both seek ASEAN as an ally or look to the pressure of the United States or the Soviet Union, it is because they want to use these third parties to help them shape a final outcome that is more to their liking. What is less clear is the weight they give to the policies of other states as these impinge upon them via Cambodia. Thus a multitude of different factors are involved, further entangled by personal obsessions on all sides that date from the Vietnam war.

In addition to the complexity and secrecy of the diplomatic tangle, any discussion in printed form is made doubly difficult by the way the situation changes with remarkable fluidity. When I first delivered a paper on this subject in the aftermath of Yuri Andropov's elevation to the leadership in Moscow, my stress on the possibility of a Sino-Vietnamese settlement seemed implausible. By October 1983, however, Nguyen Co Thach—the Vietnamese foreign minister—had been invited to the official PRC celebration of its national day at the United Nations in New York, where in the presence of China's Cambodian allies and ASEAN diplomats, he was embraced by his Chinese counterpart.[2] As we will see, considerable movement had taken place in the Chinese position. However, no thaw ensued. When the Chinese Premier Zhao Ziyang visited Washington in January 1984, he emphasized that China's and America's views on Cambodia were "identical." At the same time he said that while "conditions are not yet ripe for a political solution to Vietnam's occupation of Kampuchea . . . China would very much like to see an early solution." Zhao also repeated China's desire for improvement in Sino-Soviet relations.[3] In this context it is hard to separate the main currents from the secondary whorls and eddies. Nonetheless I will attempt to explore the prospects for a resolution of the Sino-Vietnamese conflict over Cambodia—something which would almost certainly involve a more comprehensive détente between China and Russia.

Before doing so, it will help to begin with a quick glance at the possible variations of a "Western" solution, by which I mean an outcome to the present conflict in which Cambodia ceases to be a bureaucratic-party state allied to Vietnam and instead is ruled by a national-capitalist regime.

"Western" Solutions

A Military Resolution

Short of outright war, there seems to be little likelihood that the coalition will win in a straightforward military fashion. Here, others who are much more expert than I in judging the balance of forces can read the military score. Certainly, while the Vietnamese missed the chance to destroy the Pol Pot forces completely in the immediate aftermath of the 1979–1980 invasion, the subsequent presence of the Khmer Rouge commandos strengthened the political hand of the fledgling Heng Samrin administration. Today, while the guerilla forces can lie in wait in the border regions, like vultures, they cannot themselves move in to make the kill. The coalition's military strategy must be a commitment to attrition in which success will stem from other pressures as well—the international and domestic costs for Vietnam and popular opposition in Cambodia.

So far as the first is concerned, the economic and diplomatic cold war launched by China and supported by the West has certainly had punishing consequences for the Vietnamese. But today Hanoi's economic difficulties are seen—rightly—as being determined more by internal conditions than by external forces. A slow economic improvement seems to be underway for both Vietnamese agriculture and industry, while the costs of their large military presence in

Cambodia itself may be less than the economic benefits of trade and exchange for the Saigon (Ho Chi Minh City) region. One recent analysis in the *Far Eastern Economic Review* was remarkably assured on this point: "Hanoi was almost certainly prepared to pay any price to secure Cambodia, which it regards as vital to its own security. Now, as long as Soviet aid keeps flowing, and the Vietnamese and Cambodian economies suffer no major setback, the Vietnamese leadership must feel the price is becoming increasingly bearable."[4]

Two suppositions are the key to this assessment. It presumes Soviet aid will remain at current levels and that there will be no significant economic deterioration in either Vietnam or Cambodia. If neither internal developments within Vietnam nor military or diplomatic pressure from without will break its hegemony over its smaller neighbor, then only developments within Cambodian society itself will do so. Perhaps another catastrophic famine, or an epidemic in Phnom Penh that demonstrated criminal mismanagement within the capital, might undermine completely the tentative legitimacy of the present regime. Only if some such event were to turn the indigenous population so against the present powers in Cambodia that the guerillas obtained a genuine popular base within Khmer society would the country likely be shaken from the Vietnamese sphere. Many might wish such an outcome, but few could want the Cambodian people to pay such a high price for its realization.

The key reason for the strength of the Vietnamese position in Cambodia is often underestimated. Most of the population prefers the regime Vietnam has initiated in Phnom Penh to its predecessor. A majority of the Khmer might want something different again, but a majority of mankind lives with an equivalent desire. Quite apart from the *fait accompli*, it needs to be recognized that Cambodia is a defeated country in an important sense that is rarely registered in the Western media. Pol Pot *really was* an extreme nationalist, and he showed the country the meaning of "100 percent independence." In the process he launched the border attacks upon Vietnam, as part of the dramatic mobilization entailed by his exceptional regime. The whole country has learned a hard lesson in the process. The Cambodians know that they began the war, and this lends them a certain realism and an awareness of the legitimacy of the Vietnamese presence, however reluctant or unspoken—much as the German population accepted the respective occupations by American or Soviet forces after 1945 and continues to accept their presence now. They may not want it, they may seek to end it, but they can hardly suggest that it was not Germany's historic responsibility for bringing about the state of affairs in the first place. The equivalent holds true for the presence of Vietnamese troops in Cambodia, which means that the latter has not been "colonized" by the former, if by this is meant a domination similar in form to those of historic European imperialism.[5] It follows that the Vietnamese-backed regime in Cambodia, however unstable, is unlikely to be subject to the same historic instability.

Nothing is certain in international politics, but short of another war, it seems to be pretty definite that in fact the Vietnamese have "won" in Cambodia in that the Khmer administration they have constructed in that country will not be removed by force alone. Only a political decision taken in Moscow to curtail military supplies is likely to reverse the naked balance of power.

A Political Solution

Rather than dismissing pro-Western peace feelers out of hand, Hanoi has sought agreement. Possibly it is concerned about the likelihood, rather than the improbability, of a Sino-Soviet agreement, one that might indeed mandate Vietnam to a settlement in Cambodia. Certainly, from the point of view of Vietnamese diplomacy, if Hanoi were able to reach a settlement with ASEAN this would significantly increase its bargaining position vis-à-vis Beijing. Despite their antagonisms, therefore, both ASEAN/Washington and Hanoi share a short-run desire that any breakthrough should come about through their initial agreement. Although the Vietnamese undoubtedly have a more profound difference of principle with what we loosely call "the West" than they do with China or the Soviet Union, they would undoubtedly prefer to see a solution to the current Cambodian problem initiated by a settlement between themselves and "the West," rather than one that takes its cue from an understanding between their brethren to the north.

Yet there is something implausible about the prospects for any such agreement between Hanoi and those countries that either fought in Vietnam, like the United States, or supported its intervention. A formal understanding will necessitate one side or the other making a concession of principle. Either Hanoi will indeed agree to use its leverage to dismantle the state structure it has tried to put in place in Phnom Penh, and allow Cambodia to swing over to the capitalist world, or the ASEAN countries will agree to ratify, in one way or another, the incorporation of Cambodia into Vietnam's Indochina alliance.

Apart from both sides agreeing to allow Pol Pot back, to form a pro-Chinese regime, it seems to me that these are the only two alternatives so far as negotiations between the pro-Western and the Vietnamese parties are concerned.

It follows from the military assessment that any likely settlement of the Cambodian diplomatic imbroglio that stems from an agreement between Vietnam and the pro-Western countries is going to be one that recognizes the substance of Hanoi's *fait accompli*. In other words, of the two alternatives sketched out above, so far as a Vietnam/ASEAN settlement is concerned, only one of these is at all plausible, given the situation on the ground—namely, an agreement that legitimizes a government in Cambodia akin to Heng Samrin's. It might change its superficial appearance, but the presence of some Vietnamese troops and the continued activity of the existing internal security apparatus would be confirmed. How likely is this? The United States—and in particular the Reagan administration—has little interest in seeing any such agreement. It would further antagonize

China, it would be seen as a victory for Russia, it would legitimize Hanoi, and for what? There are some very good answers, not least the healing of the wounds of war and retention of the diplomatic initiative in the region, but these are not likely to be persuasive in American national security calculations.

The only halfway plausible scenario, then, for any such settlement is if ASEAN comes to an understanding with the Vietnamese that grants Cambodia a kind of Laotian status, despite the inclination of the United States. It is important not to underestimate the reality of ASEAN's autonomy from Washington. Yet I remain skeptical of so dramatic a demonstration. Not because I think it inconceivable that the Southeast Asian alliance could defy the United States on such an issue, but because at present the Association does not have an overwhelming, collective interest in doing so. Thailand and Indonesia might benefit in certain ways, the Malaysians clearly prefer to see a settlement, but Singapore does not, and for evident reasons. The city-state benefits from the tension in the region: its own position within ASEAN is safer, its ideological prominence would suffer most from any détente with Hanoi. Whatever the wishes of any individual ASEAN country, none values an agreement with Vietnam above the relationship with its regional associates. It is therefore improbable that ASEAN will come to terms with Hanoi and Vientiane. The Association "led" the United Nations into its endorsement of Pol Pot's well-known "Democratic Kampuchea" and maneuvered to have it graced with Sihanouk himself. This caused considerable embarrassment to many a diplomat, and a U-turn will demand at least a heavy veneer of justification. The very independence of ASEAN's foreign policy, and the way in which it first made its mark politically with its policy over Cambodia, may mean that it is stuck with it henceforth. As so often in international affairs, its policy seems to have the weakness of its strength: that which first gave ASEAN definition is now a liability in that it cannot be disregarded with equivalent ease. There have been strains and creaking noises by a majority, but so far they remain imprisoned by their own collective structure.

Perhaps the impasse in Vietnam/ASEAN diplomacy can be summarized as follows: The Vietnamese have a great interest in reaching a settlement on Cambodia prior to reopening negotiations with Beijing, but this is not sufficient reason for them to make the only concession of real interest to ASEAN, namely, relinquishing control over Cambodia. In the absence of this, ASEAN has no special interest in legitimizing the presence of Vietnamese forces inside Cambodia, forces which will remain there, as Hanoi has stated more than once, so long as the threat from China continues.

In the final stages of the Vietnam war, Richard Nixon showed that by playing China off against the Soviet Union, it was possible to have a Western sphere of influence in Indochina conceded by Beijing. Washington's strategy was clear enough: its interests could be defended by dividing its communist opponents against each other. Can this strategy really be repeated today so far as Cambodia is concerned? It might be possible if the West were willing to do the fighting. But the essence of the Nixon Doctrine was to have Asians fight Asians in Asia (Africans in Africa, etc.). Well, they are certainly killing each other still in Indochina, but will they do so on such a scale that American-backed regimes can simply pick up the best pieces afterward? For that is what the Son Sann approach seems to mean. The Washington/ASEAN axis hopes to use China as the force to lever Cambodia away from Vietnam and from China as well. The idea is that at the crucial moment when victory is assured, the Pol Pot forces will be surgically eliminated.

This brings us to the question of how Beijing would view a Cambodian settlement between Vietnam and "the West." Clearly it would be hostile to one that underwrote the status quo in Phnom Penh for the same reasons that Hanoi would favor such an outcome. But how would China view a Vietnamese defeat in Cambodia that led to the "liberation" Son Sann has demanded?

The role of China in such a scenario is absolutely crucial, yet in a paradoxical way: as both prime agent and passive loser. It is Beijing that is applying direct military force to Vietnam, along its northern border. It is Beijing's embargo that is most damaging to the Vietnamese economy. Similarly, in Cambodia itself, it is the Chinese armed, trained, and organized Khmer Rouge that dominate the guerrilla forces. So the question has to be asked, why should the Chinese do all this to give Washington a Son Sann government in Phnom Penh? While the Chinese apply the pressure against Vietnam in Cambodia—the main human victims, it virtually goes without saying, being the Khmer—the West is dedicated to preventing a pro-Chinese regime emerging in Phnom Penh when the Vietnamese finally leave. A pro-Western outcome will hardly increase Beijing's leverage on Hanoi, and the diminution of the Soviet presence in the region would be small compensation for China's loss of influence. Even belligerent Americans concede that a stable Son Sann regime will only emerge as a result of negotiated concessions. Such negotiations will inevitably incorporate the recognition of Vietnam and some security guarantees for it against further border assaults. Thus the envisaged Western-style settlement may well consolidate Vietnam's independence from China.

Under what circumstances can we imagine Beijing really accepting such an outcome? It will surely demand a quid pro quo. When it was still possible for the Chinese to hope for major concessions on Taiwan, for example, or while the Russians were being seen as behaving in a directly bellicose fashion, the "loss" of Cambodia to the West might have seemed a reasonable price in terms of Beijing's global strategy, especially as it would be an even greater loss for the Vietnamese. Today, however, such an outcome might leave Beijing itself empty-handed, despite all the expenditure and odium that it has incurred.

In other words, if the United States was willing to deal China into a strategic relationship of substance, in which economic grants on a massive scale assisted its development and an eventual reunification with Taiwan was made politically and militarily credible, then Beijing might happily use its leverage to wrest

Cambodia away from Hanoi knowing that Phnom Penh would become an ASEAN capital. But there is no such juicy prospect on the horizon for Beijing.

On the other hand, the Chinese value their present relationship with the United States, which undoubtedly assists their economic development. At the same time, on most major foreign policy issues from the Middle East to Central America they now diverge, and Beijing feels obliged to stress the difference between the social systems in the two societies. In these circumstances, Cambodia and Afghanistan, but particularly Cambodia, provide a fulcrum of common interest. It is therefore handy for the Chinese to keep Cambodia on the boil, especially when this is to the liking of an American administration with a visceral hatred for Hanoi. But keeping something on the boil is different from actually cooking it and serving it up on a plate. The present unresolved Cambodian conflict is useful to China insofar as it provides a basis for common policy with the United States in which China has a degree of leverage; in which China provides the arms and pressure that the United States apparently desires. But by the same token, a *resolution* of the issue that established a pro-American regime would bring China's role to an end, and it might even diminish Beijing's influence in Washington, should the Chinese seek to preserve a place for the Khmer Rouge.

As Beijing assesses the reality of the affair, and as it comes to terms with its failures of the 1970s and its own limitations as a regional power, it will continue to rethink the nature of its friendship with Washington. Indeed, the Chinese government now openly recognizes the cynical character of the Reagan administration's attitude so far as China itself is concerned.[6] Western (and ASEAN) policy on Cambodia is to use China against China: it aims to take China for a ride. But the longer this takes, the less likely it is that the mule will stick the course. If Beijing and ASEAN/Washington were in positive agreement about the kind of regime they wished to see installed in Phnom Penh, their combined pressure might well prevail in the long run. However, if, as it seems, they concur only on the perpetuation of the conflict rather than on its outcome, then in the end they may seek to get the better of each other in a settlement with Hanoi.

It is this, the fact that China has little interest in a pro-Western resolution of the Cambodian question and has been offered no compensation of consequence for accepting such an outcome, that may make Beijing look more favorably on a settlement with its communist neighbors, provided that this confirms rather than diminishes its status.

The point is a crucial one in any examination of the long-term development of Indochina diplomacy. So it is worth further emphasis by extending it to Vietnam—which, as noted above, is what the Cambodia question is actually about. Some Western exponents of *realpolitik* suggest that Vietnam will "break"; informally, State Department experts foresee Vietnam denouncing the USSR and turning to the West, in a miniature rerun of the trajectory China took between 1960 and 1972. Skepticism is justified here. The Soviet Union has itself learned from the debacle of its Chinese policy, and it is hardly likely to push

Vietnam in the same direction. There is no source of rivalry between the leaderships in Moscow and Hanoi as there was between Khrushchev and Mao. Already the duration of the active alliance between the Russians and the Vietnamese goes back over two decades (and many "difficulties"), more than twice the period of comparable Sino-Soviet cooperation in the fifties. Furthermore, this American scenario omits the presence of China itself. There, in Beijing, an equivalent calculation has been made, that when the Vietnamese learn what it is like to be solely dependent upon Moscow, they will indeed "do a China," but will turn not to America but rather to their natural ally, China itself. The one outcome is as improbable as the other; we are unlikely to see any sort of diplomatic capitulation by Hanoi, in either direction.[7]

Yet this is of less importance than the existence of an underlying conflict between Washington and Beijing. Each may now be willing to push Hanoi into reliance on the USSR, hoping that it may turn to them for a way out. But the last thing the United States (or ASEAN) wants to see in Southeast Asia is a strong Chinese presence in a unified Vietnam. Similarly, the Chinese have no desire to see Vietnam concede in Cambodia and strike up in "regional alliance" with a Washington that remains the dominant influence in Taiwan. Thus both the United States and China may actually prefer to see the USSR remain influential in Vietnam rather than have it supplanted by the other. If so, this may prove to be of decisive advantage to Hanoi.

Chinese-Soviet Relations

Vietnam has played a significant role in the shifting relationship between Beijing and Moscow, perhaps most of all during leadership changes in one of the two major communist capitals. Stalin's death led to the end of the fighting in Korea and to the Geneva Conference at which the fate of Indochina was arbitrated. A decade later came Khruschev's fall; described as Hanoi's "biggest windfall since the Chinese Communists arrived on the border in 1949."[8] Perhaps the sharpest symbol of this critical moment was Kosygin's presence in Hanoi when President Johnson began to bomb the North.

It seems that, at that crucial moment, the new leadership in Moscow decided to forestall an impulsive Khruschevian "pullout" from Vietnam, one that would allow the Chinese a free hand there, not least because this might allow China to dominate communist relations with the United States in the Pacific. While the new Kremlin leadership sought to improve relations with the Chinese, it equally insisted upon its own right to support Vietnam; for it was not going to grant Beijing any exclusive sphere of influence in Asia.

The intensity of the dispute with the Chinese that followed can hardly be exaggerated. It was Mao's ambition to make China one of the great powers, while the Russians saw this as a threat to their leadership of the communist world. The Chinese then proceeded to denounce the escalation of the conflict in Vietnam as a double encirclement of themselves by the United States and the Soviet Union, two "superpowers" in collusion against Beijing.

The extraordinary path of triangular diplomacy that followed, and which was centered upon the maelstrom of the Vietnam war, is not for discussion here. The final outcome, however, was undoubtedly a setback for Chinese diplomacy given the "promise" of the 1972 Shanghai communiqué. After 1975 Beijing backed the new Pol Pot regime in a way that was calculated to underwrite its hostility toward Hanoi, as part of a more general effort to limit Soviet influence in Asia while expanding its own. But while this approach may have been Mao and Zhou's in its origin, it took on a different—more violent—form under their successors.

Just as changes in Soviet leadership brought new policies toward its global competitors, and thereby shifted Moscow's priorities in Vietnam, so the post-Mao leadership struggle in Beijing focused in part on Vietnam. Deng Xiaoping seems to have gained crucial Red Army support for his turn toward the United States by outbidding his domestic rivals in their belligerency toward Hanoi. Unfortunately for him, the Chinese generals were unable to deliver his promised "Indian-style" invasion, which was supposed to demonstrate a masterful strategic superiority over the area.[9] Instead, in heavy ground fighting on its northern border in 1979, Vietnam staved off the Chinese forces. By so doing it ensured its preeminent role in Cambodia.

When Yuri Andropov gained the Soviet leadership at the end of 1982, he made a highly visible effort to conciliate China at Brezhnev's funeral ceremony. In the funeral address itself Andropov proposed an order of precedence in the communist world that would probably have given China a senior rank irrespective of Vietnamese wishes, if Beijing had so desired.[10] Like many of his initiatives, Andropov's China policy lost momentum with the rapid failure of his health. Yet it was never likely that China would accede to an alliance with the USSR in which it gave first priority to its relations with the Soviet Union. For this would place at risk its newly won economic relations with the United States, Japan, and Europe.

In early 1983, however, there was a clear shift in China's position toward Vietnam, as an aspect of Beijing's Soviet diplomacy, and in response to Andropov's feelers. I will discuss some of the details below. They should be read in the context of intercommunist diplomacy on the larger scale.

It seems that Beijing would like to see a relationship with Moscow that is stronger than ordinary, normalized, state-to-state relations, yet not so close as to jeopardize the continued investment of Western companies in China. The Chinese have stated that there are three sticking points to their reaching a satisfactory agreement with the Russians. The clearest and most recent such statement came in Premier Zhao's address to the Canadian parliament, after his tour in the United States. Stressing his desire for normalization, Zhao demanded the removal of three obstacles: "First, the Soviet Union must stop supporting Vietnam in its aggression against Kampuchea; second, it must withdraw its troops from Afghanistan; and, third, it must withdraw its forces from the Sino-Soviet border and Mongolia.'' Zhao continued by saying that the Soviet side had evaded ''these three questions which threaten China's security.'' The formulation is a significant one. It means that China has drawn up its list not on the basis of principle (which is notoriously unamenable to negotiation) but in terms of threats to its own security, something that might be assuaged in a variety of ways. Furthermore, in the same speech, Zhao called for greater efforts ''by the two sides,'' rather than by the Soviet Union alone, an additional sign (because it implies the possibility of mutual concessions) of his genuine rather than merely rhetorical interest in some form of ultimate normalization.[11]

Here we have a special interest in asking why the Chinese should have placed priority on Cambodia. The question is especially pertinent because it is the only one of the three conditions that the USSR is unable to meet directly and fully of its own accord. Put at its strongest, the Chinese demand is that the Soviet Union concede to China the primary role in Indochina. Beijing wants this because it is here, and perhaps only here, that it can hope to claim its own ''sphere of influence,'' an essential qualification for an aspirant ''great power.'' The Russians have their zone in Eastern Europe; without an equivalent in Asia, how can the Chinese be their equal?

Moscow will not make such a concession. Since 1964 it has sought to prevent Beijing from establishing an exclusive zone of influence in Southeast Asia, and this policy has brought substantial dividends to the Soviet Union in great-power terms. It is unlikely to discard them now. Nor does China present any persuasive reasons why it should do so. The first and major stumbling block to improvement of Sino-Soviet relations on the basis of mutual *equality* is the actual and massive inequality between the two, which stems from China's intractable economic inferiority. Beijing has had to confront the same conundrum in its relations with Washington: it seeks mutuality but is offered only a junior partnership.

Perhaps this is why the Chinese leadership now seeks its own autonomous course, as they would describe it, one that is independent rather than neutral or ''equidistant,'' in which China can try to gain the best of both worlds so as to make up for its backwardness relative to each. Such an outcome requires the consolidation of China's existing relationship with the United States followed by the normalization of relations with the Soviet Union (in which Cambodia figures prominently for the Chinese). The possibility of the latter can now be considered, under four headings: ideological, economic, military, and geopolitical.

Ideology—State Socialism

The ideological basis for a rapprochement became a serious possibility with the introduction of martial law in Poland. The Reagan administration apparently asked China to join in the general condemnation and thereby to categorize

General Jaruzelski's coup as a Soviet-imposed, "surrogate invasion." Instead, the Chinese defined the Polish action as being within the sacred limits of a country's "internal affairs."[12] Perhaps this response may eventually be seen as reversing China's reaction to the invasion of Czechoslovakia in 1968. The latter, with its associated annunciation of the Brezhnev Doctrine of limited sovereignty within the Soviet bloc, was taken as a direct threat by Beijing to its own legitimacy. To be sure, Mao used the overthrow of Alexander Dubcek to his personal advantage in the contest for supremacy over China, and it provided an excuse to intensify a hostility toward the USSR that was already in place. In this sense the Russian invasion of Czechoslovakia in 1968 accelerated a trend that was already discernable in Chinese policy—it was not its cause. In the same way, the restraint demonstrated by Moscow with regard to Poland, and the reality of the problems posed by Poland for bureaucratic regimes everywhere, provided lessons that some Chinese leaders at least were already beginning to learn for themselves. So it would be wrong to suggest that the Polish events in and of themselves caused a new feeling of shared interest between Beijing and Moscow. But if the USSR had sent its troops into Poland, the poor relations between the two largest communist states would not now be improving.

In addition there was the nature of the Polish crisis itself. When Solidarity was founded the Chinese were showing little concern for the Soviet view of the world; they were still flirting heavily with the Carter administration. But at the very least, the Chinese were reading *Fortune*. In its September 1980 issue there appeared a brief and vivid account of "What the Bankers Did to Poland," which opened with a description of how a meeting of Western bankers in Warsaw helped to "trigger the workers' revolt," because it insisted upon an increase in domestic prices. The report went on to discuss how 92 cents of every dollar of Poland's hard currency exports went to pay interest and principle on its debts.[13] The Chinese, already beginning to go into debt themselves, were also showing signs of being unable to digest the huge capital projects that Western companies were queuing up to sell to them, on credit.

The actual meditations of the Chinese leaders upon the meaning of the Polish events remain closed to us, but their import was clear enough. Economically, China concluded that massive Western credits would not provide a quick and safe route to modernization, but would rather risk the political security of the regime itself. Within limits, technical and financial help could be most valuable; beyond them it could be destabilizing: Hence China's present policy of clamping down hard on all dissent, intellectual, industrial, and economic, while retaining a controlled degree of imported Western technology and fighting for a share of its export market.

The Polish events might seem to take us a long way from Cambodia, but they functioned to confirm China's new course, one which will bring it consciously into line with, broadly speaking, a Soviet model. The USSR is no longer "revisionist." Its "hegemonism," therefore, does not have the same economic drive to dominate as does capitalist imperialism. Moscow may be nearer, and in that sense a greater perceived threat, but there is no inherent reason for this to remain the case. U.S. bases, as well as Soviet ones, have to be removed from the Pacific according to China's current perspectives. Thus the supreme rationale for the whole of China's policy in Indochina since 1972, namely that the USSR is the greatest enemy of all, has evaporated.[14]

To underline this point, which has received almost no attention in the Western media, Chinese Vice-Premier Wan Li received the Polish minister of foreign trade in Beijing the week before Zhao went to visit Reagan. *Beijing Review* immediately followed its account of the official White House welcome with a report of Wan Li stating: "We are glad to see that the Polish people have overcome one difficulty after another. . . ." He agreed to try to establish mixed Sino-Polish commissions for economic cooperation.[15] Few more pointed demonstrations of China's ideological affinity with the communist bloc could be imagined.

Economic Coincidence

The shift in China's position creates some painful ideological conundrums for the Vietnamese communists, whose leadership has declared that the People's Republic of China is no longer a socialist country (although they have drawn back from claiming that it is a capitalist or revisionist one). As has been frequently pointed out, the economic reforms underway in China and Vietnam have many features in common and, indeed, are now being echoed in the Soviet Union as well. In all three countries there is clearly an intense debate underway between strongly opposed groups about the legitimacy, and the necessary extent, of economic liberalization. The remarkable coincidence of these debates in all three countries (the three most populous communist states) testifies to their common social character as regimes. It is hardly surprising, although nicely ironic, that the ideological postures of avowedly Marxist states should be determined by economic factors. But one factor that will continue to impel the trend toward détente, however gradual, among the communist regimes, is their economic compatibility with each other.

The limited nature of Western aid to China provides an opportunity for Russian assistance: the long frontier, the poverty of the outer regions, the mutual benefit of exchanging Chinese consumer goods, such as textiles, for Soviet machinery point to increased Sino-Soviet trade as a natural direction for the two economies.[16] For the Chinese it should be stressed that such a development could only complement, not replace, the more important trade and assistance from the OECD economies. For the Vietnamese, there can be little doubt that normalization with China would relieve some of the worst economic bottlenecks that impede the country's growth at present: it would be a decisive lifting of the siege. Paradoxically, this may mean that China has all the more reason to move quickly, for Hanoi does also seem to be showing that it can eventually do without China trade if it has to.

Military Balance

The centripetal forces drawing the neighboring communist regimes together, and thus, through its sheer momentum, including Cambodia as well, should not be painted in colors of rosy hue. The lines of military force cut across the picture in bold strokes. They may start with the Chinese failure in North Vietnam and the Vietnamese success in Cambodia. They extend to the relative restraint of Moscow in limiting itself to threats with respect to China in 1979 and Poland two years later.[17] Between these dates, however, the Soviet invasion of Afghanistan seems to have been motivated in part by a Russian fear that the Chinese would link up with a draconian Amin regime in Kabul. After Deng had bragged that the polar bear had no claws, the Afghan invasion seems to have made the Chinese more conciliatory rather than less. But for all the setbacks China may have experienced on the battlefield, its recent technical success has undoubtedly contributed to Beijing's greater willingness to come to terms with Russia. In October 1983, China successfully fired its first submarine-launched missile capable of carrying a nuclear warhead.[18] For the first time, therefore, it has a credible nuclear "deterrent," in the sense that part of its nuclear forces might hope to escape an attack and be capable of delivering a "second strike." There can be little doubt that this at least makes the Chinese feel more capable of negotiating with the Soviet Union on a basis of equality.

Geopolitical Reality

Finally, there are geopolitical factors outside the communist states that make it rational for China to seek some common terms with the USSR: above all American policy under Reagan. Beijing's hopes for a favorable resolution of the Taiwan question (favorable, that is, to itself) have dimmed considerably. The United States has proved unwilling to open up its domestic markets to Chinese goods on a decisive scale. There is therefore little sense in China pursuing a one-sided policy any longer. In the 1970s pure anti-Sovietism was a means of gaining diplomatic and economic assistance from the West. Today, this approach seems to have exhausted its potential. Indeed, by carefully publicizing its renewal of interest in good relations with Moscow, Beijing is not only testing the water with the Soviet Union; the threat of such a rapprochement is today a means of squeezing concessions from the United States.

Chinese-Vietnamese Relations

With a relatively small expenditure in arms and supplies, the Chinese can oblige

Vietnam to maintain a mobilization that is costly and obliges it to rely upon Soviet arms and aid. But the cost to China of any escalation may reverse the imbalance. A further full-scale Chinese invasion of Vietnam might well do greater proportional damage to Beijing's program of modernization than it would further ''punish'' Vietnam. The priority China rightfully accords to its enormous domestic problems will incline the pragmatic toward rather than away from all-round normalization with its communist neighbors. It is said that the main obstacle to such an outcome is the arch-pragmatist himself, Deng Xiaoping, whose animus against the Vietnamese is such that no settlement is likely while he retains his personal power. But eventually the laws of human biology dictate that younger leaders will take full command in Beijing, Moscow, and Hanoi and past obsessions may count less than the mutual advantages of a pluralistic communist diplomacy.[19] It is not inconceivable, therefore, that future—one almost writes ''forthcoming''—leadership changes in all three capitals will initiate a comprehensive settlement that would incorporate Cambodia.

Recently, there have been signs of movement in the Chinese position. Of course, this is movement from outright support for the return of the Pol Pot regime in Cambodia, a politically ludicrous and morally abhorrent policy from which there was no way out but up. One might feel relief, but there is no need to feel grateful or obliged to policymakers who are slowly ceasing to be completely and criminally foolish. Nonetheless, the present trend demonstrated by Chinese policy is encouraging. Previously it demanded an immediate Vietnamese withdrawal; now Beijing has stated that under certain conditions it is willing to talk to Hanoi while Vietnamese troops remain in Cambodia, and without a specific timetable as to their final pullout.

Thus in March 1983, China stated officially that 1) Vietnam *must* first declare an unconditional withdrawal of all its troops, 2) that the Soviet Union *should* cease supporting Vietnam's aggression (note the difference between the words I have emphasized). 3) If Vietnam should announce its withdrawal, then, as this occurs, China will take ''practical steps to improve its relations.'' 4) After the withdrawal China *wishes* to see an independent, nonaligned Cambodia (thus China no longer rules out the possibility of a Phnom Penh regime aligned to Hanoi, only that the Vietnamese Army should not be indefinitely based in Cambodia).[20]

Inevitably, Hanoi promptly rejected the Chinese position as being an interference in the affairs of Vietnam and the People's Republic of Kampuchea and as imposing unacceptable ''preconditions'' to the renewal of talks between Vietnam and China.[21] But the shift in the Chinese position can be registered clearly enough by comparing the Vietnamese statement of the end of February with that of March 3, 1983.[22] What is there today that prevents renewed talks between Hanoi and Beijing? The Chinese say that Vietnam must first announce that it will withdraw all its forces from Cambodia, and that improvements in relations will then follow. Hanoi has however repeatedly stated that it will withdraw all its troops when China ends its threats, and it is anyway withdrawing some unilaterally. There are real differences between the two parties, but the gap between their formal positions hardly needs shuttle diplomacy to get them to the table.

Cambodia is not the only issue of contention between Beijing and Hanoi. Another is the status of the islands in the South China Sea. Here, too, there is evidence of strains in PRC policy. China claims as its sovereign, territorial zone seas and atolls stretching well south of Vietnam to the shores off eastern Malaysia. It might be thought, as I have heard it said, that this is "merely historical." Yet China has never accepted the "loss" of the Spratley Islands as an unfair consequence of history, as, for example, it has the northern slopes of Siberia. So perhaps these islands will prove to be as much a matter of history as the Falklands are to Britain. At any rate, while China controls the Paracel group to the north, Vietnam has garrisons on some of the widely dispersed Spratley group, the Philippines controls others, and Taiwan has soldiers on one of the larger atolls.[23]

The islands returned as an issue in the spring of 1983. Vietnam appeared to have strengthened its position militarily in Cambodia and diplomatically at the New Delhi nonaligned conference. In April there was a publicized clash on the Sino-Vietnamese border.[24] Then, "to facilitate navigation" the New China News Agency circulated a list of the 300 Chinese names of every reef, bank, and shoal in the South China Sea.[25] The Vietnamese promptly denounced the "land grabbing claims." One explanation of Beijing's behavior is that it escalates its public rhetoric and its combative stance at the same time as it probes for an overall resolution: the technique both strengthens Beijing's hand in the event of successful discussions and provides it with cover in the event of a stand-off. However, there is a certain curiosity about the way the islands were raised as an issue vis-à-vis Hanoi. *Beijing Review* of April 18, 1983, carried a long analysis of "China's Future Position in Asia" by Pei Monong. It listed the various recoveries China would make in the course of its modernization: Taiwan, Hong Kong, Macao.[26] Its final two paragraphs then read: "Of course, China's sovereignty over islands of the South China Sea will also be recovered in the process. The fulfillment of the above-mentioned tasks will make it possible for China to step into the 21st century, a modernized and fully united socialist nation." The reference to the islands is rather curt and embarrassed, as well it might be, for it is rather difficult to plead the case for peaceful relations with ASEAN and to praise regional coexistence while making a territorial claim that is bound to unite the states of Southeast Asia in hostility toward China should any attempt be made actually to realize it. Thus whether by accident or design, the definite article was omitted: China's sovereignty over "islands" may be made good, but that is quite a different thing from "the islands," implying the totality.

Does this difference of emphasis emerging within a few days signal some disagreement among the Chinese over the realism of their maritime claims? Possibly we can be more confident that Pei Monong's conclusion was carefully scrutinized, for in it, extraordinarily, he places the normalization of Sino-Soviet relations before reunification with Taiwan.

Even more remarkable evidence of the way the Chinese were exploring the possibilities of a settlement with Moscow with respect to Vietnam came in June 1983. Beijing published a letter from China by Hoang Van Hoan, apparently written in February. Hoan was the "Chinese comrade" in the Vietnamese leadership. He defected to China itself in 1979, from whence he has issued various diatribes against the "Le Duan Clique." This time, however, his accusations have taken a new turn. Previously he had only attacked the present Vietnamese leaders for their "anti-Chinese policy," something that Ho Chi Minh never allowed, according to Hoan (although there is powerful evidence that Ho was quite wary of the Chinese). But now Hoan has explicitly praised Ho's policy of cooperating with *both* the Soviet Union and China: "President Ho stood for unity with the Soviet Union and China, but Le Duan has entered into 'all round cooperation' with the Soviet Union and has gone all out to oppose China with all the manpower and material resources at his disposal."[27]

For connoisseurs of irony this statement must rank as one of the finest pieties on record, spoken as if the last seven years had simply ceased to exist. If one thing is well known and well documented about Vietnamese foreign policy after the liberation of the South in 1975, it is that Hanoi attempted to retain Ho's balance between China and the Soviet Union. Immediately after the fall of Saigon, Le Duan went to Beijing and Moscow, and to the displeasure of each he publicly praised the aid and assistance of the other in both capitals. The Vietnamese resisted early efforts by the Russians to draw them into an anti-Chinese alliance, just as they turned down Chinese blandishments to become part of an anti-Soviet bloc. It was for this that the Chinese "punished" Vietnam in the first place, forcing them into reliance on the Soviet Union.

The factor that upset Vietnam's balancing act was Cambodia. Pol Pot was lavishly welcomed in Beijing in September 1977, while a Khmer Rouge mass raid took place secretly over the Vietnam border. Le Duan also went to the Chinese capital but was unable to persuade the Chinese to desist in their support for the Khmer Rouge regime. In December the Vietnamese launched their own invasion across the border, to demonstrate their determination, keeping it secret in the hope of preventing a public and lasting breach. According to China's own account, the Vietnamese then "asked China to exert 'influence' on the Kampuchean Government and force it to succumb to the pressure from the Vietnamese aggressors. China flatly refused. China suggested instead that Vietnam withdraw its troops from Kampuchea and then negotiate a settlement."[28] This was what the Vietnamese then attempted to do.

The Chinese did more than refuse to aid the Vietnamese. They encouraged the Pol Pot regime to announce that fighting was taking place inside Cambodia's borders and to sever diplomatic relations with Hanoi. The representatives of Democratic Kampuchea accordingly held a press conference in Beijing at the very end of 1977 to break the news, apparently taking the Vietnamese by surprise.

Vietnam then withdrew its forces and considered its position. On February 5, 1978, it proposed a peaceful negotiation of the dispute: mutual withdrawals from the frontier between Vietnam and Cambodia to create a demilitarized zone, agreement about the small differences on the delimitation of the frontier itself, and an international control commission to inspect any infringements. This offer, as was noted at the time, was addressed as much to Beijing as to Phnom Penh.

We should consider its meaning for a moment, as everything that has happened in Indochina since then really turned upon the events of those few months—from mid-1977 to mid-1978—a period of intense American activity as well. What would have happened if the Pol Pot regime had accepted Vietnam's February proposal as a basis for negotiations? There is evidence that the Chinese were becoming alarmed at the prospect of uncontrolled Khmer Rouge escalation. They might have pressed harder, or more successfully, for Pol Pot to agree to enter discussions. Hence the question, what would Hanoi have done if Phnom Penh had then nominated China as one of the parties to the international supervision? In 1980, I posed this to Hoang Tung, in the editorial offices of *Nhan Dan*. He replied:

> We have had a lot of experience of international supervision, for example after Geneva. There is the matter of having one country completely on your side, one on their side, and one acceptable to both. Then Poland was on one side, Canada was for the French, and India was acceptable to both sides, with Nehru. Pol Pot could have suggested China. Then we would have suggested the USSR. It is more difficult to find a small one acceptable to both, perhaps Yugoslavia.[29]

What the Vietnamese were offering in February 1978, it seems, was the following: that China could have its special relationship with Cambodia under Pol Pot, to the exclusion of any Vietnamese influence in that country, provided that it granted Vietnam's right to follow its own independent course, one that kept a balance between Beijing and Moscow and retained good relations with the West and Third World countries as well—in short, the Ho Chi Minh approach mentioned by Hoang Van Hoan. It was the Vietnamese who sought, it seems almost desperately, to retain a multilateral communism, and it was the Chinese who finally made this impossible through their endorsement of Pol Pot.

Perhaps we should not exaggerate the Chinese role. The Chinese leadership was divided and the main objective of the Deng group was to consolidate positive relations with the United States in order to further China's modernization. If the American price for this was a bellicose attitude toward Hanoi and Moscow, that was a price that could be paid. Secondly, the Pol Pot regime itself played its own autonomous part. Just as the Vietnamese had shown that it was possible for a small communist state to escalate a conflict and thereby oblige its larger allies to support it, however reluctantly, so it is possible that Pol Pot defied China's advice and continued with the fighting. Pol Pot's calculation was presumably that if he forced a choice upon Beijing, to support either him or the Vietnamese, then China would commit itself to his side rather than line up with a Soviet ally. If so, Beijing was trapped by its "general line."

Today that line is changing, but the peoples of Indochina have paid a fantastic cost for China's folly, Cambodians especially, while all that Vietnam gained for trying to keep a balanced course after 1975 was a general scorn for their naiveté followed by the opprobrium of being invaders. If Hanoi was right after all in its attempt to sustain good relations with both the Soviet Union and China, it will now be made to feel all the more wretched because it is unable to welcome a belated Chinese conversion to the Ho Chi Minh policy of a balanced relationship with Moscow. For the recent years of blood and bitterness will hardly be forgotten lightly. Indeed the more cynically they are ignored, as if China has nothing to apologize for, the more evident it is that little has really changed, and that for Beijing the aspirations of countries in Southeast Asia are to be manipulated in its own great-power interest.

Thus Chinese concessions may intensify Vietnamese resistance to a rapid settlement in Cambodia. The most obvious catalyst for a resolution of Sino-Vietnamese differences is Sihanouk. So perhaps it will be best to conclude this section with a consideration of his role. In 1982 Cambodian Foreign Minister Hun Sen issued a statement that in effect invited Sihanouk to return to Cambodia as a private citizen.[30] The Khmer and Vietnamese authorities were, and possibly still are, willing to envisage Sihanouk's return, but on their terms, to grace the Heng Samrin regime with his presence during the final years of his retirement. This, however, is not yet a fate to which the one-time ruler has reconciled himself. Instead, he has justified his adherence to the Khmer Rouge coalition, for example, as follows: "One journalist asked me why I chose to be with the Khmer Rouge now, because the Khmer Rouge had killed my children, my grandchildren, and many, almost two million of my people. So I told him: to be killed, to lose life, yes it is very serious, very sad, but to lose our national identity as Cambodia, to lose our national independence, it is terrible."[31]

At present the Khmer are not being killed, except through the actions of the coalition, and the Vietnamese are far from liquidating Cambodia's "national identity." For Sihanouk, however, he personally *is* the country's national identity; it has no real existence apart from him. He really does regard the Khmer as his "children"—he called them that for thirty years after he became king at the age of eighteen—and naturally, he believes that the family cannot regain its identity until its "father" has returned to take up his rightful position at its head.

Not that this means that Sihanouk is serious about the coalition. He is well aware that the Pol Pot forces will never return to power in Phnom Penh, and his contempt for Son Sann, who speaks Khmer with "a Vietnamese accent," is well known. He is using the coalition as a means of keeping himself in play. His object is to leave it behind, in the Thai border regions, while he himself is sprung back to power with the help of Chinese and, perhaps, Soviet diplomacy. The Vietnamese once described Sihanouk as being a ''cheap cosmetic'' on the face of the Pol Pot coalition, to prettify it for international consumption. But even the Vietnamese will admit that in any eventual, and for them also beneficial, settlement with Beijing, the Chinese will have to be offered some face-saving device. What could be better to save Beijing's face than a cosmetic? This seems to be the line of Sihanouk's reasoning.

But Sihanouk's return could also be destabilizing for the Heng Samrin administration, especially if the old king (then ''prince'') has influential Soviet and Chinese patrons. He is a past master at setting officials off against each other: the whisper in an ear that so-and-so is a Vietnamese stooge, that this reward awaits that good turn; the calculated interview with a Western correspondent denouncing Hanoi's influence. The Vietnamese will neither desire such an outcome nor feel, after all the fighting, that they should concede to it. Hanoi almost certainly wants time to consolidate the regime in place today in Phnom Penh. It desires a settlement with China, but in the years rather than the months ahead.

Soviet-Vietnamese Relations

How are the Russians and the Vietnamese responding to the shifts in the Chinese position? One sign of the tension in their response can be seen in the interview Le Duc Tho gave to Novosti Press, the Soviet news agency, on January 26, 1983. Although Tho is not the formal head of the Communist Party in Vietnam, Le Duan's illness and the collective structure of the Politburo leadership in Hanoi seem to have allowed him to become perhaps the most influential member of the Politburo (if still one who can be outvoted). There can be little doubt that Tho shares the general hostility of the leadership toward Beijing. The usual tone of Vietnamese comments on Chinese policy is indeed bitter and strident—after the traumatic invasion of the North the intricate politeness sustained with such concentration from 1954 to 1979 has, naturally enough, been abandoned. Thus such statements as this by Le Duan on the international situation are par for the course:

> In Southeast Asia, the Chinese expansionists in collusion with the U.S. imperialists and other reactionaries are frenziedly conducting a very wicked and intensive war of sabotage in many fields against the Vietnamese, Kampuchean, and Laotian peoples. They hope to weaken and eventually to annex Vietnam and destroy the banner of national independence and socialism upheld by Vietnam. This strategic scheme of the Chinese reactionaries has not changed an iota.[32]

This emphasis has, if anything, taken on added urgency as the Chinese position softens. "Modernized Maoism is the most dangerous of all," Hanoi has warned as it assaults China's "campaign of good will."[33] Yet this was not at all the note struck by Le Duc Tho in his Novosti interview, even though his influence is not nugatory on the standard Vietnamese denunciations. Asked about his country's diplomacy "to consolidate peace in Southeast Asia," Tho replied with restraint, rather than the usual vehemence, to his Soviet interlocutor:

> The Vietnamese people are determined to defend their independence and sovereignty, to maintain the close solidarity with the fraternal Lao and Kampuchean peoples to defeat the Chinese authorities' expansionist and hostile policy, still hoping to reestablish their long-standing friendship with the Chinese people. Vietnam has put forward just and reasonable proposals for moving toward the normalization of relations between Vietnam and China. But we have not yet had any response from the Chinese side.

Earlier in the interview Le Duc Tho at least managed to describe his country's victories over Pol Pot's and Beijing's attacks as "the blocking (at least until now) of the shady maneuvers of Chinese expansionism and hegemonism." The parenthesis can be read as a neat warning aimed at Soviet inclination to strike an agreement with Beijing.[34]

Geyder Aliyev, who is apparently a rising influence in the Soviet Politburo, visited Hanoi at the end of October 1983. The joint statement issued on November 4 by his delegation and that headed by Pham Van Dong provided an interesting barometer of Sino-Vietnamese relations. Both sides endorsed the People's Republic of Kampuchea as the only legal representative of the Cambodian people. Vietnam declared its full support for the "Soviet Union's principled line aimed at normalizing its relations with the People's Republic of China." Whereupon the statement immediately continued, "The two sides have exchanged views on numerous issues related to the situation in Southeast Asia." Where views are exchanged, they are not identical.[35]

The statement declared that the Soviet Union "totally supports" Vietnam, "in its efforts to normalize its relations with the People's Republic of China on the basis of the five principles of peaceful coexistence, in the interests of the two peoples and of peace in Asia." The Chinese, of course, also regard "the five principles of peaceful coexistence" as the basis for any settlement. What is significant in this statement is the stress upon the interests of the two peoples, something that is open to Moscow's own interpretation, as is "the interests of peace in Southeast Asia." If Hanoi is concerned to delay a settlement with China, there seems little doubt that Moscow favors concessions that might lead to a rapid improvement in its relations. In particular, the establishment of party-to-party contacts would intensify pressure on the Vietnamese.

At the time of writing, the fourth round of talks between Russia and China is due to start in Moscow. Wan Li had attended the Andropov funeral there, where he had talks with Aliyev. Both stressed the possibility of improved relations as the result of "joint efforts."[36] The fluidity created by the renewed leadership crisis in Moscow makes a dramatic breakthrough unlikely. Nonetheless, just as with China's relations with Vietnam, so also in Vietnam's relations with the USSR, there seems to be a steady pull toward an eventual settlement in the course of the decade of differences between all three communist states.

The Soviet Union, China, and Vietnam could agree to disagree about their varieties of communism if they wish; to recognize their general backwardness; to cooperate as countries in improving their standards of life. Hanoi would be more than nervous about such an understanding; it wants its independence underwritten by the West, first of all. But any formalization of their pluralism by the communist countries that was not an attempt to construct a geopolitical alliance would strengthen their hands separately as well as together.

Were this to result it would regularize the situation in Cambodia. The consequences for Cambodia would be threefold. First, if least important, the United States and Thailand would be left looking after Pol Pot. This is (were it to happen) as it should be, given their repeated and quite unconvincing assurances that they have "nothing" to do with him or his cohorts. (The Chinese hold Democratic Kampuchea by the hand, while the U.S. observers walk alongside, making sure, as they put it, to keep at "arms-length." It seems, therefore, that the Americans have "clean" hands, and by comparison with the Chinese they do. But should the Chinese cross over the road, it will be immediately evident who else has been walking out with the Khmer Rouge.) Second, a confirmation that Cambodia's fate is not going to be that of a buffer state within which East and West eternally contend for influence, rather that it is to remain within a bloc linked in the first instance in Hanoi. As such, however liberal its regime might be, it will remain in the communist sphere. Third, that such a regularization would allow multiple economic relations to be established by the regimes in Hanoi and Phnom Penh alike.

Although this scenario could certainly be interpreted as regularizing a Vietnamese "victory," it would be due in the main to the original foolishness of Chinese policy in the last years of Mao; a foolishness for which, however, the Vietnamese have paid at least as heavily as China. Furthermore, any such outcome would necessitate significant concessions from Hanoi. It cannot hope totally to banish all Chinese influence from Indochina and ASEAN alike. Above all, such a result would finally mean that the United States had lost the war, and that the inhabitants of Vietnam, Cambodia, and Laos could finally begin to live in peace.

Even to suggest such an outcome might be construed, in some quarters at least, as an apology for Stalinism. The problem in sustaining such a criticism, however, is the ethical bankruptcy of Western diplomacy with respect to Cambodia. Earlier I compared the legitimacy of the Vietnamese occupation of Cambodia with that of the allies in Germany after 1945. The point can be made more emphatically still. It is said that the Vietnamese should negotiate with ASEAN (and listen to the voice of Washington) on the terms of the UN resolutions over Cambodia. The United Nations, however, recognizes the cadres of the Pol Pot regime among its proper members. How, then, can any negotiations be conducted under its aegis? If the wartime allies were to sit down and discuss their withdrawal from central Europe, that might be fine, but who would advocate or expect them to do so with a German delegation that included a strong Nazi secretariat that also happened to control the main body of their delegation's own armed forces? The question of the genocidal policies of the past would rule this out, as would the catastrophe that the war itself had visited. Doubtless there could be an element of hypocrisy involved—after all we are discussing international affairs. But the fact that the moral record of the allies was shameful prior to 1945 and that many a crocodile tear was shed after does not alter the way that the crimes of genocide have now become irrevocable public knowledge, with all the associated political charge. For similar reasons the West—and ASEAN—should really cease to have anything whatsoever to do with the Pol Pot regime. Were this to happen, then indeed Western voices might have a greater say in any final Indochina settlement. It would mean that Washington had accepted the legitimacy of Vietnam's victory in its struggle for national independence under the flag of socialism. Until that happens the United States deserves to have no say whatever in the final destiny of Cambodia, for it was the United States, above all others, that forged the present tragic unity of Cambodia with Vietnam in the terrible fire of its intervention.

Notes

1. Videoaddress to the conference on "Kampuchea in the 1980s: Prospects and Problems," Princeton University, November 12–14, 1982.
2. *Far Eastern Economic Review* (*FEER*), December 15, 1983.
3. *Beijing Review*, January 23, 1984.
4. Paul Quinn-Judge, *FEER*, November 19, 1982.
5. Apart from the fact that no imperialist country was ever attacked on its own soil by the countries it colonialized, the other obvious reason why the term does not apply is that under colonialism sovereignty is removed, whereas the Vietnamese have gone out of their way to install the appearances of Khmer national sovereignty. Thus, despite many vivid observations and the telling description of heavy-handed Vietnamese indoctrination efforts, Elizabeth Becker's reports from Phnom Penh are mistaken in categorizing developments there as "a politics of colonialization"; *Washington Post*, February 27 and 28, 1983.
6. In its conclusion to its report on the "big success" of Premier Zhao's U.S. visit, *Beijing Review* (January 23, 1984) quotes the *Wall Sreet Journal*, which opined that the most useful outcome for President Reagan of Zhao's trip was that it would assure him a

friendly spring reception in China, which "in an election year is good politics."

7. Becker was surely right to quote "a foreign expert" in Hanoi who said that "theft is one of the biggest problems in Soviet projects," and not just Soviet. The inability of the Vietnamese effectively to utilize aid and credits is today probably a bigger problem than getting such help. The Russians are justified in complaining for their part, therefore, about the waste of resources, at least as much as the Vietnamese are about the inadequacy of assistance. At the same time it should be noted that in addition to decisive military supplies, the Soviet Black River Dam project has just completed its first stage with the successful damming of the river. When the huge project is completed, the Hao Binh hydraulic center will generate nearly 2 million kw—the largest dam of its kind in Southeast Asia. Nobody else would have helped the Vietnamese to build such a project.

8. Jay Taylor, *China and Southeast Asia: Peking's Relations with Revolutionary Movements* (New York: Praeger, 1974), p. 29. Taylor's book is an essential and underrated source for the recent history of China's diplomacy in Southeast Asia.

9. See Z. Brzezinski's *Power and Principle* (New York, 1983), pp. 409–10.

10. Anthony Barnett, "A Meeting of Giants," *New Statesman*, December 16, 1983, in which I also discuss some of the same issues.

11. *Beijing Review*, January 30, 1984.

12. John Gittings looked at what he termed China's "surprising reluctance" to use the declaration of martial law as a whip against Moscow, in *The Guardian*, London, January 22, 1982. There have been a number of reports on the rapprochement, all necessarily tentative in nature. For one that gives a clear picture of Russia's cautious optimism, see Jonathan Steele, "Peking on the Long Road to New Detente," *The Guardian*, London, March 22, 1983. Alan Jacob has suggested that Moscow may concede that China has a special "interest" in Indochina, if Peking grants the Soviet Union its "interest" in Afghanistan, *Le Monde*, November 19, 1982.

13. Juan Cameron, "What the Bankers Did to Poland," *Fortune*, September 22, 1980.

14. It is remarkable how rapid this change has been. In 1981 a series of interviews with Chinese officials by Henry Kamm revealed a unanimity of attitudes similar to Dulles's containment mentality. The Chinese believed that protracted war was necessary in Cambodia and that world pressure would bring down pro-Soviet regimes in Hanoi and elsewhere. *New York Times*, April 21, 1981.

15. *Beijing Review*, January 16, 1984.

16. See, among others, Roy Medvedev, "The USSR and China: Confrontation or Detente?" *New Left Review* (November-December 1983).

17. In both cases the Soviet Union's armed forces partially mobilized and prepared to attack, but stood down. With regard to the strange goings on when the Chinese invasion ground to a halt in Vietnam, the Chinese threatened to extend the conflict, and some sharp Soviet ultimatums were delivered, see my brief discussion in the *New Statesman*, April 6, 1979.

18. *The Economist*, March 12 and August 13, 1983, which states that China is going to build six nuclear submarines.

19. But, for the structural problems that impede any such relationship between one-party regimes, see my discussion in the *New Statesman*, December 16, 1983, and also Fred Halliday on the "involution" of the societies concerned, in *The Making of the Second Cold War* (London and New York, 1983).

20. "Foreign Ministry Statement," March 1, 1983, in *Beijing Review*, March 7, 1983.

21. "China's Deceitfulness Exposed," a VNA report of March 3, 1983, recounting a *Nhan Dan* editorial response to the Chinese statement of the 1st.

22. "Fair and Reasonable Answers of Vietnam to the So-Called 'China's Stand on the Kampuchean Problem,'" in the press release of the embassy of the Socialist Republic of Vietnam, London, February 25, 1983.

23. For a map of these dispositions and a brief discussion of the significance of the issue, see my piece on "The Chinese Invasion," in *Aftermath*, John Pilger and Anthony Barnett, New Statesman Report No. 5 (London, 1982).

24. The fighting at the border in April 1983 seems to have been "overstated" by both sides, according to reporters who visited the area subsequently; Mark Baker, *Melbourne Age*, July 4, 1983. I am grateful to Ben Kiernan for sending me a copy of this report.

25. The Chinese statement on the islands was issued on April 25, 1983. See the analysis by John Gittings, *The Guardian*, London, May 9, 1983. *Nhan Dan* published its reposte on May 12. For their part the Chinese authorities may have felt annoyed by Vietnam's "Declaration on the Baseline of Vietnam's Territorial Waters" of November 12, 1982; see *Vietnam Courier*, Hanoi, December 1982. It may be especially important for the present discussion to note that Moscow appears to have refused to extend the defense commitment it made to Vietnam after 1977 to the disputed islands; Gareth Porter, "Vietnam's Soviet Alliance: A Challenge to US Policy," *Indochina Issues*, Washington, no. 6 (May 1980).

26. Pei Monong, *Beijing Review*, April 18, 1983.

27. "Letter to the Vietnamese Compatriots" by Hoang Van Hoan, *Beijing Review*, July 25, 1983.

28. "Facts About Sino-Vietnamese Relations (2): Origin and Development of Present Differences," *Beijing Review*, October 19, 1981, p. 27.

29. Interview with Hoang Tung, Hanoi, February 24, 1980.

30. Hun Sen, interview with the SPK, September 18, 1982. But while this statement was indirect, Hun Sen used a more recent press conference to launch into an angry denunciation of Sihanouk; Paul Quinn-Judge, "Adding up the Sihanouk Solution," *FEER*, May 26, 1983

31. Interview with Bruce Palling, Bangkok, January 1983, later shown in his television film on Cambodia, "The Eagle, the Dragon, the Bear, and Kampuchea," Channel 4, London, April 28, 1983.

32. Press release, embassy of the SRV, London, July 1, 1983.

33. Jacques de Barrin, *Le Monde*, December 9, 1982.

34. Press release, embassy of the SRV, London, February 10, 1983.

35. *Vietnam Courier*, December 1983; see also the discussion of Aliyev's visit by Nayan Chanda, *FEER*, November 17, 1983.

36. *Daily Telegraph*, February 16, 1984.

David R. Hawk

INTERNATIONAL HUMAN RIGHTS LAW AND DEMOCRATIC KAMPUCHEA

Cambodia in the 1980s is a particularly unfortunate product of its past. The challenges it faces—peace, self-determination, and economic, social, and political reconstruction and redevelopment—must be viewed in the light of the preceding decade of foreign intervention, civil war, genocide, and famine.

The spillover of the Second Indochina War into Cambodia shattered traditional society. Thus it was in an already broken land that the Khmer Rouge, after their defeat of Lon Nol in 1975, perpetrated one of the most violent and massive atrocities of the twentieth century: the substantial destruction of a people by their own leaders. Previous examples of the attempted destruction of a people have involved the persecution of some nation, race, or tribe by another. The possible uniqueness of the Khmer Rouge period will require careful examination, analysis, and explanation, though the documentation for such an undertaking is not yet available. In any case, the Cambodian tragedy must not be lost to history, as was, for example, the genocide of the Armenians during the first part of this century.

I propose to document the events occurring between April 1975 and January 1979 in accordance with the standards on international human rights, a body of laws and institutions established after World War II in the hopes of deterring the reoccurrence of such atrocities as had taken place in the 1930s and 1940s.[1] Such standards not only can provide a framework for analyses by historians, political scientists, and sociologists, they also set norms for governmental conduct and are supported by law. As a member state of the United Nations, and as an accessory to some of the relevant international human rights treaties, Cambodia under various regimes has opened itself to international scrutiny.[2]

In this context the Khmer Rouge engaged in wholesale violations of international standards of human rights; but the laws, systems, and institutions designed to ensure those rights failed not just to protect the Cambodian people but even to mount a significant protest on their behalf. Unlike Chile or Argentina, there has been no international insistence that the perpetrators be held accountable for their crimes. There have been no Nuremberg Trials for the Khmer Rouge. Not only has the UN General Assembly failed to condemn Cambodia, the Democratic Kampuchean representative has continued to hold his seat since the Khmer Rouge were ousted from power in 1979.[3] This is not simply a historical question; unlike

Uganda's Idi Amin, Pol Pot and his closest associates remain at the center of vital questions relating to Cambodia's future. Politically and militarily, those responsible for the Khmer Rouge crimes against humanity are very much part of the ongoing tragedy and present predicament of the Cambodian people.

The continuing, active presence of the Khmer Rouge leadership group is the source of the tacit acceptance by the Khmer people in Cambodia of Vietnam's occupation insofar as it prevents Pol Pot's return to power. It is a central, though not the only, impediment to a negotiated solution to the international conflict in and over Cambodia, which in turn is necessary for social and economic reconstruction and development and a "durable solution" to the continuing refugee crisis on the Thai-Cambodian border. As the 1981 Human Rights Subcommission analysis noted, "the horrible genocide perpetrated by the Khmer Rouge between 1975 and 1978 . . . lies at the root of everything that has happened in Kampuchea during the last five years."[4]

The Norms, Standards, and Binding Obligations of International Human Rights

The major statement of human rights norms since World War II is the Universal Declaration of Human Rights, proclaimed without opposition[5] by the General Assembly of the United Nations in 1948 as "the common standard of achievement for all peoples and all nations."[6] The Universal Declaration upholds the right to life; security of person; freedom of movement, opinion, belief, expression, association; education; social security; participation in the cultural life of the community; equal protection of the law; etc. It prohibits enslavement, torture, arbitrary arrest, arbitrary interference with privacy of the family, etc.

Inasmuch as UN General Assembly resolutions are declaratory rather than mandatory, the Universal Declaration itself has not been enforceable as law. But by ratifying or acceding to a particular treaty or convention, a nation-state agrees, as it were, to be "legally bound" by its provisions and is obligated to the other states that have ratified the same convention. In the words of one legal expert, "Any state that adheres to an international human rights agreement has made the subject of that agreement a matter of international concern. It has submitted its performance to scrutiny and to appropriate, peaceful reaction by the other parties, and to any special procedures or machinery provided by the agreement for its implementation."[7] The conventions provide standards with which ratifying nations can bring their own legal codes into conformity. International human rights law is often an essential part of the more complex process of ameliorative pressure against violations.

The International Covenant on Civil and Political Rights and the International Covenant on Economic, Social, and Cultural Rights translate the articles of the Universal Declaration of Human Rights into the language of international law. Additionally, there are regional and particular human rights conventions, violations of which are illegal under international law. These include conventions against genocide, slavery, racial discrimination, and apartheid and conventions to define and promote the protection of refugees, religious groups, and women. The two international human rights instruments most relevant to Democratic Kampuchea are the Universal Declaration of Human Rights and the Convention on the Prevention and Punishment of the Crime of Genocide.[8] The former is the general standard of reference for all states and peoples. Democratic Kampuchea, still recognized by the United Nations as the legitimate governmental authority, ratified the International Covenants on Civil and Political, and Economic, Social, and Cultural Rights.[9] The Genocide Convention is relevant because Cambodia had long been a "state's-party" to it and because it is the human rights convention that most explicitly deals with particular aspects of the phenomenon of mass murder by governments outside of a wartime situation.

Human rights violations range from relatively minor miscarriages of justice to wholesale violations of such an extreme character that the physical existence of entire groups is threatened. Democratic Kampuchea is at the latter end of the continuum. It should not be assumed, however, that all human rights violations occurred with regularity in every locale or that everyone who suffered did so equally. There unquestionably were variations as to time and place, and certain categories of people clearly had it much worse than others.[10] But those whose treatment was absolutely bad amounted to a substantial percentage of the population. Furthermore, some violations—such as the strict prohibition against religious observance or the denial of family rights by forced communalization of eating—afflicted virtually everyone. It seems, nonetheless, that there were district and regional sections of the Khmer Rouge that resisted the thorough implementation of the most extreme Khmer Rouge policies. But some such areas (e.g., the Eastern Region), where conditions were initially better, later became the site of perhaps the worst and most widespread massacres perpetrated by the regime.[11] Khmer Rouge dissenters were brutally subjugated by soldiers and cadre loyal to the Cambodian Communist Party policy-making center. When the evidence is in, it is probable that the continuities will outweigh the discontinuities, and that the consistencies in policy and practice will exceed the significance of variations of place.

Refugee accounts are one important source of information, especially those from the 1979–80 group, which largely corroborate, but are more detailed than, earlier testimony. Of particular note are the horrifying pictures painted by children in the refugee camps. With the help of psychologists, children too terrified and disoriented to put their experiences in words objectified their trauma by drawing. There is no appreciable difference between these refugee accounts and either testimony from Cambodians inside the country who have told their stories to Khmer-speaking scholars, journalists, or relief workers or the accounts gathered in the records of the Phnom Penh Tribunal of August 1979.[12] Other documentary evidence includes the extraordinary archives of the S-21 Khmer Rouge extermination center in Phnom Penh, left behind when its officials fled the Vietnamese. The S-21, or Tuol Sleng, documents—entire records, photographs, prisoner confessions, execution lists, etc.—are a chilling microcosm of the political pathology of the Democratic Kampuchean regime.

There is additional evidence of the Khmer Rouge attempt to extirpate the cultural heritage of Cambodia: destroyed Buddhist pagodas, smashed statues and icons, and ransacked or destroyed libraries. Finally there are the mass grave sites each containing the remains of hundreds, sometimes thousands, of execution victims, whose skeletal remains are sometimes blindfolded and bound.[13]

The Universal Declaration of Human Rights and Violations by Democratic Kampuchea

In 1978 the governments of Australia, Canada, Great Britain, Norway, and the United States, along with Amnesty International and the International Commission of Jurists entered submissions documenting violations by Democratic Kampuchea to the United Nations Human Rights Commission.[14] The allegations were analyzed by the chairman of the Subcommission on the Prevention of Discrimination and the Protection of Minorities according to the articles of the Universal Declaration of Human Rights.

The subcommission report begins by summarizing the events that were reported to have taken place in Cambodia subsequent to the assumption of power by the government of Democratic Kampuchea on April 17, 1975:

> 1. The forceable and precipitant deportation immediately after April 17, 1975, from Phnom Penh and other urban centers of the country of all residents, totalling an estimated three to four million persons—without regard to age or physical condition.
>
> 2. The compulsory resettlement of the population of the country in rural areas, frequently uninhabited, and the organization of the population into collective work brigades.
>
> 3. The imposition of a draconian discipline upon the entire population with respect to their work as well as private conduct, and of strict controls over their freedom of movement.
>
> 4. The launching of a systematic program aimed at the physical elimination of various categories of persons formerly associated with the previous regime or belonging to higher social or educational categories.
>
> 5. The launching of a sustained program designed to destroy traditional religious and family life and previous economic or social values and practices.[15]

Reviewing the allegations of violations of the Universal Declaration of Human Rights, the report continues:

> *Article III. Everyone has the right to life, liberty, and the security of person.*
>
> (i) The forceable and precipitant evacuation of the population of Phnom Penh and other cities and towns ordered by the Kampuchea authorities immediately upon assuming power, in the absence of adequate arrangements by the authorities to provide food, water, and medical care to the evacuees while in transit or at their destinations, caused the loss of many lives, particularly among the aged and the young and among the many sick and wounded persons who did not endure the rigors of the evacuation;
>
> (ii) Within the first few days after the assumption of power by the Kampuchean authorities, a large number of former military officers, senior officials, policemen, intelligence agents, country officials, and military police were executed in various parts of the country as part of a systematic campaign of extermination, and in a very large number of cases the wives and children of such categories of persons were also executed;
>
> (iii) Many persons belonging to such categories who had initially succeeded in concealing their identities or former occupations were subsequently systematically sought out and were also executed, especially since early 1976;
>
> (iv) That although the treatment of lower level personnel associated with the previous regime, such as minor officials, noncommissioned officers, soldiers, headmen, and members of paramilitary units appears to have been different from region to region, many personnel in such categories were also executed immediately following the takeover or subsequently, particularly since 1976;
>
> (v) So-called ''intellectuals'' such as doctors, engineers, professors, teachers, and students have also been summarily executed, especially since 1977;
>
> (vi) Many ordinary persons have died as a result of being forced to perform exhausting manual labor, under a strict regime, without being provided with sufficient food, rest, or medical care;
>
> (vii) Many ordinary persons whose attitudes had not been deemed satisfactory by the new authorities, or who had committed minor infractions (such as being late for work,

losing their tools, etc.) have also been frequently punished by execution after one or two verbal admonishments;

(viii) During 1977 and 1978 large numbers of Kampuchean administrative and military personnel of various levels and their families were also executed in a series of internal purges, and that in a number of instances even the villagers and peasants who had been working under the authority of such purged officials were also executed.

Article V: No one shall be subjected to torture or to cruel, inhuman, or degrading treatment and punishment.

(i) The population at large has been subjected to degrading treatment through being forced to live and work under an extremely harsh regime of discipline and fear of summary punishment;

(ii) Numerous persons have been subjected to summary execution through cruel and barbaric methods including disembowelment, pole axing, and beating to death, etc.

(iii) Prisoners have been kept bound and chained for long periods or tortured.

Articles VI through XI (concerning recognition of persons before the law, entitlement to equal protection of the law, right to effective remedy, freedom from arbitrary arrest, detention, or exile, right to a fair and public hearing, right to the presumption of innocence, etc.)

The constitution of Democratic Kampuchea contains two brief articles concerning the administration of justice, appointment of judges, and forms of punishment. In fact, however, there was a total absence of judicial process and punishment of offenses tended to be entirely of a summary character. DK Prime Minister Pol Pot is quoted from September 1977 to the effect that "reactionary elements" who constituted "one or two percent of the population" and who are regarded as "enemies" should be "educated," "neutralized," or "irradicated."[16] The report also refers to the Canadian government's assertion that "the majority of the population of Democratic Kampuchea are subject to an arbitrary system of justice . . . in addition to crimes normally subject to sanction, . . . punishable crimes in Democratic Kampuchea include lateness for work, complaints about work, complaints about lack of food, requests for improvement in living conditions and having been an official or soldier in the former government. The punishment for these kinds of offences is frequently death."

Article XII: No one shall be subjected to arbitrary interference with his privacy, family, home, or correspondence, nor to attacks upon his honor and reputation.

Everyone has the right to protection of the law against such interferences or attacks.

The imposition of a communal system of life and a system of surveillance by informers (*chlops*) in the villages created an atmosphere of suspicion and fear and destroyed any semblance of privacy. Refugees also describe work brigades organized on the basis of sex and age that often entailed separation of family members for long periods, or even permanently. These kinds of complaints were voiced most frequently by refugees coming from the northwestern part of Democratic Kampuchea; there are indications that in some of the southern and eastern provinces the infringement of these rights was less widespread.

Article XIII: (1) Everyone has the right to freedom of movement and residence within the borders of each state; (2) Everyone has the right to leave any country including his own, and to return to his country.

The movement of the population within Democratic Kampuchea was strictly controlled by the authorities; the population did not enjoy the right to leave the country. Refugee accounts contain numerous descriptions of persons being summarily executed when found to be travelling without specific authorization or attempting to leave the country.

Article XVI: (concerning the right to marry; the free and full consent of the intending spouses; the entitlement of the family unit to protection by society and the state).

The accounts allege that the right to marry was consistently violated; several refugees have claimed that prior to 1976 marriages were prohibited altogether in the communities where they lived. Specific allegations cite the strict prohibition of courting, the need to obtain permission to marry and the arbitrary handling of such applications, and the infringement of the right of free choice on the part of females, particularly in cases where local officials or members of the Khmer Rouge wish to marry them. The separations arising from the establishment of work forces according to sex and age in certain parts of the country, and the practice of separating children from their parents, are also cited in some refugee accounts as infringements of the right of the family to protection by the state.

Article XVIII: (concerning the right of freedom of thought, conscience, and religion, etc.).

The materials contain numerous allegations that Buddhism—the former state religion practiced by an estimated 85 percent of the population—as well as the practice of other religions such as Islam and Christianity was almost completely suppressed. According to the materials, a number of Buddhist leaders were executed immediately after the victory of the revolutionary forces and many thousands of Buddhist monks were turned out of their pagodas to work in the rice fields or on public works projects along with the rest of the population. Many refugees report displays of strong hostility on the part of authorities to religious belief or observance of any kind, and the destruction of religious buildings, symbols, or artifacts.

Article XIX: (concerning the right to freedom of opinion and expression, the right to seek, receive, and impart information, etc.).

Refugee accounts consistently allege that the right to freedom of opinion or expression was systematically and absolutely suppressed throughout Democratic Kampuchea. Numerous instances of summary execution of persons who expressed even minor complaints about the conditions of life or who voiced any criticism of government policies are recorded in the materials.

Article XXI: (concerning the right to take part in the government of one's country; the provision for periodic and genuine elections based on universal and equal suffrage, secret balloting, etc.).

The materials as a whole convey the impression that the population played no role in the formulation of government policies, was generally ignorant of such policies, and did not know how policies were formulated or who formulated them.

Articles XXIII through XXV: (including rights to free choice of employment, to just and favorable conditions of work, to rest and leisure including reasonable limitations on working hours, to a standard of living adequate for health and well-being, etc.).

Refugee complaints are particularly frequent regarding the following practices: imposition of excessively heavy work quotas, long working hours, and an exhausting pace of labor; lack of provisions for adequate rest (until 1977 when the system of one day of rest every ten days was generally introduced, the population was apparently obliged to work every day); insufficient food rations; and inadequate or unavailable medical care.

Article XXVI: (concerning the right to education . . . directed to the full development of the human personality and to the strengthening of respect for human rights and fundamental freedoms, etc.).

According to the materials submitted, the educational system that existed in the country prior to April 1975 was abolished and was not replaced by any system of general education. Children from the age of six upward were customarily required to work in the fields or at other tasks and were given little or no opportunity for formal education. To the extent that education existed, children were grouped together for short periods during the day and given rudimentary literacy training or were taught revolutionary songs and slogans.

In short, the report of the chairman of the subcommission, Mr. Bouhdiba of Algeria, indicates that in Democratic Kampuchea there was wholesale violation of the rights enumerated in the Universal Declaration. In his report to the UN Human Rights Commission meeting in March 1979, Mr. Bouhdiba concluded that the situations described in the submissions were "the most serious that had occurred anywhere in the world since Nazism" and constituted "nothing less than autogenocide."[17]

The Genocide Convention and the Policies of Democratic Kampuchea

The international human rights law that most directly relates to the mass murder in Cambodia between 1975 and 1979 is the 1948 Convention on the Prevention and Punishment of the Crime of Genocide, which has been ratified by ninety nations.[18] "Genocide" is an elastic concept, formed in the wake of Hitler's "Final Solution." It has come to include cultural assimilation and atrocities for which more precise but less emotive or condemnatory language was not deemed adequate. The word itself is of recent vintage, coined during the Second World War by Raphael Lemkin, a jurist and scholar whose efforts led directly to the 1948 Convention.[19] Its literal meaning is the killing (Greek, *cide*) of a people, race, or tribe (Latin, *genos*).

The term genocide has frequently been applied to the situation in Cambodia between 1975 and 1979.[20] Yet, when U.S. Senator William Proxmire wrote to a number of leading American international law experts during this period inquiring if Cambodia constituted a case of genocide in the terms of the Genocide Convention, he received, on the basis of what was then known, a mixed and muted response. Since this is the most detailed law dealing with large-scale murder and other phenomena related to the actual or threatened destruction of a group of people, it is important to examine, *on the basis of what is now known*, whether Khmer Rouge policy constituted acts of genocide in the legal sense prescribed in the 1948 Convention.

Article II of the convention specifies that genocide is any of the following acts committed with the intent to destroy in whole or in part a national, ethnic, racial, or religious group, as such: a) killing members of the group; b) causing serious bodily or mental harm to members of the group; c) deliberately inflicting on the group conditions of life calculated to bring about its physical destruction in whole or in part; d) imposing measures intended to prevent births within the group; e) forcibly transferring children of the group to another group. Conspicuous by its absence is any reference to economic, social, and political groups. Included in earlier drafts, they were removed in the process of negotiations between nations in the UN Economic and Social Council (a subsidiary body of the General Assembly).[21] Reading through accounts of the successive drafts and arguments of the *travaux préparatoires* a generation later, it is difficult not to conclude that the strained legalistic arguments were politically motivated. These arguments were acceded to, however, because it was feared that otherwise the many states unwilling to limit their means to suppress political opposition would block overall ratification.[22]

Cambodia's ratification of the Genocide Convention in 1950[23] was never renounced or abrogated by the Democratic Kampuchean regime. Thus there are no legal impediments to the application of its provisions to the situation in Cambodia between 1975 and 1979. But because the convention does not cover political and economic groups (such as former Lon Nol government officials or army officers), the execution of the Tuol Sleng prisoners and the murder and mistreatment of commercial and professional classes cannot technically be regarded, ipso facto, as acts of genocide. These acts are more properly referred to in international human rights and legal terminology as extrajudicial executions—murder by government outside the process of law.

Nonetheless, human destruction on so vast a scale as occurred in Pol Pot's Cambodia did include the destruction of groups, in whole or in part, that *are* specifically covered by the convention. Such is the case with the Cham ethnic minority, which had its social structure eliminated and much of its population killed, and the once dominant Buddhist monkhood, which was virtually destroyed.

The Cham

The Cham are a non-Khmer Islamic minority who once lived in the central coastal region of what is now Vietnam. They were conquered by the Annamese (Vietnamese) in the fifteenth century. From the fifteenth through the eighteenth century the Cham migrated into Cambodia, where they settled along the rivers and Tonle Sap. The Cham concentrated in the province of Kompong Cham and around Phnom Penh and often engaged in fishing trades. Living apart from the Khmer in their own hamlets and communities, they were recognizable by their distinctive dress and language and their Islamic religious practices.

The Khmer Rouge divided the Cambodian population into three categories of rights and obligations, and this categorization allowed no room for ethnic minorities. They instituted a racist, chauvinistic policy of forced assimilation, or "Khmerization," of the Cham (and other ethnic minorities). In decrees circulated to provinces it was declared that:

> There is one Kampuchean revolution. In Kampuchea there is one nation, and one language, the Khmer language.
>
> From now on the various nationalities [listed according to province] do not exist any longer in Kampuchea.
>
> Therefore [Cham] individuals must change their names by taking new ones similar to Khmer names.
>
> The Cham mentality [Cham nationality, the Cham language, Cham costume, Cham habits, Cham religion] are

abolished.

Those who do not abide by this order will reap all consequences.[24]

In many areas Cham communities were broken up and dispersed among the general Khmer population, sometimes two or three families to a commune. Community and religious leaders were executed. Cham women were required to cut their hair short. Men were not allowed to wear their distinctive caps or sarongs. Mosques were destroyed or turned into warehouses. According to the *Hakkim* (religious leader) at Chaing Chamres, a suburb of Phnom Penh, of 113 mosques prior to 1975, only 20 remain. Cham burial grounds were desecrated and Cham who died were defiled by being buried "upside down" (i.e., not facing Mecca). Some Cham were forced to eat pork, forbidden to Moslems, as a test of loyalty, and some who refused were executed.[25]

According to Mat Le, now an official in the Agricultural Ministry in Phnom Penh, numerous Cham communities resisted the forced "Khmerization"—which like other repressive policies was not applied everywhere at once—and rebelled. In response to this resistance, massacres of entire villages ensued.[26] Early escapees reported wholesale slaughter of Cham at Trea and at Kroch Chhmar in Kompong Cham province.[27] Mat Le mentioned massacres at Koh Thas village and Svey Klieng in Kompong Cham.[28] Up to 40,000 are reported to have been killed in the hamlets in the districts of Kong Meas and Kompong Siem in Kompong Cham.[29] A survivor from the Kong Meas district reported to a Japanese journalist that 2,000 people of 500 families of Khach So community, Peam Chikang village, Kompong Cham province were killed in one evening in April 1978.[30] Thousands were thrown into a ravine known as Chros Stung Treng; other Cham massacres reported at the Phnom Penh Tribunal took place in Khlong district, Kratie province, O Russay district, Pursat province, and Kompong Trolek district, Kompong Chhang province. By August 1979 only 60 of 1,200 pre-1975 Cham households had returned to the Chrui Changvar area near Phnom Penh.[31] Mat Le lost his father, wife, three married children and their spouses, three grandchildren, and an unknown number of nephews and nieces.[32] The *Hakkim* of Chaing Chamres, Him Mathot, lost eight of his fourteen family members.[33]

According to Francois Ponchaud, a French priest who lived many years in Cambodia, the Cham had long been "regarded with suspicion by Cambodians because of their cultural and religious separation."[34] Ben Kiernan notes that the Cham, "as an autonomous national group were considered a 'weak point' in the state."[35] As a former Khmer Rouge cadre expressed it, "The Cham are hopeless. They abandoned their country to others. They just shouldered their fishing nets and walked off letting the Vietnamese take over their country."[36] Kiernan also notes that even Cham who had joined the Khmer Rouge revolution and had become minor officials were removed from their positions.[37]

Estimates vary on the number of Cham existing in Cambodia before 1975.

Cham inside Cambodia and in exile claim 700,000, one-tenth of the Cambodian population.[38] This figure is based on a survey of households connected with Cambodian mosques.[39] Thai Moslems figure their Cambodian coreligionists to number approximately half a million.[40] Pre-1975 Khmer accounts regarded the Cham population as much smaller, as low as 150,000, and some earlier texts did not count the Cham at all because they were regarded as foreigners and not Cambodian.[41] Father Ponchaud places the Cham population before 1975 at 250,000.[42]

A 1981 Cambodian census counted 200,000 minority group members, including Cham, after the overthrow of the Khmer Rouge.[43] Even using the 250,000 figure for the pre-1975 population, it is clear that there was a substantial massacre of the Cham. If, on the other hand, the 500–700,000 figures are more accurate, then fully three-fifths to five-sevenths of the Cham did not survive the three and a half years of Khmer Rouge rule. (These figures all include Cham massacred and those who died of a combination of induced, unnecessary starvation, disease, and exhaustion.) This scale of destruction clearly falls under the jurisdiction of paragraphs IIa, IIb, and IIc of the convention, as listed above. Paragraphs IId and IIe may not apply because, although there was a decline in births and transfer of children, there is no specific evidence that this was intended by the Khmer Rouge. The convention, as noted earlier, requires proof of intent. These phenomena among the Cham population may merely have been the consequence of other Khmer Rouge policies such as dispersion of ethnic minorities, communalization, segregation of men and women, and separation of families. In addition, these policies, which had consequences for the Cham birthrate and placement of children, were not directed specifically at the Cham, but at the Cambodian population as a whole.

Even without the actual killing, the treatment of the Cham can be considered genocidal in character because of the dispersion of the group and the prohibition on its language, customs, and religion. And even if the paper record of the policy decree to abolish ethnic minorities did not exist, there would still be evidence of "intent": the acts of dispersal, prohibition, and destruction were carried out too repeatedly and thoroughly to have been random. That the policy was not carried out fully only indicates that there may have been dissenting elements within the Khmer Rouge regional and district structure. But that these policies and practices continued throughout the period of Khmer Rouge rule, even through 1978, does show that this was the intent of the political organization (the *Angkar*), which finally brought most regions and districts under central control.

Other Minority Groups

Less is known about the treatment of other ethnic minority groups. Extrapolations based on 1980 interviews conducted by Stephen Heder, an American Cambodia scholar, with ethnic Chinese in refugee camps in Thailand indicate that as many as half of the urban ethnic Chinese died during the Pol Pot years from murder, induced starvation, disease, or exhaustion from forced labor and marches.[44] This treatment may be more a function of economic class (excluded from the Genocide Convention) than a policy directed toward the ethnic Chinese as a group.

Tens of thousands of ethnic Vietnamese left or were expelled from Cambodia in 1975. The ensuing years, as the relationship between Vietnam and Cambodia deteriorated, saw repeated massacres not only of the remaining ethnic Vietnamese, but also of Cambodians of partial Vietnamese descent. A middle-aged Cambodian peasant, Heng Chan, at Tur-A-Kur village in Pursat province related that his first wife, of Vietnamese descent, was killed one night along with five of their sons, three daughters, three grandchildren, and sixteen members of his wife's family.[45] Massacres of Vietnamese and mixed Khmer-Vietnamese would also seem to qualify as genocidal acts within the terms of the convention.

There were also massacres of ethnic Thai in Cambodia, particularly in the Kos Kong area near the Thai border, where ethnic Thai were actually a majority in some districts. This area had been dominated by left-wing Thai forces during the Lon Nol regime. It is not clear if massacres of ethnic Thai groups were directed against the Thai minority as such or were part of the factional struggles within the Khmer Rouge that took place in numerous areas, irrespective of ethnicity, as the center attempted to extend control.

Even less is known about other ethnic minorities, including the Khmer Leou or "hill tribes"; Shan or Kula, who lived in the Pallin area and once dominated the gem-mining business; Indians; and tribal minorities, long the object of discrimination, who lived in the mountainous regions of Preah Vihear, Strung Treng, Ratanakiri, Mondulkiri, and Kratie provinces. Although some had supported the Khmer Rouge revolution, they were allowed no autonomous existence or protection in revolutionary Cambodia's class-denominated structure. Testimony at the Phnom Penh Tribunal states that some minorities were separated from former dwelling areas and relocated in the plains.[46] Other Cambodian hill tribes suffered relatively less because of their remoteness and the difficult terrain of their habitat. Minority group cadres within the Khmer Rouge were purged in 1977 and 1978. The tribunal in Phnom Penh also noted that as of August 1979, seven months after Pol Pot's ouster, there was no trace of the Kula from the Pallin area,[47] although it is not clear that the Vietnamese and Heng Samrin forces have complete control of that area or that the Khmer Rouge have not dragged these people to the Thai borders as they did others still under their control. A Buddhist monk, formerly of Battambang province, told me that ethnic Indians there were taken away for execution.[48]

The Buddhist Monkhood

The destruction of the Buddhist monkhood constitutes another case of genocide according to the terms and provisions of the Genocide Convention. The Khmer Rouge had a general animus against religion. The Democratic Kampuchea constitution (Article XX) supposedly provided "the right to worship according to any religion," but it "strictly prohibited all reactionary religions that are detrimental to Democratic Kampuchea and the Kampuchean people." In effect, all religions and religious practices were regarded as reactionary and were strictly prohibited.

The Catholic cathedral in central Phnom Penh, presumably a symbol of French colonialism and Vietnamese colonization (most Cambodian Catholics were ethnic Vietnamese), was dismantled stone by stone, including the floors and central foundations. There is now a weed-filled vacant lot next to the central railroad station on Monivong Avenue in Phnom Penh. The religious prohibitions against the Muslim religion have already been discussed. The antireligious animus, however, fell most heavily on Buddhism, which before 1975 had been the established state religion.

Buddhism was in fact much more than the religion sanctioned and supported by the state; it was integral to the traditional Cambodian way of life. While educated urbanites were not immune to the secularizing tendencies of modernization, Cambodians were a highly religious people. Like Roman Catholicism in Poland, Theravada Buddhism in Cambodia embodied and represented Cambodia's tradition, culture, and identity. For centuries it was the main source of learning and education. As one scholar noted, "Buddhism, the village pagoda, and its monks provided continuity in the Khmer state."[49] The Buddhist monkhood (*sangha*) along with the monarchy were the institutional authorities around which national life revolved. It had been the focus of life in the villages, where most Cambodians lived until the U.S. bombing and the civil war drove them into the cities in the early 1970s. In fact, the Cambodian village (*khum*) was defined as a group of houses supporting a particular temple.[50] The rhythm of life of the Cambodian peasant revolved around the monsoon cycles, which determined rice production and the calendar of Buddhist festivals. Father Ponchaud notes that "until April 1975 the word for "race" and "religion" in Cambodia were the same, and in everyday language "Khmer" implied "Buddhist."[51]

According to a refugee Buddhist monk, during the struggle against Lon Nol, the Khmer Rouge respected the monk and the Buddha. Many of the "country monks" had been sympathetic to the Khmer Rouge because of its alliance with Prince Sihanouk and because their ignorance of Communist doctrine."[52]

After the April 1975 victory things were quite different. According to Laura Summers' analysis of the Cambodian class structure as seen by the Khmer Rouge, monks were regarded as a special class apart from the regular social structure, a classification reserved for selected residual elements of the "old regime" including "reactionary intellectuals and military men and policemen of the old regime"—all of whom were singled out for elimination.[53] A former Khmer Rouge regimental political commissar said that on April 17, 1975, the day of victory over Lon Nol, Pol Pot laid out an eight-point Khmer Rouge program at a special assembly for DK ministers and regional secretaries that included as point 4 the order to "defrock all Buddhist monks, and put them to work growing rice."[54]

The policy of the Khmer Rouge toward Buddhism was as follows.

> 1) Buddhism is a reactionary religion and as such must be eliminated.
>
> 2) The bonzes (monks) who have been exploiting the population for more than 2,000 years do nothing but eat and sleep. They are leeches who suck the blood of the people.
>
> —The bonzes must discard their robes, abandon their spiritual way of life, and return to a secular existence. Those who refuse to obey will be regarded as enemies and be dispatched to the next world.
>
> —Believers of both sexes must refrain from all incantations, genuflections, and pious offerings in both the pagoda and at home. It is strictly forbidden to offer food to the bonzes.
>
> —Persons failing to comply with these orders will be regarded as enemies who undermine from within and to whom a policy will apply "to leave them alive brings no advantage; to kill them incurs no loss."[55]

In Theravada ("lesser vehicle") Buddhism, defrocking or disrobing is the equivalent of disordination or disbarment. Without the saffron robe, a monk is not a monk. Nor should monks engage in regular work. The Buddhist monkhood is a mendicant contemplative and teaching order. Monks obtain food for their two meals a day by circulating at dawn with their begging bowls and from offerings brought by villagers to the temple to obtain "merit." After 1975 people were prohibited from feeding the monks and were told that "extra rice is for the nation not the bonzes."[56] Early refugee accounts relate that in propaganda meetings the Khmer Rouge explained their policy by denouncing monks as "leeches and bloodsuckers."[57] As Timothy Carney noted in 1977 after discussions with early escapees, the production cooperative replaced the local temple as the ritual and ceremonial center and the focus of social organization.[58]

The mere disrobing of the monkhood was deemed insufficient. The Buddhist leadership was destroyed through execution. Some, such as Samdech Preh Vannarat, Preh Krou, and Siri Seltakoev Sangh, were summoned in Battambang province ostensibly to "receive" Prince Sihanouk but were instead taken away for execution.[59]

Samdech Sangh Huot That, leader of the Mohanikai sect, and Venerable Krou Thomabal Khieu Choum, a leading modernizer of Khmer Buddhism, were also executed.[60] Other senior notables, such as the Venerables Thamma, Lithet Kinto, and Preh Pothivong Schai, disappeared.[61] The Phnom Penh Tribunal reported massacres of fifty-seven monks in the Ok forest near Chanta Sor, Seam Riep province, and some 200 monks from Phnom Penh at the Chompukack pagoda in Kien Svay district, Kandal province.[62] Surviving monks inside Cambodia and in refugee camps report that monks who refused to disrobe were beaten to death or executed by other means.

Other Khmer Rouge acts against the Buddhist religion included the closing of the Buddhist Institute complex in Phnom Penh and the destruction of religious literature including ancient texts inscribed on palm leaves. Temple libraries and the Pali schools (the language of Buddhist scripture) were ransacked. Temples were destroyed or desecrated and converted into warehouses, workshops, or prisons and occasionally execution centers.[63] One monk I interviewed listed the temples destroyed in Ba Phnom district in Prey Veng province: Koh Sanvek, Viher Thom, Tang Kok, Wat Thamei, Wat Theay, Trong Khoa, Kompong Seung, Kompong Trabek, Angkor Tret.[64]

The organized monkhood was disrobed, disbanded, and destroyed. Prior to 1975 there were 40–60,000 monks—the number varied because it was customary for young men to become monks for several years before marriage, or even for several months during adolesence. By late 1979, almost a year after the Khmer Rouge were ousted from Phnom Penh, only 800 monks had returned to their former monastery sites. The Buddhist leadership, and those who refused to disrobe, had been executed. Thousands more died from some combination of starvation, disease, and exhaustion. Having previously eaten only two vegetarian meals a day, and having been part of a profession of meditation, not labor, the monks were particularly susceptible when pressed into forced labor and restricted rations. Other monks were forced to marry. In 1978 Yat Yen, Democratic Kampuchea's minister of culture and education, told a visiting Yugoslav journalist that Buddhism was incompatible with the goals of the revolution and was "a relic of the past, forgotten and surpassed."[65]

Like the elimination of the Cham, the disrobing of the monkhood was not carried out with equal thoroughness in all regions: in some places it was carried out fully and immediately, while elsewhere monks were not defrocked until late 1975, mid-1976, or even later. Again, this indicates that some Khmer Rouge district authorities may have dissented from central policy.[66] In some areas former monks, once disrobed, were kept isolated, while in other regions they were dispersed among the population. But it does seem that there was no place that monks continued to practice for the duration of the regime. This indicates that the destruction of the monkhood was a policy to which the Khmer Rouge central leadership group was firmly committed, and which was finally implemented with devastating thoroughness.

The folk-level revival of Buddhism after 1979 is one of the encouraging signs in post-Khmer Rouge, postfamine Cambodia. But there is still no Buddhist literature for the devotion and the instruction of the faithful. Most importantly, there are hardly any monks. Many were sent back to their villages or to work camps under the DK regime, and their fate is unknown. There are, no doubt, surviving former monks who will not reveal themselves out of fear of the Vietnamese and the Heng Samrin regime, or out of shame because of the transgressions forced upon them.

The National Group

The attempted, and partially successful, destruction of the Cham, an ethnic group, and the Buddhist monkhood, a religious group, has been examined, but a more vexing legal question arises in reference to whether the Khmer Rouge committed genocidal acts against the national group itself. In other words, did so many Cambodians die during the three and one half years of Khmer Rouge rule that their policies can be generally characterized as genocidal? The answer to this question depends on three factors: the question of numbers—how many did not survive; the extent to which the acts enumerated in the convention took place in Cambodia between April 1975 and January 1979; and the interpretation of "intent" in international law.

Almost all relief workers or journalists who have spent time in Cambodia since 1979 or the refugee camps in Thailand are aware of the high proportion of Cambodians who lost family members and friends due to the policies of the Khmer Rouge—through either execution or some combination of exhaustion, starvation, and disease. Yet there is still no adequate study of the demography of death during the rule of Democratic Kampuchea.

The best available statistical survey remains *Kampuchea: A Demographic Disaster*, a 1980 publication of the U.S. National Foreign Assessment Center based on research completed in January 1980. The CIA estimates that "the savagery of the regime caused an actual drop of between 1.2 million and 1.8 million people" between April 1975 and January 1979 (p. iii). This study has been criticized on various grounds,[67] but its most serious deficiency regarding the 1975 to 1979 period is that actual death rates (by massacre) in 1978, and probably 1977, were much higher than the CIA assumed. Also, inasmuch as there were births during the Pol Pot years, the number who died would be higher than the CIA figures, which measure projected gross, or absolute, decline.

In 1977 Barron and Paul estimated that 1.2 million had died in the first eighteen months of Khmer Rouge rule.[68] Their figures were criticized for being worst case projections based on refugee accounts from atypically bad areas.[69] Perhaps the estimates of total deaths resulting from the first forced exodus from the cities and early massacre, starvation, and disease were high; unfortunately, as "worst case" conditions spread, Barron and Paul's not unreasonable assumptions turned out to be not very wide of the mark.

A different approach to statistical quantification was undertaken by a Japanese reporter, Katuiti Honda, from *Asahi shimbun*. For three weeks in August and September 1980, Katuiti toured Phnom Penh, four Cambodian provinces, and the Sakeo refugee camp in Thailand, conducting in-depth interviews with survivors documenting the deaths within 216 families. His results have been published in English, along with diagrams reconstructing each family tree and descriptive accounts of the death or survival of family members.[70] While not a scientific or statistically representative sample, his data confirm the experience of relief workers and other journalists that it is difficult to find a family that did not suffer deaths. Katuiti found only four among his sample. The 216 families, accounting for 1,896 individuals (excluding deaths before 1975 and after 1979), lost 827, leaving 1,069 survivors. The mortality rate for this group of families would be 44 percent. Even if a small number might have died during this four-year period irrespective of Khmer Rouge policy, and even if his survey sample is biased toward "new people" (those not living in Khmer Rouge-controlled areas before April 1975), his results are still staggering.

Khmer-speaking Cambodia scholars have independently interviewed larger, more representative samples of Cambodians inside Cambodia and among Khmer coming to the Thai border for refuge or famine relief. More precise statistical quantification of their data, when competently analyzed, will provide much more definitive information. Their preliminary and provisional results, however, indicate very large-scale mortality. Ben Kiernan estimates that roughly 500,000 were killed by execution and massacre alone.[71]

Stephen Heder's interviews with more than 1,500 Cambodians—classified by place of residence, political affiliation (if any), ethnicity, prewar social status, age, sex, time, place, and circumstance of death—yield data on over 15,000 individuals and show a mortality rate of roughly 50 percent among urban ethnic Chinese, with deaths divided almost evenly among execution, starvation, and disease; a mortality rate of roughly 25 percent among urban Khmer, with deaths again divided almost equally among execution, starvation, and disease; and a mortality rate of about 15 percent among rural Khmer, with approximately half dying by execution, a quarter by starvation, and a quarter by disease. Projected to cover the pre-1975 population, these ratios would indicate that, allowing for varying demographic assumptions, 1.5 to 2 million Cambodians died from execution, starvation, and disease.[72]

Extremely limited resources and the very difficult economic, social, and political conditions in post-Khmer Rouge, postfamine Cambodia preclude census-based determination of political mortality between 1975 and 1979. It is hoped that macropopulation and micropopulation studies will yield more precise quantification than exists at present. In the meantime, it is a reasonable working hypothesis that somewhere between one-seventh and one-third of the Cambodian people died as a result of Khmer Rouge policy and practice.

The following is an analysis of genocidal acts against the Cambodian people as a whole in terms of the articles of the Genocide Convention.

Killings: Article IIa. The policy and practice of execution has already been discussed in reference to the Cham and the Buddhist monkhood. Beyond this distinctions can be made between several categories of massacres and executions, these often accompanied by torture. Massacres took place during the evacuation of the cities, against leadership groups, "enemies" or perceived enemies, undesirables, and those "insufficiently disciplined." During the evacuation of the cities, some of those who refused, procrastinated, or showed signs of opposition were killed or beaten. Others were killed by Khmer Rouge troops in order to keep the outward marches moving or to maintain discipline.

In some sectors, Khmer Rouge soldiers pulled people suspected of being associated with the former administration out of the column and summarily executed them. This was the beginning of a general purge of officers, senior officials, and high-ranking civil servants from the previous Lon Nol regime. These purges were at first carried out against high-ranking officials and officers, but by late 1975 and early 1976, they included lesser officials as well. In some instances wives and children were also executed, presumably to prevent them from becoming opponents or taking revenge against the new regime.

There was also a policy of executing members of the bourgeois or middle classes and the Cambodian intelligentsia. In addition, thousands of individuals were executed during the Khmer Rouge years for minor infringements of work discipline: "illicit" sexual relationships, complaining about conditions of life, and even laziness. In 1977 and 1978 massacres were directed against dissenting elements within the Khmer Rouge itself and those who were affiliated with or living under the jurisdiction of what the center leadership considered to be dissenting elements inside the Khmer Rouge movement. And, as has been discussed, throughout the entire Khmer Rouge period there were executions and massacres directed against ethnic minority and religious groups (particularly their leadership), which should be included in the category of killings directed against the Cambodian people as a national group.

There was also a prison-execution system operating throughout Cambodia at national, regional, and district levels, which provided for the deaths of scores of thousands of additional Cambodians. At its apex was the Tuol Sleng or S-21 prison in Phnom Penh.[73] At S-21, Khmer Rouge officials kept meticulous records of the murders. These documents were left behind and discovered by the Vietnamese who captured the city in January 1979. There are documents on more than 14,000 people. However, because the materials for June to December 1978 are missing, it is estimated that 15–20,000 were exterminated at Tuol Sleng. The execution schedules show that, "on October 15, 1977, 418 were killed; on October 18, 179 were killed; on October 20, 88; and on October 23, 148. The highest single figure was 582 recorded executions on May 27, 1978."[74]

S-21 is a former school consisting of four three-story buildings. Some classrooms were used as common cells where forty to one hundred persons were lined up and shackled in leg irons. Other rooms were subdivided into tiny cinderblock cubicles, sixteen or eighteen cubicles to a room, where important prisoners or those undergoing interrogation were isolated and shackled to chains cemented to the floor. S-21 held an average of 1,000–1,500 prisoners at one time. Forty categories of men, women, and children—mostly Khmer Rouge officials and cadres suspected of dissidence, but also captured foreigners, workers, and intellectuals—were brought to be photographed, interrogated, and tortured until they confessed to being CIA or KGB agents, naming their contacts, meetings, and accomplices. The prisoners were then executed. According to Stephen Heder, in the case of allegedly disloyal Khmer Rouge cadre, purges included not only region secretaries themselves but whole strings (*khsae*) of their supporters, allies, friends, relatives, associates, and colleagues reaching very far down into the party, administrative, and military structures.[75]

The Khmer Rouge prison officials in some instances took photographs of the emaciated, bruised, and deceased victims to send on to party higher-ups to demonstrate that the prison guards and officials had been doing their job. The written confessions indicate in marginal notations when and how often torture was applied. The documents and archives now housed at Tuol Sleng provide a remarkable insight into the pathology of the regime and reinforce the testimony and evidence of the handful of survivors.

People were brought to Tuol Sleng because their names had appeared on previous confessions or were forwarded from district or regional prison and execution centers. Some were "dissidents" captured by the central security forces (*Sontebal*). The six prisoners who survived did so only because their technical skills were useful in the administration of the prison.[76] Throughout Cambodia there are, however, survivors of regional and district-level prisons.[77] Their testimony indicates that there were scores of executions conducted weekly in each of the smaller prison-execution centers. What remains to be ascertained are the variations in time and place and the political dynamics of the cycles and waves of murders. These may possibly be reconstructed from prison records at Tuol Sleng, refugee accounts, and testimony provided by Cambodians in Cambodia.

Destruction resulting from conditions of life: Article IIc. Far larger numbers of people died due to the conditions to which they were deliberately subjected than due to execution and massacre. These conditions include starvation or malnutrition, disease, and exhaustion from forced labor and marches.

Press reports and 1975–76 refugee accounts of the evacuation of Phnom Penh reveal that it caused a large number of deaths among the young, the old, and the sick. It was originally argued that the evacuation of Phnom Penh was rational—even humanitarian.[78] The initial rationale seemed plausible: a shortage of food in a capital swollen by those who had fled the U.S. bombing, the Khmer Rouge-Lon Nol civil war, and early Khmer Rouge repression in areas long under their control. However, not only Phnom Penh was evacuated, but provincial cities and large towns as well. And no exceptions were made for those who would not survive the march, including hospital patients, belying any supposed humanitarian intent. It is now clear that the cities were evacuated for ideological reasons: to make a "proletarianized" nation of peasants, the Khmer Rouge instituted a policy designed to destroy the bourgeois and capitalist classes that had resided in the city. The Khmer Rouge also sought to maintain political control over any large groups of people who might not be sympathetic to the revolutionary policies that were to follow the Khmer Rouge victory.

Throughout Khmer Rouge rule there continued a series of forced marches of large numbers of these evacuees. For example, there were forced marches across Cambodia from Prey Veng province in the south to Battambang province in the northwest; from Kompong Speu in the south central region to Battambang; from northern Kompong Thom to Siem Reap; from Kompong Cham to Kompong Thom, Pursat, and Koh Kong.[79] Additionally, after the initial evacuation those who could not be accommodated in the areas to which they were sent were then sent back to their native villages, giving rise to additional clusters of travellers for whom little or no provision had been made. These migrations resulted in deaths, especially of the old, the young, and the sick.

In Khmer Rouge theory, the elimination of exploitive social classes and the old social order, along with the collectivization and communalization of agriculture, were supposed to unleash the productive power of the peasantry. Making the "new people" (city dwellers and those not living under Khmer Rouge control) into peasant laborers was supposed to bring new areas under cultivation. This new system amounted to virtual slave labor; refugee accounts make numerous references to extremely long and gruelling work weeks of six to seven days. The strict regimen and rigidly enforced work discipline led to exhaustion, particularly on the part of the former city dwellers who were not used to the back-breaking work of rice production. Few of the new dams or water works built by the "new people" to bring new land under production seem to have worked. Agricultural productivity declined. In many areas rations became more and more meager.

The Khmer Rouge continued to export rice in order to import oil and military supplies despite the fact that Cambodians were succumbing to malnutrition and exhaustion.[80] While people starved, the Khmer Rouge proudly relied on their own efforts to solve food problems "without importuning the international community."[81] Democratic Kampuchea declined to avail itself of the food aid available internationally from public and private sources. The Khmer Rouge leadership is thus responsible and accountable for malnutrition and starvation that might otherwise have been avoided.

The large portions of the population weakened by forced marches, forced labor, and food shortages were obviously susceptible to disease. Khmer Rouge policy was to use "the healing methods of our ancestors," and "to take nothing from the Western medicine."[82] Many doctors and nurses were executed because of their educational level or training abroad. Other doctors hid their identities. The destruction of Cambodia's limited system of modern medicine meant that many Cambodians succumbed to the reemergence of preventable diseases for which medical science has found cures: tuberculosis, malaria, dysentery, diarrhea, beri-beri, and fever, among others.

Articles II(b), (d), and (e). Khmer Rouge policy certainly caused serious mental harm to surviving members of the Cambodian national group as well. Many Khmer in refugee camps or in Cambodia say "their head isn't right" since Pol Pot, or that some family members have become deranged due to the suffering and loss of loved ones. Psychologists who have worked in the Cambodian refugee camps in Thailand found trauma and mental disorder in those numbed by the sight and experience of prolonged terror.[83]

As we have seen, children were separated from their parents in the course of urban evacuations, dispersion of ethnic minorities, communalization of agricultural production, and creation of mobile youth brigades. Because of such separation and the overall living conditions, there was a severe reduction in the birth rate during the Khmer Rouge period. This was apparently judged to be such a serious problem that forced marriages were instituted in many areas. If women did not find someone of their choice to marry, Khmer Rouge cadre would match up men and women randomly. But it would seem that the transfer of children away from their parents and the reduction in the number of couples giving birth was a consequence of other policies rather than a deliberate attempt to destroy the Cambodian people through "imposing measures intended to prevent births within the group" (article II [d]) and "forcibly transferring children of the group to another group" (article II [e]).

"Intent" and the Partial Destruction of the Cambodian National Group

Adequate consideration of "intent" in the international legal sense is beyond the scope of this chapter and the expertise of the writer. A few observations may be made, however.

The intent requirement was introduced into the Genocide Convention to distinguish between genocide and homocide[84] and exclude large-scale murder for legitimate self-defense or from accident.[85] It seems that in any situation where the evidence of large-scale death is well-established or acknowledged, the lack of specific intent is an obvious line of defense against the accusation of genocide. Indeed, when the governments of Brazil and Paraguay were accused of genocide and complicity with genocide against their indigenous Indian populations, the "intent" to destroy was denied.[86]

In Cambodia, the intent of the Khmer Rouge policy-makers to dissolve and destroy the Cham as a distinct ethnic minority group and the Buddhists as a religious group seems reasonably well established. In regard to acts of genocide against the Cambodian national group as a whole, the situation is more complex.

The drafters of the Genocide Convention anticipated that one national group, or the leaders of one national group, might destroy or seek to destroy in whole or in part *another* national group. Autogenocide adds another eventuality to the scope of the convention.

The language of the convention itself does not preclude responsibility or accountability because the destruction of part of a national group was carried out by the nation's own leaders. As one legal expert noted, "The Kampucheans cannot escape responsibility for genocide simply because it was practiced on members of their *own* nationality since this massacre did not take place as part of a civil war, but as a calculated, systematic elimination of entire segments of the national population."[87]

The terms of the convention require only the intent to destroy *part* of the national group as such. It is not required that a "final solution" to eliminate the group in its entirety be intended. It was certainly the intent of the Khmer Rouge policy-makers to dissolve and if necessary eliminate the part or parts of the Cambodian nation that were perceived to stand in the way of the revolutionary transformation of society. This is exactly what happened. In practice this came to include additional and ever-growing parts of the nation as the revolutionary transformation failed to work.

Some aspects of the Khmer Rouge mentality are revealed in their claim to have "won these great victories thanks to our decision to track down and liquidate enemies in a systematic way,"[88] or that "to rebuild our new Cambodia, one million men is enough. Prisoners of war [people expelled from the cities and villages controlled by the government as of April 17] are no longer needed, and local chiefs are free to dispose of them as they please."[89]

Intent in the legal sense, however, can be something that is determined on the basis of a rendering of the evidence of what actually took place. One legal scholar has attempted to apply to genocidal intent the descending order of the elements of culpability: purpose, knowledge, recklessness, and negligence, reserving the most serious charges and penalties for offenses involving purpose or knowledge.[90] Others extend liability to "culpable negligence."[91] The attitudes displayed in the above quotes from Khmer Rouge sources would seem to encompass purpose and recklessness.

How legal scholars ultimately resolve the question of intent in regard to the partial destruction of the Cambodian national group cannot be known. However it would indeed be disturbing if accountability for the massive death and destruction were removed because DK policies were based on an analysis of class divisions or because the ultimate goal was an egalitarian society.

The exacting and sometimes trivializing nature of a legal vocabulary may seem an inappropriate one with which to describe one of the worst examples of genocide in recorded history. Measuring specific aspects of the atrocity against legal statutes almost mocks the enormity of the crime. But it does not. If we are ever to move beyond the paralyzing rage that such inhumanity inspires, specific steps must be taken to identify the perpetrators of these crimes and to work against such massive breaches of civilization in the future. The architects of the Kampuchean holocaust are not only alive and at large, they are diplomatically supported by most of the free nations in the world. In anticipation of the day when Pol Pot and his lieutenants can no longer depend on the vagaries of geopolitics for their survival, it is important to establish the framework within which they might be brought to justice.

Notes

1. See Louis Henkin, *The Rights of Man Today* (Boulder, Colo.: Westview Press, 1978), chs. 1 and 3.

2. See Louis Henkin, ed., *The International Bill of Rights* (New York: Columbia University Press, 1981), p. 15.

3. Crimes against humanity, including genocide, have no statute of limitations.

4. UN Doc. E/CN.4/1437, January 19, 1981.

5. There were no votes against this resolution, although the Soviet bloc, Saudi Arabia, and South Africa abstained. The Soviet bloc subsequently accepted the Universal Declaration of Human Rights.

6. As opposed to a "moveable ceiling." See Stanley Hoffman, *Primacy or World Order* (New York: McGraw Hill, 1978), pp. 287–88.

7. Henkin, *International Bill of Human Rights*.

8. There are others. Cambodia also ratified the Supplementary Convention on the Abolition of Slavery, the Slave Trade, and Institutions and Practices Similar to Slavery; it signed the International Convention on the Elimination of All Forms of Racial Discrimination. Also relevant would be the UN Draft Code of Offenses Against the Peace and Security of Mankind and the Crime Against Humanity Doctrine that emerged from the Nuremburg trials.

9. This was part of an effort to improve the international image of the Khmer Rouge. Other such actions included dropping socialism as a programmatic goal, a supposed (but not widely believed) dissolution of the Communist Party of Kampuchea, and the substitution of Khieu Samphan for Pol Pot as the head of the rump state of Democratic Kampuchea. Pol Pot remains head of the Khmer Rouge armed forces and presumed chairman of the again clandestine Communist Party of Kampuchea.

10. See Michael Vickery, "Democratic Kampuchea—Themes and Variations," ms.

11. See Ben Kiernan, "Wild Chickens, Farm Chickens, and Cormorants: Kampuchea's Eastern Zone," in David Chandler and Ben Kiernan, eds., *Revolution and Aftermath in Kampuchea* (New Haven: Yale University Press, 1983).

12. The August 1979 tribunal, formally entitled "People's Revolutionary Tribunal Held in Phnom Penh for the Trial of the Genocide Crime of the Pol Pot-Ieng Sary Clique," was held by the National United Front for the Salvation of Kampuchea, a self-proclaimed patriotic mass organization led by the leaders of the People's Republic of Kampuchea, the successor Cambodian regime installed by the Vietnamese after 1979. This tribunal, which *in absentia* condemned Pol Pot and Ieng Sary to death, was not without the attributes of a communist show trial. The "defense" of Pol Pot was a particularly shoddy propaganda piece even considering that little can be said in his defense. The evidence and material organized and brought forward at the tribunal is, however, consistent with refugee accounts and testimony provided to journalists and scholars by Cambodian survivors inside Cambodia. The unique aspect of this evidence is the internal Khmer Rouge memoranda and directives provided by Khmer Rouge dissidents and defectors. Khmer-speaking Cambodia scholars regard this material as authentic. The Vietnamese have submitted some of the tribunal records to the UN Human Rights Commission.

13. Australia, UN Doc. E/CN.4/Sub.2/414/Add.8, September 20, 1978; Canada, UN Doc. E/CN.4/Sub.2/414/Add.1, August 14, 1978, and E/CN.4/Sub.2/414/Add.7, September 8, 1978; Great Britain, UN Doc. E/CN.4/Sub.2/414/Add.3, August 17, 1978;

Norway, UN Doc. E/CN.4/Sub.2/414/Add.2, August 18, 1978; United States, UN Doc. E/CN.4/414/Add.4, August 14, 1978; Amnesty International, UN Doc. E/C-N.4/414/Add.5, August 15, 1978; International Commission of Jurists, UN Doc. E/CN-.4/Sub.2/414/Add.6, August 16, 1978.

14. These submissions included John Barron and Anthony Paul, *Murder of a Gentle Land* (New York: Reader's Digest Press, 1977); Francois Ponchaud, *Cambodia: Year Zero* (New York: Holt, Rinehart and Winston, 1978); preliminary report of "The International Cambodia Hearing," Oslo, Norway, April 21–23, 1978; *Human Rights in Cambodia*, hearing before the Subcommittee on International Organizations of the Committee on International Relations, House of Representatives, Ninety-fifth Congress, First Session, May 3, 1977, and July 26, 1977.

15. UN Doc. E/CN.4/1335, January 30, 1979.

16. Ibid.

17. UN Doc. E/CN.4/SR.1510.

18. See *Multilateral Treaties in Respects of Which the Secretary-General Performs Depository Functions: Lists of Signatures, Ratifications, Accessions, etc.*, ch. 5, Human Rights.

19. Holocaust scholar Helen Fine describes genocide as "organized state murder," Helen Fine, *Accounting for Genocide* (New York: Free Press, 1979). Sociologist Irving Lewis Horowitz defines genocide as a "structural and systematic destruction of innocent people by a state bureaucratic apparatus," I. L. Horowitz, *Genocide: State Power and Mass Murder* (New Brunswick, N.J.: Transaction Books, 1976), p. 18. Lemkin himself defines genocide as "a coordinated plan of different actions aiming at the destruction of the essential foundation of life of a national group. . . . The objectives of such a plan would be the disintegration of the political and social institutions of culture, language, national feeling, religion, and economic existence of national groups and the lives of the individuals belonging to such groups," Raphael Lemkin, *Axis Rule in Occupied Europe* (Washington, D.C.: Carnegie Endowment for International Peace, 1944), p. 79. The 1946 UN General Assembly Resolution 96(1), which led to the drafting of the Genocide Convention, defined genocide as "the denial of the right of existence of entire human groups" and noted that "such crimes have occurred when racial, religious, political, and other groups have been destroyed, entirely or in part."

20. For example, Barron and Paul's *Murder in a Gentle Land* is subtitled *The Untold Story of Communist Genocide in Cambodia.* Former Secretary of State Henry Kissinger routinely characterizes Khmer Rouge policy as genocidal in his memoirs, *Years of Upheaval* (Boston: Little, Brown, 1982), vol. 2. The UN Human Rights Subcommission rapporteurs used the terms "auto-genocide" and "genocide" in describing Khmer Rouge human rights violations.

21. It was argued that social and political groups, unlike religious groups, had no stable, permanent, and clearcut characteristics and do not constitute an inevitable homogenous group independent of the will of the individual. See Nehemiah Robinson, *The Genocide Convention: A Commentary* (New York: Institute of Jewish Affairs, World Jewish Congress, 1960); Leo Kuper, *Genocide: Its Political Use in the Twentieth Century* (Harmondsworth, England: Penguin Books, 1981); and the UN *Study on the Question of the Prevention and Punishment of the Crime of Genocide*, E/CN.4/Sub.2/416, July 4, 1978.

22. Robinson, *Genocide Convention*, p. 59.

23. ST/LEG/SER. D/13 at 85.

24. Phnom Penh Tribunal document No. 2.4.02. Submitted to the UN Human Rights Commission A/C.3/34/2 in slightly different translation (from French to English) than the tribunal document translated from Khmer to English.

25. Interview with Hakkim at Chaing Chamres, March 1981, and with Mat Le, Phnom Penh, April 1982.

26. Interview with Mat Le.

27. Ponchaud, *Cambodia*, p. 133.

28. Interview with Mat Le.

29. UN Doc. A/34/491, September 20, 1979.

30. Katuiti Honda, *Journey to Cambodia* (Tokyo: Committee of ''Journey to Cambodia,'' 1981), p. 135.

31. Phnom Penh Tribunal Doc. No. 2.4.02.

32. Interview with Mat Le.

33. Katuiti, *Journey to Cambodia*, p. 133.

34. Ponchaud, *Cambodia*, p. 132.

35. Kiernan, ''Wild Chickens,'' p. 27.

36. Kiernan interview with Nao Cha, Smong village, Treang district, Takeo province, August 26, 1980. Cited in Kiernan, ''Wild Chickens,'' p. 27.

37. Ibid.

38. Interviews with Mat Ly, the Hakkim at Chaing Chamres, March 1981, and others. Dr. Abdul Gaffar Peang-meth, an American-educated Ph.D., reports that he participated in the census of Moslem families attached to mosques in Cambodia. Interview, November 1982.

39. This figure would also include the Moslem Cambodians of Malay origin.

40. Interview with Mr. Derek Kulasiriwasd, a leading Thai Muslim scholar, Bangkok, April 1981.

41. Phnom Penh Tribunal Document No. 2.4.02.

42. Ponchaud, *Cambodia*, p. 132.

43. This figure would not include several thousand Cham refugees now in Malaysia or the Cham who live in the settlements along the Thai-Cambodia border.

44. Interview, October 1982.

45. Interview, April 1982.

46. Phnom Penh Tribunal Document No. 2.4.02.

47. Ibid.

48. Interview with Ven. Pin Sem, Nong Samet, a settlement along the Thai-Cambodian border, April 1982.

49. Milton Osborne, *The French Presence in Cochinchina and Cambodia* (Ithaca: Cornell University Press, 1965), p. 5.

50. Milada Kalab, ''Study of a Cambodian Village,'' *Geographical Journal* (London) 134, 4 (December 1968):523, 528, as cited in Timothy Carney, ''Communist Power in Kampuchea,'' data paper no. 106, Southeast Asia Program, Cornell University, January 1977, p. 20.

51. Ponchaud, *Cambodia*, p. 126.

52. Interview with Ven. Pin Sem.

53. Laura Summers, ''Cooperatives in Democratic Kampuchea,'' ms., 1981, cited in Timothy Carney, ''The Organization of Power in Democratic Kampuchea,'' ms., 1982.

54. Kiernan, "Wild Chickens," p. 42.

55. Phnom Penh Tribunal Documents submitted to United Nations. Doc. No. A/34/569, October 12, 1979.

56. Ponchaud, *Cambodia*, p. 130.

57. Ibid.

58. Carney, "Communist Party Power," p. 20.

59. Ponchaud, *Cambodia*, pp. 127–28.

60. Ibid., p. 127.

61. UN Doc. A/34/569, October 12, 1979.

62. Ibid.

63. Viewed by author on ten-day visit to Cambodia (Kandal, Takeo, Prey Veng, and Svay Rieng provinces), March 1981, and three-week visit (Kandal, Takeo, Kampot, Pursat, Battambang, and Siem Riep provinces), March-April 1982.

64. Interview with Ven. Nuon Sing Siphy, Khao-I-Dang Holding Center for Kampucheans, Thailand, April 1982.

65. Dispatch by Tanjug's correspondent, Slavko Stanic, in BBC *Survey of World Broadcasts* FE/5773/B/5, March 25, 1978, as cited in UN Doc. E/CN.4/Sub.2/414/Add.5.

66. Carney, "Communist Party Power," p. 20.

67. Michael Vickery, "Kampuchean Demographics—Cooking the Books," forthcoming.

68. Barron and Paul, *Murder of a Gentle Land,* p. 206.

69. Noam Chomsky and Edward Herman, *After the Cataclysm: Postwar Indochina and the Reconstruction of Imperial Ideology* (Boston: South End Press, 1979), p. 245.

70. Committee of "Journey to Cambodia," international edition (Tokyo: Heiwa to Rodo Kaikan, 1982).

71. Interview, November 1982.

72. Interview, October 1982.

73. The best available account of Tuol Sleng is Anthony Barnett, Ben Kiernan, and Chanthou Boua, "Bureaucracy of Death," *The New Statesman*, May 2, 1980.

74. Ibid., p. 671.

75. Stephen R. Heder, "Kampuchea: From Pol Pot to Pen Sovan to the Villages," *Indochina and Problems of Security and Stability in Southeast Asia* (Bangkok: Chulalongkorn University Press, 1981), p. 23.

76. Interview with Ung Peck, one of the survivors and present curator of the Tuol Sleng archives, March 1982.

77. In March 1982 I interviewed survivors from prison execution centers at Ta Mon and Tonli Bati. Other journalists and scholars have encountered survivors of numerous district-level prison execution centers.

78. See George Hildebrand and Gareth Porter, *Cambodia: Starvation and Revolution* (New York: Monthly Review Press, 1976).

79. *Kampuchea in the Seventies: Report of a Finnish Inquiry Commission* (Helsinki, 1982), p. 15.

80. See Elizabeth Becker interview with Ieng Sary, *Far Eastern Economic Review*, August 7, 1981.

81. Submission of Democratic Kampuchea to UN Human Rights Commission, E/CN.4/1343, p. 4.

82. Testimony of Chau Trey to the International Cambodia Hearing, Oslo, Norway, April 1978, as cited in UN Doc. E/CN.4/Sub.2/414/Add.2, August 18, 1978. There are conflicting reports as to whether or to what extent the Khmer Rouge imported medicine or accepted offers of medical aid.

83. See Neil Boothby, "The Horror, the Hope," *Natural History*, January 1983, p. 64.

84. M. Cherif Bassiouni, *International Criminal Law* (Alphen san den Rijn, Netherlands: Sijhoff and Noordhoff International Publishers, 1980), p. 73. See also UN Doc. A/C.6/SR.73.

85. Robinson, *Genocide Convention*, p. 60. Also Kuper, "International Action," p. 5.

86. Kuper, "International Action," p. 5.

87. Letter of Professor David Weissbrodt to Senator William Proxmire, April 19, 1979, p. 5.

88. Internal Khmer Rouge directive.

89. Statement of Khmer Rouge cadre reported by Ponchaud, as cited by Barron and Paul, *Murder of a Gentle Land*, p. 197.

90. Roger Clark, "Does the Genocide Convention Go Far Enough?" *Ohio Northern University Law Review* 8 (1981):324–25.

91. Peter Drost, *The Crime of State, Book II: Genocide* (Leiden: Sijthoff, 1959), as quoted in ibid., pp. 324–25.

POLITICS

Serge Thion

THE PATTERN OF CAMBODIAN POLITICS

It might seem paradoxical to consider the evolution of modern Cambodian politics as a continuous and at times repetitive process, because it appears mainly as a succession of brutal changes, involving merciless replacements of ruling elites.[1] Successive regimes abhored the preceding ones and tried to stamp out any leftover influence. But these total and abrupt changes, sometimes labelled revolutions, occurred in a distinctively Khmer way that calls for some reflection on what Cambodian politics is really about. The most crucial question for Cambodia is not so much its international position as its ability to rebuild a political system of its own.

When King Sisowath of Cambodia first visited France, at the beginning of this century, he provoked quite a sensation in Paris. *La belle époque* was already fond of exoticism, but the display of Oriental luxury surrounding the king, his royal ornaments, his retinue, and his ballet—which amounted more or less to his harem—sparked a cultural shock. This shock recurred later, in 1930, with the *Exposition coloniale*, the biggest colonial exhibition in France, at which a huge cardboard rendering of Angkor Wat struck the fantasies of a whole generation. In these fantasies, Cambodia would always be a beautiful and graceful country, shrouded in Oriental mystery and ruled by a divine king, flanked by dancing *apsaras*. No attention was paid to the fact that the bill for the show was footed by the colonial administration and that the king was pretty much its servant.

This illusion had its local counterpart in Cambodia itself. The French intrusion into the country had also been a cultural shock, triggering a violent insurrection in 1886–87, after the colonial authorities had at gunpoint secured the allegiance of King Norodom (Sisowath's brother and predecessor) to their own particular concept of "protection." Although the French had several able administrators who could quickly grasp and analyze the basic functioning of the Khmer kingdom (see, for instance, the works of Leclère and Moura), colonial authorities were for the most part unable to understand, and even less adapt to, the basic working concepts of the Khmer political system. The repeated and fruitless efforts to build up a system of private land ownership, based on notions of Roman law, bears interesting witness to this enduring misunderstanding. The Khmer, for their part, ensconced in their own culture, naturally made few efforts even to grasp the new rules of the political game. They stuck very much to their traditional leaders and ways of exercising a power that was now under the control of a foreign administration. In some ways, they behaved with the deep lack of realism that has up to now been a permanent factor in Cambodian politics. For a long time, they somehow pretended that the French went about their own business, and this business was unrelated to the real power struggle. That struggle mainly consisted of obtaining honorary positions, along with titles bestowed by the king—titles which the French, using a word out of its original context, likened to those of mandarins. There was in fact nothing in Cambodia like the Chinese mandarins, civil servants climbing up sheerly by dint of their literary knowledge and managerial abilities.

The more appropriate reference here is the tradition of the Theravada pali Buddhist states of Southeast Asia, where literate men, until the last century and to a degree still now, were usually elected to join the *sangha* (monks' community). They did not join the civilian or military upper administration, which was reserved for the sons of the aristocracy or, up to a point, those of the merchant class.

Politically, Cambodia should be first understood as part of a cultural world that makes it foreign to both Western values and Chinese (or sinicized) political and administrative traditions. Just ask the Vietnamese. This bitter lesson in cultural difference was first brought home to them in the decade of the 1830s, when Emperor Minh Mang had to withdraw his protectorate after a complete failure to change the nature of politics in Cambodia. The same may be expected today, as soon as the military situation stabilizes and brings the question of political reorganization to the forefront. This is only prevented now by the continuous turmoil at the border, which is almost completely fueled from abroad.

To grasp fully the present situation, we have to go back to the nineteenth century. To simplify very much, I would say that Cambodia was then a part of the Siamese political system. Being "part of" should be understood not as what it means today (a region included in a state, e.g., Sicily being "part of" Italy), but as what it meant at the time: an area under a local ruler—a country, a kingdom, whatever its official name—having a specific relation to another that is considered by both parties as the center, and which is ruled by a paramount king to whom some kind of personal and renewable loyalty is owed by the local ruler. This local ruler maintains a court, an army, a corvée labor force in much the same manner as the paramount king but on a less grand scale. I would hesitate to call this arrangement feudalism because that might confuse rather than clarify the issue. Yet there is something of feudalism here: the personal bond of loyalty, usually referred to as the "drinking of allegiance water." Tribute was paid, but this was not the same as the tributary system of the classical Chinese empire, where tribute was first and foremost an economic affair. It was perhaps more like the Aztec federation, the way the Spaniards found it, with Cortes playing the periphery against the center.

The Siamese kingdoms of Ayutthia and Bangkok were so constructed as to include peripheral states, which were ever-poised to seize upon the opportunity of a weakened central influence in order to set themselves free from that influence and attempt to establish themselves as competing centers. These peripheral states at times included culturally related areas like the Chiangmai kingdom in the North or the Laotian principalities in the Mekong valley, as well as foreign entities, such as the Malay sultanates in the South, large slices of Khmer-speaking areas of the former Angkorian empire, and the kingdom of Cambodia itself, the *srok khmer*. A detailed analysis of these polities and their roots in Buddhist conceptions of religion and kingship, particularly in the Indian empire of Ashoka, is to be found in S. J. Tambiah's book.[2] He calls the system a "galactic policy," which seems suggestive enough. This also is the backdrop for the feeling shared by most Thais that Cambodia should somehow belong to a Thai sphere of influence. At the first opportunity, i.e., the submission of the French Vichy regime to the Axis powers, the Thais managed to seize several western provinces of Cambodia, which Thailand had to give back after the downfall of its Japanese protector. Again, in 1970, in the wake of the March coup, the Thai military intervened as the Lon Nol policy was opening a new opportunity. And nowadays the Khmer resistance groups, the Khmer Rouge included, are under the complete control of the Thai military.[3] Thai politicians are quite reluctant to accept the idea of an independent Cambodia because it challenges their own "legitimate" claim to some kind of overlordship.

For reasons that probably have more to do with the shifting of international trade routes than the exhaustion of the soil or Thai military attacks, the Angkorian empire fell into oblivion. By the eighteenth century, for all practical purposes, Siamese influence dominated every area of political life, including administration, court rituals, education, kinship alliances, and royal chronicles. But the kings in Oudong, the royal capital before the coming of the French, had a much smaller economic base and were relatively poor. The basic Theravada conception of the ruler was that of the *dharmaraja*: the righteous ruler, embodiment of the *dharma*, maintainer and supporter of the *sasana* (religion), benefactor of his people, and source of the moral law. Closely connected was the notion of *chakravartin* or world ruler (with the palace representing the center of the world), as a necessary link between the several levels of cosmic existence, animals, nature, mankind, gods, etc. The divine nature of kingship, inherited from a very early Hindu influence, was not altogether stamped out by Theravada Buddhism, as can be seen by the role played, up until 1970 in Cambodia and still today in the Thai court, of Brahmin families attached to the royal cult. In a way, all this removed the king from the direct management of affairs, which was entrusted to powerful ministers and their wives, not to mention royal wives and consorts, thus precluding the establishment of fixed rules of succession and leaving the door open to much internal fighting among princes and other royal offspring. Siamese, Burmese, and Khmer history is replete with internecine struggles for succession and long periods of decline due to political instability.

The colonial impact completely changed the way this system operated, but it did not suppress it altogether, at least not in Thailand, Cambodia, or Laos. In Burma and Ceylon, the system was decapitated with the disappearance of the monarchy, but its background in the local culture has not yet disappeared.[4] Politics in Theravada societies is supposed to be chiefly concerned with the morality of the leaders and the welfare brought by their action. Righteous leadership and mass welfare are the basic requirements for the growth of religion, which allows individuals to accumulate merit through donations and good deeds. Power is supposed to reward a good kharmic "credit" for past merits, accrued from former lives. Legitimacy stems mostly from the *sangha*, which is in turn honored and protected by the powerholder.

Sihanouk became king on the basis of this mental framework, though he was selected and put on the throne by the French. Since the death of King Norodom (1904), the French had favored the Sisowath branch of the royal family, which they deemed more pliable to French interests. Palace intrigues to win the governor general's favor replaced open struggle by contending families. Colonial interference thus brought swift successions, but old cliques and factions were still playing very much the same game as before, even though the power of a Cambodian king was more apparent than real.

It has been said that French colonial authorities enhanced and refurbished the status of kingship in Cambodia. There is some truth in this, but the French closely watched and always controlled the administration. Nevertheless, this added prestige caused more competition among "powerful" families, some of commoner origin, to fill court and local administrative positions.

The role of family links seems to be highly contradictory in this respect. Although the concept of the extended family, well established throughout Asia, is not unknown in Cambodia, genealogical ties are remembered only in some families, when they have some link to the royalty.[5] Solidarity is far from being automatically deducible from kinship relations. There are many examples of ordinary people, before the war, trying to take advantage of their claim to kinship solidarity and being mistrusted or rebuffed by those to whom this claim was addressed. There is a tendency to model kinship ties on the patron-client pattern and to use them to fulfill individual ambitions.

What the French had in mind was to *conserve* kingship. Calling the king powerless may not imply a great departure from traditional realities. In the last century, he was surrounded by powerful lords, like the *obareach* (pali *uparaja*, "underking"), who shared part of his political and religious privileges, and other royalty who had control over large provinces and could "eat" them, i.e., collect taxes and revenues. Although his religious role made him unique, the king was far from being an absolute monarch; his will was often checked and rebellion was always waiting in the wings. Rebels would often seek foreign help in order to win power. With French dominance came administrative interference and tax control.

But for a long time, the Khmer elite could conveniently ignore it, as modern administration did not attract much interest. No Khmer king had really thought of attempting reforms in the way King Chulalongkorn had in neighboring Siam. So the "bureaux" were conveniently filled with trained Vietnamese freshly imported from Saigon.

The introduction of a parliamentary democracy in the aftermath of World War II provided an entirely new political arena. Old courtiers and traditional administrators had now to compete with wealthy businesspeople and young French-educated intellectuals, who heralded the emergence of a still tiny local middle class. The new political game had rules that were entirely foreign and directly derived from the French Third and Fourth Republics. But there was nothing like the big, established sociopolitical blocs that shared power in the French Assemblies. The nationalist movement in Cambodia, born in the late 1930s, was torn apart by diverging loyalties toward its prime leaders, like Son Ngoc Thanh, or the king, or the Viet Minh. It largely disintegrated and could not build a regular electoral constituency despite its rather broad base in public opinion. Other parties had only a narrow social base, if any. It was thus only normal to see parliamentary life as a new opportunity for factional and individualistic power contests. Naked self-interest was the rule for a host of small leaders and chieftains.

We know of course of other fissiparous polities, working along the same lines, in Thailand for instance, or even in Japan. There, however, some kind of political device ensures the overall unity of the system. In Cambodia, as in Thailand, this role obviously fell to the monarchy as the sole embodiment of all values that could tie the people together. The modern concept of nation had not yet reached the outer fringes of village life, although, of course, the feeling of Khmer cultural identity was firmly established. Nation as a mold for unity and action was slowly trickling down along with educational progress and was still a rather new and somewhat puzzling concept in the countryside before the war in 1970. The nationalist "bourgeois revolution" which erupted in the urban population, and mainly among students, after Sihanouk's overthrow in March 1970, was largely confined to the main cities. In many places, it provoked an angry rejection by large segments of the rural populace, which rallied to the older concept of kingship, even though it implied at the time the acceptance as allies of foreign (Vietnamese) troops.

In the 1950s, Sihanouk, as king, was confronted with a difficult choice. He could remain a constitutional monarch, the paramount symbol of unity, his actions mainly directed to ritual observance of an elaborate religious protection of the kingdom, in much the same role as the kings of Thailand after the 1932 "revolution"; or, he could spring into action and make political use of his charismatic appeal to further his own policies. Feeling it impossible to expose the throne to the hazards of daily politics, he stepped down and had his father, an obscure and shadowy figure, replace him. This arrangement did not really work because, with characteristic megalomania, Sihanouk wanted to have his cake and eat it too. Although formally relinquishing the regalia to his father Suramarit, he nevertheless retained as many royal trappings as possible, and there was no question for the population as to who the real king was. Entering politics, he moved quickly to build his own party, soon to become the only one in the country. Factional struggles thus took place right at the heart of the party, and the traditional pattern of individual squabbling petty leaders, including princes and other aristocrats, could go on unimpeded. The victim was not so much Sihanouk as politician—he was successful for a rather long time—but rather kingship itself. After Suramarit's death in 1960, Sihanouk, drifting toward a sterner autocracy, did not dare have a successor appointed. A new king would have drawn support that he, Sihanouk, needed and commanded. Such a change would have destroyed an ambiguity that was very profitable to him because it allowed his mistakes to be erased by the magical aura he drew from his position as former king and lord protector of the *sangha*.

The French parliamentary tradition thus had no chance to graft itself onto the Khmer body politic, except for a short period in 1955 with the rise of the Democrat party, which was aborted when Sihanouk stepped down. He was the perfect "Oriental" ruler, with harem politics, secret police, widespread corruption, costly festivals, and even a touch of buffoonery in the shape of his absurd dabbling in singing and movie-making. (Even in 1982, while attending the UN session in New York, Sihanouk could not refrain from offering lavish parties and jumping on the stage to sing for his guests. He had a Khmer band specially flown in from Paris. As a Khmer singer, he is indeed not bad.)

But when, under the pressure of the outside world, the internal crisis got out of hand and Sihanouk's ubiquitous demonstrations failed, the whole system broke down. The coup of March 18, 1970, created a gap between politics, concentrated in the towns, and kingship, abolished by law but still a central point of reference for most of the peasantry. Republican politics, without a democratic tradition to lean on, could not fare much better and was very soon heading back to the old pattern of squabbling petty leaders. Prominent among them was Lon Nol's own brother. As is usual in a case such as this, these leaders tried to improve their hand by relying on foreign influence and money. Thus American, Thai, South Vietnamese, French, and even Russian help was solicited and sometimes given. These contending politicians appeared typically as patrons commanding a more or less important retinue of clients whom they had to pay off with gifts and posts. They had to control resources and redistribute wealth as a means of keeping and enlarging their crowd of clients. These resources could come from illegal business through Chinese middlemen, gross corruption, or foreign help. The pattern did not look very different from that of the old days; only the economic opportunities and environment had changed.

By cutting all ties with kingship, even allowing the queen mother, Kossomak, to join her son Sihanouk in Beijing, the republican regime lost its chance, not only to unify the country, but to gain legitimacy, even among the mass of urban dwellers. The *sangha* was deeply divided and could not fill the gap. Lon Nol's personal mystique, though strange, appeared as a dull and unattractive compound since he was never invested with any recognized supernatural power, as kings "naturally" are. His appeals for struggle against the *Thmil* (for Tamils, i.e., nonbelievers) could not match Sihanouk's appeal for struggle against the "rebels."

In his Beijing exile, Sihanouk did not feel he should change his political demeanor. Still surrounded by his court, though on a somewhat reduced scale, he tried to muster as much personal allegiance as he could. The Sihanoukist and communist groups in Beijing lived in completely separate quarters and had almost no contact with each other. Usually Chinese officials would act as intermediaries. For the communists, Beijing was a training ground for many students who were being prepared for the honor of going back to fight inside their country. They were thus set in the classical party life, which applied to sympathetic nonparty members as well. Obviously there was no possible reconciliation between traditional and communist ways of thinking and acting politically. This situation gave rise to Sihanouk's widely reported remark that, after victory, the communists would spit him out, "comme un noyau de cerise" (like a cherry pit).

Interestingly enough, the communists after they came to power were more cautious than the would-be Jacobins of 1970 had been in not disposing entirely of kingship. They kept Sihanouk under a rather liberal house-arrest, in spite of the boiling hatred that most Khmer Rouge veterans harbored toward him. He was allowed to celebrate Queen Kossomak's funeral in the palace in September 1975, some months after she had died in Beijing. He was several times allowed to travel in the countryside and shown off to startled tillers. All in all, he got better treatment than the last of the sons of heaven in China, mainly because the Pol Pot regime thought his royal appeal should be kept as a potential asset, and they used it when they saw that their policy toward Vietnam was growing sour. This regime, possibly the most radical eraser of "old-regime" politics, thus held on to a chance to manipulate royal symbols (the Angkorian past, the Sihanouk charisma) that had proved so useful during the 1970–75 war. It was to be useful again, as the establishment in July 1982 of an unnatural coalition exile cabinet was to prove. This all goes a long way to showing how influential the traditional political pattern still is and why no other political tradition, whether communistic as represented by both Pol Pot and Heng Samrin, or Western democratic, as heralded by Son Sann, has been able to establish a stronger foothold in the country. The obvious caution demonstrated by the Vietnamese regarding the person of Sihanouk is another illustration of this point.

It is obvious for most observers that, whether they like it or not (and I do not), Sihanouk still has a large appeal to broad sections of the Khmer population. But it would be mistaken to assume that this appeal is an approval of the prince, his person or his politics, as was shown by the rather cool reception he got in

Khao-I-Dang in July 1982. The older people have many causes of resentment for his past mistakes and the younger people, say under twenty, know nothing about him. His strength is thus much less due to his own debatable skill as a politician than it is to his charisma as former king, the embodiment of *dharmaraja* and of the unity of the *srok khmer*. His resumption of power would mean—in the realm of Buddhist ideology—the reestablishment of the kind of harmony between nature and world that has been so badly shattered for the last fifteen years.

What were the means whereby Sihanouk or his successors could claim legitimacy while an empty throne was denying them its religious sanction? The nationalist credo was not very efficient because its basic tenet had not yet permeated the entire society. Chauvinistic appeals to the preservation of Khmer "race" or "blood," successively launched by Sihanouk, Lon Nol, and Pol Pot, failed to transcend the boundaries of the educated class. The related manipulation of the image of the "hereditary foe," the Vietnamese, also failed to produce spontaneous action or commitment. Measures against the Vietnamese civilian population settled in Cambodia were taken by, again, Sihanouk, Lon Nol, and Pol Pot, but they never aroused any visible public passion. Seen from the village level, the peaceful coexistence and codwelling of local Khmer and Vietnamese peasants and fishermen is a too ancient fact, both in the Mekong Delta and in Cambodia proper, to stir up homicidal feelings. The dirty job of massacring Vietnamese civilians was carried out under military orders by the troops alone (Lon Nol, 1970; Pol Pot, 1977–78). The Khmer are not given to pogroms, no matter what officials may say about the Khmer hating the Vietnamese.

It is fairly logical, I think, to relate the question of the Cambodian borders to these attempts to raise nationalist feelings and to promote the successive regimes' legitimacy. The most fascinating feature of this question is how these regimes exactly repeat each other's posture. The basic legal fact is that Cambodia has no border freely agreed upon with its neighbors. Frontiers with Vietnam, Laos, and Thailand were drawn by the French, and only with Thailand was the line defined in an international treaty signed by both sides—the French acting as "protector" of Cambodia. This situation is far from being unique in the aftermath of the colonial era. But what is striking is the refusal of the successive Cambodian regimes to commit themselves to any precise boundary. The most clearly articulated definition was reached by Sihanouk when, in 1968–69, he appealed for international recognition of Cambodia "inside its present borders" ("à l'intérieur de ses frontières actuelles").

This vague formula can be understood as an implicit renunciation of former Khmer territories lost in the course of past centuries to the expansion of the Thai and Vietnamese states, and an implicit acceptance of the Brévié administrative line as an international boundary.[6] This contradicted earlier claims made by Sihanouk at the time of the Geneva Conference to the effect that the lost territories, now in southern Vietnam, had never been formally renounced by any Khmer king and that, on the contrary, actions taken by the French to subsume Khmer territory under Cochinchina colonial administration had always drawn angry protests from the Khmer sovereigns. He knew well, of course, that Cambodia would never regain these provinces, now heavily populated with ethnic Vietnamese, but he thought this nonrecognition gave him some leverage with Saigon.

There is also a wide body of evidence from refugees that Pol Pot troops, when they were hurled across the border in murderous raids in 1977–78, were told by their commanders that these operations were aimed at regaining lost provinces "as far as the *thnot* (sugar palm trees) grow," this being a sign of Khmer land occupancy. Sihanouk reported the same thing from his conversations with Democratic Kampuchea officials while he was in Phnom Penh. Unbelievable as it might seem in retrospect, this territorial ambition should be taken as a fact. It reflects the survival of the concept of a "galactic" state with no fixed border line, whose influence extends as far as it can go without meeting a force of equal strength emanating from another central polity, like Bangkok or Hue in former times.

Of course, though they may be unrecognized by treaty, borders do exist. In practice, only a few tiny spots still seem litigious along the land border; the problem of delimitation is somewhat more complex at sea. Old principles concerning territorial waters were swept aside when technical advance permitted offshore exploration for, and exploitation of, oil. The Gulf of Siam has quite a considerable potential for oil and gas. The concept of a 200-mile exclusive economic zone has further sharpened conflicts between neighboring countries. The Lon Nol and Nguyen Van Thieu governments negotiated an acceptable border line dividing the continental shelf, but they failed to agree and succeeded only in establishing respective sea-patrolling areas. The Khmer called the Vietnamese claims abusive, and the student opposition in Phnom Penh voiced strident protests against the very idea of an agreement with Saigon. Later, in April 1976, the newly installed Pol Pot government abruptly broke off a round of talks with the newly unified Vietnamese government, claiming they would not even discuss border delimitation. The main bone of contention was again the sea line. In July 1982 the Heng Samrin government held talks with Hanoi. Although it seemed that the prevailing political conditions would ensure a quick outcome, the two delegations were unable to reach a settlement on the sharing of the continental shelf. Details have been agreed upon that are very similar to those in the Thieu-Lon Nol arrangement. According to the official statement, the question of sovereignty was to be solved "later." No further details were given.

It seems, then, that no Khmer government, whatever its political stance or degree of independence, is able to reach the point of no return, i.e., the establishment of a clear border line that would mean the relinquishing of any potential claim on land or waters once under undisputed Khmer sovereignty. It is to be noted, for instance, that neither the Khmer Rouge nor Sihanouk has, to my knowledge, criticized the Heng Samrin-Hanoi agreement. In the view of the particular situation of Cambodia, it might thus be said that an essential attribute of legitimacy is the refusal to sign any agreement recognizing that lost territories are, effectively, lost forever. This is of course purely symbolic, since borders actually exist, but the fact that this symbol has appeared necessary to all successive Khmer governments since the beginning of the colonial era is all the more significant.

Equally indicative may be that the most learned and articulate exponent of the Khmer border case, Sarin Chhak, who had strong ties with Prince Sihanouk and had worked with the Pol Pot regime, although he was far from being a communist himself, is now detained in secret by the Vietnamese. He was apprehended when Phnom Penh fell in January 1979 and allegedly refused to cooperate with the Heng Samrin government. As he is known to have had no part in the Khmer Rouge atrocities, it seems likely that only his expertise in the border question has prompted his continued detention and "reeducation" in a camp where he is reported to be in extremely poor condition.

What we know of the life of Khmer Rouge elite in 1975–78 provides some glimpses into something very ordinary in Cambodian politics: nepotism and corruption. This is more astonishing when actual practice is compared with official ideology and with the treatment of the population and average party members. Everyone was supposed to relinquish family ties and feelings for relatives. Children were taken care of by the organization and parents were supposed to transfer their love to the party and the country.

It is thus all the more remarkable that Pol Pot himself, who had no children, saw to it that several of his nephews and nieces had good careers, particularly in the Foreign Affairs Ministry, run by another uncle, Pol Pot's brother-in-law, and in the Central Committee offices he personally controlled. One of these nephews, called Hong, "Ieng Sary's right arm, Pol Pot's nephew and special commissar for Phnom Penh,"[7] could travel abroad with Ieng Sary. He managed to get a sister of his late wife appointed head of the office for general affairs of the ministry. This lady was known to have a private flat and to live entirely out of the community system; to maintain her status, she later married a highland tribesman, which in Phnom Penh at the time was seen as a final proof of unconditional loyalty to the party. Khieu Samphan is also said to have married a tribesgirl and to have had his son Kroch ferried about in a Mercedes limousine.[8]

Ieng Sary had several children, in their teens at the time. His eldest daughter was appointed director of the former Calmette hospital and the second one director of the Pasteur Institute, although neither had finished her secondary education. The third and youngest daughter was a military officer. His son had the privilege of going to China for a pilot training course, usually reserved for poor peasants' sons. His nephew (the son of another Khieu lady, sister to Madame Ieng Sary and Madame Pol Pot), strangely named Noeu ("stupid"), was one of the rare Moscow students to survive; he was even authorized to leave Beijing and join the guerillas before 1975 (where he married a tribesgirl). He was later made president of the Civil Aviation Company. His sister, Da, was employed as an

English translator and a news reporter for the radio, although she was notoriously ill-equipped for these duties. Their mother was receiving, with each bimonthly plane from Beijing, fruit, vegetables, and fabrics from China. All these relatives enjoyed privileges in lodging, food allowances, and even some luxuries. Along with Nuon Chea's nieces, they were part of the tiny crowd that appeared at every official party for foreign guests.

Until February 1977, the Foreign Ministry was colonized by another powerful family. The head of the Office for Trade in the Industry Ministry, Doeun, had his wife, Roeun, nominated head of the Diplomatic Shop, a profitable business in any communist regime. Since the shop was also connected with the Foreign Affairs Ministry, Roeun could expand her influence there through a nephew, named Roun, who had wide access, and through a niece, Moeun, married to a petty official called Pich Chheang. Roeun won the intense struggle that followed the decision to reopen some embassies abroad by having her niece and her husband nominated for the Beijing embassy, the most desirable position then. She later introduced another niece, Noeun, as head of the ministry's children section, and other relatives took control of the moving section (in charge of controlling and emptying houses and properties in Phnom Penh), the manual work section, the sewing section, and the kitchen section. These posts were particularly important at a time when offices were rather understaffed and real power mostly lay in the control of food and other basic necessities. These people were thus not only in control of other people's material needs, but they could enjoy private banquets with the best food, servants, and so on. The whole family, including children, with the exception of the Beijing couple, was later purged and disposed of.

Because the facts are well known, I shall only make passing mention here of the family nexus composed of Ieng Sary and his wife Khieu Thirith, sister of Pol Pot's wife, and the lesser family of Son Sen and his wife Yun Yat, also both ministers. Son Sen (or possibly his wife) had a younger brother (or half-brother) named Kon, who had a military position until he went to the Foreign Affairs Ministry and soon became chief of protocol. (He is the Kan described in Y Phandara's book.) Upon the fall of Phnom Penh, he returned to military duty and is now one of the most prominent field commanders in Phnom Malai (the Mount Malai area), near the Thai border. Kon, or Nikon, came into the limelight because he was implicated in the murder of Malcolm Caldwell by the "confession" extracted from "the contemptible Chhaan," apparently a member of the murderers' gang. This document comes from the Tuol Sleng interrogation center. It is not known if Kon was actually part of this plot or if this confession has been fabricated in order to implicate Son Sen and bring about his removal. Kon was probably involved in some kind of factional activity. Son Sen was the obvious target of the next high-level purge planned by the Security to purify the party and eliminate the "traitors" hidden at the top. In the ensuing controversies in the Western press, the fact that Kon is a relative of Son Sen has usually been overlooked.

Among other families who had played a role in Cambodian politics under previous regimes, we should mention some who were also present in the upper strata of the DK regime, such as Nuon Chea's relatives, or the Thiounn family, who managed to survive rather well, or the Thouch or Pok families, who did not do so well.

It should be remembered that this ''family strategy'' was far from being systematic. Relatives could be judged as useless or even nonentities by powerful figures. It was said in Phnom Penh that Ieng Sary refused, in late 1978, to meet a brother of his who had come from Preah Vihear province to Phnom Penh for medical treatment. In 1981, the *Vietnamese Courier* carried an interview with an elder half-brother of Pol Pot who explained that a large part of Pol Pot's most direct relatives were just ignored and suffered the common fate of ordinary people.

It seems again that kinship solidarity was far from being automatic but was used as a convenient way to reach for more power. Y Phandara made the following observation regarding Ieng Sary and the formation of a new committee in the Bung Trabek camp in late 1978: ''He did not want husband and wife to sit in the same committee, but a couple could take part in two different committees.'' The reference was to his leading the Foreign Affairs Ministry while his wife Ieng Thirith led the Social Affairs Ministry. He offered an indirect justification, saying that ''if a high-ranking cadre gives important tasks to members of his family, the reason is that he cannot find other people whom he can trust.'' Y Phandara was led to the following conclusion: ''Norodom Sihanouk and Lon Nol have been accused of giving privileges to their relatives. But the Khmers Rouges, under the cover of principled ideas, did the same.''[9]

As for corruption, refugee testimony shows that it was very common, the most favored currency being not money but food. Pin Yathai provides an excellent description of this traffic with a wealth of detail reminiscent of the living conditions and ways to buy off guards in the German concentration camps of World War II.

All these details, for all their triviality, can help us understand some aspects of the tragedy that took place in the DK period. The fact that removing or purging an individual cadre meant that the same fate was met by his relatives, often his close associates, and sometimes most, if not all, of the people under his authority, had some root in social reality. People tended to look for a patron who would help them and promote them. Among the motivations of most young educated people who returned to Cambodia after April 17, 1975, the craving for power was certainly foremost. Peasant motivations in joining the Khmer Rouge are not well known, but individual ambition certainly ranks quite high. This would be reminiscent of similar motives in Vietcong recruitment in South Vietnam, which is better known.

But in such a context, one does not compete all alone; the best chance is found with a powerful leader who can provide protection (and the Security service was a deadly threat to anyone) and promotion. These bonds of loyalty, however, were not absolute; they could be switched to another, more powerful patron. But it would be quite normal for a cadre, when replacing another who had been purged, to bring along with him his own dependents and clients, leaving thus few chances for the clients of the former boss to fit into the new network of personal affinities, for lack of "trust." The likely outcome for the follower of a losing patron was that he would share his patron's fate. In this highly collectivized society, no one could survive outside a strongly hierarchical group.

Along with other authors, I once argued that the KCP was a loose federation, composed mainly of three contending trends with diverging political affiliations.[10] After more research and more discussions with Khmer participants in this period, I would now tend very much to downplay the role of ideology as a divisive factor. Take, for example, the fall of the Eastern Zone command in April-May 1978: I see much less evidence of an ideological split between So Phim's politics and that of the Center. I believe that the Eastern Zone had been basically loyal to the party line throughout. But the urge to take possession of this quite autonomous "fortress" was strong enough, I think, to prompt a coalition of Pol Pot (Center), Ta Mok (Southwest), and Pok (North) forces to smash what had been one of the oldest strongholds of the party. Party loyalties there were oriented toward So Phim, an old veteran, and in order to reorient these loyalties toward the sole legitimate power, the Center, So Phim and the whole party chain of command of the zone, together with a large chunk of the local population, had to be removed.[11]

The Heng Samrin regime is certainly not immune to this kind of problem. Its ruling layer is composed of very diverse groups: Khmer Viet Minh returnees, former local Khmer Rouge, and former Sihanouk and Lon Nol military and civil servants. But there are other sides to it. I would suggest that the removal of Secretary General Pen Sovan in early December 1981 is a case in point. From random conversations with various officials in Phnom Penh in the summer of 1981, I got the feeling that Pen Sovan was rather isolated, or perhaps was isolating himself, and that he had "his" people. He had been placed in the top position in the new party because he was a choice for the future. The bulk of the party, at least of its surviving members, had followed Pol Pot in his retreat to the Thai border. A new party had, in the Vietnamese view, to be rebuilt, as a safeguard for the future, out of the remnants of several generations of activists, former Khmer Viet Minh, former Khmer Rouge who had followed Pol Pot or some other local leader, and even left-leaning intellectuals and administrators of Sihanouk and Lon Nol vintages. An experienced and able leadership is the key to the development of the party and, again in the Vietnamese view, the sole guarantee that the Khmer party would not fall again into criminal errors. But no obvious personality was available for this crucial task. The most trusted elements being already quite old, the Vietnamese probably settled for the younger Pen Sovan, who would slowly mature, during maybe ten years, into a suitable and efficient secretary general. In the meantime, the task of reorganizing and expanding the party would fall to older, more experienced hands, like Bou Thong and Say Phuthang. What probably went wrong was that Pen Sovan was unable or unwilling to be just a learner. He apparently started to play politics and build up a faction of his own followers, exactly the opposite of what was expected of a secretary general, whose main task is to hold the party together, continually reinforcing unity. If he was not a unifier, then he was not suited for the job. I think this fits well with the comments of some Vietnamese diplomats expressing their "disappointment" with Pen Sovan's performance in his job. Speculation about his playing the Russians against the Vietnamese does not have much substance and is in fact must less to the point than the very crucial question of the party's reconstruction. The Vietnamese were reacting not so much to his political alignment as to the fact that he was apparently trying to create a political faction. One may surmise that the Vietnamese, who belong to the only party in power that has so far avoided schisms and bloody purges, are particularly sensitive to any trend in the new Khmer party that might lead to a renewal of the factional struggles that so strongly undermined the party under Pol Pot's leadership and finally wrecked the DK regime. Factionalism is not a Khmer monopoly, of course, but it is a choice specialty in the local political tradition.

Nepotism, corruption, and factionalism, as means of governing, are too well documented for the Sihanouk and Lon Nol periods to deserve more comment here. The same can be said about the present situation at the border. The bloody fighting among petty warlords in 1979–80 has been widely reported in the press, although sometimes in a rather unclear way. Now the surviving warlords have gained a respectable cover by allying themselves to some prominent politicians. But, all the same, corruption, nepotism, and killings are still prevalent in most camps outside of the Khmer Rouge areas. The last such killing prompted the return of Son Sann from the United Nations and the demotion of General Dien Diel from his KPNLF military command. So wide is the gap between the politicians and the civilian administration of the camps on one hand, and the warlords, turned military leaders who control the border markets, on the other that only a very strong Thai pressure can prevent a complete decomposition of the camps. There, in this unstable setup, it is clear that nothing has been learned from the past and nothing credible is proposed for the future.

Southeast Asian Theravada societies never had an institutional framework of hierarchical relations, like the Hindu caste system, but they had other means for building up pyramidal relations, namely, patronage and slavery, which complemented each other. The very elaborate system of slavery has formally disappeared, but it has left a deep historical imprint. It is fascinating to consider how quick and easy it was to turn Khmer society into a system having so much in common with the old slavery, with its hierarchical layers of slaves. Such very oppressive regimes always had a very low economic productivity, contrary to what some ignorant ideologists expect. But my point here is concerned mainly with the political system and its continuities, as they persisted through a series of different regimes. I think it is fair to say that factionalism is the most recurrent tendency. It is the modernized version of the old patronage system. What we call nepotism, favoritism, graft, are just the means and ends by which this system is expressed and fed with people, money, and other required resources. There is no judgment here. This political pattern obviously caters to social needs. The Khmer society, compared with others, appears loosely structured. Before the war, the presence of the state was far from being felt everywhere. In other words, political integration had not been completely achieved. The same can be said for Cambodia today, with the state appearing even weaker than it used to be, the Vietnamese presence notwithstanding.

These continuities should be the basis of any projection of a political future for the country. They are certainly bounds within which a political reorganization is thinkable. Any proposal going beyond these bounds, calling for some kind of modern representative system, had better be forgotten as unrealistic and probably dangerous. It is a time now in Cambodia when, having gone through the most excruciating experimental reforms, the people can think only of one thing: the past, how good it was, how peaceful, how unmurderous. The criticisms of this past may not be forgotten, but they are hardly relevant. The Khmer dream of the advent of a *dharmaraja*, maybe without the name, but a just and righteous ruler, a source of harmony, a fountain of merit. The only candidate for such a wishful dream would be a king. This is not a matter of any particular person; nobody is looking for another politician. On that level, they are all bankrupt. But let us for one moment entertain a sociological dream, and picture the royal palace with a great festival going on for the crowning of a young descendant of Ang Duong, the royal barges on the river, the monks chanting, the villages rejoicing, the dawn of a new era. . . .

The decisionmakers in our world do not usually indulge in dreams, even if they be sociological. But who cares for the dreams of the Khmer?

Notes

1. I would like to thank Donald Nicholson Smith for his help in revising the text.

2. S. J. Tambiah, *World Conqueror and World Renouncer* (Cambridge: Cambridge University Press, 1976).

3. Thailand's view of Cambodia as a buffer zone, which should be open to some form of Thai influence, is well presented by Khien Theeravit, professor at Chulalongkorn University and adviser to Premier Prem, in a paper presented to the pro-Democratic Kampuchea Tokyo Conference in June 1981, "Thailand's Response to the Vietnamese Aggression in Kampuchea."

4. See Tambiah, *World Conqueror,* and Immanuel Sarkisyanz, *The Buddhist Background of the Burmese Revolution* (The Hague: Nijhoff, 1965).

5. The Saloth family of Pol Pot, for instance. See Serge Thion and Ben Kiernan, *Khmers rouges! Matériaux pour l'histoire du communisme au Cambodge* (Paris: J. E. Hallier/Albin Michel, 1981), p. 367.

6. The Brévié line was established in January 1939 to allocate to Cambodia and

Cochinchina islets and islands on the coast of the Gulf of Siam, for administrative purposes only: "The matter of territorial dependence of these islands remains entirely reserved." The Khmer governments have insisted on saying that the present border was established mostly by the French colonial authorities, without proper and free acceptance by Cambodian authorities, to the advantage of Cochinchina, which at the time, as a colony, was French territory. Sarin Chhak, *Les Frontières du Cambodge* (Paris: Dalloz, 1966), contains all the documents, including the Brévié letter, together with maps and the Khmer judicial point of view.

7. Y Phandara, *Retour à Phnom Penh* (Paris: A. M. Métailié, 1982), p. 176.

8. Ibid., p. 102.

9. Ibid., pp. 149–50.

10. See Thion and Kiernan, *Khmers rouges!*

11. Ben Kiernan provides the most detailed description to date of the Eastern Zone crisis in his essay "Wild Chickens, Farm Chickens, and Cormorants: Kampuchea's Eastern Zone Under Pol Pot," in Ben Kiernan and David P. Chandler, eds., *Revolution and Its Aftermath in Kampuchea: Eight Essays* (New Haven: Yale University Southeast Asia Studies, 1983).

David P. Chandler

A REVOLUTION IN FULL SPATE: COMMUNIST PARTY POLICY IN DEMOCRATIC KAMPUCHEA, DECEMBER 1976

The People's Republic of Kampuchea (PRK) in many respects is a postrevolutionary society. Democratic Kampuchea (DK), nearly two years after the abolition of the Communist Party of Kampuchea (CPK), disavows any revolutionary intentions. For its part, the PRK regime has so far been unwilling and probably unable to bring the people of Kampuchea to heel to serve its avowedly socialist goals. For millions of people in Kampuchea, such loyalty as they feel toward the regime is connected directly to the contrast between PRK behavior toward them since 1979 and DK behavior over the preceding years. It seems likely that *internal* politics in the People's Republic of Kampuchea will be acted out against this background for the next few years at least. For these reasons, as well as the intrinsic interest of Kampuchea's recent past, it seems worthwhile to examine the DK revolution in full spate, using a fifty-eight-page document prepared by the CPK for private circulation at the end of 1976.[1]

We must be especially grateful for a document of this kind (and there are not very many of them) because almost nothing was printed in Democratic Kampuchea between April 1975 and the end of 1978. The scarcity of documentation, which reflects official policy, is compounded by the regime's contempt for paperwork and research, and its much publicized preference for practice at the expense of theory. To a Palestine Liberation Organization visitor in 1976, Ieng Sary remarked, "We did not act with the guidance of definite theories, but followed our feelings and carried out the struggle in a practical way. . . . What is important is the determination and faith of the principal revolutionaries. We did not study in ideological schools, but practiced a struggle in the light of the concrete situation."[2]

In a speech commemorating the eighteenth anniversary of the founding of the CPK in 1978, Pol Pot was even more schematic, remarking that "theory helps practice, and practice helps theory, but practice remains the basis. In this way the masses can accomplish their goal. Theory alone is too much for the masses. Practice alone will not go far. For these reasons we must combine theory and practice."[3]

These passages raise questions about the "feelings" mentioned by Ieng Sary, and the "determination and faith of the principal revolutionaries," as well as about whose practice and whose theory were to be combined and, more importantly, what sorts of practice "[remained] the basis" of the Kampuchean revolution.

Another characteristic of Democratic Kampuchea that deflects inquiries is its preference for being clandestine, for verbal ambiguity, and for subterfuge. This cast of mind meant, for example, that the existence of the CPK itself was kept a secret from the Kampuchean people until September 1977, and that the secretary general of the party, "Pol Pot," preferred to reveal to nobody that for the first fifty years of his life he had been known as Saloth Sar. Examples such as these could easily be multiplied. In 1976, a party spokesman, writing in a CPK journal, went so far as to suggest that secrecy itself was to be taken as the "basis" of the revolution.[4] Certainly the number of facts kept from the Kampuchean people between April 1975 and the end of 1978, and the methods Democratic Kampuchea employed to keep anyone from sharing or receiving information, suggest a conscious choice, by DK leaders, to run the country secretly, and to explain themselves, if at all, on the basis of a person's "need to know."

Moreover, Ieng Sary's phrase "determination and faith," like a great many DK declarations, suggests that the attitude of the regime toward information was ambiguous, to say the least. The preference for being clandestine encouraged the DK leaders to listen, for the most part, to themselves. The decision to carry their conspiratorial style of government over into the late 1970s, after the liberation of Phnom Penh, is difficult to assess. Did it reflect enormous self-confidence on the part of the leaders, or none at all? Had the revolution succeeded so spectacularly because of clandestinity, or was it on the verge of collapse unless clandestinity was maintained? Who trusted the party? Whom could the party trust? These are fascinating questions, and they have been tackled by several scholars. What interests me at this point is the attitude toward information that prolonged isolation and clandestinity had fostered among these men and women at the center of the party. As William Willmott has shown, it is clear that not much serious research lay behind the CPK's ham-fisted "analysis" of Kampuchean rural sociology, or its decision to single out "feudal landlords" for special blame, rather than local Chinese and Sino-Kampucheans, who dominated the economic life of Kampuchea.[5] It seems likely that information per se was of little interest to them, until it could be harnessed to their ideas of strategy and tactics. Despite lip-service in texts and speeches to studying Kampuchea's "objective conditions," it is unlikely that much studying took place, and while this is unsurprising in the grueling civil war of 1967–1975, the self-assurance of the CPK after 1975 reflects their frequently stated view that *victory* was identical to "practice."

Pol Pot and his colleagues were entranced by the exercise of power, and by the praxis of prolonged and unrelenting warfare. To bring about a millenial transformation of the nation and its people, through internal and external war,

DK leaders depended on the hard work and loyalty of everyone in the country. They sought to insure these by controlling everyone's access to food, leisure, movement, and information. The balance of evidence, and the document I will discuss, suggest that the leaders knew very little about the "people" and distrusted them en masse. Information reaching them that class warfare and national transformation were proceeding slowly, painfully, or expensively seems to have made little difference to their 1976 decision to accelerate these two procedures. Even though regional differences of fertility, personnel, and leadership (to name only three) were often extreme, as Michael Vickery has shown, CPK policies, as internal documents suggest, aimed at a full-scale, consistently *paced* revolution. In 1976–77 at least, what was actually happening to the "new" people, in terms of their own health and sanity, made little difference, as the widely circulated slogan about "new people"—"sparing them's no profit; losing them's no loss"—suggests. On the other hand, information that conveyed the impression to the leadership that everyone's loyalty was open to question—i.e., that everyone was potentially disloyal—appears to have been accepted at the party center in a comparatively uncritical way.

The regime spoke often of "contradictions" (*tumno'h*) and suggested that the clash of opposites itself, rather than any synthesis between them that might develop, was what gave life to the revolution. The regime's own relations with the majority of the people, of course, represent a contradiction of another kind. On the one hand, the masses (*mahachon*) were the ones on whose behalf the revolution had ostensibly taken place. On the other, revolutionaries must always be on their guard against them, at least on an individual basis. In the war with Vietnam that began in 1977, the people were ordered to "fight to the death" but the regime gave them no weapons to fight with. And although the regime claimed in late 1975 that there was only one stratum in Kampuchean society—the worker-peasants—everyone was enjoined to wage warfare against "class enemies." As far as we can tell, these enemies were never defined by the regime with any precision. Although "99 percent" of the masses ostensibly supported the revolution, those who opposed it were thought to permeate every level of the society, including the ranks of the CPK itself. Pulverizing this embedded segment gave the revolution a *raison d'être* (and the party center, incidentally, carte blanche). In other words, forming all of Kampuchea's people into collective life, swiftly and thoroughly, was not intended merely as a means of transforming society in a rewarding way. Rather, as some documents assert, the main purpose of collectivization was to expose its enemies and prevent the old society from reemerging: "Why do we make revolution? We have not come to make revolution blindly. Nobody has forced us to do it. We have consented to all sacrifices for this cause because we want to shake off the domination, oppression, and repression of the enemy and help liberate our worker-peasant class."[6]

The CPK held power in Phnom Penh for barely forty-one months. In the first year or so, operating as a front with Prince Sihanouk as chief of state, the party concentrated on transferring "nearly four million people" (its own estimate) from the cities to the countryside, and on tackling the "problem of water" so as to achieve national self-sufficiency in rice. Less openly, the party center used this time to consolidate its hold on the army, and on the military regions that had operated with a good deal of autonomy during more than five years of civil war. Although it was clear to foreign observers, such as Timothy Carney, that the CPK was identical to the "organization" (*angkar*) supposedly running the country, the party kept its existence secret from the masses throughout 1975 and 1976.

When Democratic Kampuchea assumed state power in April 1976, the party center took several steps to inaugurate what it referred to secretly at first as a "socialist revolution" (*padevat sangkum niyum*). At the end of August 1976, an economic development plan for the years 1977–1980 was discussed at a three-day meeting sponsored by the party center. The plan, which was instituted at the beginning of 1977, called for the collectivization of industry and agriculture and set targets for agricultural production. Shortly afterward, the CPK set in motion the first of many purges against "traitors" (*kbot*) allegedly embedded in its ranks. Veteran members of the party, like Keo Meas and Non Suon, were arrested, interrogated, tortured, and killed. I have argued elsewhere that the purges were probably connected with a controversy connected with the birthdate of the CPK and its implications for relations with Vietnam, and with arguments within the party about making its existence known.[7] The party center decided to set 1960, rather than 1951, as the birthdate of the party, and to keep the party's existence a secret from nonmembers.

On December 20, 1976, the party center completed a fifty-eight-page document, for internal distribution, entitled "Report on the Activities of the Party Center: Political Tasks, 1976."

The typescript has survived, and it forms the basis for most of what follows. This text offers a rare glimpse of Democratic Kampuchea examining itself, evaluating its performance and setting its priorities at a turning point in its short career.

The report is divided into three equal parts, following a one-page introduction. The first of these, entitled "Revolutionary Tasks, 1976," opens with a discussion of the party's plan to introduce and intensify the socialist revolution in Kampuchea. Three aspects of society still impede such a revolution: exploiting classes, individualism, and privilege. In 1976, however, exploiting classes fell "even farther down" than in 1975, but remnants remained to be "beaten, pulverized, and uprooted," and class enemies existed even in the ranks of the party. As for individualism, the key to obliterating it was to abolish private and even personal property. Privilege, which still existed "especially inside the party," needed to be eradicated as well.

The report goes on to make a case for an all-embracing socialist revolution, claiming that "if our revolution is not all-embracing, the party will be confused, the army will be confused, the people will be confused, officers and ministries will be confused, and the 1976 development plan will be defeated.''

But the diffusion of clarity is not sufficient reason to wage a revolution, as the next three pages of the text make clear. A major reason for doing so is to extirpate its enemies, compared to ''microbes'' (*merok*) embedded inside it. The ''people's'' and ''democratic'' revolutions of the 1960s and early 1970s had allowed some of the microbes to take shelter inside the CPK itself:

> The heat of the peoples' revolution and the democratic revolution were not enough. . . . The level of people's struggle and class struggle meant that [when] we searched for evil microbes inside the party, we couldn't find them. They were able to hide. Now that we are advancing on an all-embracing socialist revolution . . . in the party, in the army, among the people, we can find the evil microbes. They emerge, pushed out by the true nature of the socialist revolution.

Throughout the text, ''enemies'' and ''traitors'' inside the party and abroad are never identified by name or in terms of ideology. The tendency to leave readers and listeners guessing about such things was widespread in DK texts before a change in 1978, when all enemies became generically Vietnamese. This 1976 text, however, fails to clarify who the ''traitors'' were who have lingered on from the earlier stages of revolution and must now be pushed aside. Perhaps they were Sihanouk and his entourage; perhaps they were those like Keo Meas who argued against the party's continued clandestinity (or some of its policies); perhaps they were the DK ambassadors recalled from their posts in November 1976 for ''consultation'' and arrested for treason the moment they arrived. The text, in other words, gives us no criteria for judging ''friends'' or ''enemies.'' Categories of people to guard against, and the kinds of offenses they are likely to commit, are not made clear. As in so many DK documents, one gets the impression that a major purpose is to provide the party center with freedom to maneuver.

There is no need for more precision, the text contends, because a socialist revolution provides party members with correct behavior on its own: ''We must resolve the errors that we encounter in our ranks, but no ministry of state security will reveal them. The socialist revolution [itself] will do so, and will seep into the party, the army, and the people, sorting good from evil.''

The revolution proceeds by the collectivization of food, supplies, the means of production, workers, and peasants. In 1976, we are told, the process had already made enormous progress, against enormous odds: ''We had empty hands. We lacked food. We lacked tools. Cattle and buffaloes were sick, and many died. Enemies from outside, and enemies within, continued their activities. We had no assistance from anywhere in the world.''

At the risk of laboring a point (particularly for those who have immersed themselves for any length of time in DK documents), there are no answers provided here for the important questions: Who were the enemies? What were they trying to do? Moreover, nothing is said of any human deaths connected with the acceleration of social change.

The report goes on to suggest that the only way to counter the hostility the revolution has met inside Kampuchea and elsewhere is to prepare for a people's war: guerilla warfare inside the country, and a full-scale war along the frontiers, particularly since there are secret agents (*phtai knong*) working for unnamed powers everywhere.

The second section of the report deals with what it calls "Two Key Tasks for 1976": organizing collectives into larger units than villages, and solving the problem of water. The expansion of collectivization was delayed because there were insufficient party members to supervise the process—an issue taken up in more detail at a later stage of the report. Solving the water problem means, in essence, providing all collectives with enough water to deliver the three-ton per annum per hectare quota, or to exceed it. According to the report, the population of Kampuchea, estimated at 7.7 million people, consumed thirteen *thang* (i.e., approximately 780 pounds) of *padi* per capita per year. At 1976 production rates, calculated in some detail, the party center estimated that this would already allow for an exportable surplus of milled rice of between 100,000 and 150,000 metric tons.

The report goes on to review the production of *padi* throughout Kampuchea in 1976. Four zones (Northwest, East, Southwest, and West) are deemed to have produced sufficient rice for food and seeds, and to maintain enough reserves for some of the surplus to be exported. Exportable surplusses in the East and Northwest, traditionally the "rice-baskets" of Kampuchea, are set at 50,000 tons apiece. The less favored West and Southwest are asked to export 20,000 and 30,000 tons respectively. The North (an area including Kompong Thom) was short of *padi* for food to the extent of 12,117 tons in 1976, according to the report; the Northeast was similarly a deficit area. Region 106, finally, which included Siem Reap, enjoyed a 10,000-ton surplus of *padi* and was asked to set nearly all of it aside for export.

One point that emerges from this passage is that the Northwest (largely liberated in 1975, and then repopulated with hundreds of thousands of "new people") is asked by the party to export less than half of its alleged surplus of *padi*, while the East, with its healthy workforce, its competent cadres (and, perhaps, its proximity to Vietnam) is levied more than two-thirds of its surplus.

With exports out of the way, the report goes on to discuss five problems that may impede the acceleration of agriculture in Kampuchea. The first of these is the problem of water. More work is needed, the report asserts, on water shortage tanks, feeder canals, ditches, and reservoirs. The second problem, treated in rather less detail, involves the labor force: "In 1976 the labor force was feeble. It was only in the East that it was not feeble. Thus in the coming year we must arrange things properly and persuade people to follow the party's plan, and to rest according to the party's plan. Resting three days a month is sufficient for health, but when people work nonstop, it has ill-effects on them."

There are two things of interest in this passage, aside from the righteousness of its tone. The first is that it confirms the conclusions of several scholars, derived from interviews, that conditions in the DK Eastern Zone were somewhat more benign, and that the labor force was handled somewhat more humanely than elsewhere in the country.[8] One reason for this is that the region had a long tradition of radicalism and an unusually high proportion of indigenous peasants. Many CPK cadres operating in the East had become affiliated with the Vietnamese communists during the war against the French (1945–1954), and this long association gave them not only some sympathy for the revolution but also some experience of working—both as cadres and as farmers—among the local population. The passage itself implies that inexperienced cadres elsewhere in Democratic Kampuchea, feeling pressure from the party center and indifferent to the fate of inexperienced newcomers, were prepared to work some of them to death.

The remaining problems obstructing the achievement of CPK goals have to do with seed-rice, fertilizers, tools, and medicine for the peasants. Throughout this section, the party center deals rather mechanically with what it perceives to be a mechanical problem. There is nothing here about increasing one's awareness of contradictions. Instead, the "labor force" is seen as an ingredient of the problem, rather than its solution; it is akin to tools and water. Like livestock, the labor force needs food, rest, and medicine. In the context of the passages being discussed, the aspirations and "liberation" of people inside the work force are neither here nor there.

The report then makes a *tour d'horizon*, setting forth plans for the future in other areas than self-sufficiency in rice. These involve increasing production of secondary and plantation crops, raising more animals for food, producing more medicine, repairing transportation facilities, and so on. Interestingly, "culture" is mentioned in this context, briefly and last of all; the passage is worth quoting in full.

> Progress in the work of abolishing illiteracy has been good. Some places manage to set some time aside, specifically for studying. But other places have failed to do this. Arrangements should be made carefully. Books should be prepared that are identical throughout the country. It is important that these books be easy, so that people can learn [from them] to follow the revolutionary path.

Here again, the people are treated glancingly, as objects rather than subjects, and education (a word that is not mentioned in the report) is defined as the absorption of simple, standardized texts that enable otherwise uneducated people to follow the revolutionary path.

The report's final section opens on the following page, and it carries the title "Tasks of All Sorts Which Are Subservient to the Political Central Tasks of 1976." The first of these is seen as the problem of developing the party and its central organization. The report claims that all party members have been informed by now of the decisions of the party center that led to the policy of accelerating a socialist revolution in Kampuchea. Once again, as in the first section of the report, the main obstacle to this object is a lack of understanding, and therefore, "We must grasp the essential character of Kampuchean society more clearly, so as to know who friends are, and who enemies are, and in what ways the class struggle in Kampuchea must be carried out. This is a real necessity, especially in forming cadres at every level."

The report goes on to give an "example" of what is required:

> In order to defend the country well, one must grasp the situation inside the country firmly, the situation all around us firmly, and also the situation of the world. If these are not grasped firmly, the enemies' weaknesses and the enemies' strengths cannot be seen; nor can our strengths and weaknesses. Wrong measures will be taken, leading to disorder.

To avoid "disorder," cadres must be trained carefully to be in tune with party strategy for the next "ten, twenty, thirty . . . hundred years." The report admits the difficulties of inculcating so much subtlety at the pace required by the revolution, but it also stresses—in a brief paragraph—the need for unspecified, intensive study.

A major problem the party center faced at the end of 1976, which the report discusses next, was the need to expand party membership to provide cadres capable of understanding the principle of collectivization on the one hand and of administering collectives on the other. Although the number of party members already in place in each collective is not given (and would be impossible in the report), the CPK suggests that this figure be increased in each collective by 40 percent in the first half of 1977, and by 60 percent in the second half, so as to double by December 1977.

How were new members to be selected? Each *sruk* was to provide "four to ten" (i.e., several) candidates per quarter, chosen in terms of "life history" (*pravatt'rup*) and from recommendations. Candidates would be told that "higher authority has called them away to study." When the (unspecified) studying was done, candidates would be made to prepare their own life histories. If these failed to conform with information received from the candidates' base areas, they would be asked to reexamine the profiles, and then the candidates would rewrite them until the two reports converged.

People's life histories, the report contends, are the strongest recommendation for new members of the party—far better than someone's deceptive record of activities, which might allow enemies to appear eligible for inclusion. What one did was important during the war against Lon Nol, says the report—the closest it comes to saying that the war was fought by a united front—but people's class background—i.e., whether they were born poor or not—is now more crucial. Collectives in the hands of these properly screened people, under the leadership of the party, will provide a bulwark (*kampeng*) of defense.

The next section takes up the question of whether the party should "come into the open or remain clandestine." The section is worth quoting in full, perhaps as an example of the ambiguous prose that characterizes so many DK documents from this period:

> The situation inside the country and outside the country is sufficiently developed for the party to come into the open.
>
> Friendly parties express the wish that our party come into the open. For one thing, they see that the situation would allow the party to come into the open. For another thing, they need our support.
>
> Enemies also want to see us come out, so as to see clearly; this would enable them to pursue their own ends.
>
> But if the party came into the open, this would pose problems of protecting the leadership of the party. During September and October 1976, we were on the point of coming into the open. But since that time, documents have revealed that the enemy has sought to beat us, more than ever.
>
> Still another problem is that if the party came into the open, difficulties would be created with certain people. We have therefore decided to defer a decision on the issue of the party's coming into the open.

It seems likely that while pressure to bring the CPK into the open probably came from China, and perhaps North Korea, further pressure came from within the party, particularly as the numerologically significant twenty-fifth anniversary of the founding of its predecessor party, the Khmer People's Revolutionary Party (KPRP), approached on September 30, 1976. Pressure to celebrate this anniversary probably came from elements in the party opposed to its anti-Vietnamese stance. The KPRP had been founded, after all, under the patronage of the Indochina Communist Party (ICP) soon after that party's "dissolution." The "documents" referred to here may be confessions extracted from some of these long-serving party members, like Keo Meas, who had been arrested in late September; one of his confessions in the archives at Tuol Sleng is entitled "1951 or 1960?"—a reference to the alternative founding anniversary (1960) preferred by Pol Pot and his colleagues.[9]

While it is likely that the text refers to this sequence of events, we are given no clear assurances as to who the "enemies" were, or what their "activities" involved. And how would bringing the party into the open have endangered the party center? It is possible that what is meant is that the pressure to bring the party into the open was accompanied by pressure for a coup d'etat; certainly DK spokesmen implied as much, in 1977–78, when referring to September 1976. But there is no way of knowing this from the passage as it stands.

The report turns next to the problem of inducing the people to strengthen their syndicates and collectives "so as to develop the worker-peasant alliance under the leadership of the party."

Previously, the report admits, political training had been ignored or insufficient. In many places distrust had built up between the people and CPK cadres. By improving cadres' political consciousness, the report declares, the people's confidence in them will increase. Cadres who live apart from the people, and don't suffer from the same shortages, after all, are not respected, so they must live among the people and serve them unafraid: "Don't be afraid of enemies buried among the people. Whenever we serve all of the people, even though there are one or two enemies among them, they cannot stay buried for long."

These three pages reveal lyrical confidence in the capabilities of ordinary people, unique in the report; in fact, they encourage an open dialogue (about a limited number of subjects, to be sure) which the party center itself appears to shy away from. In a revealing passage, the report goes on to say that "some comrades seem to believe that all 'new people' are enemies and so they do not pay attention to expanding their awareness of politics, and do not pay attention to solving problems of their livelihood. This is a very great error, for it would mean not gathering everyone up on the side of the revolution."

One way of "stirring up the people," and of increasing their understanding of the revolution, is to hold short, frequent meetings to explain the party's plans to them. There are problems even here, however:

> If we stir up the masses in this way, one crazy fraction will take advantage of us, and will raise this difficulty, that difficulty, beg for this, beg for that. But we shouldn't be afraid. Our path is correct, our rationale is correct, we serve the people, the people support the party. In the beginning, one or two people [presumably cadres] are able to mistreat the people, but as time goes on, the people gain awareness, they understand.

The next section deals with the expansion of the party's control of state power. The "very poor class" nourishes the party and protects its dominant position by uprooting "secret networks of the enemy." These can be found in collectives and in factories. The frightened and menacing tone of this passage contrasts sharply with the benevolent one before it. Why is this so? Perhaps because the "enemies,'' unlike the ''people'' are unspecified, and, because they are ''buried,'' hard to find. If ''the people'' are those who believe in the revolution (as defined by the party center), enemies are those who actively oppose it, and those who only *seem* to give it their support. How is hypocrisy to be ferreted out? Primarily by verifying everyone's class status, and by taking advantage of the heightened clarity of vision that accompanies one's understanding of the revolution.

The ''enemies'' of the party center at the end of 1976 appear to have been a transitional group. On the one hand, in the past, lay the Americans, the Lon Nol ''lackeys,'' capitalists, and big landlords of the ''old society,'' making up ''new people'' who had been—in the words of the report—''rounded up'' into collectives, particularly in the Northwest. On the other hand, perhaps, in the future, but clearly hinted at in the report, are the Vietnamese. The enemies of the moment, in other words, are those who might tend to support those of the past, those of the future, or both. By keeping all of their identities imprecise, the party center maintained its freedom of maneuver to attack the enemies it liked, and its ability to do so was enhanced by its assumption of state control.

At the end of 1976, the report suggests, however, that there were parts of Kampuchea where the CPK was not sufficiently in control:

> In some places, particularly in the Northwest, the proportion of new people is far too high. For this reason, former soldiers and former government employees have insinuated themselves into positions of responsibility in many collectives. For this reason the party has been unable to develop sufficient influence, so far, and so a number of combatants and a number of revolutionary youths must be chosen to take charge of these collectives to prevent enemies from taking over.

No one can transform himself into a revolutionary, in other words, unless his class origins allow the transformation to take place. The report makes this point, by warning against accepting people from the ''upper stratum'' for positions of responsibility in the collectives, because they may belong to ''enemy networks,'' while ''as for people from lower strata who have just come out of the cities, don't choose them yet [for positions of responsibility]. They are too diverse.''

Turning to DK foreign relations, the report makes the point that the CPK has established friendly relations with a number of ''Marx-Lenin'' parties, including parties in and out of power. In the future, the report goes on, DK plans to increase the number of noncommunist countries with which it has diplomatic relations, partly to overcome the insults it has received from ''American and French imperialists,'' who have taken unfair advantage of Democratic Kampuchea's inability to defend itself abroad. Characteristically, no other friendly or unfriendly countries—and particularly China and Vietnam—are mentioned at this point. The section closes by stating,

> Our only failing overseas has been a shortage of cadres in the revolutionary movement who are able to work in the field of foreign affairs. Some [foreign affairs] personnel have come back from outside the country, and thus are separated from the movement. Others have acceptable life histories, but lack the skills to serve the movement. At the moment we need four to ten new ambassadors, and we beg those attending this meeting to put their minds to solving this problem.

I have quoted this passage in full because the last line indicates that the report was delivered orally for discussion, and more importantly because some members of the party center would have been aware that on the very day this report was tabled—December 20, 1976—DK ambassadors and diplomatic personnel, summoned home for "consultations," were arrested at Phnom Penh airport and taken off for interrogation, torture, and death—thus creating some personnel vacancies in DK missions overseas. Interestingly, foreign affairs was the only area in which Democratic Kampuchea was required, by other people's conventions, to behave recognizably as a government, with accredited diplomats, published policies, and so on.

The following section of the report takes up the issue of implanting the ideas of revolution through Democratic Kampuchea: "We are not thinking merely of forests and mountains, we are paying attention to the entire country: the countryside, the cities, the collectives, the factories, offices, and ministries, all of them can see how the revolution has progressed. . . ."

An issue of concern to the party center, however, was that "certain places, some in the interior, others along the frontiers, are not yet firm, or reliable enough . . . there are problems here. People are running off to Vietnam, to Siam [*sic*], to Laos. The places are not reliable, for one thing, because of new people. And yet there are also base areas of long standing, from which people are fleeing to other countries."

To combat this problem, the report suggests "toughening" the army and expanding the role of the CPK. In practical terms, people who lack "good characteristics" should be moved from frontier areas, and their places taken by "old, reliable people" (i.e., hardened revolutionaries).

The remainder of the report (pp. 57–58) exhorts listeners to protect the party, by remaining vigilant, "without making their alertness known," and by working in secret to ferret out enemies wherever they may be found.

The report concludes by pledging its listeners' obedience to the tasks imposed on them by the party center, "no matter how much initiative or creativity you may show."

Inadvertently, the report tells us a good deal about the mentality of the DK regime and the ways in which it governed Kampuchea from 1975 to 1979. The stress the document places on clandestinity and "enemies," for example, and the ways in which it seeks rather than avoids ambiguities of language, probably derive in large part from the experience and preferences of the men and women who made up the party center.

Throughout the history of the CPK and its predecessor parties, the membership of the Central Committee, party policies, and everything else about the party had been kept a secret. Such clandestinity was normal for a party intent on seizing power and was forced on the CPK by Sihanouk's anticommunist policies. Moreover, it worked: before 1970, Sihanouk himself seems to have had no idea of the membership of the party center. After 1963, when this group of people, led by Saloth Sar (Pol Pot), sought refuge in the remote northeastern province of Ratanakiri, the disjunction between them and the Cambodian political process became even more pronounced. From then on, the center seldom explained itself, even to party members, or argued the feasibility of the policies it was developing in secret, on its own. No party congresses were held between 1963 and 1971.[10] Instead, at the core of the party was an anonymous "organization" (*angkar*) impervious to attack (or consultation) and indifferent to information. Anonymity served the leaders well during the civil war that broke out in 1967–68, but why did Saloth Sar's secretiveness about himself, and the CPK's reluctance to come into the open, persist after April 1975? Three explanations suggest themselves. One is that Pol Pot, who assumed this pseudonym at about this time, felt no obligation to reveal himself to his enemies, who had lost the civil war. From 1975 to 1979, indeed, the "organization" treated survivors of the Khmer Republic as if they were citizens of another nation, rather than fellow Khmer. A second explanation is that the leaders of the CPK, by keeping under cover, were not required in public to honor the internationalist components of communist ideology, or to express fraternal solidarity with other parties, particularly with Vietnam. The Vietnamese, of course, knew who was in charge of the CPK without being told, and they were aware of the anti-Vietnamese policies the CPK had pursued since 1971. Moreover, the mere expression of solidarity between two communist parties does not preclude contrary behavior. Thus it seems likely that the CPK kept its identity a secret primarily in order to convince people inside the country that the "organization" owed nothing to communist history, foreign aid, or foreign models. It served *Cambodian* ends. Because it was incomparable, the Cambodian revolution was idiosyncratically national, rather than recognizably communist. Enemies were identified as foreigners, or those who served them.

The question then arises: why did Pol Pot and his colleagues persist in calling themselves communists at all? They did so, perhaps, to retain the support of China and North Korea, to retain the services of CPK cadres tempered by the civil war, and because they saw themselves not as schismatics but as communist revolutionaries of a particularly pure sort whose victory offered a model for revolutionaries elsewhere in the Third World.

By remaining clandestine and anonymous, the party center maintained its freedom of maneuver. By using ambiguous language, it was able to keep its "enemies"—who changed from day to day—off-balance. The idea that the Cambodian revolution was exemplary and incomparable, coupled with the notion that the party center was always right, may have led the CPK to force the pace of the revolution in 1976–77, with little regard, if any, for the human costs involved. In December 1976, when the report was prepared, the leaders of the CPK may well have been uncertain of Chinese support, following Mao's death; uncertain about relations with Thailand, following the right-wing coup of October 1976; and uncertain about Vietnam's long-term plans. These uncertainties probably intensified the party center's search for "enemies" inside Kampuchea's borders, and even inside the CPK. It is possible that a sense that time was running out encouraged the CPK's leaders in their belief that the revolution in Kampuchea had to be accelerated and *completed* as rapidly as possible. Once again, the absence of historical precedents for such a transformation does not seem to have worried the party center; in fact, if their statements are to be believed, the absence of models was what encouraged them to push ahead.

As things turned out, the party center had little cause to be alarmed about China or Thailand in 1977 and 1978. On the other hand, encouraged by Beijing, the CPK soon chose to embark on a provocative policy toward Vietnam, to preserve Kampuchea's independence and to focus the virulent energies of the revolutionary process in general, and the army in particular, on a new target. It is impossible to say whether those in the party center believed that Vietnam could be defeated militarily, or that Vietnam would never dare to invade Kampuchea in strength. In the process of "purification" that swept through the country in 1977–78, as the war was waged, hundreds of thousands of Kampucheans were executed as "enemies" or died of malnutrition or disease as they were mobilized for all-out war. There is no way of telling how far this process would have continued had the Vietnamese postponed or cancelled their attack, but it seems likely that the policies of 1977–78, many of them foreshadowed in the 1976 report, laid the basis for millions of Kampuchean survivors turning against an indigenous nationalist regime and welcoming a foreign invasion. In this sense, the CPK's estimate that its enemies were everywhere may turn out to be one of its very few accurate assessments.

Notes

1. *Reaykar sakommapheap reboh mochumpak taam pear'kech niyobay ruom chanm 1976*. I am very grateful to Ben Kiernan for his comments on an earlier draft of this chapter and for allowing me to study this document. The original typescript has been preserved by officials of the People's Republic of Kampuchea, who allowed Kiernan to photocopy it in the course of his research in Phnom Penh in 1980.

2. Quoted in Hamad Abdul Aziz al Aiya, "Modern Kampuchea," *News from Kampuchea* 1, 4 (October 1977):35.

3. See United States, *Foreign Broadcast Information Service (FBIS), Daily Broadcasts*, October 2, 1978.

4. *Tung Padevat* [Revolutionary Flag], (September-October 1976), p. 27.

5. William E. Willmott, ''Analytical Errors of the Kampuchean Communist Party,'' *Pacific Affairs* 54, 2 (Summer 1981):209–27.

6. *FBIS*, September 4, 1978.

7. See David P. Chandler, ''Revising the Past in Democratic Kampuchea: When Was the Birthday of the Party?'' *Pacific Affairs* 56, 2 (Summer 1983).

8. See the essays by Ben Kiernan, Michael Vickery, and Anthony Barnett in David P. Chandler and Ben Kiernan, eds. *Revolution and Its Aftermath in Kampuchea: Eight Essays* (New Haven: Yale University Southeast Asia Studies, 1983).

9. I am grateful to Ben Kiernan for this information about Keo Meas's confession, which I discuss in more detail in ''Revising the Past.''

10. Ben Kiernan, personal communication. According to Vorn Veth's confession in 1978, many policy decisions flowing from the 1971 congress were amended, cancelled, or initiated by the party center after the congress had adjourned.

Timothy Carney

THE HENG SAMRIN ARMED FORCES AND THE MILITARY BALANCE IN CAMBODIA*

Introduction

The question of a military force for the People's Republic of Kampuchea (PRK) bears directly on a key aspect of the argument about Cambodia's future.* Without a military force of their own, the Heng Samrin authorities could not credibly claim legitimacy internally. Nor could their international supporters hope to postulate that accepting the status quo would result in an eventual recrudescence of Khmer nationalism and consequently greater independence from Vietnam. The creation of a PRK military force would not, of course, automatically guarantee legitimacy and eventual independence. But a viable PRK military wing might in time relieve the People's Army of Vietnam (PAVN) of some of the burden of fighting.

Enough information is available from Vietnamese and PRK media, and from refugees and defectors, to begin to examine what appears to be a serious effort to create a loyal Khmer military arm and to assess its weight in the balance of forces now fighting in Cambodia. PRK national-level or "main-force" troops have been built to three divisions and a brigade, which may be in the process of developing into a division. Provincial forces include at least two battalions for each province. The latter help defend important points and operate with the PAVN on pacification missions. The main-force units are being groomed for combat against the larger resistance forces near the border with Thailand. A militia ensures order in the villages.

Historical Background

Vietnam's first attempt to create a Khmer military force dates to the late 1940s. At that time the anti-French Khmer Issarak independence movement evolved a wing that was oriented toward the Viet Minh. Up until 1951, the Vietnamese and Khmer dissident strategy was to establish "mobile units" of mixed Khmer-Vietnamese forces, generally with one company operating in a sector of one or

*The analysis and conclusions here are mine alone and not necessarily those of the Department of State and government of the United States.

two provinces. This changed in 1951 as Khmer began to form in separate units.[1]

The original Vietnamese-trained Khmer force was probably no more than 3,500 men, speculates Kiernan,[2] and was so new that it lacked combat experience and full training. A military school had only been opened in August 1950.[3]

The Geneva Conference cut this effort short, and at least 3,000 Khmer, both leaders and rank and file, regrouped to North Vietnam, boarding Polish ships in the guise of Viet Minh troops under the eyes of the International Control Commission.[4]

The People's Republic of Kampuchea sees its military roots in that early Issarak period. On June 6, 1980, Heng Samrin, chairman of the Revolutionary Council of the People of Kampuchea, signed a directive naming June 19, 1951, as the date of the founding of the Issarak army to fight the French. It was to be celebrated as the anniversary of the Revolutionary Armed Forces of the People of Kampuchea (KPRAF).[5] A succeeding circular explained that the Issarak army was "founded and presented to the masses" at the subdistrict of Kompong Tramung in Srei Ambel District of Kampot province.[6] As are other components of the People's Republic of Kampuchea, the army is seen as the continuation of traditions forged during the First Indochina War. In 1981 Radio Phnom Penh broadcast "Directives for Study on the Occasion of the 30th Anniversary of the Unity Day of the KPRAF," explaining that one of the study aims of troops is to "heighten the sense of responsibility of the revolutionary combatants who are continuing the role of the 'Issarak' army."[7]

The Issarak army itself came to nothing as a result of the Geneva Accords and the subsequent development of a more independent-minded party leadership in Cambodia. Vietnam's own objectives soured its relationship with this emerging group, which by the mid-1960s included most of the senior members of the current leadership of the Khmer Rouge, now a component of the resistance against the Vietnamese: Pol Pot, Nuon Chea, Ta Mok, Ieng Sary, and Son Sen. Ideological and foreign policy differences strained the relationship. The Khmer Rouge have, probably accurately, claimed that the Vietnamese opposed the Khmer party's class analysis of Cambodian society, arguing that, as in Laos, Cambodia had not yet developed "conditions of a society divided into classes."[8] The Khmer Rouge claim that this difference meant that the Vietnamese opposed the Khmer party line.

Nor were the Vietnamese willing to aid a Khmer party effort, which began in 1968, to overthrow Cambodian Chief of State Prince Norodom Sihanouk by force of arms. The Prince had accepted Vietnamese military installations on Khmer soil, and Khmer facilities were given over to resupplying them. The Khmer Rouge were so small at that time that the Vietnamese would have seen no hope for their struggle and only risk in provoking the Prince by any important assistance to his internal enemies. One former Khmer Rouge senior officer said in 1980 that the party had asked Vietnam for radios in 1968 and was refused. This must have seemed an especially sharp blow. Khmer Rouge bases were then widely scattered in isolated areas. Their organization was suffering from Phnom Penh's reaction to the opening of armed hostilities in January 1968, from which the Khmer Rouge date the foundation of their armed forces. As it turned out, this refusal paid dividends for the Khmer Rouge in the war against Lon Nol because it forced them to develop a courier system to ensure security. A farther-reaching effect was the formation of a corps of dependable young men to act as couriers. They themselves came into leadership positions through the patronage of their powerful military or party employers.

This is not to suggest that the Khmer party received no Vietnamese assistance. In 1965 Party Secretary Pol Pot visited Vietnam as part of a trip to China. The Khmer Rouge leadership subsequently criticized Vietnam for pressuring them to change their line.[9] The next year, however, the Vietnamese set up an office under Politburo member Le Duc Tho's authority to assist in training Khmer cadre.[10] The Vietnamese likely assisted the Khmer Rouge in more ways than simply escorting them up the Ho Chi Minh trail in 1965 and again in 1969 when Pol Pot led another delegation for talks in Hanoi and Beijing. The Khmer seem to have mistrusted Vietnamese motives then, but evidence for serious differences does not appear until later.

If Hanoi's goals in the 1960s seem to have been to constrain the Khmer movement and keep Vietnamese material assistance to a minimum, the March 18, 1970, coup changed their tack completely. Control over the Khmer Rouge was apparently hoped for, but given up as unrealizable fairly early on. The PAVN offered radios shortly after the coup, but only with operators. The Khmer Rouge, claimed a former senior officer, declined. Once Prince Sihanouk joined the National United Front of Kampuchea (NUFK), the Vietnamese invoked contingency plans to operate against the Phnom Penh coup regime. The plans incorporated the use of Prince Sihanouk's appeal to recruit a peasant army. Hanoi also returned the Khmer who had spent the previous sixteen years in training in Vietnam. By mid-1971 they repatriated and took training roles or became small unit commanders, or at most deputy commanders of larger units. The PRK party secretary for Kandal province, Hem Samin, who trained in mechanical engineering during his stay in Vietnam, told an academic interviewer that he and his colleagues were aware that all was not well between themselves and those who had stayed on in Cambodia after 1954.[11] In fact, the regroupees were not made to feel welcome, and they began to be purged in 1972. A number of current PRK figures saw the likelihood of their own demise and dropped out, fleeing to Laos or back to Vietnam.

Even in 1970, the Khmer Rouge had enough indigenously raised and trained cadres to make themselves relatively independent of the regroupees. As they expanded their party structure behind the screen of the NUFK, the regroupees were increasingly seen as a liability, as the thin edge of a Vietnamese wedge. Evidence contemporary to that period is scanty; however, two Hanoi-trained Khmer Rouge defectors told an interviewing American embassy officer in late

1971 that they rallied to Lon Nol because they believed the Vietnamese were only using the Khmer; because of the extent of Vietnamese control over the Khmer movement; and because of the excessive violence toward the people of the Khmer party itself.[12] Conspicuously absent among their reasons at that point was a fear of being purged.

In 1981 Hem Samin noted that, of those who joined the party in the Issarak days, only a hundred or so survived.[13] Military personnel figure prominently among the survivors: to judge by the candidates' official biographies, approximately a quarter of the PRK National Assembly selected in March 1981 elections was from the military. Some of the former military include regroupees. Chan Si (currently prime minister), for example, returned and worked in the Khmer Rouge ranks until 1972, the earliest date a regroupee cites for leaving the Khmer Rouge movement. Most of the other military members of the Assembly served in Khmer Rouge ranks during the 1970–75 war.

Thus, the military outlook of the PRK elite must have been deeply colored by its Khmer Rouge experiences. The KPRAF acknowledges that its roots date to the alleged reorganization of Issarak forces in Ratanakiri on April 17, 1970.[14] The military history of one important PRK commander, Nhek Huon, born in 1945, original chief of the first PRK unit to be upgraded to division structure, is illustrative of Khmer Rouge experience. According to Vietnamese sources, which refer to him as Nhat Huon,[15] he was "former deputy political commissar of the autonomous company of Region 21" of the East Zone, the same area of Kompong Cham province from which Hun Sen fled to Vietnam in 1977.

The army the Khmer Rouge built from 1970 to 1975 rapidly developed a rigidly disciplined organization. Reward and punishment were strictly based on performance. The People's Republic of Kampuchea retains elements of that organization. In 1970, just before the coup, the Khmer Rouge force included only three companies. They expanded into battalions, three companies each, by the end of the year. At the end of the war, the Khmer Rouge had 230 battalions divided among 35–40 regiments and 12–14 brigades. Regiments held three battalions. The brigades grouped more than a dozen battalions.[16] The command structure of Khmer Rouge units then, as now, centered on a three-man committee for all units from company up. A political commissar held authority over the military commander while a third committeeman was usually listed simply as a member of the command committee or as the deputy military commander. This force promoted and punished savagely, as Hem Samin later noted,[17] on the basis of results, and it forgave little, least of all indiscipline. It ruthlessly sacrificed men for little apparent gain, using frontal attacks against entrenched positions. It pushed forward toward Pochentong Airport in the face of aerial war. It was a discipline many of the current PRK commanders fled when it turned on them in 1977 or 1978, and they have yet to install anything like it among their own forces now.

Building a Dissident Army

The Vietnamese took some time, probably until late 1977, to decide to create an armed Khmer resistance force from among refugees in Vietnam. In the meantime, at least one organized, armed group of former Khmer Rouge was resisting the Pol Pot regime from bases along Cambodia's southwest border. Say Phuthang, an ethnic Thai, had escaped the 1974 purge of Region 11 (Koh Kong) and waged a distant and low-key battle against the party center from 1974 to 1979. He was so far away he was unable to get to the party reorganization congress PRK leaders held in Memot along the Vietnamese border as Phnom Penh fell in January 1979.[18]

The Vietnamese did not initially seem to know what to do with either the Khmer party defectors or former Lon Nol military who fled to them. One former Lon Nol officer who defected in 1981 said that he had been jailed in Vietnam in 1976 after fleeing from eighteen months in a Khmer Rouge prison. In mid-1977 the Vietnamese sent him to study political theory at Long Thanh near Xuan Loc, east of Ho Chi Minh City. After a year at this Khmer refugee center, in July 1978 he became a company commander in what became the PRK forces. Hem Samin himself was more or less ignored during his first year in Vietnam.

The fleeing Khmer Rouge party people began to arrive in Vietnam in 1977. Hun Sen left his regiment in July and arrived with five others. Hem Samin said that he joined them, and with the cooperation of Vietnamese security agencies began work; his first activity was with a Vietnamese force that entered Cambodia in September 1977.[19] Samin told the *Vietnam Courier* in 1978 that dissidents "all put their hope on So Phim," head of the Eastern Zone and senior member of the Khmer Rouge party Standing Committee.[20] The interview and publication seem to have taken place before he and the Vietnamese were aware that Phim had died when the center smashed the Eastern Zone in May 1978. The SRV apparently then had hope of using him, or possibly of spreading a little disinformation to spark the anger of the Khmer Rouge center. The latter interpretation would fit Samin's comments about So Phim in 1981 which showed hatred for Phim, describing him as ruthless, typical of the Khmer Rouge, and unworthy to live.[21]

Hun Sen and the others Hem Samin identified remain important among PRK military and civilian leaders today:

—Hun Sen himself is a member of the Politburo and deputy prime minister charged with foreign affairs;

—Hem Samin is party secretary for Kandal province;

—Nuch Tan, after participating in the December 25, 1978, invasion as a commander, became head of the Youth Organization;

—Nhek Huon rose to command the 196th Division; he was apparently relieved to become "representative" of the Phnom Penh military command by June 1982, and then identified as deputy chief of staff and Phnom Penh garrison commander on December 22;

—Meas Kroch is political commissar for the KPRAF;

—Ung Phan was, from early 1979, chief of the Office of the Chairman of the Revolutionary Council and continues as minister, director of the Cabinet of the Council of Ministers;

—Peng Path was commander of Phnom Penh Municipal Forces in March 1980 and was subsequently identified as commander of the First Brigade, presumably for the capital region; Nhek Huon has replaced him;

—Meas Huon has not done so well: he was named as deputy director of Kompong Som port and vice-chairman of the People's Revolutionary Committee in March 1980 and may be the same man identified as acting *chargé d'affaires* of the PRK embassy in Bulgaria in September 1982.

Other elements of the initial PRK leadership came from former regroupees who defected from the Khmer Rouge in Cambodia's Northeast. Perhaps the most prominent is the minority tribesman Bou Thong, who became minister of defense in January 1982, replacing the purged Pen Sovann. Thong became a dissident in 1974, and he captained various armed bands against the Khmer Rouge zone leadership thereafter.[22] Another minority member, Bun Mi—a member of both the December 1978 and September 1979 Front Central Committees—has apparently ceased to be active since the latter appointment. Some regroupees fled to Laos where, according to Chea Soth, they just farmed.[23] Soth became a member of the secret party politburo in January 1979 and was the first PRK ambassador to Vietnam. He is now minister of planning.

The group of exile Khmer best known to Hanoi would have included Pen Sovann, who regrouped while very young. He became vice-chairman of what was essentially a provisional government when the Revolutionary Council was formed on January 8, 1979. He did not figure in the front created a month earlier, nor did his fellow regroupee ministers, Keo Chanda (Information), Nu Beng (Health), and Mok Sankun (Economy). The initial front consisted of men who were either old-line Issaraks who broke early with the Khmer Rouge or were known to Vietnam through close association. Pen Sovann had been part of the Cambodian propaganda apparatus in Hanoi in the early 1970s; he has been purged. Mok Sankun died in mid-1979. Nu Beng lost his ministerial portfolio and was relegated to a vice-chairmanship of the National Assembly. Keo Chanda, after numerous vicissitudes, which included a period in late 1981 that saw him suspended as Phnom Penh party secretary and party central committee chief of cabinet as well as minister of industry, finally resumed public life in January 1982, but only in his party positions.

The Political Aspects of KPRAF Development

In late 1979 and early 1980, the KPRAF began its development toward a modern armed force. Pen Sovann's January 1980 description of the force's ideals, given to the first course for primary class regular cadre at the Infantry Graduation

School, noted "only one ideal: to liberate the nation and motherland from the danger of extermination and from a life of tears and blood."[24] He emphasized six points: loyalty, solidarity, and internal unity; international solidarity with Vietnam, Laos, and other socialist countries; discipline and organization of the KPRAF; learning, study, and self-respect; active work of food production; and correct behavior in daily life. By October 17, 1980, Pen Sovann was emphasizing an expanded concept when he told the "first national high and medium-ranking cadres political course" that "If we want to build up a strong and capable army to ensure the defense of our territory, waters, and space, we must have a strong KPRAF. . . . Our KPRAF constitutes an elite tool for the defense of our state power."[25] Developing this theme, in April 1981 then Vice-Defense Minister and Chief of the KPRAF General Political Department Chan Si listed armed forces tasks for 1981 as "defense of the territorial sovereignty of the motherland"; maintenance of political security and social order; maintenance of security for the holding of general elections and during important holidays; protecting the people's lives and both private and state property; and, finally, joining with other forces to improve the mass agitation movement, to build genuine revolutionary forces at the local level, and to stimulate production and improve the livelihood of the people.[26]

At Army Day on June 19, 1982, Defense Minister Bou Thang brought these themes together, stressing the need to maintain vigilance; to recognize party leadership of the army; to serve the people; to improve itself to earn popular respect; to maintain solidarity and unity at all levels; to eschew favoritism, bureaucratism, militarism, liberalism, and disrespect for authority; to heighten "international proletarian" solidarity, especially "close and militant" solidarity with Vietnam and Laos (though only "close" solidarity with the USSR and others); and to maintain vigilance and heighten spirit in deflecting Chinese maneuvers undertaken "in collusion with the imperialists and their lackeys."[27]

The KPRAF had to concentrate on cadre education and party-building in order to achieve these goals and to counter tendencies for men to drift away from the forces to less disciplined and dangerous pursuits, simply letting the PAVN do the dirty work. Among the earliest efforts was a drive to build the "core organization" in KPRAF regular and provincial units. The use of such a body dates to the Khmer Rouge period. Its members, referred to simply as "core," are drawn by the party from the most "progressive" masses to be "leading links between the party and the masses." Candidates for party membership are picked from the "core" group. Politburo member Chea Soth has said that the PRK core organization formed only in 1980.[28] The various units have regularly reported expansion of their core membership. Among the earliest was the 41st Battalion of Stung Treng province, which noted successful election of core in a November 4, 1980, broadcast. Building the core organization in the military only accelerated after the May 1981 Fourth Party Congress. A broadcast Revolutionary Army newspaper editorial explained that thousands of core had been admitted since the Congress and hundreds of core had become party members.[29]

Although an increase in numbers of core was welcomed, among shortcomings that figured in editorials were criticisms that core were insufficiently widespread, of inadequate quality, and not active enough. Mass proselytizing was weak. Party organs had not grasped the role of the core. Even those who were engaged in combat were criticized for failing to recruit good cadre and combatants to fill core groups. There were other deficiencies as well. At the party congress in May 1981 Chan Si decreed a low level of KPRAF political consciousness, as well as lax discipline and inadequate facilities for "building spirit."[30] As an indication of lack of progress in cadre development, high-ranking military cadre have recently come in for criticism. The Armed Forces Radio Program charged some of them with failing to maintain a close relationship with troops and of being unaware of "unhealthy phenomena in their units" that might affect readiness.[31] Possibly this criticism of commanders is a reaction to the continuing high level of KPRAF desertions. Pen Sovann had, two years earlier in October 1980, urged graduates of the First National High and Medium-Ranking Cadres Political Course to focus on living conditions of junior cadre and combatants.[32] He noted then that building cadre was necessary for the formation of an all-around conventional infantry, navy, and air force.

Structural Development of the KPRAF

Hem Samin said that by the end of 1978, the Khmer dissidents still had just a few battalions.[33] These were enough to form a plating on the PAVN spearhead, but little more. The battalions were organized into three or four brigades. In 1980 Pen Sovann noted that the 2d Brigade had participated in the January 1979 victory.[34] The 1st and 3d brigades also operated with the PAVN then. The former, founded on April 22, 1978, appears to have been the original Vietnamese-created element.[35] The units actually fought as battalions rather than as brigade-sized entities. It is doubtful if adequate brigade staffs existed. Indeed, the People's Republic of Kampuchea does not appear to have developed a military General Staff until mid-1979. In June 1979 the radio listed a chief of staff, Soy Keo; a deputy chief of the Political Department, Meas Kroch; and the head of the Signals Department, Lim Nay.

Immediately after victory the army seems to have undergone a crisis, with massive desertions and uncoordinated efforts at expansion. While volunteers to fight the hated Khmer Rouge were not lacking, revenge did not make for good discipline. Moreover, the PAVN was doing the job and seemed loath to let the Khmer get in the way of its fast moving combined-arms assault. In truth, the Khmer were neither sufficiently disciplined nor equipped to be of importance.

The end of 1979 and the beginning of 1980 seems to have marked a turning point, a fresh start toward creating a military establishment. One element forcing this change was the massive desertion of PRK units at both national and provincial levels. The initial Khmer military enthusiasm died as the Khmer Rouge entrenched themselves in the malarial hills. Another reason for the reorganization was that the Vietnamese drive to wipe up "Pol Pot remnants" stalled in the face of Khmer Rouge tenacity, Vietnamese logistical inadequacies, and a terrain ill-suited for pursuit. Finally, the beginning of a major border relief effort in October 1979 meant an international presence and spotlight on the border, which may have stayed the PAVN from undertaking either hot pursuit into Thailand or enveloping moves through Thailand to outflank Khmer Rouge border bases. The problems were both logistical and political. Vietnam's brief incursion into Thailand in June 1980—and attempt to encircle a noncommunist border center—cost them dearly in political capital, as did PAVN attacks against a Khmer Rouge base in March 1983.

By the beginning of 1980 signs of progress began to appear. Radio broadcasts were giving increasing attention to KPRAF development. On January 23, 1980, then Commander-in-Chief Pen Sovann opened the first course for "primary class regular cadre," presumably noncommissioned officers, at an "Infantry Graduation School." The army weekly newspaper *Revolutionary Army*, a product of the Army Printing House, had begun to appear. The printing house, formally opened on May 6, 1980, was said to have begun work in October 1979. The first army logistics training course with forty-five trainees began on April 30, 1980, and lasted almost four months. In Kompong Cham, on August 23, 1980, a PAVN unit turned over logistical self-management to KPRAF "Division B."[36] An armed forces Logistics Department has existed, at least in name, since mid-1979, when its chief was identified. It now has a handful of Soviet advisers. Radio appeals for army volunteers greatly increased in March 1980. The army itself received a founding date, June 19, 1951, only a few weeks before the 1980 celebration.

The year 1980 also saw the creation of other services. The air force began its first political course in June, graduating on July 31. By September 25, 142 trainees were attending classes. In November, 100 air force recruits went to Vietnam for training. Sign-up lists for air and naval training in the Soviet Union were posted in Phnom Penh at the end of 1980, according to Cambodian travelers to the border. A separate Air Department of the KPRAF has yet to be described in detail, though Vietnamese pilots are known to fly Soviet helicopters given to the People's Republic of Kampuchea in early 1981. Deputy Minister of Defense and Deputy Chief of Staff Tea Banh, in charge of civil aviation, may be commander of air force units. A KPRAF naval arm exists, but as of 1982 it was under the General Logistics Department of the General Staff, perhaps indicating that its primary function is water transport.

The main emphasis is on infantry and supporting arms such as armor and artillery. Only the military skeleton survived the confusion of the 1979 period. The four "regular" brigades have been in the process of expansion and development into infantry divisions. These units seem assigned to designated areas of operation for use as main-force infantry units operating on their own. They have, however, as many as forty PAVN advisers, and they work in close coordination with main-force Vietnamese units to attack resistance strongpoints.

Once again, 1980 was the key year in the formation of large infantry units. By the end of the year, some senior military leaders had been in the Soviet Union for three months of military studies. Among them was Khang Sarin, currently minister of the interior and member of the party Central Committee. He was made vice-minister of defense and deputy chief of staff in 1981 after returning from the Soviet Union. In December 1981 Radio Phnom Penh identified him as a member of the Central Committee Secretariat, bringing that body up to full strength by filling the slot Pen Sovann's purge vacated. Sarin, born in 1935, is a regroupee who returned in 1970, but, according to his official biography, left the Khmer Rouge in 1972 to organize dissident forces. In 1979 Radio Phnom Penh variously described him as chairman of the Phnom Penh People's Revolutionary Committee, deputy chief of staff, and commander of the First Brigade. This First Brigade is not to be confused with the main-force unit of the same designation. The former developed after January 7, 1979, and seems to have been specifically formed for the capital's defense. After March 1980 command of this unit shifted to Peng Path, commander of Phnom Penh Municipal Forces. Sarin may have made his way to the Soviet Union about then. He had returned by April 1981 to be listed as a vice-minister of national defense.

The main-force First Brigade received priority in expansion. In January 1981 a tenth regiment was associated with the brigade, suggesting that three battalions were being grouped under regiment staffs for greater ease of command. On April 26, 1981, Radio Phnom Penh announced that four days earlier, on the brigade's founding anniversary, it had been upgraded and redesignated as the 196th Division. Its commander, Nhek Huon, first identified in mid-1980, attended the May 1981 Fourth Party Congress as a military delegate. He spoke following the Lao and Vietnamese party secretaries, suggesting both his importance and the importance of military matters in the People's Republic of Kampuchea. By mid-1981, his unit had been assigned to protect Route 10 from Battambang to Pailin. It did not distinguish itself, suffering a high desertion rate and exhibiting a lack of aggressiveness. Toward the end of the year a mutiny among division elements broke out, possibly over poor logistical support. At least one battalion marched on Battambang town. By June 1982, however, Huon had been transferred and promoted. Radio Phnom Penh described him as "representative of the Phnom Penh Military Command," and then, in December 1982, as "deputy chief of staff and commander of the Phnom Penh garrison." He replaced Peng Path in the latter capacity. Huon's former deputy division commander, Ham Kin, took command of the 196th in early 1982.

The Second Brigade has received less domestic media attention. It began to be upgraded in 1981 and late in the year, or in early 1982, moved from Kompong Speu to Pursat province to back up PAVN units in action against the Khmer Rouge in the Cardamom Mountains. On June 15, 1982, Radio Phnom Penh called the unit the "Second Division" and identified its commander as Mao Chhem. Chhem, as late as August 1981, had been deputy to Heng Samol (alias Mean Samol, alias Nhim Samol).

Receiving by far the largest share of publicity has been the Third Brigade, which operates in Siem Reap and Oddar Meanchey. In December 1981 its commander, Un Bien, along with the 196th division commander, became a member of the National Advisory Council of the Front for the Protection and Defense of the Motherland, the current title of the Salvation Front formed in December 1978.[37] The brigade was trying to recruit throughout 1981 and apparently reached its target; in mid-1982 the unit was formally upgraded to a division, designated 286th, and Radio Phnom Penh confirmed the information in a December 29 broadcast that praised the unit's Fifth Regiment.[38] In January 1982, Prom Samen, formerly division deputy commander, was publicly named commanding officer.

There is no information on the fate of either the old Third Brigade commander, Un Bien, or the old Second Brigade chief, Heng Samol. Considering the low military educational level of senior PRK leaders in general, both have possibly gone for further training rather than being purged.

The Fourth Brigade, headquartered at Kompong Cham, has yet to be described as a division, and most likely it is still expanding. Activities of two of its battalions, the 16th and 32d, received mention on Radio Phnom Penh in mid-1982.

In general, the performance of PRK military units is poor. While a Vietnamese advisory staff of about forty is assigned, with members located at all regiments and battalions, PAVN main-force units have sheltered PRK forces from combat. Military skills are low. KPRAF units are not aggressive enough, they still suffer high desertion rates, but they are well equipped. New Soviet equipment, delivered in 1982 to regular army units, includes plastic stock AKM assault rifles, mortars, 122-mm howitzers for artillery battalions, and single-barrel antiaircraft guns. Units that figured in annual honors lists included the First Brigade in 1980, the Third Brigade in 1981, and some of their component battalions in 1982.

These units vitally need trained manpower. As enthusiasm for service against the Khmer Rouge waned, the People's Republic of Kampuchea was forced to institute an increasingly coercive military conscription system. Up through 1980 volunteers kept PRK units staffed, although an editorial in the army newspaper noted at mid-year that "some difficulties still exist this year,"[39] including enemy sabotage of recruitment efforts, inadequate propaganda efforts, and "lack of economic stability" (low salaries?). New measures were instituted to solve the problems. Takeo established an Army Recruitment Council, which included the Youth League, by the end of September 1980. Kompong Thom and Pursat established Recruitment Councils "at all levels" by July 1980. District and village authorities began to draw up lists of youth in Prey Veng and Svay

Rieng. By the end of the year, fear of a draft colored refugee accounts.

The draft began in earnest in 1981. Young men were called to meetings at subdistricts, entertained with speeches and, sometimes, artistic performances. Trucks waited to take the harvest to training centers. A serious effort in Siem Reap subdistricts met with indifferent success as many young men began to flee to the border. But enough answered the call to raise the strength of the Third Brigade significantly, presaging its restructuring as a division. To enlist those who failed to answer the summons to the subdistrict meeting, officials went to the villages, met with the village chief, and escorted the offender from his house to a training center.

Salary may be an important incentive for those who stay in the ranks. Wages range from about 80 to 135 riels per month with a rice ration of 16–22 kg deducted at the concessionary rate of 1 kg for 1 riel. That would leave the equivalent of three or four dollars pocket money at the free-market rate of exchange.

Training primarily takes place in Cambodia, but it is supplemented with study abroad for advanced courses and artillery and other specialty arms training. Centers exist in at least five of the seven PRK military regions. These facilities seem to have originated as PAVN training schools. In a review of 1980, Army Logistics Chief Dy Phin reported that more than 2,000 students had trained in five military schools.[40] Thirteen military schools were open in Cambodia as of mid-1982. Radio Phnom Penh has identified the Region One training school, "Ban Lung" at Ratanakiri; the Region Three school is near Phnom Penh. The H-30 school at Siem Reap received honors on Army Day in 1981. That school was then, in fact, entirely Vietnamese run. According to a former staff member it trained political officers for both Vietnamese and KPRAF units. Specialty training that takes place in Cambodia includes the October 1, 1982 graduation of sixty-one radiomen and seventy-four telegraph operators after a 300-day course of intensive training.

Study abroad begins with training in Vietnam. Promising military students go to the Long Giao Military Training Center in Dong Nai province. Others study at the Thu Duc headquarters of the PAVN Fourth Corps. Armor, air, and naval training, including training in maintenance and repair, takes place in the Soviet Union. Infantry commanders have studied there as well.

The KPRAF Provincial Forces

The KPRAF formally includes three categories of troops: "regular," "regional," and "guerrilla." The regional forces are made up of standing battalions at province level with companies at the districts and platoons in some subdistricts. "Guerrillas," actually militia, are drawn from among subdistrict and village dwellers. Provinces are grouped into military regions, the exact boundaries of which are unknown to outside observers. Regular and provincial forces may total

Base Areas of the Resistance to the P.R.K.

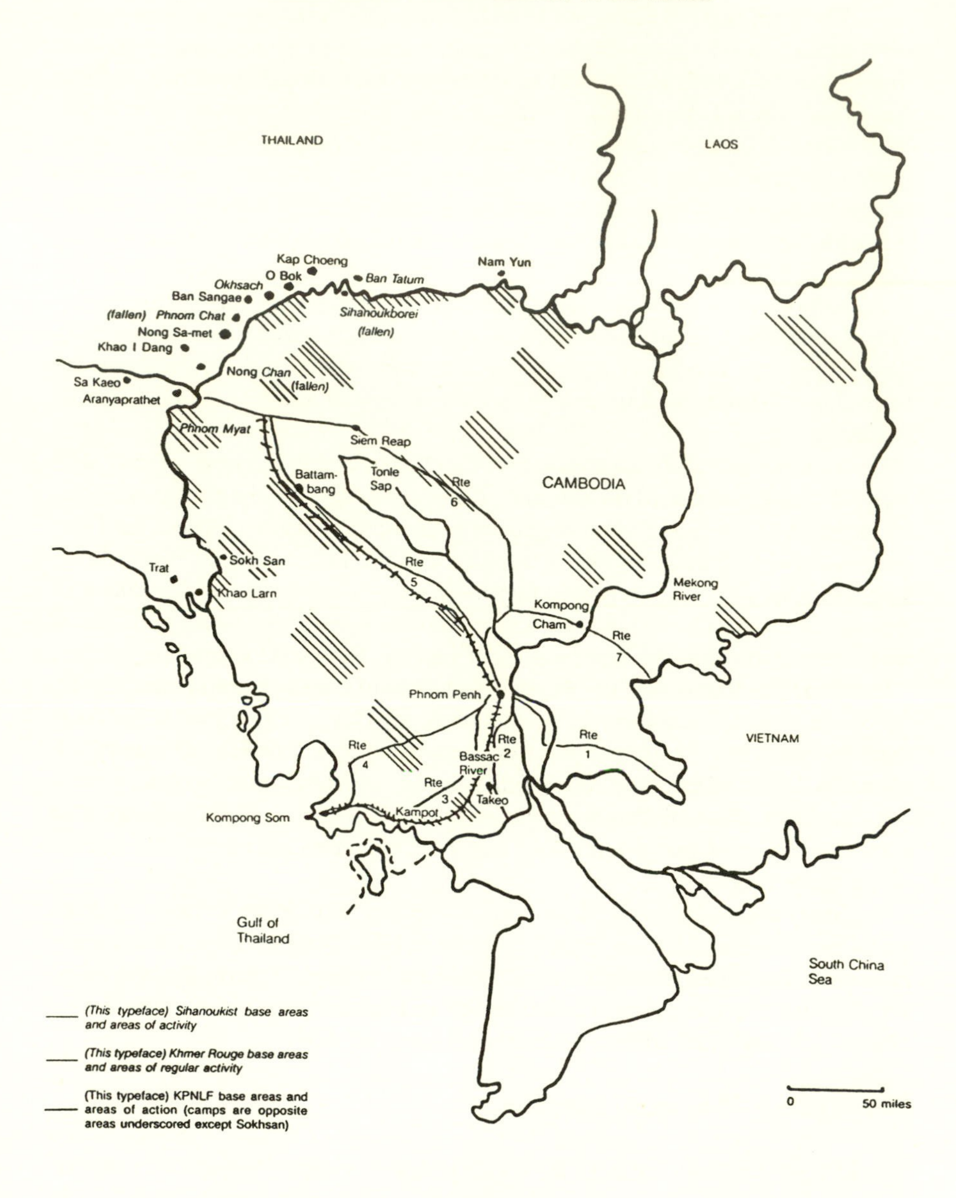

(This typeface) Sihanoukist base areas and areas of activity

(This typeface) Khmer Rouge base areas and areas of regular activity

(This typeface) KPNLF base areas and areas of action (camps are opposite areas underscored except Sokhsan)

as many as 30,000 men.

The chairman of the Provincial Military Command has authority over provincial units. He is probably a deputy chairman of the province People's Revolutionary Committee as well. A deputy seconds him in the military command, and each province has a military General Staff. Senior province military commanders are also members of the National Assembly: Battambang, Kompong Cham, Kompong Chhnang, Siem Reap, and Takeo. In two large provinces, Kandal and Takeo, 2,500 troops received pay. The early effort to develop regional forces required two battalions for each province. By late 1982, provinces in some regions had built three or more battalions with additional companies at the district level.

Provincial military goals, as the chairman of the Kompong Cham command outlined them in July 1982, were to build forces through political and military training under Defense Ministry and province Party Affairs Control Committee guidance, as well as with "the wholehearted assistance of the comrade Vietnamese experts" and the PAVN; to undertake "mass agitation work," educating people on the revolutionary line and exposing enemy psychological warfare actions; and to foster national and international solidarity with the local inhabitants and with PAVN units.[41]

Political aspects of mass agitation work were strongly emphasized throughout 1981 and 1982. Mass agitation "task groups" in rural Battambang selected capable people for "local state power," according to a January 1981 broadcast. Province political schools hold courses for "chiefs of mass agitation units." In Takeo, eighty such students had finished a course by early October 1981. Specific to the task of agitation and propaganda are standing "armed propaganda units" whose existence is reported from a variety of provincial subdistricts. These provincial units all fit into a pacification role coordinated by Vietnamese military entities in the regions.

The Vietnamese Military Role

Vietnam is spending enormous amounts of blood, treasure, and talent in Cambodia. The task of consolidating the People's Republic of Kampuchea and ensuring its loyalty is under the purview of three Vietnamese party Politburo members. Vietnam Communist Party (VCP) leader heir-apparent Le Duc Tho has long been the foremost personality dealing with Khmer issues. According to the former concubine of Ros Samay, an early PRK leader, Tho remains in charge of Khmer affairs.[42] Tho has been a regular visitor to the country since 1979. On at least one occasion in mid-1979 he lectured to PAVN field-grade officers on measures to build PRK state institutions. On the military side, Vietnamese Defense Minister and Politburo member Senior General Van Tien Dung directly concerns himself with, and is said to regularly visit, the Fourth Corps headquarters, the main center of military operations in the early years of the occupation. In charge of day-to-day operations is Colonel General Le Duc Anh, former deputy commander of the Central Office for South Vietnam during the Vietnam war. He was raised to vice-minister of defense in May 1981 and became alternate member of the Politburo at the March 1982 Fifth VCP Congress.

Only the skeleton of the Vietnamese political/military organization operating in Cambodia is clear. It has undergone expansion and change since 1979, primarily to improve logistics for all units and command and control for elements engaged in pacification. About 180,000 PAVN troops garrisoned Cambodia at the end of 1982. The initial organizational structure included a number of division headquarters, leapfrogging their way across Cambodia. A forward headquarters for the Fourth Corps opened in Phnom Penh quickly, and an entity designated "Front 479" established itself in Siem Reap to control operations in Northwest Cambodia.

Early on a group of Vietnamese military experts arrived to advise and monitor the KPRAF, but they were not publicly identified until late in 1980. In celebration of Vietnamese Army Day in December, the then head of the KPRAF General Political Department visited the PAVN "478th Division."[43] A more specific reference nearly a year later conveyed the KPRAF chief of staff's thanks to "Vietnamese Expert Team No. 478" for helping Army Academy instructors train the first class of high and medium ranking cadre.[44]

Specifically political aspects fall under the VCP Politburo office in Phnom Penh. Vice Foreign Minister Vo Dong Giang, the first SRV ambassador to Cambodia, probably held the supreme political job. His deputy, Ngo Dien, replaced him in December 1979. He would seem to lack the stature for the senior political job. The Politburo office in overall charge is designated B-68, according to defectors.[45] Two sections of it are A-40, which oversees the PRK administration, and A-50, responsible for Phnom Penh municipality. How Expert Team 478 relates to B-68 is not yet clear.

In the provinces, the Vietnamese civil/military advisory and pacification effort seems somewhat separate from but complementary to the duties of the main-force divisions prosecuting the war against the resistance. Both functions are under the aegis of regional "fronts," such as Front 479 at Siem Reap. By the end of 1981 the Vietnamese had created similar entities for all other areas in Cambodia, but they are under the authority of the Vietnamese military regions roughly adjacent. For example, by February 1981, the Third PRK Military Region came under PAVN Military Region Nine. The Third Region's Training School benefited from attentions of the Military Region Nine Command.[46] PAVN Region Nine forward headquarters is likely in Kompong Speu province.

What these fronts actually do is suggested by a *Nhan Dan* article of April 1982.[47] The SRV party daily described "Multiregimental Unit [Binh Doan] C" as a combat unit in Kampuchea. It took heavy criticism for its inadequate 1981 training activities, presumably as reflected in poor performance in the 1981–82 dry season. Its exact identity is deliberately obscured, but its three main components include integrated mobile units, units operating near the border, and units providing "assistance to friends." The term used to describe the unit (*binh doan*) carries the sense of a rear area corps, rather than a front-line fighting corps (*quan doan*). The "Cuu Long Group of Divisions," some of whose elements were promised to be among a partial withdrawal of PAVN troops from Cambodia in May 1983, is called a *binh doan* in Vietnamese.

The People's Republic of Kampuchea calls PAVN units assisting it and engaged in pacification, including the building of local government, "task forces." October 1981 reports from Kompong Thom province of military actions against small Khmer Rouge bands referred to PRK district commandos and military units cooperating with "Vietnamese task forces."[48] By early 1981 operational entities were designated by four digits, beginning with the number of the Vietnamese military region controlling the area of operations. Thus, Military Region Nine operates in southwestern Cambodia, the Third PRK Military Region. Its Front 979 has a group of experts to advise the Khmer, designated "978," and a number of operational bodies, such as Task Force 9904 for Kampot province.

This structure grew out of decisions taken in mid-1980 when the Vietnamese agreed on the need to use military specialist teams to train and strengthen the PRK forces in order to facilitate the protection of key areas such as bridges and lines of communication. These task forces were formed in October, and their elements operate throughout PRK provinces. Duties are as diverse as approving candidates for the March 1981 PRK subdistrict elections or candidate members of the Khmer party or, as in Kampot in mid-1981, selecting a handful of candidates for air and armor training in the Soviet Union. The military striking power of these task forces is built around a core of a Vietnamese regiment detached from PAVN main-force units or Vietnamese provincial regiments. PRK provincial units such as the Fifth Battalion in Kompong Chhang or the Twentieth Battalion in Siem Reap operate jointly with the PAVN and are employed as a blocking force, while the PAVN sweeps to ensure security in the more settled areas of Cambodia.

PAVN main-force units bear the brunt of the war in the border areas near major resistance troop concentrations. Although PRK regular infantry units took up road and bridge defense in remote parts of western Cambodia in mid-1981, they did not engage in important fighting until the 1981–82 dry season when a few PRK battalions moved with main-force PAVN units against the Khmer Rouge and Khmer People's National Liberation Front (KPNLF) forces along the western border.

Thai National Security Council Secretary General Prasong Sunsiri estimated that there were eight to ten Vietnamese divisions in western Cambodia as of April 1982.[49] More than a dozen PAVN divisions are assigned to Cambodia. A partial list, beginning with the Gulf of Siam and curving around to the triborder point, includes the 4th, 7th, 59th, 75th, 72d, and 307th divisions. The Thai Supreme Command spokesman identified three of these units in July 1982.[50]

While in theory a Vietnamese division has 10,000 men, in fact PAVN units in Cambodia have until recently been greatly under strength. A Vietnamese division usually includes three infantry regiments of three battalions each and supporting arms. However, especially in western Cambodia, PAVN divisions had added extra infantry regiments. All the infantry regiments have been so weakened by disease and desertion that, even with the extra regiments, divisions have been under strength. The presence of extra regiments does provide the potential for oversize divisions, if the Vietnamese command is willing to send in the manpower. The 20,000 new troops rotated into Cambodia during June and July 1982 filled out many regiments, adding strength rather than merely replacing the troops mustered out at the end of their military service in the same period. Vietnamese combat power seemed slightly enhanced at the beginning of the 1982–83 dry season compared with their situation at the same time a year before.

The Resistance

Noncommunists

The earliest noncommunist resistance groups sprang up in the wake of the Vietnamese invasion and destruction of Khmer Rouge control. Throughout 1979, colorfully named groups like the "Khmer Soul," the "White Elephant," "Discus," or "Black Eagle" vied for recruits, supplies, and recognition in the interior. One band was organized under the banner of Prince Sihanouk's son, Norodom Chakrapong. He was actually in France. Along the Thai-Khmer border, long-time anti-Khmer Rouge resistance leaders and new arrivals began to set up what became essentially warlord-led bands, engaging more in trade and internecine fighting than opposing the Vietnamese.

Several of these border figures amalgamated into the most successful of the pro-Sihanouk movements. At a meeting held along the border on August 31, 1979, under the chairmanship of Kong Sileah (alias Sell Soeng), representatives voted to unite the resistance under the "Movement for the National Liberation of Kampuchea" (MOULINAKA), which then picked Kong Sileah as provisional chairman of its Central Committee.[51] He died in August 1980 and the group came under the leadership of his deputy, Nhem Sophan, who has commanded it since 1983. Figures for its military strength varied, but estimates put it at around 500 men until recruitment began in late 1981. Two smaller organizations based on the northern Thai-Khmer border also voiced their support for the Prince.

With Prince Sihanouk's resumption of political life in February 1981, his political organization, the Front for a United, National, Independent, Neutral, Peaceful, and Cooperative Cambodia (FUNCINPEC), dispatched Khmer elder statesman In Tam to the border. He had the mandate to weld the Sihanoukists into a single force. He arrived in April 1981 and became commander-in-chief of the Prince's troops. A chief of staff, former career officer and Khmer Republican governor of Kompong Thom province, Teap Ben arrived in August 1981. The armed forces were styled the *Armée Nationale Sihanoukienne* (ANS), having formally come into existence on June 6, 1981, when Sihanouk signed FUNCINPEC's statutes. The ANS included MOULINAKA and two smaller groups, one under former air force officer Toun Chay, based opposite Thailand's Buriram province. In Tam himself moved to the base of the third component, Svy Thoeun near the Chong Chom pass between Surin and Siem Reap-Oddar Meanchey provinces. In 1983 the ANS claimed 5,000 men under arms. They received a shipment of Chinese-supplied light infantry weapons in March 1982, at which time their armed and trained troops exceeded 3,000 men, according to the Prince.[52] In May 1983 he claimed to have an additional 5,000 men trained and ready to receive arms.

Military activities of the ANS center on dispatching teams of political and military cadres into the interior.[53] Most of the units, organized as battalions, are deployed in defense of their base areas. Refugees and travelers to the border have reported arrests of individuals accused of being MOULINAKA agents; MOULINAKA had a courier network that reached as far as Phnom Penh at least throughout 1980. PRK and Vietnamese security services have since tightened up.

Larger and better organized is the Khmer People's National Liberation Front (KPNLF) under Son Sann. This movement grew out of a Paris-based exile group. Its first military leader, Dien Del, returned to the border in early 1979. He organized several bands that had fought along the border during the Pol Pot period into a Khmer People's National Liberation Armed Force on March 5, 1979.[54] The political front formally proclaimed itself on October 9, 1979, and "Decision No. 3" appointed Dien Del as chief of the General Staff.[55]

From the outset the KPNLF recognized its "numerical and material inferiority."[56] Its goal was to build a large enough armed force to make it a credible player in a political solution while concentrating on building a political infrastructure for eventual competition with the Khmer Rouge and the People's Republic of Kampuchea. The KPNLF grew slowly from 1979, when it claimed to have had 2,000 men in March. By the end of 1980, it said it had over 6,000 armed men. By the end of 1981, the organization claimed about 9,000 men under arms. In a videotaped statement to the conference at Princeton University in November 1982, Son Sann claimed 11,000 men under arms. Two shipments of weapons for 3,000 men each came from China. The April 1981 delivery was highly publicized. KPNLF spokesmen have refused to confirm reports of other sources of their arms, despite Prince Sihanouk's public statement in early May 1983 that Singapore had given Son Sann weapons for 3,000 men.[57]

The KPNLF began to train cadre in December 1979. On December 4 the front opened courses of "political, psychological, popular warfare and education of the people," the main objective of which was to "defeat a stronger enemy" by dividing and weakening through undermining his morale and discipline and sowing dissension in his ranks.[58] A military school opened in January 1981 at the

Front headquarters opposite Ban Sangae, mainly to train platoon-level officers.

The KPNLF is organized into battalions at its headquarters and at the larger camps under KPNLF control north of Aranyaprathet. KPNLF forces so far operate conventionally, in classical units of platoon, company, and battalion. At the more distant KPNLF bases, such as Sokh San opposite Thailand's Trat province, units are broken down into guerrilla teams, which the KPNLF says have penetrated as far as the railway line in Pursat and Kompong Chhang provinces.

In general, the KPNLF does not fight on a large scale. Its heaviest offensive actions took place at the end of September 1982 opposite Nong Chan border concentration, and in late December 1982. In September a week-long series of attacks about a dozen kilometers east of the border cleared away PAVN checkpoints and forced PAVN/PRK elements to withdraw temporarily. The KPNLF claimed a modest number of Vietnamese casualties (under a hundred), and the seizure of a small quantity of supplies.[59] In December the Nong Chan force ventured into villages up to 25 km from the border. Surprise attacks caused PAVN units to flee. The KPNLF slowly yielded to PAVN counterattacks, abandoning its last gains in mid-January.

The KPNLF command structure was reorganized in mid-October 1982 when Chief of Staff Dien Del resigned. A four-man Interim Military Committee replaced him, chaired by former career officer Sak Sutsakhan. He had returned to the border in early 1982. The other members included former Special Forces Commander Thach Reng, who joined the KPNLF in August 1982; KPNLF Executive Committee Delegate Hing Kunthon, who had survived the Pol Pot period; and Nong Chan Commander Chea Chhut, who has been on the border since 1975. A strategy emphasizing guerrilla actions deeper in the interior has been stressed. But a reality demonstrated during the 1982–83 dry season is the need to hold off large attacking forces long enough to enable the population to flee, as when Nong Chan was finally destroyed on January 31, 1983. This dichotomy will continue to require the ability both to put together a conventional force for population defense and to break up into small units to operate inside Cambodia.

Khmer Rouge

In the wake of the September 1979 creation of a new Khmer Rouge political front, their armed forces dropped the name "Revolutionary Army of Kampuchea." Revolution was out, as the "Patriotic and Democratic Front of Great National Union of Kampuchea" emerged with Khieu Samphan as its provisional chairman, pending the willingness of Prince Sihanouk to accept leadership. On December 18, the Khmer Rouge announced their new cabinet, in which Khieu Samphan replaced Pol Pot as prime minister. Also announced was the Supreme Committee of the "National Army of Democratic Kampuchea," as the forces

Enlargement of Border Area

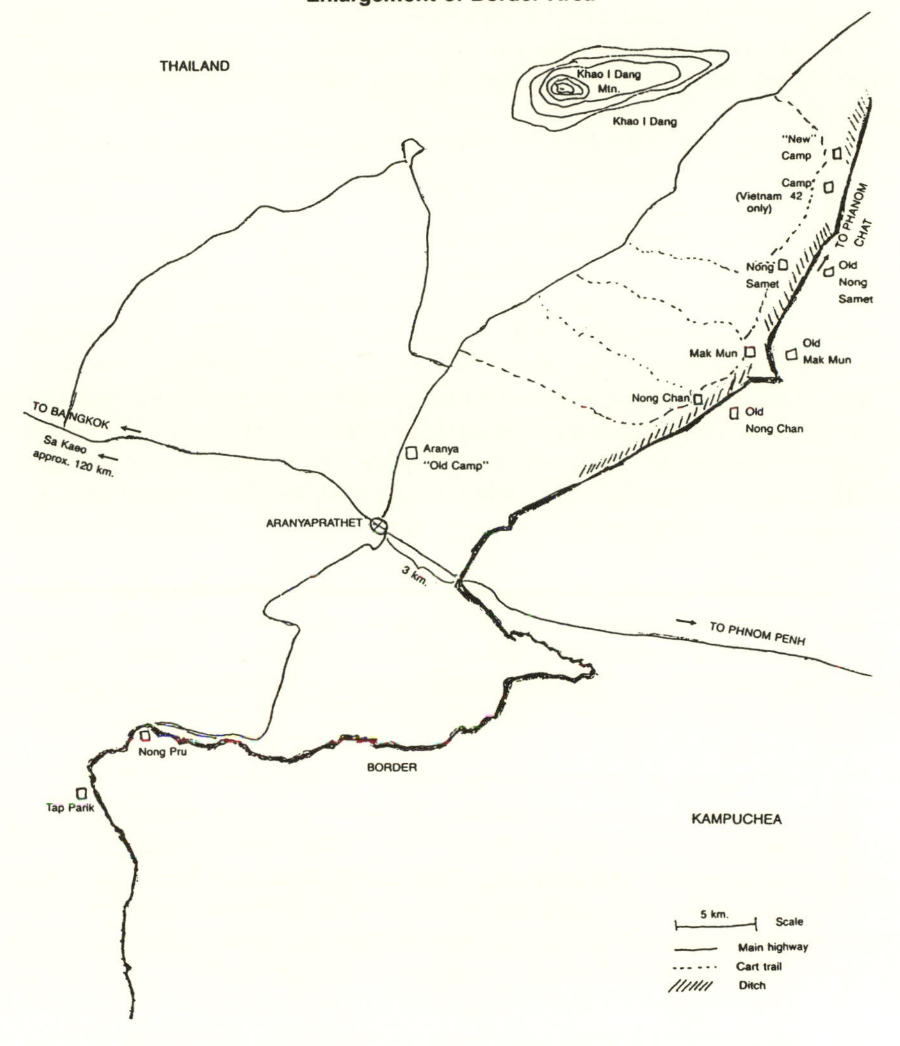

were restyled. This leading military body seems to be the public manifestation of the party military committee. Its composition includes Pol Pot, chairman and commander-in-chief; Chhit Choeurn, alias Ta Mok, vice chairman and chief of the General Staff; Son Sen, secretary general; and Ke Pauk, undersecretary general.[60]

When the December 1978 Vietnamese strike crumbled Khmer Rouge resistance, their units fled helter-skelter toward the hilly and thickly wooded border areas, taking what population they could with them. The Vietnamese provided a short breathing space in February and March 1979. By the end of March, the PAVN had resumed their offensive in western Cambodia, smashing Khmer Rouge resistance and driving farther up against the border where Khmer Rouge units piled up. In September 1979, for example, at least three Khmer Rouge divisions were packed into the Phnom Melai area. Some Khmer Rouge divisions were cut off in eastern Cambodia and forced to trek out. Division Commander (of the 801st Division) Saroeun told Chinese journalists in the spring of 1982 that his forces remained in the region east of the Mekong after the Vietnamese invasion, losing contact with their superiors when Phnom Penh fell. The division split up, and Saroeun eventually reported to the border in May 1980, gradually transferring his troops there.[61]

Elsewhere, military strategy failed, as the Khmer Rouge themselves recognized by mid-1979. Western Battambang Sector Commander Ni Kan told a visiting journalist in March 1981 that in 1979, "The Vietnamese almost wiped us out. They drove most of our forces to the border."[62] Son Sen explained why:

> At the initial stage of resistance from January to June in 1979, though guerrilla warfare was rather effectively carried out in some places, the National Army was then still accustomed to engaging the enemy in major battles in which large numbers of troops took part. This neither stopped the enemy's advance nor protected our own forces while the supply of weapons and ammunition became more and more limited. Such practice could hardly sustain a protracted war.[63]

Contemporary Khmer Rouge documents confirm this analysis and set forth their revised military strategy.[64] A June 29, 1979, analysis outlined a new direction for the "Military Line of the Party in Making People's War." Central to this document was recognition of PAVN superiority and the danger of meeting the Vietnamese head on, which might lead to success of the PAVN's strategy, dubbed by the Khmer Rouge "quick strike, quick victory." As long as Vietnamese forces remained preponderant, the Khmer Rouge would implement a period of "strategic defense," in order to protect their forces. This would be a period of guerrilla war, with "central troops" broken down into small units.

Entering the second phase, the period of "strategic contestation," would require that the balance of forces be about equal. In 1979, the Khmer Rouge saw this as likely to take a long time, despite the fact that they described themselves as "not addicted to delay." In mid-1979 the Khmer Rouge wrote that they controlled one quarter of the country, with guerrilla war in another half, leaving their enemy with control of only a fourth of Cambodia. This geographic equality of control, however, was not seen as matched by equality of strength.

Thus, the urgent tasks were to expand force levels and quality. The number of village guerrillas was to be increased, and guerrilla actions were to be heightened around the nation. At the same time, central forces were to be built, but units were to be broken into squads for guerrilla war and not employed as main forces except as needed. Areas of control also needed to be expanded. When all the conditions of increased size, improved activity, and enlarged areas had been met, then the forces could be regarded as approximately equal. At the same time, work to create secret sympathizers in enemy zones was to go forward. Fighting would increasingly see units combining to hit their enemy with the goal of progressing to the period of "strategic assault." Politically, the document asserted that the Khmer people were increasingly enraged against the Vietnamese, who were suffering diplomatically, too.

The Strategic Assault phase was seen as developing after the balance of forces was reached. Entering it required that guerrilla war increase and that central forces were sufficiently developed to attack PAVN platoons, companies, and even battalions.

This scheme was known to Khmer Rouge defectors from western Cambodian division bases. At the end of the 1980 rainy season, they were saying that the army was entering the second strategic phase. During the 1980–81 dry season, Khmer Rouge military action sharply increased with larger scale attacks in the interior against outposts and lines of communication. The traffic on Route 6, for example, was reduced to heavily armed PAVN convoys in early 1981. During 1980, the Khmer Rouge were building main-force units as well, upgrading the regional units that had fled to the border, to create "central" or main-force divisions.

The military structure described by defectors from Khmer Rouge main-force units in 1979 included a squad of twelve men, a platoon of fifty men, a company of three platoons, a battalion of three companies, a regiment of three battalions, and a division of three regiments. The theoretical infantry strength of a division was thus over 4,500 men, adding somewhat for command staff as units combined. Additional supporting arms would increase the overall manpower total. In February 1981 Khieu Samphan told the press that the Khmer Rouge had ten "skeleton divisions of infantry."[65] As of late 1982, more than a dozen such divisions were said to exist. Few were at full strength, and although estimates varied, a division of 2,500 to 3,000 men was above average. Khmer Rouge unit designations shifted regularly, but moving north from the Gulf of Thailand along the border the publicly cited order of battle began with the Third Division (then the 175th), the 30th and 474th at Phnom Melai, the 519th, and around on the northern border to the 616th, 612th, and, in the triborder area, the 801st Division. Khieu Samphan seems to have suggested to the press that his divisions totalled 30,000 guerrillas and that the Khmer Rouge had an additional 20,000 militia guards. Most estimates, however, put total Khmer Rouge armed strength at 30–35,000. In addition they have an unarmed transport corps, mainly young women, who carry food and ammunition to forward areas,[66] and a similar body of married men and women who carry supplies in base areas.

The Combat Balance

The initial plan for the December 1978 invasion, as PRK officials understood it, was to capture the area east of the Mekong River.[67] Only regional PRK forces were being built, noted Hem Samin. The late December PAVN attack started against Cambodia's Northeast. Fronts quickly opened along the length of the border, including from the areas the PAVN had cleared earlier around Snoul and Memot to enable the Khmer Salvation Front to claim a base in Cambodia for its December 2, 1978, founding. Although individual Khmer Rouge units fought ferociously in places, generally the units were weakened by combat since late 1977 and by the brutal purges of senior command committees during the years 1976–78. The Khmer were unable to compete in what they described as "full-scale" war.[68] The Vietnamese pushed on, probably invoking contingency plans as the unexpected happened and the Khmer Rouge resistance proved manageable.

The PAVN ran out of steam in the Siem Reap area toward the end of January 1979 and was forced to pause. Resupplied, the Vietnamese resumed their drive in western and northwestern Cambodia in late March. In mid-April journalists watched some of the result, as tens of thousands of Khmer Rouge-controlled villagers and troops, who were at least temporarily disarmed, moved across Thai territory to reenter Cambodia around Phnom Melai. The arrival of the monsoon helped limit PAVN military gains in this second-phase offensive. The Vietnamese had beaten, but failed to destroy, their enemy. This period and subsequent rainy-season months of deprivation and disease severely strained loyalties to the Khmer Rouge leadership and their control over population.[69]

Relatively limited Vietnamese military action during the 1979–80 and 1980–81 dry seasons (December-April) suggest that Hanoi's priorities lay elsewhere. Their effort to protect and build a PRK bureaucracy and then, in late 1979-early 1980, to make a fresh start at creating a Khmer military required enormous Vietnamese energy. The object seems to have been to hold larger Khmer Rouge units away from the main populated areas and vital lines of communication while the Vietnamese advisory staff and PRK figures built national institutions. Initial concentration on building the village militia sought to ensure order in the countryside without bringing PAVN units into unnecessarily close contact with the Khmer population, which would risk igniting traditional

Border Camps of the Anti-P.R.K. Forces

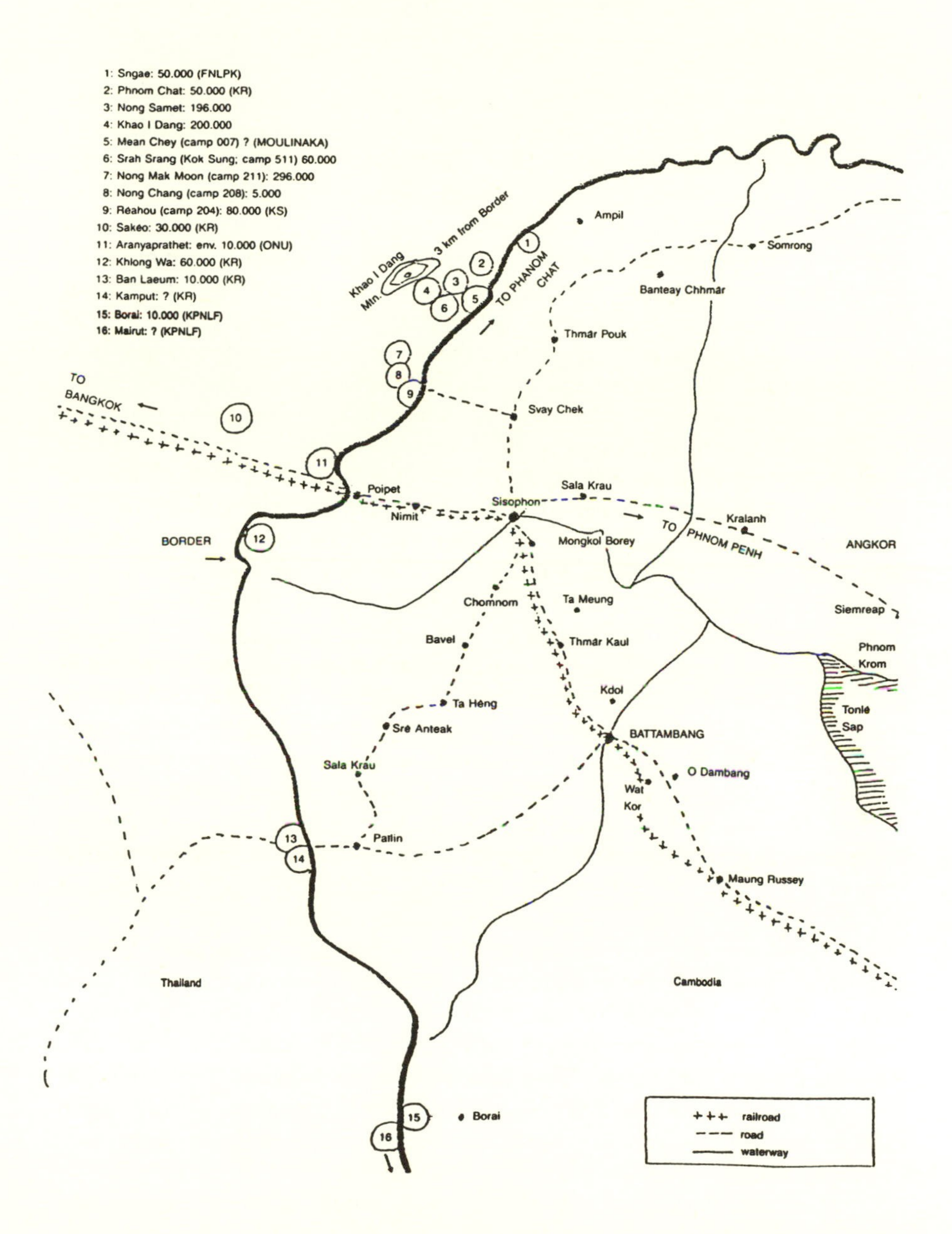

hostilities. In the development of national institutions, early 1981 saw the public unveiling of a Communist Party, elections at local and national levels, inauguration of a parliament, and promulgation of a constitution. All of this took place before the end of June.

Militarily, the 1980–81 dry season marked a resurgence of Khmer Rouge activity. They considered this to be the second strategic military phase. They had recovered from the beating and deprivation of 1979 and rebuilt by reorganizing their force structure and moving commanders around. Vietnamese units had, at the same time, pulled back from positions along the border, consolidating in larger garrisons closer to population centers and major lines of communication. Khmer Rouge communiqués from the 1980–81 dry season and 1981 rainy season betrayed an exultation even beyond that warranted had their grandiose claims been true. They seem to have believed that their pressure rather than PAVN strategy had caused the Vietnamese withdrawals. In fact, the Khmer Rouge made some military gains inside Cambodia, especially in Kompong Thom province, interdicting Route 6, which runs north of Tonle Sap. Forces from Phnom Melai expanded their zone of control to within 30 km of Route 5, west of Monkolborei.

After a rainy season period of troop rotation and the delivery of new equipment, the PAVN took back the initiative in the 1981–82 dry season. A series of probes in November 1981 at defense lines of noncommunist resistance camps north of Aranyaprathet and a solid, bloody punch at the Khmer Rouge 801st division prefigured the real action. In mid-January 1982, under the heaviest artillery barrage since 1979, major elements of two Vietnamese divisions tried to destroy the Khmer Rouge garrison at Phnom Melai, a governmental headquarters and the focus of world press attention. A series of infantry, artillery, and armor attacks, including the alleged use of chemical weapons, regained the ground the PAVN had ceded in the previous dry season. Vietnamese forces seized a number of important training bases and villages as the Khmer Rouge fell back. But the Vietnamese campaign failed either to pin down and destroy large Khmer Rouge units or to drive the Khmer Rouge into Thailand. PRK units did not participate in this effort.

The number of casualties from the three months of fighting is not known. The Khmer Rouge refused to let ICRC medical teams into their hospital areas near the fighting, probably to restrict knowledge of their losses. They clearly fought harder for this territory and suffered more casualties as a result than guerrilla theory would dictate. They reinforced themselves and later claimed to have succeeded in attacking PAVN supply lines. The Vietnamese, however, seem to have ended the attack in early April, as much because of early rains as because of Khmer Rouge tactics.

A similar attack in mid-March, involving major infantry elements of the PAVN Seventh Division supported by PRK units, dislodged the KPNLF's strongest garrison from its two-year-old border base at Sokh San near the Kranhung River in Pursat province. This operation was interpreted widely as an attempt to discourage ASEAN support for the KPNLF. Neither side appeared to have suffered significant casualties. The KPNLF garrison held back the Vietnamese for just long enough to allow the noncombatant population to flee, with cooking gear, into the Thai-Khmer border area. PAVN and PRK units stayed only a few weeks, and the KPNLF population began to return in July.

The 1982 rainy season followed the pattern of earlier years as the PAVN rested, trained, and refurbished. A large rotation of troops in July permitted demobilization of draftees who had been in the ranks for four years. Up to 20,000 new troops came in to complete basic training with front-line divisions. The Khmer Rouge mounted more guerrilla strikes than in the previous rainy season, especially in western Cambodian border areas and along some interior lines of communication. The KPNLF became more active on a larger scale, catching press attention opposite Nong Chan as the noncommunist resistance in general modestly enhanced its military arm. Heng Samrin's KPRAF began to assume a stance closer to the front, promising a more active role.

The PAVN continued to hold the initiative during the 1982–83 dry season, but it changed the pattern of its attacks. The Vietnamese operated two coordinated campaigns during the dry season. They took a break from mid-February to the end of March, when major political events, and the need to resupply, dictated a stand-down in fighting. In short, sharp, multiregimental attacks, the Vietnamese destroyed bases of all three resistance factions: Nong Chan of the KPNLF at the end of January 1983; Phnom Chat of the Khmer Rouge at the end of March; and Sihanoukborei of the Sihanoukists in early April. The last action included one documented atrocity by PAVN troops.[70] An attacking PAVN element called civilians out of a bunker, forced them back in, and then threw two hand grenades in, wounding eight. Other actions against Khmer Rouge areas failed, although a modest success against the Khmer Rouge headquarters just west of Pailin on the western border did temporarily force a Khmer Rouge retreat at a high cost in Vietnamese casualties. The Vietnamese failed to kill or capture any significant number of resistance troops. In general, the resistance forces held out long enough for their noncombatants to flee.

The KPRAF was minimally engaged during the dry season. Elements of its 286th Division participated in the attack against the Sihanoukist headquarters, but PAVN regiments did the major work.

The Vietnamese showed skill in their use of combined arms, artillery, and armor, where the terrain permitted, along with infantry. The PAVN did not use air, nor were there reports of the use of lethal chemical weapons, "yellow rain," as prevalent as during the two previous dry seasons.

Increased KPNLF aggressiveness was a new development. Son Sann's forces at Nong Chan drew on lessons learned during rainy season operations opposite the camp. In late December they attacked Vietnamese 5th Division elements up to 25 km into Cambodia. The PAVN responded poorly, retreating from a number of villages. By the time they built a reaction force, the KPNLF had established a solid line at an abandoned village just east of Nong Chan. The Son Sann forces held that line against Vietnamese artillery and armor for more than two weeks before retreating to Nong Chan, which the PAVN ultimately destroyed during their offensive of January 31 to February 2.

The Khmer Rouge performed poorly at Phnom Chat (March 31-April 7). Several hundred troops broke when they feared that the Vietnamese had flanked them through Thailand. They ultimately refused to return to the ranks. The Vietnamese captured supply caches, but again they failed to destroy significant numbers of the enemy. This attack also included PAVN attacks on two Thai units on Thai soil. Vietnamese forces tried to dig in on the Thai side of Phnom Chat hill as well. Thai artillery and air strikes, combined with worldwide diplomatic pressure, forced the Vietnamese to withdraw. International pressure, waning morale among Vietnamese units engaged in the fighting, and, perhaps, logistics requirements along with the possibility of rains by mid-April caused the PAVN high command to desist. The PAVN did not follow up with strikes against the two adjacent major KPNLF centers. Vietnamese troops were in their rainy-season posture, with regimental headquarters well back from the border by mid- to late April. Some battalions as usual did stay close to the Khmer resistance bases.

As the Vietnamese pulled back, the Khmer Rouge—relatively untouched by dry-season activities—began a series of harassment operations, infantry attacks, sabotage of the rail line and bridges, and stand-off artillery attacks. These continued into the early rainy season (June 1983). KPNLF patrols and clashes in western Battambang and Preah Vihear provinces were also reported. Thus, the PAVN, despite a higher level of dry-season fighting and destruction of larger resistance camp infrastructures, failed to put the resistance off balance.

Prospects

A number of military lessons, some with political implications, can be drawn. First, despite its size and firepower, the Vietnamese garrison in Cambodia is neither large enough nor strong enough to destroy the Khmer resistance. This is not to deny the weakness of individual resistance military bases and civilian camps. The Khmer Rouge, however, have demonstrated that they can fight the PAVN to a standstill, making it too costly, as at Phnom Melai in early 1982, to capture a base. Even the noncommunists have the ability to delay the PAVN long enough for civilians to move to sanctuary. Should the noncommunists receive supplies of adequate antitank munitions, the Vietnamese tactic of the 1982–83 dry season, which saw infantry advancing behind armor, would fail.

Second, the KPRAF, after four years of effort, including receipt of modern Soviet arms and artillery and training of large number of officers and men in Vietnam and the Soviet Union, remains inadequate to cope with the resistance, even in an auxiliary role. The SRV does not have confidence in the loyalty or discipline of KPRAF main-force units. The KPRAF, considering its resources, has failed to develop as well as the noncommunist resistance. This powerful evidence undercuts notions that the Heng Samrin regime is succeeding in consolidating its hold in Cambodia. The resistance continues to view the KPRAF as a potential source of converts rather than as implacable foes.

Third, the Khmer Rouge remain a tough, disciplined fighting force. Nevertheless, they are not monolithic and weaknesses of discipline and morale became clear from August 1982 to April 1983. While the Khmer Rouge can recruit, it is not clear that they can do so in sufficient numbers to replace losses, much less grow.

Fourth, the noncommunist resistance organizations have shown they will fight. The Sihanoukist group is still faced with the need to improve its organization if it is to capitalize on the Prince's remaining appeal. The KPNLF argues that it can produce recruits to carry all the weapons it is given and that, with more supplies, including food, it can greatly enhance actions inside Cambodia on the basis of its new guerrilla-oriented strategy.

A military balance in Cambodia still centers on the contest between the Soviet-equipped People's Army of Vietnam and the Chinese-equipped National Army of Democratic Kampuchea—the Khmer Rouge and PAVN. The Vietnamese have improved their command structure, logistics, and firepower. The resistance has increased its arms. The noncommunists are more aggressive. The Heng Samrin army seems far from significantly contributing to easing the PAVN's burden by augmenting its operational capability. The military situation is stalemated, but the level of violence increases every year. The enhanced capabilities of all the actors raise the potential for serious consequences if one side or the other miscalculates on the battlefield.

Notes

1. Ben Kiernan, "Origins of Khmer Communism," *Southeast Asian Affairs*, Institute of Southeast Asian Affairs (Singapore: Heinemann Asia, 1981), p. 171.
2. Ibid., p. 172.
3. Tan Kim Huon, "The Indochinese Communist Party" (1972), translated from French by Timothy Carney, ms., p. 4.
4. U.S. Embassy, Phnom Penh, "Conversation with Khmer Rouge Rallier Ieng Lim," decontrolled airgram, A-179, November 24, 1971, and "Khmer Rouge Rallier Keoum Kun," decontrolled airgram, A-5, January 7, 1972.
5. U.S. CIA, *Foreign Broadcast Information Service* (*FBIS*), 070652GMT, June 1980.
6. *FBIS* 080936GMT, June 1980.
7. *FBIS* 200615GMT, June 1981.
8. Democratic Kampuchea Ministry of Foreign Affairs, "Black Paper: Facts and Evidence of Acts of Aggression and Annexation of Vietnam Against Kampuchea," [Ministry of Foreign Affairs] Group of Khmer Residents in America edition, 1978, p. 24.
9. Ibid., p. 26.
10. Timothy Carney, "Unexpected Victory," in a forthcoming book to be edited by Karl D. Jackson.

11. Stephen Heder, Interview with Hem Samin (1981), Phnom Penh, ms.
12. U.S. Embassy, "Conversation with Ieng Lim" and "Khmer Rouge Rallier."
13. Heder, Interview with Hem Samin.
14. *FBIS*, 190702GMT, June 1980.
15. "Kampuchea Dossier III: The Dark Years," *Vietnam Courier*, 1979, p. 53.
16. Carney, "Unexpected Victory."
17. Heder, Interview with Hem Samin.
18. Ben Kiernan, "National Rehabilitation in the Eye of an International Storm," *Southeast Asian Affairs 1982*, Institute of Southeast Asian Affairs (Singapore: Heinemann Asia, 1982), p. 170.
19. Heder, Interview with Hem Samin.
20. "Kampuchea Dossier II," *Vietnam Courier*, 1978, p. 72.
21. Heder, Interview with Hem Samin.
22. Heder, personal communication.
23. Stephen Heder, Interview with Chea Soth, 1981, ms.
24. *FBIS*, 271144GMT, January 1980.
25. *FBIS*, 191010GMT, October 1980.
26. *FBIS*, O31955GMT, April 1981.
27. *FBIS*, 221428GMT, June 1982.
28. Heder, Interview with Chea Soth.
29. *FBIS*, 041145GMT, November 1981.
30. *FBIS*, Bangkok, 031955GMT, April 1981.
31. *FBIS*, 130808, October 1982.
32. *FBIS*, 191010GMT, October 1980.
33. Heder, Interview with Hem Samin.
34. *FBIS*, 240935GMT, June 1980.
35. *FBIS*, 271014GMT, April 1981.
36. *FBIS*, 070337, September 1982.
37. People's Republic of Kampuchea, List of Names of Candidates Inscribed for Election, Phnom Penh, Khmer language, ms., December 24, 1981.
38. *FBIS*, Bangkok, 300914GMT, December 1982.
39. *FBIS*, 300736GMT, July 1980.
40. *FBIS*, 270944GMT, February 1981.
41. *FBIS*, 041139GMT, July 1982.
42. *Bangkok Post*, June 18, 1980, pp. 1, 3; see also *FBIS*, 180313GMT, June 1980.
43. *FBIS*, 211402GMT, December 1980.
44. *FBIS*, 221449GMT, November 1981.
45. *New York Times*, October 9, 1982, p. 5.
46. *FBIS*, 100449GMT, February 1981.
47. See *FBIS*, 201025, April 1982.
48. *FBIS*, 231136GMT, October 1981.
49. *Straits Times*, Singapore, April 30, 1982.
50. *FBIS*, 161502GMT, July 1982.
51. MOULINAKA, "Procès-Verbal du Congrès des Délégués et Représentants du Différents Mouvements du Résistance du Kampuchea Tenu le 31 Août 1979 à Soerng (?) dans le Srok de Sisophon (Battambang), 1979," photocopy.
52. Norodom Sihanouk, "Mise au Point de Norodom Sihanouk du Cambodge,"

Pyongyang, March 15, 1982, ms.

53. Jacques Bekaert, "The Khmer Coalition Who Wins, Who Loses?" *Indochina Issues*, Center for International Policy Indochina Project, Washington, D.C., September 1982, p. 7.

54. Khmer People's National Liberation Front, declaration, 1979.

55. KPNLF, proclamation, 1979.

56. KPNLF, Son Sann speech, 1979.

57. *Straits Times*, May 6, 1983.

58. KPNLF, statement, 1980, p. 8.

59. *Paris Herald Tribune*, October 13, 1982.

60. Democratic Kampuchea, "Decisions of the National Assembly," 1979.

61. *Renmin ribao*, May 20, 1982, p. 6; see *FBIS*, Hong Kong, 020600GMT, May 1982.

62. Daniel Burstein, "The Long Campaign of the Khmer Rouge," *Boston Globe Magazine*, July 5, 1981.

63. *Renmin ribao*, April 1, 1982, p. 6; see *FBIS*, Hong Kong, 060938GMT, April 1982.

64. Democratic Kampuchea, "Some Guiding Directions of the Military Line of the Party in Making People's War to Strike the Contemptible, Invading, Expansionist, Land-Grabbing, Genocidal Vietnamese Enemy to Defeat Them Completely and Finally," Khmer language, June 29, 1979, ms.

65. Claire Hollingsworth, "Khmers Seek Allies in Cambodia's Anticommunists," *Daily Telegraph* (London), February 9, 1981, p. 5.

66. *Beijing Review*, May 10, 1982, p. 17.

67. Heder, Interviews with Hem Samin and Chea Soth.

68. Democratic Kampuchea, "Decisions," p. 9.

69. Stephen Heder, "Kampuchean Occupation and Resistance," Chulalongkorn University, Department of Asian Studies, monograph no. 27, Bangkok, 1980.

70. *New York Times*, April 11, 1983.

AGRICULTURE AND ECONOMIC DEVELOPMENT

Map 1. **Medieval Agriculture**

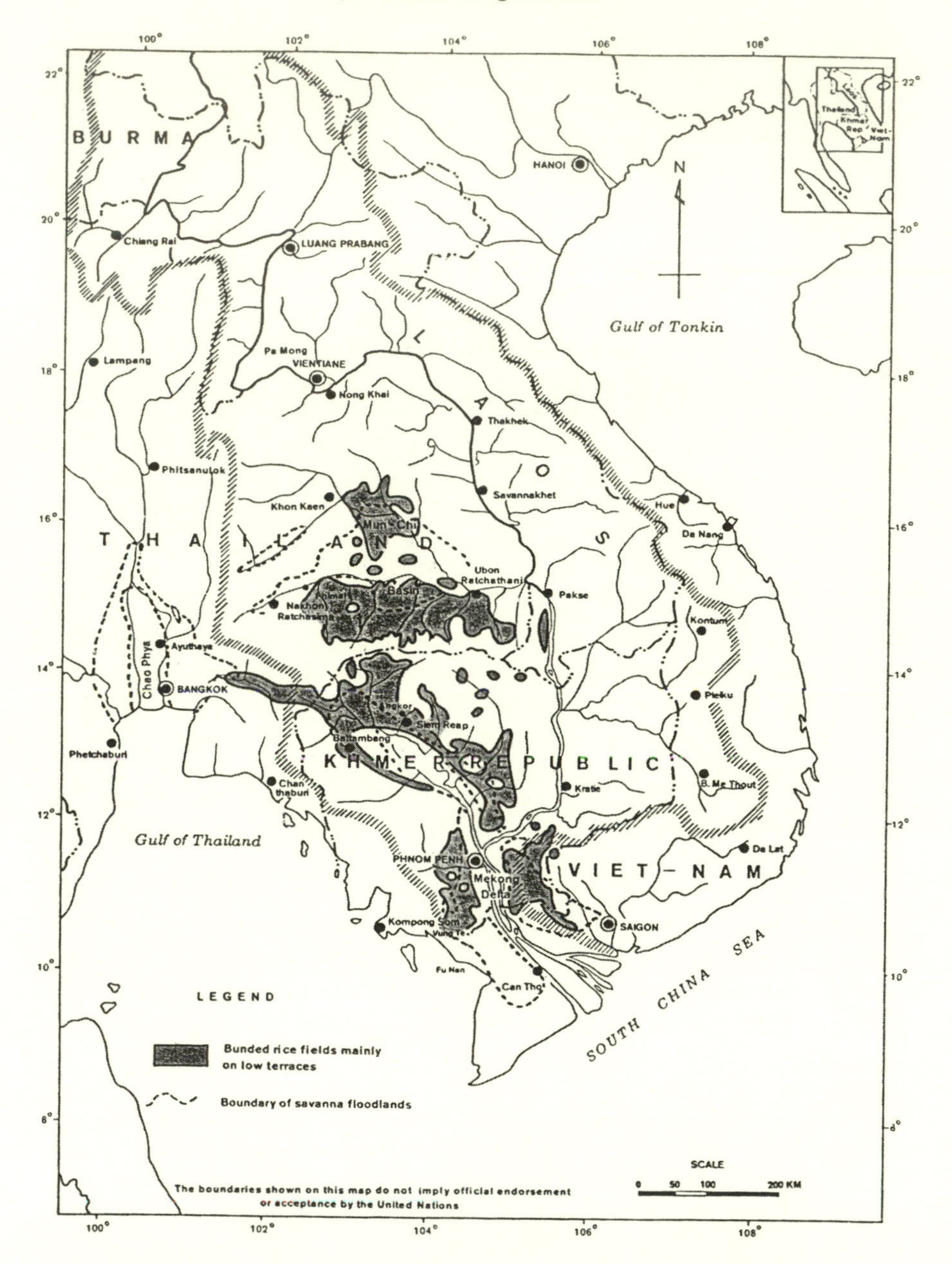

Source: Mekong Committee. 1973 Archeology and the Mekong Project. WRD/MKG/INF L 574

John V. Dennis, Jr.

KAMPUCHEA'S ECOLOGY AND RESOURCE BASE: NATURAL LIMITATIONS ON FOOD PRODUCTION STRATEGIES

Introduction

As in other countries, agricultural development in Kampuchea has always entailed a struggle by man to obtain greater control over his physical environment and resource base. The irony of this struggle is that the more successful man is in gaining control over his environment, the more rapidly the resource base tends to be depleted. This is readily apparent in Kampuchea. Within a central zone around Tonle Sap and Kampuchea's major rivers, annual flooding during the wet season months of July to October rejuvenates the soil but makes careful water control impossible. The mixed result is that this area can produce low rice yields indefinitely. By contrast, irrigation schemes on higher terrace soils outside this zone have allowed periods of superior productivity that ended with permanent deterioration of the soil. According to several contemporary development strategies, it is now time to curtail the extent of the annual flooding within Kampuchea's zone of flooding.

The annual flooding, however, is crucial to the maintenance of soil fertility and high rice and fish production. A greater awareness of the ecological relationships involved may encourage aid donors and Kampuchean planning officials to develop improved food production strategies that take advantage of the existing strengths of the natural resource base but do not destroy them at the same time. This strategy would lessen the perceived need to implement a standardized package of Green Revolution rice technology. Such a package requires costly petrochemical imports and a high level of water control that would be both very difficult to achieve and incompatible with the rejuvenating floods.

Contrasting Strategies of Water Use

The rice agriculture of the Khmer empire, which flourished from the ninth to the twelfth century, was based on Pleistocene terrace soils north of Tonle Sap in what is now Siem Reap province. This agriculture was located beyond the zone of flooding (see maps 1 and 2) and depended on gravity-flow irrigation systems to divert water from the Puok, Siem Reap, and Roluos rivers. The location of all major Kampuchean towns and cities, with the exception of Phnom Penh, just beyond the zone of flooding suggests that both agriculture and political power were based outside this zone for many centuries. However, the high agricultural productivity that supported the building of Angkor Wat was not sustainable, and today the rice soils of Siem Reap are some of the worst in Kampuchea.

The French decision in 1866 to establish the political capital at Phnom Penh in the very center of the flooded zone suggests an implicit recognition that the old terrace soils were exhausted and that consequently the focus of both agriculture and political power had shifted to the zone of flooding.[1] The French evidently recognized the relationship between flooding and the maintenance of soil fertility, for they supported a practice called *colmatage* in which canals (*prek*) were dug through the natural levees of the rivers in order to allow silt-bearing flood waters to penetrate behind the levees to agricultural land or areas where farmers wished to create new arable land (See Figure 1). By the early 1940s some 370 colmatage canals existed in Kandal province.[2] It is estimated that these canals deliver a layer of fresh silt about 20–25 mm thick each year to an area of about 17,600 hectares.[3] Upstream from Kandal province the Mekong River tends naturally to flood its banks in the wet season, so *colmatage* is rarely used except where roads block the flood waters.

Although colmatage rice lands have never comprised more than 15 percent of the total rice lands in Kampuchea, the adoption of the practice in the mid-1800s was probably a response to the irrevocable ecological degradation that deforestation and farming had caused in the nonflooded zone in the preceding centuries. As deforestation proceeded, the arable land in the nonflooded zone became increasingly drought-prone as well as infertile. Just as the bamboo tree in Asian proverbs survives by bending with the wind, colmatage was a pragmatic response that derived impressive benefits from the disruptive annual flooding.

By contrast, some food production strategies of the Khmer Rouge regime were like the oak tree which stands straight in the wind but is eventually broken by it. Farmers in Kandal province were ordered to close off *colmatage* canals to the wet-season flood waters and in some cases to fill them in as well.[4] Under the *colmatage* system the land grows no crops during the months of flooding and a rice crop is transplanted either as the flood waters recede or early in the year to allow the crop to be harvested before the flood waters rise.

The Khmer Rouge regime sought to grow more than one rice crop per year on the *colmatage* fields, evidently determining that the only way to do this was block the flood waters from the fields altogether. The plan failed, not because of an immediate loss of soil fertility, but because flood water no longer reached the inland lakes (*beng*) where it was needed for dry season irrigation. Whether or not this particular project succeeded in the short run, it suggests that Khmer Rouge leaders either were not aware of the critical role flooding plays in maintaining soil

Map 2. **Major Rice-Growing Areas in Kampuchea**

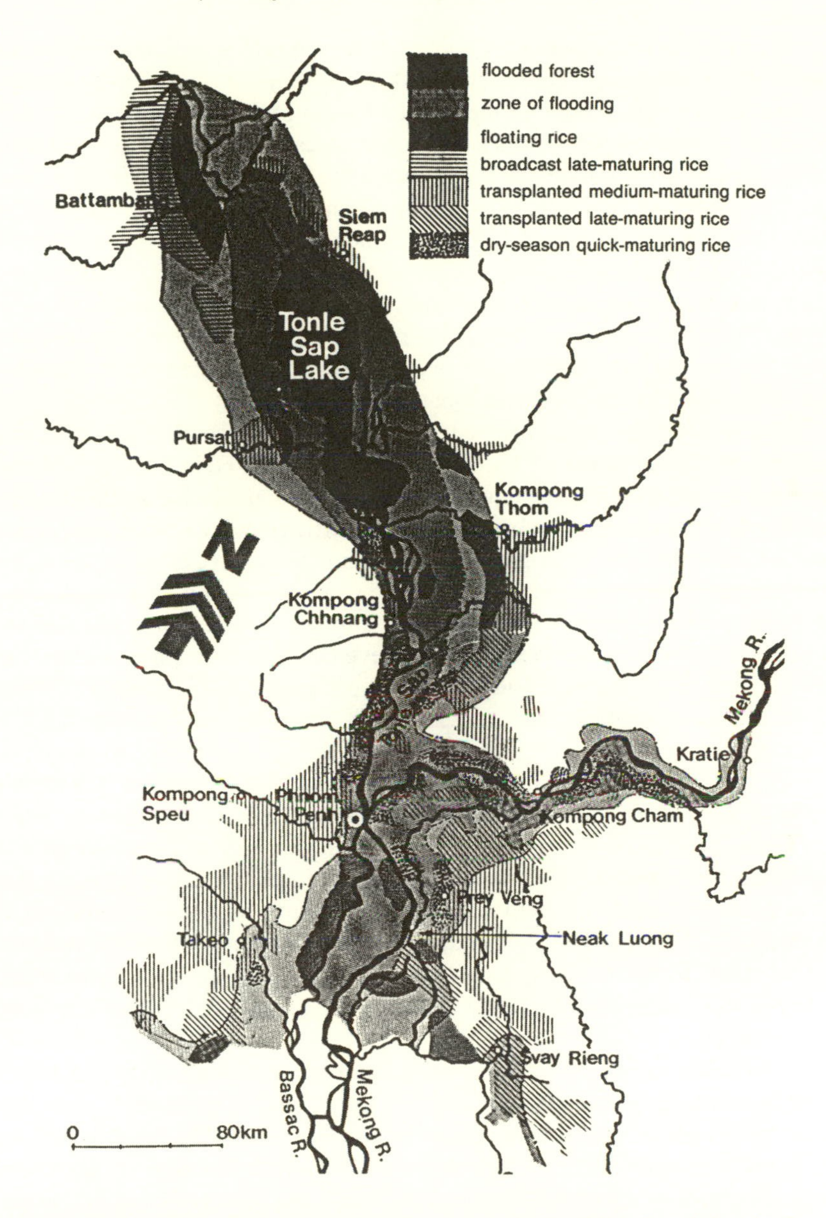

Source: Adapted by Ho Kwoncjan from four maps (no. 1, 3, 5, and 9) in Jean Delvert's, *Le Paysan Cambodgien*, Mouton & Co., Paris, 1961.

Note: See Tichit, 1981, and United Nations, 1968, for similar maps of the rice-growing regions of Kampuchea fertility or were confident that they could replace the silt deposits with Chinese-style fertilization practices.

The present Heng Samrin regime's policy on the management of the annual flood water is ambiguous. On the one hand, it has sought to reestablish practices and structures in place prior to the Khmer Rouge regime. Farmers have been allowed to reopen *colmatage* canals, in some cases assisted by the government and Western aid. A long dike that broke during the Khmer Rouge regime has been repaired, allowing a large area in Kompong Cham and Prey Veng provinces to be returned to wet-season rice cultivation. On the other hand, three years after aid agencies began sending in Green Revolution-type rice seed and chemicals, the regime continues to emphasize these types of materials in its requests to foreign donors. Heng Samrin officials admit that most of the Green Revolution-type rice inputs are used for the much smaller dry-season rice crop when water control is more often sufficient to use these inputs successfully.

These officials are confronted with a policy choice that has profound implications for the future of Kampuchean agriculture. Food and Agricultural Organization (FAO) and some other agricultural aid donors have argued that the best way for Kampuchea to achieve food self-sufficiency is to replace local rice varieties and low-cost cropping practices with input-intensive cropping using the short-stemmed Philippine rice variety, IR36.[5] In my opinion, it is misleading simply to point to the fact that IR36 and similar varieties have enabled the Philippines, Indonesia, and Malaysia to achieve rice self-sufficiency. All three of these countries have export-oriented capitalist economies that either can afford to import the energy-intensive inputs essential to Green Revolution technology or have the resources to produce these inputs themselves. And unlike Kampuchea, none of these countries is dominated by a central floodplain that is annually refertilized by flood waters. At present, less than 10 percent of Kampuchea's agricultural production teams have access to land with sufficient water control to implement the Green Revolution technology. A food production strategy that focuses on such a small area will result in either increased social inequality or, if the surplus is rigorously redistributed, political discontent within the more productive areas. To date the Heng Samrin regime has shown no signs of embarking on any massive campaign to greatly increase the amount of rice land with good water control.

The other, and in my opinion, more appropriate strategy open to Kampuchean agricultural planners is not to fight the annual hydrological cycle, but rather to launch an intensive program to improve the yield potential of the numerous local rice varieties which over the past few thousand years have been adapted to the diversity of soil, water, and disease conditions existing in Kampuchea. Such a program could still incorporate various desirable genetic traits already identified in rice germplasm in neighboring countries or at the International Rice Research Institute (IRRI) in the Philippines, but in ways compatible with the existing hydrological cycle in Kampuchea.

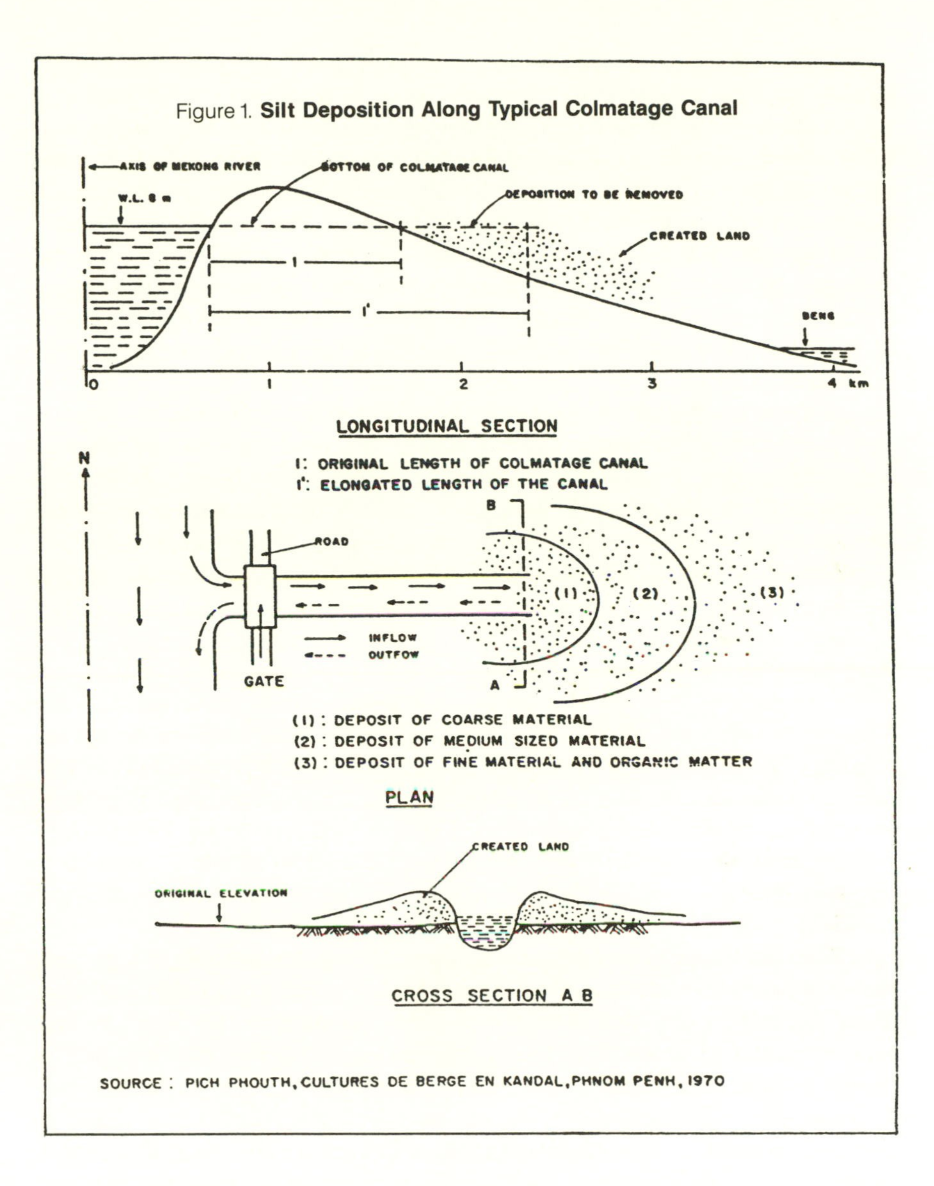

Figure 1. **Silt Deposition Along Typical Colmatage Canal**

SOURCE : PICH PHOUTH, CULTURES DE BERGE EN KANDAL, PHNOM PENH, 1970

The Mekong Secretariat based in Bangkok has planned a more radical strategy for water control in Kampuchea. This plan, whose principal objective is the production of hydroelectricity, calls for a massive dam across the Mekong River between Thailand and Laos at Pa Mong. This dam would reduce both the extent and the duration of the annual flooding in Kampuchea by at least 15 percent.[6] According to one study sponsored by the Mekong Secretariat, Kampuchean agriculture is "at the mercy of natural hydrologic conditions" and the reduction of "this violent water regime" would allow more intensive cropping within the zone of flooding, though at the same time it would increase the extent of drought-prone areas.[7] Fish production and soil fertility would also be adversely affected.[8]

Careful analyses sensitive to long-term effects (100–500 years) on Kampuchea's soils and fish-producing capacity need to be carried out before such a plan is acted upon. The reduction of silt deposition would probably help prolong the existence of Tonle Sap, which is already very shallow, but otherwise the Pa Mong scenario suggests that it would achieve short-term gains (50–100 years) at the cost of permanently degrading crucial portions of the natural resource base.

The Land Resource Base

Kampuchea is a small tropical country with generally poor soils and almost no mineral resources. Located between 10 and 15 degrees north latitude and 102 and 108 east longitude, the country is smaller than any of its three neighbors: Vietnam, Laos, and Thailand (see table 1). The sum of its arable and permanently cropped land is considerably greater than that of the Lao People's Democratic Republic (PDR), about half that of Vietnam, and about one-sixth that of Thailand.[9]

The most striking contrast shown in table 1 is the remarkably low proportion of population to arable land in Kampuchea. If one assumes a 1979 population of six million people, then Kampuchea probably has less than two people to feed for every hectare of arable land. By contrast, Vietnam must feed over nine people from every hectare of arable land, the Lao PDR about 3.7 people per hectare, and Thailand about 2.6 people per hectare. In effect, per hectare rice yields in Kampuchea could be only 25 percent of those in Vietnam and the Kampucheans would still have more rice per person. Agricultural techniques are more intensive in Vietnam than in Kampuchea, but it is not clear whether such techniques led to the higher population density or vice versa. The causes of the markedly different population densities appear to predate the colonial period.

Some evidence suggests that the population-carrying capacity of Kampuchean soils is quite low. Kiernan, for example, cites a survey of the paddy soils of all tropical Asian countries as showing that Kampuchean soils sampled were, on average, the least fertile of all.[10] The Kampuchean soils sampled were poorest in

Table 1

Some Comparative Statistics

Country	Surface area[a] (1000 ha)	Arable land & perm crops[a] (1000 ha; 1980)	Arable land surface area Ratio	Population (millions, mid-1979)[b]	Persons/ ha of arable land	Rice yield (kg/ha) (1980)[a]
Kampuchea	18,104	3,046	17%	6	1.97	967
Lao PDR	23,680	880	4%	3.3	3.70	1,562
Vietnam	32,956	6,055	18%	52.9	9.09	2,239
Thailand	51,400	17,970	35%	45.5	2.56	2,079

Sources:
[a]Food and Agriculture Organization, *1980 Production Yearbook*, vol. 35 (1981), pp. 51–52, 98.
[b]World Bank, *World Development Report, 1981* (1982).

four of fourteen soil qualities, second or third poorest in seven others, and below average in the remainder. An initially lower geochemical potential of Kampuchean soils, rather than the nature of farming practices over the centuries, may have exerted a greater check on population growth in Kampuchea than existed in Vietnam.

The topography of Kampuchea has often been likened to a tilted saucer ringed by highlands and mountains to the west, north, and northeast. The saucer is "cracked" by the Mekong River, which flows in from the north to Phnom Penh at the center and then continues southeast to Vietnam and the South China Sea. The deposition of sediments by the Mekong River in the Quaternary period formed the saucer-like basin of central Kampuchea when it was still part of the sea. Sediments eventually filled in the basin except for what became a shallow inland lake, the Tonle Sap. This was originally drained by a single river flowing southeast. However, at some point the larger Mekong River shifted its bed southward, breaking into the bed of the other river. This act of "stream bed piracy" resulted in the "four-armed" or *quatre-bras* confluence of rivers which was to be crucial to both fish production and the renewal of soil fertility.[11] This confluence also became the site of Phnom Penh.

The Hydrological Cycle

Kampuchea has a tropical monsoon climate. Mean annual rainfall is 1,385 mm (55 inches) at Phnom Penh and 1,955 mm (77 inches) on the coast at Kampot.[12] Most of this rainfall occurs between May and mid-October, leaving the country distinctly arid during the rest of the year.

The onset of the monsoon season on the Indochina peninsula coincides with the melting of winter snows in Tibet and other upper reaches of the Mekong

River. As a result of this double input of run-off water, the Mekong River rises at Phnom Penh from a low of about two meters above its bed in April and May to a height of between nine and eleven meters above its bed in the months of August through October.[13] The volume of water carried in the Mekong increases about twenty-eight times from 1,760 cu.m/sec in the dry season to 52,000 cu.m/sec at flood level. Speed of flow increases about seven times from 0.2 m/sec to 1.4 m/sec.[14]

Neither the main, eastern bed of the Mekong nor its western addition, the Bassac River, can contain all of the annual rise in flood waters. The result is an extraordinary natural mechanism for the renewal of food productivity in Kampuchea. When the flood waters reach a height of about eight meters above the river beds, the waters begin to flow inland through small rivers and canals which drain the flood waters into basins or reservoirs called *beng*, from where the water may in turn flood onto surrounding agricultural land. Delvert estimates that a total of about two million hectares is flooded in Kampuchea every year.[15] By far the most dramatic example of this mechanism occurs when the flood waters actually reverse the flow of the Tonle Sap River, which drains the lake.[16] The river generally stops flowing downstream in early June and flows inland about 110 days a year from mid-June to early October. During this time the area of the lake roughly quadruples from a surface of about 270,000 hectares to about 1 million hectares. The volume of the lake at peak size is estimated to be 65–80 billion cubic meters, more than half of which originates from the Mekong River.[17] The external origin of this water helps explain the paradox that such a productive lake exists within a local watershed of low geochemical potential.

The lake acts as a buffer or flood regulator of the lower Mekong Basin, drawing off water during the peak of the flood and then adding it back to the Mekong-Bassac River system later in the year.[18] Using LANDSAT imagery of the lake and data on Mekong flow rates, an engineer at the Mekong Secretariat estimates that the Tonle Sap increases main stream flows by about 16 percent during the dry-season months. This contribution to dry-season flows is crucial to limiting problems of salinity intrusion in delta areas of Vietnam.[19] On a smaller scale water flooding the left bank of the Mekong at Kompong Cham province is carried by the Tonle Toich River to a large basin in front of Prey Veng City. Similarly, the Prek Ambel delivers flood water to the Angkor Borrey Basin in Takeo province.

Flooding and Maintenance of Soil Fertility

The ability to maintain paddy soil fertility for century after century has been a major test of Asian civilizations. It appears that the Chinese were able to overcome long periods of deforestation and soil erosion by perfecting a meticulous system for recycling organic waste. Japan, the Philippines, and Java are fortunate to have a rich base of volcanic ash soils on which to farm. Good rice crops, for example, are still grown around the temple of Borobodur in central Java, much as they were when the temple was built in the ninth century. By contrast, the fields near the twelfth-century temples of Pagan in northern Burma are now arid and windswept, supporting only meager crops of peanuts.

Within Kampuchea, long-term shifts in population settlement patterns are quite probably related to regional changes in soil fertility. The most reasonable supposition is that vigorous cropping lowered soil fertility wherever it was practiced. However, within the zone of annual flooding, soil fertility did not drop below a critical level needed to support permanent settlement. On the other hand, beyond this zone there was no natural mechanism to prevent the soil from becoming permanently degraded and therefore much less attractive to permanent settlement.

In his book, *A History of Cambodia*, David Chandler discusses the dynamics of population shifts over the centuries. He suggests that the lack of an institutionalized ancestor cult may be one reason why Kampuchean farmers have been less tied to particular areas than farmers in Vietnam. Population levels appear to have been quite low—probably less than 750,000 in the 1840s.[20] The distribution of pre-Angkorian inscriptions indicates that in the seventh and eighth centuries settlements were based along the Mekong and the lower Tonle Sap. From the ninth century to the decline of Angkor in the fourteenth and fifteenth centuries, the center of population and agricultural production was north of Tonle Sap, much of it beyond the zone of flooding.

The shift away from Angkor back to the Phnom Penh region and areas to the east and south began in the 1440s. Chandler suggests that the rise of Ayutthaya in the Chao Phraya valley to the west, the need to be closer to trade routes to China, and the shift from Hinduism and Mahayana Buddhism to Theravada Buddhism may all have been factors related to the move.[21]

But concentrations of population eventually result in lower soil fertility. This may be the crucial factor that has necessitated various population shifts back to the center of the flooded zone. Chandler (p. 100) cites Ba Phanom district in Prey Veng province and Bati in Kompong Speu province as two of the few rural areas with relatively high population densities prior to the 1860s. Both these areas are now poorer than average, suggesting a higher rate of soil degradation.

A better studied area is the site of Angkor itself. The soils of Siem Reap province, alluded to earlier, once formed the agricultural center of the Khmer empire, but now yield on average only 800 kg of paddy per hectare.[22] During the period of the Khmer empire, gravity irrigation systems were used to fill large man-made reservoirs, which in turn supplied enough water for two or more rice crops a year over a period of several hundred years. Groslier describes the silt carried down from the Kulen hills by the three rivers, which supply the systems as a ''poisoned gift'' that led to an ''arteriosclerosis'' of the canal and reservoir network. Silt carried by the river water did renew soil fertility, but it eventually built up the level of the fields to such an extent that it became difficult for irrigation water to reach them by gravity flow. Groslier also maintains that the frequent cycles of wetting and drying led to a progressive "laterization," or hardening of the soil, which greatly reduced its productivity.[23]

The fertility of most soils is renewed by the gradual weathering of minerals contained within the soil particles. However, due to the alluvial origin of Kampuchea's rice soils, the soil particles had already been subject to considerable weathering before the Mekong River deposited them during the Pleistocene era as part of an earlier delta. Soils that are farmed are routinely stripped of their vegetative cover and exposed directly to the sun and rain, which causes them to weather more rapidly than unfarmed soils. It is not clear whether the present low-nutrient status of many of Kampuchea's rice soils is due more to centuries of farming or to natural weathering in earlier geologic periods.

In either case, the present-day reserve of plant nutrients tends to be very low in those rice soils not receiving annual deposition of silt. A Japanese team which analyzed fifty-nine alluvial soil samples from lowland Kampuchea in the 1950s described the soils as weakly acid or neutral clayey loams with good nutrient retention capacity, but deficient in nitrogen, phosphorus, and especially potassium.[24] Due to the low nutrient status of these soils, the annual flooding is much more important for maintaining soil fertility than it would be for younger, less degraded soils. In addition to supplying moisture, the flood waters contribute to the soil's productivity in three other ways: neutralization of soil acidity, silt deposition, and supply of soluble nutrients.

Under normal circumstances an attempt to determine the difference between soils affected by flooding and nonflooded soils farther inland should begin by taking a series of soil samples along transects perpendicular to the river. However, the extent to which the flood waters enhance this productivity is suggested by two pairs of soil samples I collected on a trip in Kandal and Prey Veng provinces in 1981. In each pair, one sample was gathered within the zone of flooding and the other just beyond the zone of flooding. In both provinces the sample collected within the zone of flooding was considerably richer in available potassium, magnesium, calcium, and phosphorus. Nitrogen and organic matter were low or moderate in all samples. In addition, the long-term potential to supply certain nutrients was markedly higher in the samples from within the zone of flooding. The reserve phosphorus was three to five times higher, reserve potassium three to seven times higher, reserve magnesium four to eleven times higher, reserve copper one and a half to four times higher, and reserve zinc two to six times times higher. (See appendix.)

Soil pH of one sample, taken from a field beside the Mekong, was neutral, while the other three, all sampled farther inland, were moderately acid. This is consistent with the fact that Mekong River water is slightly acid to alkaline (pH 6.4–8.3) and exercises a neutralizing influence on its silt load and the soils it floods.[25] Most soils and local river water in Kampuchea are considerably more acidic (pH 4.5–6.0).[26] Saeki et al. found the average pH of forty-seven alluvial soils to be 5.7, or more than ten times as acidic as Mekong River water. Thus, the effect of Mekong flood waters is to neutralize soil acidity and maintain soil pH in ranges more favorable to plant nutrition. By contrast, annual applications of some chemical fertilizers can make some soils progressively more acidic, and this eventually lessens their productivity.

A more tangible effect of the flood waters is the deposition of fertilizing silt. Delvert, in his invaluable book *Le Paysan Cambodgien*, describes the central flood plain around Phnom Penh as a "Mesopotamia."[27] The comparison is an apt one. After the Mekong River has left Laos and passed between the mountains of northern Kampuchea, the slope of its bed flattens out considerably and in the wet season the waters overflow onto the adjacent flood plain. The resultant slowing of the flood waters causes suspended materials to be deposited. Between Kratie and Kompong Cham, coarser sediments are deposited, frequently in the riverbed itself upstream from various islands. Between Kompong Cham and Neak Luong, medium-sized silt is deposited, whereas farther downstream alluvial deposits tend to be mostly clay, resulting in dense but fertile soils.

The same deposition pattern occurs on a smaller scale as flood waters flow inland from the riverbank: the larger, heavier sediment is deposited first on the immediate river bank, while finer materials are carried farther inland. The total amount of silt deposited is immense. Delvert estimates that in one year a net amount of three million tons enters the Tonle Sap River from the Mekong. The rate of deposition decreases as the water moves downstream. According to a Mekong Committee study, flooded areas upstream from Phnom Penh receive 5–10 mm of silt in normal flood years and 20–30 mm in high flood years; the Phnom Penh zone receives 2–3 mm and 5–9 mm respectively, and farther downstream land receives almost nil and 1 mm per annum respectively.[28] As mentioned earlier, *colmatage* systems are considerably more efficient. The silt deposition patterns within them are clear enough to be seen in satellite imagery.[29] A study of silt deposition on farm land in the Mekong Delta of Vietnam found that while the nutrient content of the silt was higher than that of the soils it was deposited on, an estimated deposition rate of only 2 mm per year would be unable to sustain high levels of production without the addition of fertilizer. The study also noted that problematic, acid-sulfate soils only existed *beyond the zone of annual silt deposition.*[30]

Silt deposition is not a gradual process occurring throughout the wet season. According to the Mekong Committee study, the most silt is delivered to low-lying farm land when the Mekong reaches peak flood levels. As the Pa Mong dam is designed to smooth out the flow in the river, it is estimated that the average peak height of the river that now occurs annually would occur once every ten years if the dam were in operation. Given that the silt load of the below dam portion of the Mekong would be only that added between Pa Mong and lowland Kampuchea, the resultant silt deposition might be less than 10 percent of present levels.

Silt brought into Kampuchea by the Mekong should not be emphasized to the exclusion of hydrological processes within local watersheds. Charles Crocker, an American agronomist who conducted a thorough soil survey in Kampuchea, describes the soils around Tonle Sap as lacustrine alluvial soils formed by "colluvial-alluvial outwash" from the uplands bordering the lake plain and by "alluvial silts and clays carried in by the Mekong flood waters."[31]

The dissolved minerals in Mekong River water, though present in very dilute form, represent a fertilizer resource that is both continuous and free of charge. Weathering minerals release soluble forms of four plant nutrients (potassium, calcium, magnesium, and sodium) through a process geochemists call cationic degradation. Water draining from watersheds transports these and other plant nutrients to rivers and then onward to the oceans where these nutrients are apt to be removed from terrestial food production systems for hundreds of thousands of years. Irrigated rice paddies are an ingenious and remarkably durable invention of ancient man, designed not only to supply water to the rice crop, but also to short-circuit this relentless loss of dissolved nutrients to the sea.

The net balance of dissolved nutrients contributed to the soil (or directly to the rice roots) by river water varies with the fertility of the water and the infertility of the soil. A poor soil will tend to gain more than a rich soil, and a rich soil may lose more nutrients than it gains, especially to low-nutrient (oligotrophic) waters.

Among all of the domesticated grains, rice, which is a semi-aquatic plant, appears to be an especially capable forager of nutrients when soil fertility is low. Subsistence rice farming on poor soils with no fertilizer inputs tends to "bottom out" at annual yields of about 1.2 mt of paddy per hectare. However, sophisticated nutrient balance studies are required to determine the respective amounts of the nutrients derived from the irrigation water, soil nutrient reserves, and biological nitrogen fixation. One indication of the significant contribution of irrigation water is the fact that floating rice, which sprouts many additional roots from nodes along the stem, can continue to grow in the flood water alone if it becomes uprooted from the soil. Another indication is the fact that irrigated rice grown in mainland Southeast Asia rarely shows any response to potassium fertilizer even when grown on potassium-deficient soils. Ho Tong Lip, chief of the Agronomy Division in Phnom Penh in the 1960s, wrote of the role of nutrients transported to rice fields from local watersheds: "In our sandy areas where rice fields have good production, it is primarily because of neighboring forests which supply nutrients from runoff water."[32]

The discussion in this section has not been meant to imply that Kampuchea must continue to rely solely on traditional methods of rice soil fertilization, while other Asian countries steadily increase their consumption of chemical fertilizers. On the contrary, a mixed strategy relying on chemical fertilizers, recycling of organic wastes, and the fertilizing value of flood and irrigation water should prove the most cost-effective and reliable. Although the Pa Mong dam would greatly limit the extent of the natural fertilization process, smaller-scale water-management projects may actually be used to enhance the efficiency of the process. Kaida points out that ''soil dressing by *colmatage*'' has been used in Japan to improve excessively permeable rice fields, but that the process is only feasible where an intensive terminal canal system is already in place.[33]

The Mekong Secretariat has done some very impressive work in their study of the resources of the lower Mekong basin. However, they must avoid the temptation of deemphasizing the disruptive effects in any one country in order to enhance the chances of securing approval for large, multinational projects such as the Pa Mong dam. When one reads in the Mekong Committee study that the low terrace soils of Kampuchea have ''deteriorated severely because of *yet unknown reasons*'' [my emphasis], one senses an unwillingness to draw a connection between the poverty of these soils and their location beyond the zone of flooding.[34] According to the committee's computer projections, the Pa Mong dam would remove from the flood zone about 184,000 hectares around Tonle Sap and a further 200,000 hectares along the Mekong and Bassac rivers. The committee's Working Paper No. 6, published in 1976, suggests that this reduction in the flooded area might cause a decrease in the fish catch due to decreased flooding; expansion of the ''drought-susceptible areas''; or reduction in fertilization by natural silt on agricultural lands. All three possible negative impacts are critically related to the well-being of the ecology and economy of Kampuchea, and yet none was mentioned in a briefer version of the same study, which appeared in 1979 in the Kyoto University journal, *South East Asian Studies*. The following sentence, however, had been added: ''Once this violent water regime has been relieved to some extent, it will be relatively easy and almost costless, in the low alluvial plains, to develop the land into fairly productive agricultural farms.''[35] Neither version of the study mentions that almost half of the ''flood-relieved'' land would be within the ''flooded forest,'' a unique natural resource that is essential to the high productivity of Kampuchea's inland fisheries. Further study and discussion of these issues are needed.

The Rice Variety Question

The most productive zones within the biosphere occur at the interfaces between air and soil, air and water, and water and soil. Rooted in soil, partially submerged in water for much of its growing season, and spreading its leaf canopy in the air, lowland paddy rice is unique among the world's major food crops. During its growth cycle it grows across all three interfaces, and appears to gain various advantages from each. The aquatic environment, for example, favors the growth of certain blue-green algae that fix nitrogen from the atmosphere, and some of this nitrogen is eventually passed on to successive rice crops.[36] Under flooded conditions, various nutrients in the soil such as phosphorus tend to become more available to the rice roots than they would have been under nonflooded conditions.[37] And, as already stressed, the water column may contain a supplementary supply of plant nutrients transported from higher up on the watershed.

But, even if one accepts the idea that rice is the most appropriate major food crop for countries like Kampuchea which have only meager supplies of fertilizer, there remains the issue of which types of rice varieties governments should stress in their food production strategies.[38]

In the fall of 1979, Kampuchea was stricken by famine. Relief agencies were informed that farmers were eating their rice seed stocks and consequently large shipments of rice seed would be needed from other countries to avert a cycle of hunger and malnutrition.

This sad situation appeared to offer an unusual opportunity. To the extent that war and then the isolationism of the Khmer Rouge regime had prevented Kampuchean farmers from taking advantage of the high-yield-potential varieties developed by modern rice-breeding programs, these farmers could now catch up with their contemporaries in other countries in the 1980 growing season, using seed and other inputs donated by Western and socialist countries. Farmers who would otherwise be too conservative to try more than a small plot of a new variety would, due to the lack of local seed, be compelled to adopt the new varieties on a large scale. In the spring and summer of 1980, millions of dollars of relief funds were spent to ship in rice seed, much of it the modern, short-stemmed variety, IR36, and a full complement of fertilizers, pesticides, knapsac and mist sprayers, water pumps, and tractors. Unquestionably, these inputs played an important role in the increase in food production that followed. However, the above scenario of a leap forward into the Green Revolution was not realistic for a number of reasons.

First of all, Kampuchea was not a virginal backwater waiting anxiously for the seed of the first modern varieties to be planted. IR8 and other modern varieties had been sent to Kampuchean rice research stations in the late 1960s and early 1970s. More recently, the totalitarian Khmer Rouge regime had sought to extend various short-stemmed, modern varieties developed in China. In the dry season of 1983, farmers in Kandal province were still growing one of these varieties, renamed *sro matthaphiap* (friendship rice).

Another misconception concerns the extent of the famine and the lack of local seeds. Stephen Heder, an expert on Kampuchean politics, has written that the starving civilians who shocked the world as they straggled over the border Rouge had evacuated, often forcibly, to mountainous areas. That their extreme malnutrition was not representative of the majority of Kampucheans who remained in the lowlands. During trips to Takeo and Kompong Speu provinces in August 1980, I was surprised by the number of draft animals still in the villages and by the large proportion of farmers interviewed who were growing rice from local seed stocks. Asked to describe conditions during the preceding year, villagers along main roads spoke of hunger and hardship, but not of starvation. However, Molyda Symusiak provides a harrowing account*

*1986. The Stones Cry Out: a Cambodian Childhood, 1975–1980. Hill and Wang. New York: 245 pages.

showing that large pockets of severe malnutrition and starvation existed during much of the Khmer Rouge regime.

Despite the fact that the Heng Samrin regime has offered a 50 percent higher price for IR36 within the mandatory state marketing system, IR36 has failed to spread as widely in Kampuchea as it has in the Philippines and Indonesia. The different spectrum and variability of ecological conditions in Kampuchea accounts, in large part, for the low rate of adoption. The Philippines and Indonesia have topographically well-drained insular environments with large areas of fertile volcanic soils. Kampuchea, as Mekong Secretariat publications stress, is dominated by a large continental river. Considerable variation exists in the extent and duration of the annual floods and in the beginning and consistency of the monsoon rains. Within the zone of flooding, a short-stemmed, modern variety is riskier to grow in the wet season than a tall-stemmed, traditional variety. Yet beyond the zone of flooding soils tend to be infertile and rainfall unreliable. Here, too, the modern variety proves riskier unless a stable environment can be ensured through the use of irrigation and fertilizer.

Various types of rice have been in long use in Kampuchea. Although each is adapted to a particular set of soil and water conditions, they nonetheless have broader environmental tolerances than IR36-type varieties. The average farmer or solidarity team is apt to grow rice in fields that span at least two different sets of soil and water conditions and to often grow several varieties within each set of conditions. The effect is to diminish risk. Farmers in rain-fed areas frequently face the problem that the rains come late, or start and then stop for several weeks.

Late-maturing varieties, which by modern standards include everything maturing in over 140 days, are better adapted to these variable rainfall conditions than the quick maturing, modern varieties. As can be seen from table 2, more than 90 percent of the varieties traditionally grown in Kampuchea mature in 150 days or more. If the rains come in sufficient amounts, the farmer can proceed to transplant his seedlings when they are thirty to thirty-five days old. Otherwise, the farmer can, if necessary, hold his seedlings in the seedbed for as long as two months without the yield potential of the plants being damaged. IR36 and most other modern varieties mature in less than 130 days and are therefore much more sensitive to stresses that interrupt their tight growth schedules. If the rains are delayed and the farmer cannot transplant on schedule, the modern variety will begin its reproductive stage in the seedbed and yield will be greatly depressed. Once transplanted the modern variety is also sometimes more vulnerable to stress from lack of nutrients, drought, and in some cases insect attack.

On the other hand, the rice breeders have not been idle. Modern varieties, as they are much shorter, tend to put a greater proportion of the nutrients they obtain into the grain portion of the plant. On fertile soil, the modern variety will put almost all the additional nutrients available into grain production, whereas the traditional variety is apt to grow much taller with little increase in grain yield above roughly four metric tons per hectare (4 mt/ha). When plied with chemical

Table 2

Types of Rice Production in Kampuchea in 1967

	Hectares	% of crop
Early varieties (90–150 days)	78,000	3.1
150–180 day varieties	313,700	12.5
180–210 day varieties	1,080,500	43.0
Late varieties (over 210 days)	487,600	19.4
Floating rice (about 270 days)	390,000	15.5
Dry-season rice (90–140 days)	163,900	6.5
Total:	2,513,700	

Source: L. Tichit, *L'Agriculture au Cambodge* (Paris: Agence de Cooperation Culturelle et Technique, 1981), p. 79.

fertilizers, the traditional variety may grow taller yet and lodge. As the plant is much less efficient at photosynthesis when lying on the ground, yields may be even less than if no fertilizer were applied.

The plant breeders have also given IR36 and similar varieties considerably more resistance to most insect pests and diseases of rice. Ironically, the resistance has been so good that certain insects have responded by evolving new "biotypes." Biotype III of the brown planthopper is now able to feed on IR36 and transmits a serious rice disease called Tungro virus in the process. For this reason, the useful lifespan of the average modern variety is apt to end very abruptly within less than a decade. If a particular modern variety is cropped on a large proportion of a region's rice land, the region may experience sudden food shortages when the insect resistance of the variety suddenly breaks down. The susceptibility of the modern variety, IR26, to biotype II of the brown planthopper resulted in food shortages in certain Mekong Delta areas of Vietnam in the late 1970s.[39] Ragged Stunt, a new virus disease discovered in 1977, could pose a serious threat to the more than ten million hectares of Asian rice fields that are now planted with IR36.

Although a modern variety such as IR36 is apt to have a yield potential twice that of most traditional varieties, there were serious questions as to whether the potential of IR36 could be realized in the 1980 wet season in Kampuchea. And even if it had been successfully extended to all of the dry season and early rice areas shown in table 1, this still only equalled 13 percent of the estimated 1.32 million hectares of rice that were reportedly planted in 1980. The question of which varieties of seed were most appropriate to send into Kampuchea in the spring of 1980 was thus very difficult to answer satisfactorily.

The floating rice areas appeared to be the easiest to bring back into production for a country scarred by warfare and with a shattered infrastructure. These areas made up 15.5 percent of the rice area planted in 1967 and thus tentatively could make up almost 40 percent of the one million hectares needed to feed the post-Khmer Rouge population.

Floating rice is a hardy subsistence crop that needs minimal soil preparation, very little labor, and no fertilizer or other chemical inputs. It does have some drawbacks: low yields (0.8–2.0 mt/ha), a high seeding rate, and inferior eating quality. It is troublesome to harvest. It is also critical that the crop be planted sufficiently in advance of rising flood waters—i.e., April-June in Kampuchea—so that the chosen varieties can grow tall enough to survive in the maximum water depths possible in each area. However, given the existing shortage of labor, draft animals, and chemical inputs and the need for early planting, it was the common opinion of farmers, Kampuchean officials, and relief agency personnel in the spring of 1980 that the importation of floating rice seed should be given top priority.

As a result, more than 6,000 tons of floating rice seed were unofficially sent across the Thai border, and a smaller amount was exported under license by air and sea from Bangkok. *Khao khiew*, a rugged floating variety adapted to water depths of 3–3.5 meters, began to be distributed across the Thai border in April. If necessary, it could simply be broadcast on unplowed soil and covered lightly with soil. When the FAO finally received a license to export rice seed in June, an airlift was hastily arranged due to concern that floating rice seed sent by sea could not be distributed to farmers in time for planting. Two aircraft, each carrying forty tons of seed, flew daily flights from Bangkok to Phnom Penh via Ho Chi Minh City. Surprisingly, most of the initial deliveries by air were then shipped by truck and by train westward to Battambang province despite information that this province had already received floating rice seed in excess of projected needs. By this move, the Heng Samrin regime established more of a basis from which to claim credit for what was subsequently to be a good rice crop in that province. The airlift carried mostly the improved floating varieties, *lep mue nang* and *pin gaew 56*, both of which are more specific in their cultural requirements than *khao khiew*.

As shown in table 2, not even including the floating rice crop, 75 percent of the 1967 rice crop consisted of varieties that matured in more than 150 days. Relief agencies in effect took two very different positions on the question of what outside varieties were appropriate for the nonfloating rice areas. Care and World Relief, both operating in Thailand, were responsible for most of an additional 7,000 tons of traditional-variety seed, which was distributed under International Red Cross and UNICEF supervision to Kampuchean farmers at the Thai-Kampuchean border.[40] The tall, late maturing (more than 150 days) varieties, *ta haeng* and *met lek*, were sent from the central plain of Thailand, while more drought-resistant varieties such as *dawk mali* and *khao khao* were sent from Northeast Thailand. As with the floating rice shipments to the border, all seed bags were marked in Khmer to indicate variety name, plant height, and approximate harvest date.

By the end of June, most agencies delivering rice seed to the government in Phnom Penh began purchasing IR36 or similar short-stemmed varieties from Asian countries other than Thailand. This policy was due both to the difficulty of exporting rice from Thailand by air or sea and to the belief that modern rice varieties were the best type to send. Thus the OXFAM/NGO consortium of thirty-five Western aid agencies purchased high-priced "IR36-equivalent" from China. However, in the opinion of one IRRI rice breeder, none of the varieties available in China (e.g., *guang lou AI4*, *zhen zhaui*, *qing gan huang*, and *hong 410*) was well adapted to conditions in Kampuchea.[41] Various agencies, including the FAO, bought out the IR36 seed supply in the Philippines, and at least one agency shipped the short-stemmed variety, *jaya*, from India.[42] A further reason for the emphasis on IR36-type varieties was that the Ministry of Agriculture in Phnom Penh was impressed by high yields obtained with IR36 in some dry-season areas.

Unfortunately, there is very little information on the respective performances of the various varieties sent in. After three years of IR36 shipments to Kampuchea, however, government figures in 1982 indicated that IR36 accounted for 66,000 hectares, or only 4.1 percent of the annual crop.[43] IR36 has not been the panacea that FAO and some other agencies expected it to be.

Trips to Kompong Speu and Takeo provinces in August 1980 revealed that a surprisingly large amount of rice land was still fallow and that at least 90 percent of the on-going planting was being done with local, tall-stemmed varieties. Villagers interviewed said that they had not consumed their own seed stocks. The "little dry season," which usually occurs in part of July and August in central Kampuchea after an earlier period of rain, was more severe than usual. Officials said that of 15,000 hectares planted in Takeo province, 2,500 had already died due to drought, although they did not know what proportion of this area had been planted with modern varieties. An area of floating rice near the Vietnamese border planted with improved-variety seed sent by the OXFAM/NGO consortium was reported to be growing well. Similarly, IR36-type varieties were growing well on fertile, well-watered soils near the village of Tra Hieng on the main road west of Takeo City. (Losses occurred in a 3,000-hectare area of floating rice in Battambang because the floating rice seed sent in also contained nonfloating variety seed.)

The risks of planting IR36-type varieties were evident elsewhere, particularly along the route south from Phnom Penh into Kompong Speu province. The soils grew sandier and less fertile with increasing distance from the Mekong. It appeared that chemical fertilizer would be rapidly leached from the root zone of such soils if added. Due to drought, soil preparation was delayed and farmers were transplanting with forty-five-day-old seedlings. An equally long delay with IR36-type seedlings would have lowered yields significantly. In one area of particularly poor soils, a crop of IR36 was not worth harvesting; the plants were only about thirty-five cm high at maturity. But only a few kilometers down the same road farmers at a threshing site were delighted with IR36 yields of about three mt/ha and said they planned to plant IR36 again as their wet-season crop.[44] In Sroc Bati of Kompong Speu province, farmers had reportedly been prompted by the distribution of irrigation pumps and modern-variety rice seed to plant a dry-season crop where none had been grown before. However, fuel for the pumps ran out in mid-season and the crop was lost.[45]

During the rice harvest in January 1982, it appeared that farmer solidarity teams in Kandal province had spread their risks by planting numerous varieties. One team interviewed within the zone of flooding about ten kilometers south of the Bassac River in Kandal province had planted seven varieties on 12.5 hectares of land.[46] On two hectares with maximum water depths of two meters, they had planted two floating varieties. A local variety, *ga sala*, had yielded about 0.8 mt/ha, whereas a Vietnamese variety donated by a ''sister province'' in Vietnam had yielded about 1.8 mt/ha. On half of their fields, the team had planted three local, tall-stemmed varieties (*sro saw buon guar*, *ga kaniey*, *jong bonla*), which had yielded between 1.5 and 1.8 mt/ha. And on their fields with the shallowest water, the team had grown IR36 and a variety shipped from India, *mahsuri*, with yields of 5.5 and 4.5 mt/ha respectively. But due to the low eating quality of IR36, they planned to plant only *mahsuri* on the shallow-water fields the next year. However, *mahsuri* was not adopted in Thailand due to inadequate eating quality, and this may also have been the fate of *mahsuri* introduced to Kampuchea in 1969.[47]

By 1983, the Ministry of Agriculture in Phnom Penh was supporting the use of both IR36 and IR42 in the dry season and appeared to be committed to improving local varieties. Soviet agronomists were reportedly working on varietal improvement at the Tuol Samrong Rice Station in Battambang. At the Prey Phdau Station in Kompong Som the Department of Agronomy was testing about 120 varietal accessions of traditional Khmer rice which had been returned from the International Rice Germplasm Center in the Philippines. It is expected that, by use of the pure-line selection method, yields of some local varieties can be improved 20–30 percent.

Pure-line selection is based on the fact that local varieties passed on among farmers tend to contain a great deal of genetic variability and varietal mixtures within the population. Furthermore, yield potential and competitiveness tend to be inversely correlated within a given population of a rice variety. As a result, those plants within the farmer's field that have higher yield potential will tend to be crowded out by leafier, more vigorous plants. If researchers plant out the same population using wide spacing, it then becomes possible to select the plants with the highest yield potential. The cost of selecting that part of the population which is highest yielding but less vigorous is that it then requires greater environmental stability. From these relationships stems the view that the adoption of energy-intensive rice agriculture in tropical Asia is historically inevitable.[48]

In my opinion, the introduction of new varieties must proceed cautiously. It is important not to replace the risk-minimizing and location-specific diversity of the traditional varieties with a system that relies on only a few relatively stress-sensitive modern varieties. For Kampuchea, it is crucial that the effort to create the improved water control required by the modern varieties through the building of dams and levees does not undermine the natural renewal of soil fertility within the zone of flooding. Furthermore, the use of higher yielding varieties results in the more rapid use of soil nutrients. Unless steps are taken to renew these nutrients through a program of organic cycling or the use of chemical fertilizer, soil fertility will be lowered and the initial increase in food production will not be sustainable.

Fish Production in Tonle Sap

The first farmers to grow rice in paddy fields in Kampuchea may have been emulating the natural processes of the Tonle Sap. The enduring productivity of both aquatic environments is based on the annual destruction by flooding of a terrestrial ecology followed by rapid growth fueled by the decay of the terrestrial biomass. Whereas farmers plant rice in the shallow water of the fields, natural spawning cycles bring fish to the newly flooded portions of the lake. At the end of the wet season, the aquatic environments of both retreat. The paddy fields dry up entirely; the lake shrinks to a fourth of its wet-season area.

Although varying in extent from year to year, it is the annual flooding of nearby plains that accounts for the Tonle Sap's remarkable production of fish. Though sea water is generally more productive of fish than fresh water, the Tonle Sap was reported in 1934 to be, on a unit area basis, nearly ten times as productive as the best fishing grounds in the North Atlantic.[49] Since then the lake has remained famous for both the quality and quantity of fish it produces. The ecology of this marvelous productivity is not completely understood, but various scholars have described some of the most salient aspects. It is hoped that the description provided here will promote a greater awareness that Tonle Sap is at the heart of Kampuchea's productivity and must be carefully protected as one of its most valuable resources.

Tonle Sap is not a postcard lake with sandy beaches dropping off into deep blue waters. Its waters are turbid throughout the year. With an average depth of only 1.5 meters in the dry season, slight breezes are sufficient to keep water stirring against the muddy bottom. In the wet season shallow tributaries are transformed to muddy torrents. The dry season shore of the lake is rimmed by a thicket generally 30–40 kilometers wide of shrubs and small trees. This is the "flooded forest" (*forêt inondée* in French and *prey roniam* in Khmer), which provides the famous spawning ground when it is flooded each year from about June to October.

The lake still retains some traces of its marine origin. At least sixteen species of marine fish continue to inhabit the lake.[50] During the period of lowest water level in March and April the level of the lake is only sixty cm above sea level. Though separated from the South China Sea by about 355 km of river, the tides there are reported to vary the level of the Tonle Sap River and the lake itself by about ten cm during the period of lowest water.[51] In 1925 it was discovered that during this same low-water period, the water draining from the Tonle Sap was so rich in nitrogenous products that every year temporary banks of marine fish formed in the ocean several kilometers off the coast of Vietnam to feed in the plumes of the Mekong and Bassac rivers.[52] Chevey thought that it was due to differences in nutrient levels that a species of fish that grew to seventeen cm in length in two years in the Tonle Sap River grew to thirty-two cm in length in the same time period in the lake.

Subsequent researchers have stressed the role the flooded forest has played, both around the lake and in low areas closer to Phnom Penh, as a refuge and spawning ground for many of the fish in the Mekong River south of Phnom Penh. According to Pantulu, "the life cycles of the vast majority of the fish in the Mekong depend prominently on the annual inundation cycle."[53] About 90 percent of the fish in the Mekong system spawn not in the rivers themselves, but in adjoining lakes, submerged fields, or the flooded forest areas.[54] Most species find the swifter currents, the sandy beds, and the more turbid water of the rivers less suited for spawning.

When the river level rises and current speed begins to increase in May, many fish begin to swim upstream in search of spawning grounds. Due to the Tonle Sap's slower speed and lower turbidity, many species from as far away as the Mekong delta in Vietnam proceed up the Tonle Sap River to spawning grounds in newly flooded portions of the lake. The spawning peaks in July, whereas high water levels do not peak until September-October. This allows adequate time for the fry to develop before water levels begin dropping. Many species migrate downstream from the spawning grounds on moonlit nights in December, January, and February. The excellent fishing opportunities on the Tonle Sap, particularly during spawning and down-migration runs, help explain why Vietnamese fishermen have also migrated to the Tonle Sap.

The fish fauna in the Mekong-Tonle Sap system are quite diverse. The main categories are carp (54 percent), catfish (19 percent), and murrel (8 percent). Among the remaining groups are herring, climbing perch, and gourami.[55] Fishermen in Kampuchea group most species as either "white fish" (*trey saw*) or "black fish" (*trey kmao*).[56] "White fish" have silvery scales, are intensely migratory, and feed upon plankton. Among the most economically important species are three carps, *Cirrhinus auratus*, *C. jullieni*, and *Thynichthys thynnoides*, and five river catfish, *Pangasius macronemus*, *P. sutchi*, *P. larnaudi*, *P. micronemus*, and *Pangasianodon gigas*. *P. gigas* sometimes grows to weights of about 300 kg.

The "black fish" are an ecologically complementary group. They feed near the bottom on insects and smaller fish and do not migrate from the flooded forest when the waters first begin falling. Examples of the most economically important "black fish" are two carnivorous murrel, *Ophicephalus striatus* and *O. micropeltus*, a climbing perch, *Anabas testudineus*, and a fourth species, *Oxyeleotris marmorata*.[57] The murrel and climbing perch are adapted to breathe in low-oxygenated water and to travel for appreciable distances over wet land, both useful adaptations when feeding in the stagnant waters of shallowly flooded floodplains.[58]

As the annual flood waters spread out through an estimated 750,000[59] hectares of "flooded forest," the fish population follows. The broad diversity of fish species that migrate into the flooded forest is probably essential to maximizing the efficient use of the food supply created by the masses of decaying organic matter. Fish yields are at least sixty-five kg/ha/yr in the lake if calculated from the dry-season area of the lake. This compares to an average yield of twelve kg/ha/yr in tropical rivers and thirty-two kg/ha/yr in a carefully managed reservoir in Thailand. The high Tonle Sap yields are due to the presence of a large proportion of plankton- and detritus-feeding fish which crop the lowest levels of the food chain and to the huge amount of external (allochthonous) organic matter which is suddenly submerged in an aquatic environment. The subsequent decomposition of the foliage and other organic matter produced in the preceding dry season provides a surge of soluble nutrients for the growth of phytoplankton and larger aquatic plants on which many "white fish" feed.

The French ichthyologist, Chevey, described the water within the flooded forest as a "veritable vegetable and animal broth. The color of . . . tea."[60] Feeding on the rich supply of dissolved nutrients, phytoplankton rapidly multiply and form the primary food source of the "white fish." In a study of the lake in 1950, the limnologist Blache observed that by April-May the blue-green alga, *Microcystis*, had built into a dense surface bloom called *duk proam* in Khmer.[61] Blache also found that the number of plankton in the major waterways (Mekong River, Tonle Sap River, and Tonle Sap) varied inversely with the speed of the water current and the amount of silt it contained.

Also thriving in the still waters of the flooded forest and various *prek* and *beng* are floating water plants of the genera *Eichornia*, *Pistia*, *Salvinia*, *Azolla*, *Lemna*, and *Wolffia*.[62] While some of these plants and various algae can be eaten directly by fish, grazing invertebrates such as chironomoid larvae live on the plant populations and are an important source of fish food.

After the flood waters begin to recede in October, the "black fish" tend to remain in the flooded forest or in the open lake. Many take shelter in large brush piles fishermen place in the water. It is then simple for the fishermen to enclose these brush piles and catch large quantities of fish. Some "black fish" remain stranded in shrinking pools of water, and these are also easily caught. The cycle of fish production and its ecological base are described in a Kampuchean proverb: "When the waters flood, the fish eat ants; when the waters recede, the ants eat fish."

The preservation of the flooded forest is essential to the maintenance of high fish production in the lake. The major trees growing in this scrub forest are *Crudia* (*sdey*), *Homalium* (*roteang*), *Terminalia chebula* (*taour*), *Strychnos* (*sleng*), *Barringtonia acutangula* (*reang tuk*), *Gmelina asiatica* (*anchehn*), and *Hydnocarpus* (*krabau*).[63] Luckily for fish production, only one tree, *Hydnocarpus*, has any significant commercial value.

The former French administration and several subsequent Kampuchean governments recognized the value of the flooded forest by classifying much of it as an "ichthyological reserve." However, the total area of the flooded forest has tended to diminish because fishermen cut large areas for branches to make the brush-pile fish traps and farmers grow beans in some areas before the flooding and floating rice during the flooding. Much of the flooded forest that traditionally grew in low areas behind river banks was removed prior to 1900 to provide land for the production of red corn.[64]

A variety of other threats to fish production in Tonle Sap either already exist or loom on the horizon. Raw sewage from the city of Phnom Penh threatens the annual migration of fish from the Mekong up to the lake. Over-fishing was reportedly a problem in the 1960s and again in 1983. Probably the most intractable, long-term problem, however, is the gradual silting up of the lake. A reasonably large lake area must remain throughout the dry season if there are to be sufficient "black fish" to populate the flooded forest in the wet season. If the lake's average dry season depth is about 1.5 meters, a siltation rate of 0.4 mm per year measured in 1962–63 suggests the lake's remaining half-life is about 900 years.[65] However, the increasing deforestation of the Mekong watershed in the past twenty years may have greatly increased this siltation rate. In the 1960s the Mekong Committee planned to construct a barrage at the mouth of the lake to raise the dry-season level of the lake two meters. Such a project, however, would have lowered fish production by reducing the area of the "flooded forest" and by hampering fish migration to and from the lake.

The construction of the Pa Mong dam or other mainstream dams on the Mekong River upstream from Phnom Penh would have the effect of shortening the flooding season by an estimated two to four weeks and lessening the extent of the floods by 10 to 30 percent.[66] If fish production is assumed to be proportional to hectare-months of inundation, then fish production in the lake would be reduced by 10 percent or more. Due to the earlier recession of the flood waters, less organic matter and phytoplankton would be available for the fish to feed on and more fish eggs and fry would become stranded and dessicated.[67]

The total annual fish catch and its subsequent distribution have varied in part according to the type of government in power in Phnom Penh. Under the French in the early 1930s, the annual catch from Tonle Sap was estimated at 100,000 tons, over 90 percent of which was exported to Java as dried fish.[68] Under the Sihanouk government in 1966, the total national fish catch (including 30,000 tons of marine fish) was 188,000 tons of fish, but only 10.6 percent of this was exported.[69] Prior to the Khmer Rouge regime, the majority of the fishing community consisted of ethnic Vietnamese and Cham. Both these groups were harshly suppressed during the Khmer Rouge regime, and the consequent dearth of fishing expertise was reported to have resulted in ''underfishing'' of the lake and the buildup of an abundant standing crop. The resumption of more organized fishing in the spring of 1980 probably accounts for the ironic fact that dried fish from the Tonle Sap was traded in the black markets along the Thai border and appeared in restaurants in the Thai town of Aranyaprathet even as relief workers continued to distribute rice and Thai dried fish to Kampuchean famine victims who continued to come to the border for aid.

The amounts of the different fish products produced in Kampuchea in 1967 are shown in table 3. In 1968, a cannery for marine fish in Koh Kong province also began production of fish meal for use as animal feed.

In summary, the unusually large fish population in Tonle Sap depends on a unique hydrological cycle that involves the flooding of a vast area of scrub forest surrounding the dry-season lake area. There appears to be little that man can do to enhance this system of natural production. On the other hand, the presence of growing human populations within the Mekong watershed threatens to reduce the productivity of the lake. Silt eroded mainly from agricultural lands will eventually fill in the lake, but shorter term threats include pollution, overfishing, and destruction of the flooded forest.

Conclusion

Arid and bleak in the dry season, the landscape of Kampuchea miraculously comes to life again each year. The monsoon rains soak the soils. The water of the brimming Mekong backs up the smaller rivers until Tonle Sap has expanded four times. The annual flooding fills and renews the rice fields and fisheries of the country. Fish and rice have been the staple food of the Khmer for centuries, but there are obvious signs of wear and degradation within the resource base. The yields of large drought-prone areas invite pity. Deforested slopes have filled in much of Tonle Sap with their soil. The annual transition from desert to swamp to desert again appears to call out for some modifying modern intervention.

One-hectare checkerboard fields linked by badly made canals were one response stillborn in violence. The planned construction of Pa Mong dam is a response of international dimensions and yet uncertain consequences. The complex food chain supporting Tonle Sap fish production is a long-term resource that might be sacrificed by such changes. Further questions remain as to whether the Kampuchean economy can support a full shift to the energy-intensive technology of the Green Revolution. A more location-specific response involving the improvement of traditional rice varieties and the retention of some means to deliver silt to the fields would seem to be a practical alternative, at least in the beginning.

Table 3

Fish Products in 1967

Product	Production
Fresh fish	48,250 tons
Dried fish	9,800 tons
Smoked fish	1,115 tons
Boiled fish	1,090 tons
Prahoc (fish paste)	2,970 tons
Fish sauce	4,060 1,000 liters

Source: Tichit, *L'Agriculture au Cambodge,* p. 359.

Notes

1. The French policy of encouraging Vietnamese and Chinese immigrants to settle on the fertile riverbank lands, and in the process displace ethnic Khmer to the periphery of the flooded zone, may help explain the exceptionally brutal treatment of these immigrant groups and residents of Phnom Penh and nearby areas by the Khmer Rouge.

2. Than Sina, "Alternative Strategies of Irrigation Development in Cambodia," M.A. thesis, Cornell University, 1982; J. Delvert, *Le Paysan Cambodgien* (Paris: Mouton, 1961), p. 39.

3. Committee for Coordination of Investigations of the Lower Mekong Basin, *Pa Mong Downstream Effects on Hydrology and Agriculture in Democratic Kampuchea,* Working paper 6, MKG/41 (Bangkok, 1976), annex 2, p. 8.

4. Interviews with farmers in Kandal province, February 1983.

5. Conversations with FAO and other aid officials, Bangkok and Phnom Penh, 1980.

6. Y. Kaida, "Effect of Mekong Mainstream Flood Regulation on Hydrology and Agriculture in the Cambodian Lowland (I): Rice Culture," *Southeast Asian Studies* (Kyoto University) 16, 4 (1979):597, 606.

7. Ibid., p. 596.

8. Committee for Coordination of Investigations of the Lower Mekong Basin, *Pa Mong Downstream Effects*, pp. 5–6.

9. FAO, *1980 Production Yearbook*, vol. 35 (Rome: FAO, 1981), pp. 51–52; United Nations [Mekong Committee], *Physical, Climatic, and Social Atlas of the Lower Mekong Basin* (New York: UN, 1968).

10. B. Kiernan, "Introductory Note," in B. Kiernan and C. Boua, eds., *Peasants and Politics in Kampuchea, 1942–1981* (Armonk, N.Y.: M.E. Sharpe, 1982), p. 31.

11. The confluence is called *jatumuk,* meaning "four-faced." Conceivably, the four-faced statue of Bayon at Angkor could be symbolic of the river confluence.

12. M. Y. Nuttonson, *Climatological Data of Vietnam, Laos, and Cambodia* (Washington, D.C.: American Institute of Crop Ecology, 1963), annex p. 50.

13. FAO, *Kampuchea: Food and Agriculture Situation; Report on the 1983 Monsoon Season*, Office for Special Relief Operations, Report No. 16/83/E, 1983, annex 3, p. 18.

14. V. R. Pantulu, "Fishery Problems and Opportunities in the Mekong," in W. C.

Ackermann, G. F. White, and E. B. Worthington, eds., *Man-made Lakes: Their Problems and Environmental Effects*, Geophysical Monograph Series, vol. 17 (1973), p. 674.

15. J. Delvert, *Le Paysan Cambodgien*, p. 83.

16. Ibid., p. 78.

17. Ibid., pp. 69, 80; FAO, *Kampuchea*, p. 10.

18. Nuttonson, *Climatological Data*, p. 4.

19. Conversation with J. E. Collins, Mekong Secretariat, Bangkok, January 1984.

20. D. P. Chandler, *A History of Cambodia* (Boulder: Westview Press, 1983), pp. 100, 102.

21. Ibid., pp. 25–27, 72, 77–79.

22. L. Tichit, *L'Agriculture au Cambodge* (Paris: Agence de Cooperation Culturelle et Technique, 1981), p. 109.

23. B. P. Groslier, "La cité hydraulique Angkorienne: exploitation ou surexploitation du sol?" *Bulletin de l'Ecole Francaise d'Extrême-Orient* (December 1979), p. 194.

24. H. Saeki et al., "Physical and Chemical Properties of Cambodian Soils," in T. Sato and T. Takayama, eds., *Report on a Scientific Expedition to Cambodia* (Hyogo University of Agriculture, 1958), p. 21.

25. Committee for Coordination of Investigations of the Lower Mekong Basin, *Fisheries and Integrated Mekong River Basin Development* (Ann Arbor: University of Michigan School of Natural Resources, 1976), p. 115; Delvert, *Le Paysan Cambodgien*, pp. 82–113.

26. Delvert, *Le Paysan Cambodgien* p. 84; Saeki et al., "Cambodian Soils," p. 28.

27. Delvert, *Le Paysan Cambodgien*, p. 78.

28. Committee for Coordination of Investigations of the Lower Mekong Basin, *Pa Mong Downstream Effects*, summary, p. 7.

29. Ibid., annex 2, figs. 6a, 6b.

30. G. Uehara, M. S. Nishina, and G. Y. Tsuji, "The Composition of Mekong River Silt and Its Possible Role as a Source of Plant Nutrient in Delta Soils," submitted to the Mekong Committee, 1974, pp. 40, 43, 45.

31. C. D. Crocker, *A Preliminary Reconnaissance of Cambodian Soils* (Phnom Penh: USAID in cooperation with the Soil Conservation Technical Committee, Ministry of Agriculture, Cambodia, 1963), p. 49.

32. Ho Tong Lip, "Special Report II: Organic Fertilizers and Mineral Fertilizers," appendix to ibid., p. 49.

33. Committee for Coordination of Investigations of the Lower Mekong Basin, *Pa Mong Downstream Effects*, annex 2, p. 8.

34. Also see Committee for Coordination of Investigations of the Lower Mekong Basin, *Pa Mong Optimization and Downstream Effects Study, Main Report*, vol. 1, *Text* (1978), p. 2810.

35. Kaida, "Mekong Mainstream Flood Regulation," pp. 596–97.

36. P. A. Roger and S. A. Kulasooriya, *Blue-Green Algae and Rice* (Los Banos, Philippines: IRRI, 1980).

37. S. K. DeDatta, *Principles and Practices of Rice Production* (New York: Wiley, 1981), pp. 123–24.

38. There are two major ecogeographical races of the rice plant, *Oryza sativa* (DeDatta, *Rice Production*, p. 173). Most of the traditional, tall-stemmed varieties indig-

enous to tropical Asia belong to the indica race and have thick, deep-penetrating, drought-tolerant roots. The modern, short-stemmed varieties tend to have the characteristics of the more temperate range, japonica rice, including fine, shallow roots, which result in drought sensitivity. Another physiological factor is that the traditional varieties usually mature over a longer period of time (see table 2), whereas modern varieties mature in less than 140 days. This allows the traditional varieties the "vegetative lag phase" during which the plant waits for a shorter day-length before beginning its reproductive stage. However, if the plant experiences stress in the form of drought, disease, pests, or soil infertility, it can use the lag phase to "catch up."

39. T. T. Chang, "Genetic Resources of Rice," *Outlook on Agriculture* (UK) 12, 2 (1983):60.

40. Inevitably some traders and residents of Nong Chan refugee camp joined the rice seed distribution lines. Two men, Robert Ashe of Britain and Khem Sos, an American of Khmer descent, worked tirelessly to organize distributions to as many as 25,000 people per distribution. Oxcart caravans from the more distant provinces of Pursat, Siem Reap, and Kompong Thom were given first priority because of the longer return trip. Khem Sos sought to verify all claims to be from these provinces by listening to regional accents and asking detailed questions about the route and length of the trip.

41. Conversation with W. R. Coffman, Ithaca, New York, October 1980.

42. The fact that each bag of *jaya* contained a packet of toxic mercury fungicide caused some concern, for it was known that villagers were husking and eating some of the imported seed.

43. J. Charny, "Report on a Trip to Southeast Asia," Boston, Oxfam America, 1982, ms., p. 5.

44. Dennis, trip notes, August 1980.

45. Conversation with Niels Olsen, Oxfam/NGO Consortium agriculturalist, Bangkok, June 1980.

46. Yield figures are farmer estimates, not actual measurements.

47. Tichit, *Agriculture*, p. 82.

48. P. R. Jennings, W. R. Coffman, and H. E. Kauffman, *Rice Improvement* (Los Banos, Philippines: IRRI, 1979), pp. 5–9.

49. P. Chevey, "The Great Lake of Cambodia: the Underlying Causes of Its Richness in Fish," *Proceedings of the Fifth Pacific Science Congress, Canada* 5 (1934):3812.

50. T. Mori, S. Nogusa, and E. Fujita, "Fishes in Great Lake (Tonle Sap), Cambodia," in Sato and Takayama, eds. *Report*, p. 52.

51. J. Blache, "Aperçu sur le plancton des eaux douces du Cambodge," *Cybium* 6 (1951):63.

52. Chevey, "Great Lake," p. 3815.

53. Pantulu, "Fishery Problems," p. 675.

54. Committee for Coordination of Investigations of the Lower Mekong Basin, *Fisheries*, p. 72.

55. Pantulu, "Fishery Problems," p. 675.

56. Japan, Overseas Technical Cooperation Agency, *Sambor Project Report: Lower Mekong River Basin*, vol. 6: *Fishery*, 1969, C4.

57. Ibid., C2–4; R. La Font and D. Savoeun, "Notes sur la pisciculture au Cambodge," *Cybium* 6 (1951):54.

58. Pantulu, "Fishery Problems," p. 676.

59. This figure by Tichit (*Agriculture*, p. 359) includes flooded forest along the Tonle Sap River.

60. Chevey, "Great Lake," p. 3810.

61. Blache, "Le plancton des eaux douces," pp. 69, 75–93.

62. Ibid., p. 65.

63. Tichit, *Agriculture*, pp. 335–38; Blache, "Le plancton des eaux douces," p. 64.

64. Tichit, *Agriculture*, p. 335.

65. J. P. Carbonnel and J. Guiscafre, *Grand Lac du Cambodge* (Paris, 1965), p. 194.

66. Committee for Coordination of Investigations of the Lower Mekong Basin, *Pa Mong Downstream Effects*, annex 3, p. 3.

67. Interim Mekong Committee, *Environmental Impact Assessment: Guidelines for Application to Tropical River Basin Development* (1982), p. 31.

68. Chevey, "Great Lake," p. 3812.

69. Tichit, *Agriculture*, p. 359.

Appendix

Soil Analysis of Four Soil Samples from Kampuchea

		Within flood zone		Beyond flood zone	
Location		Neak Luong	Prey Veng flood plain	Ksach Sar Prey Veng	Chamcar Doung
Sample No.		1	2	3	4
Moisture % o.d.s.		1.6	3.6	1.8	0.8
Bulk density g/cm^3 a.d.s.		0.93	0.95	1.12	1.22
pH 1:5 H_2O		7.0	5.5	5.7	5.4
E.C. ms/cm 1:5 H_2O		0.37	0.05	0.03	0.05
Exchangeable[1] Cations me/100 g a.d.s.	Na	0.1	0.2	0.3	0.3
	K	0.2	0.3	0.1	0.1
	Mg	2.5	2.6	1.3	0.5
	Ca[2]	18.9	11.1	5.3	1.5
T.E.B. me/100 g a.d.s.		21.7	14.2	7.0	2.4
C.E.C. me/100 g a.d.s.		11.2	21.2	9.0	4.3
Base saturation %		100	67	78	56
Total N[3] %		0.12	0.22	0.12	0.08
Organic C[3] %		1.02	1.65	0.96	0.67
Available P (Bray) ppm		—	20	4	15
Available P (Olsen) ppm		10	10	3	7
Perchloric acid digestion ppm[5]	P	610	440	140	120
	K	16,250	20,850	5,500	2,900
	Mg	8,350	7,050	1,750	750
	Cu	30	40	20	10
	Mn	960	340	100	100
	Zn	90	110	40	20
Particle size analysis (% of dry mineral soil)	2000–500	0	0	1	1
	500–250	0	0	1	7
	250–100	0	0	12	20
	100–50	8	0	12	20
	50–20	33	2	17	19
	20–2	39	30	23	19
	<2	20	68	34	14

Source: Analysis by Tropical Soils Analysis Unit, Land Resources Development Center, Coley Park, Reading, England.

Notes:

1. Exchangeable cations determined by leaching with 1M ammonium acetate at pH 7.
2. For soils containing free calcium carbonate, lime, or gypsum, the figure for exchangeable calcium will include some nonexchangeable calcium.
3. Not including nitrate or nitrite.
4. Organic matter = 1.72 × organic carbon.
5. Ppm divided by 10,000 = %.

This table gives in very general terms the range of expected values for the soil analysis based on the results of over 20,000 samples collected worldwide.

Determination		LOW less than	MEDIUM	HIGH more than
pH 1:5 H_2O		5.0	5.0–7.0	7.0
Cond m.mhos 1:5 H_2O		0.5	0.5–1.5	1.5
Sol Na	m.eq 100 g ads	1.0	1.0–3.0	3.0
Sol K	m.eq 100 g ads	0.2	0.2–0.6	0.6
Sol Mg	m. eq 100 g ads	0.3	0.3–1.0	1.0
Exch Na	m.eq 100 g ads	0.3	0.3–1.0	1.0
Exch K	m.eq 100 g ads	0.2	0.2–0.6	0.6
Exch Mg	m.eq 100 g ads	0.5	0.5–4.0	4.0
Exch Ca	m.eq 100 g ads	4.0	4.0–10.0	10.0
TEB	m.eq 100 g ads	5.0	5.0–15.0	15.0
CEC	m.eq 100 g ads	6.0	6.0–25.0	25.0
% Sat		20	20–60	60
% Total N	ods	0.1	0.1–0.3	0.3
% Org C	ods	1.5	1.5–4.5	4.5
Total P ppm ods		200	200–1000	1000
Total K ppm ods		4000	4000–10000	10000
Total Mg ppm ods	$HClO_4$ Digests	2000	2000–10000	10000
Total Na ppm ods		200	200–600	600
Total Cu ppm ods		20	20–100	100
Total Mn ppm ods		200	200–2000	2000
Total Zn ppm ods		40	40–150	150
Avail P (Olsen) ppm ads		5	5–15	15
Avail P (Bray) ppm ads		15	15–50	50

Joel R. Charny

APPROPRIATE DEVELOPMENT AID FOR KAMPUCHEA

At the conclusion of his masterwork on rural Kampuchea, *Le Paysan Cambodgien*, Jean Delvert summarizes the status of efforts to modernize and improve a system of agricultural production unchanged for many centuries. He notes the paucity of functioning irrigation systems, the lack of rural credit institutions, and the very limited mechanization of the means of production. Delvert calls for modernizing agricultural implements, promoting more intensive cultivation patterns, providing practical rural education, and training more extension workers to help the rice farmers of Kampuchea break their bondage to a fickle climate and long-standing traditions.[1]

In the twenty-five years since Delvert conducted his research, Kampucheans have experimented with more radical solutions to the enslavement of the Khmer peasantry. The great tragedy of the past thirty years of Kampuchean history is precisely that after decades of war and revolution conducted in the name of the peasants, the rice farmers of Kampuchea today are worse off than the previous generation. The familiar litany of problems facing the peasantry remains the same. Now, however, solutions must be found in the context of a recent national experience that includes civil war, mass enslavement of the population in the name of Khmer nationalism and self-reliance, and invasion and occupation by a feared traditional enemy. The ensuing famine closed a decade that witnessed the death of two million Khmer and the near-complete destruction of Kampuchean society. In a period of rebuilding and healing after this immense human disaster, technical development approaches are inadequate unless they acknowledge the Khmer people's needs as victims and survivors.

This chapter will suggest elements of an appropriate development aid strategy for Kampuchea as the country emerges from the emergency period. I want to emphasize at the outset that my perspective on development in Kampuchea is a limited one. As an aid worker, initially for the Oxfam/Non-Governmental Organization Consortium and now for Oxfam America, I have been closely involved with the aid effort by Western private voluntary agencies on behalf of the people of Kampuchea.[2] While this work has afforded me the opportunity to travel extensively inside the country and meet with numerous rice farmers and government officials at the provincial and national levels, this experience is not the same as conducting a systematic study of the rural sector in post-Pol Pot Kampuchea. Since 1979, the few outside scholars fluent in Khmer with access to the rural areas have tended to focus on political questions arising before, during, and immediately after the Pol Pot period rather than on issues of development and agricultural production. The Kampucheans, understandably, have been unable to allocate very scarce human resources to *studying* problems; the need for solutions has been too pressing. Thus, I will have occasion to indicate when key elements of my analysis are based on educated speculation and will attempt to point toward critical areas for future research when and if opportunities for this research become available.

The Famine and the International Response

The Vietnamese invasion in December 1978 in support of a small group of Kampuchean dissidents and their followers liberated the Khmer people from the oppressive Pol Pot regime. The invasion ushered in a period of chaos, however, which resulted in the disruption of the rice production cycle and widespread famine, prolonging the nearly continuous food problems experienced by the majority of the rural population since the onset of the civil war in 1970.

It requires little imagination to picture the suffering of the Khmer people in early 1979. Exhausted and malnourished survivors, liberated with shocking suddenness by their traditional enemies, began to trek across Kampuchea to return to their homes and to search for missing relatives. As they walked, fighting between Pol Pot and Vietnamese forces swirled around them. Recapture by Khmer Rouge forces meant instant death for some or, at best, a forced march west with the retreating troops. The Khmer Rouge pursued a scorched earth policy, destroying what they could not carry with them in their flight. In areas controlled by Vietnamese and Heng Samrin forces, hungry people looted stores and granaries, eating their fill of rice for the first time in months.

The trek of the survivors across the length and breadth of Kampuchea took many weeks. When people reached their villages they often found nothing left: no houses, the *wat* destroyed, not a trace of family and friends. Draft animals roamed the countryside, there for the taking, but could not be put to use; there was little rice available for food, much less for seed for the main rainy-season rice crop. Peasants devoted what little energy they had to scavenging for food and constructing shelter for their families. In the words of a village leader from Kbal Damrei in Pursat province, "In 1979, when we returned to our village, we had nothing, absolutely nothing."[3]

The 1979 main-season rice crop, which normally supplies about 90 percent of Kampuchea's grain needs, produced only 285,000 metric tons (mt) of food rice on 700,000 hectares (ha) of land.[4] By way of comparison, rice production in the late 1960s averaged about 1.5 million mt on 2.4 million ha of land. Even the

Khmer Republic in 1974, when it controlled little of rural Kampuchea, managed to cultivate 500,000 ha. To make up for the resulting food deficits, the Lon Nol government received 280,000 mt of U.S. food aid as agricultural production plummeted.[5] Food aid of a comparable level only reached the people of Kampuchea by mid-1980, nearly a year after the most serious famine conditions prevailed.

It is probably impossible to determine how many people died of starvation in Kampuchea in 1979. Indeed, the question of who died when and how since 1970 has become such a complex political issue that population data for Kampuchea have lost meaning altogether outside of the political context in which they are presented. The fact is that despite some bilateral emergency aid in the first half of 1979,[6] including food aid from Vietnam, which was then having its own serious food problems, the first representatives from Western agencies to reach Phnom Penh in July and August 1979 were overwhelmed by the gravity of the emergency facing the people of Kampuchea. Jim Howard, Oxfam's technical adviser with years of experience in disaster relief, described the situation as one of the worst he had ever seen and predicted that two million Khmer would die in three months unless there were an immediate and massive international humanitarian response. That response was forthcoming as a result of the perception by individuals, donor agencies, and even governments hostile to the Vietnamese-backed Heng Samrin regime that the Khmer people were the ultimate victims of political forces beyond their control.

The initial focus of the relief effort was the supply of food, medicine, and other necessities required for immediate survival. By early 1980, however, enough stability prevailed inside the country to allow relief agencies to turn their attention to the revival of food production in Kampuchea, the crucial component of self-sufficiency for a people who had been unable to control productive resources for almost a decade.

Emergency aid in the agricultural sector tended to be the result of a mechanical, rather short-sighted approach to the revival of food production. The urgency of Kampuchea's food needs made this approach inevitable: indeed, a disaster assistance program that takes long-term development needs into account is probably a contradiction in terms. In Kampuchea, the consequences of the human disaster of the past decade heightened the traditional agricultural problems described by Delvert: lack of irrigation facilities, shortage of mechanized means of production, and low yields as a result of poor soils and the use of traditional rice varieties without chemical or even natural fertilizers. In addition, the human and draft animal populations, formerly adequate to work enough land for rice production, had dropped drastically in the past decade. Kampuchea, never overpopulated, now lacked the human resurces to feed itself. Even hand tools and wood plows were scarce. Finally, the overall food shortage meant that seed stocks, normally maintained from harvest to harvest, were either seriously low or depleted altogether.

In response to these problems in 1979 and 1980, the international aid agencies and bilateral donors imported more than 300,000 hoe heads, 5,000 small diesel irrigation pumps, 1,300 tractors and power tillers, 55,000 mt of rice seeds—most of them high-yielding varieties—and many thousands of tons of the chemical fertilizers essential to proper growth and production.[7]

In themselves the quantities have little meaning, but in perspective they are truly astounding. In 1968–69, when Kampuchean rice farmers planted 2.4 million ha, only 1,300 diesel pumps were in use in the entire country. Some 1,500 tractors contributed to land preparation, most of them in Battambang province. While no data are available regarding the extent of the use of international high-yielding varieties, Kampuchean farmers applied only 5,000 mt of urea to their rice paddies.[8]

Table 1 summarizes the comparison between 1969–69 and 1980–81:

Table 1

Material Inputs and Rice Production

	1968–69	1980–81
Rice cultivation 000 ha	2,427	1,320
Paddy production 000 mt	2,503	1,580
Diesel irrigation pumps	1,300	5,000
Tractors	1,500	200
Chemical fertilizer imports mt	5,000	29,800
Hybrid rice seed imports mt	figure not available	42,000

Sources: World Bank for 1968–69 data; FAO for 1980–81 data.

In effect, therefore, the international agencies were demanding that Kampuchean rice farmers employ technologies they had never used before, even in times of stability, when they had barely emerged from a decade-long period of famine and destruction. They were expected to use these imported technologies to grow the rice crop upon which their survival depended! Furthermore, they had to take this risk in the absence of a competent, trained extension service. Traditions are simply not that easily broken, particularly in a disaster situation when peasants are most likely to rely on the customs and practices that have ensured survival of their people and their culture for centuries.

This goes a long way toward explaining why, despite the tons of imported rice seed that were inundating Kampuchea, it was practically impossible to find these varieties in farmers' fields in 1980.[9] The risk was too great that something would go wrong: flooding of the short-stemmed hybrids in low-lying areas, lack of fuel for the diesel irrigation pumps essential to ensure adequate water distribution to the varieties in rain-fed areas, shortages of fertilizer when application was required to ensure promised yields. High-yielding varieties function in a clockwork fashion, and the absence of an input at each phase of the rice cycle seriously jeopardizes the yields at harvest time. Only when no other seed was available did it make sense for a Kampuchean peasant to risk the loss of his entire crop. If circumstances permitted, it was much better to eat the international seeds in anticipation of customary yields with traditional Khmer varieties.[10]

Unfortunately, in the absence of extensive *farm-level* evaluations of the relief effort in the agricultural sector, this analysis must remain speculative, based largely on limited, empirical evidence. Monitoring of the use of agricultural supplies ended with the assurance that the Ministry of Agriculture had received the aid at either Phnom Penh or Kompong Som ports. After that, agencies accepted the distribution figures of the national ministry or provincial and local agricultural committees. Relief workers fleshed out these statistics with experiences gained on field visits that rarely involved interviews with peasant farmers. Thus, the Food and Agriculture Organization (FAO), which assumed primary responsibility for leading the relief effort in the agricultural sector, has yet to go much beyond compiling statistics of aid imported in evaluating the Kampuchean experience with 1980s rice technology. Oxfam/NGO Consortium reporting throughout the emergency period worked from the premise that more hybrid rice seeds plus chemical fertilizers and diesel irrigation pumps ensured greater rice production in Kampuchea in 1980. No person or agency, however, has verified that results in the field vindicated this approach.

In the past two years, with less pressure on the agencies to justify millions of dollars in expenditures on imported technologies, more evidence has emerged to support the case for the fundamental unsuitability of massive doses of modern technology to Kampuchean rice production. The field representative of an international agency based in Phnom Penh estimated in October 1982 that 80 percent of the small diesel irrigation pumps imported during the emergency period were, by then, inoperable due to lack of proper maintenance and spare parts. Many of the remainder cannot function for lack of fuel at the village level. In Battambang, despite a superbly organized repair facility, lack of fuel also prevents complete use of the existing fleet of the over one hundred tractors necessary to meet the tillage requirements of this land-rich province. In Kandal, the director of the provincial agronomy office acknowledged that hundreds of hectares of IR36 were flooded out in 1980 because it was planted in areas unsuited to this short-stemmed variety.[11] The Kandal agricultural committee and the rice farmers themselves learned through bitter experience the areas best suited to modern varieties; they are cautiously applying this knowledge at present.

The 1980 rainy-season rice planting yielded 1.46 million mt of paddy on 1.23 million ha, an average yield of 1.2 mt/ha.[12] This crop more than doubled the figures for 1979 and represented a fantastic achievement for the government and people of Kampuchea. While falling well short of creating a surplus, Kampuchea's rice farmers reached a subsistence level, which they have maintained, despite severe weather problems in 1981, in subsequent harvests.

Contrary to the dire predictions of many observers, the people of Kampuchea survived 1980 until the harvest without a recurrence of the famine conditions of 1979. Observers in Phnom Penh in 1980 neglected to take into account the food resources of the Kampuchean countryside that did not depend on outside aid to be exploited. Thus, fruit trees, vegetable gardens, and the vast fishing reserves of Kampuchea, from rice paddies and ponds to rivers and Tonle Sap, provided essential supplements to the Khmer diet. The crucial factor was that villagers were once again free—which they had not been during Pol Pot's rule—to exploit the natural resources of Kampuchea. Dr. Carmel Goldwater, who was one of the few representatives of Western aid agencies afforded continuous access to the provinces, has reported that villagers in Takeo province identified the freedom to travel beyond their villages to look for supplementary food as the crucial factor differentiating life under the current regime from that under Pol Pot.[13] In the emergency period, manioc and maize, abhorred by Khmer peasants, substituted for rice, while numerous fruits, vegetables, and small fish provided further calories, vitamins, and protein to the Khmer diet.

Thus, crops dependent neither on outside inputs nor on extensive labor revived remarkably during the emergency period. By November 1981, vegetable production had reached 1968–69 levels. Maize production attained the 100,000 ha level in 1980, only to drop off in 1981 due to flooding of riverbank land and the low prices paid to maize producers. Fish production still lagged at less than 50 percent of prewar levels, but fishing on the Tonle Sap is both labor-intensive and dependent on resources not readily available locally (outboard motors, large fishing nets, and boats).[14]

By early 1981, the threat of famine on a massive scale had ended in Kampuchea. The euphoria prevailing in Phnom Penh on January 7, 1981, the second anniversary of the liberation of Kampuchea from the Khmer Rouge, reflected the joy and wonder that the Khmer people felt as they realized that they had survived yet another trial by fire. The joy they experienced on that day has since given way to doubt and anxiety as the intractability of the long-term problems they face has become clearer. The survivors had much to celebrate in January 1981. How much they owed their survival to the international relief effort and how much to their own skills, strength, and resilience cannot be precisely determined. In the rural areas, however, the balance was decidedly in favor of the strength of traditional Khmer food production methods and cultural practices.

Toward an Appropriate Development Strategy

Implicit in the above discussion is the judgment that the revival of the food production systems of Kampuchea is the key to the reconstruction and future development prospects of this battered country. The tragic irony of a once food-exporting country experiencing widespread famine suggested this focus immediately during the emergency period in 1980. Yet a "food first" strategy, to use the phrase coined by Frances Moore Lappé and Joseph Collins of the Institute for Food and Development Policy,[15] is equally compelling in the present reconstruction and development phase of Kampuchea's recovery. Rice is not only the staple of the Khmer diet, but as a commodity it was also the country's major foreign-exchange earner in the prewar, independence period. Rice garnered as much as 60 percent of export revenues in bumper crop years.[16] In the long term, the revival of Kampuchea's food export capacity, assuming the existence of formerly reliable markets for its products, will provide the impetus for the overall growth and development of the Kampuchean economy.

A development strategy focusing on food security for the people of Kampuchea neither minimizes the importance of aid in other sectors nor addresses the restraints on Kampuchea's economic relations with its Asian neighbors and the rest of the world as a function of the present political impasse in Indochina. Thus, the relief effort in the agricultural sector could not have taken place without millions of dollars in assistance to the country's shattered infrastructure. Trucks, river ferries, and barges provided by the international agencies and bilateral donors constitute the basis of Kampuchea's internal transportation and distribution network. Relief agencies supplied the spare parts to revive important consumer-oriented industries in Phnom Penh, particularly textiles, and have supported the efforts of Khmer technicians to assure adequate power and clean water to the residents and workers of the capital. Even in 1969, the World Bank study team noted that the electricity grid in Phnom Penh was hopelessly antiquated and that breakdowns disruptive to the city's industries were an increasingly frequent problem.[17] The modernization of Phnom Penh's water and power systems, now seriously damaged by war, lack of maintenance, and the creative repair efforts of the present civilian population, would be important major projects for willing bilateral or multilateral donors.

Medical aid has been especially critical. The disastrous attempts by the Khmer Rouge to base health care exclusively on traditional medicines have thoroughly discredited these remedies in the eyes of the people. Thus, the demand for modern, imported medicines is great.

Development aid in the health sector, largely the province of bilateral donors, should focus (which to some extent it has) on the creation of Kampuchea's first extensive rural health-care network through training of personnel and clinic-building programs. Reconstruction of a minimal level of modern medicine processing capacity is another priority. Kampuchea will have to import virtually 100 percent of its modern medicine requirements for the foreseeable future, either through grants or by using up scarce credits or foreign exchange. Thus, in the absence of large-scale aid to the health sector, the chances are great that the people of Kampuchea will have to rely on medicine smuggled from Thailand and sold at exorbitant prices in markets throughout Kampuchea.

No aid strategy developed outside Kampuchea can address the independence issue, which is primarily political. At the moment, however, it is an economic as well as political fact that the opportunities for Kampuchea to choose its own trading partners and terms of trade do not exist. The Soviets, for example, have clearly established a neocolonial hold on Kampuchea's remaining rubber resources. Reports from Phnom Penh indicate that the Heng Samrin regime has agreed not to export the Tonle Sap fish catch through 1985 *except* for 15,000 mt per year to Vietnam.[18] Given the precarious, albeit improving, food situation in Vietnam, no genius is required to predict the destination of Kampuchea's initial rice exports. Japanese businessmen have visited Kampuchea to discuss fish and timber development, but probably to no avail in the present political climate.

The prospects for economic independence are indeed bleak. Reinforced by the near total destruction of the economy since 1970, the Heng Samrin regime will have to own up economically to its huge political ''debt'' to the Soviet bloc. Assuming a basic political continuity, the price that Kampuchea will pay for Soviet and Vietnamese support will be a near complete integration into the Soviet-bloc economic system.[19]

The Khmer Rouge disaster had the effect of thoroughly discrediting self-reliant approaches to development that depend on mass participation. The Khmer people *worked* during the Pol Pot time, worked as a mass, on rural development projects that, were they found in a consultant's report, would win the approval of a wide cross-section of the development community: digging irrigation channels, clearing new land for rice production, mixing biofertilizers, growing medicinal plants, exploiting timber reserves—all tasks involving minimal use of fossil fuels and virtually no imported resources.[20] As we now know for certain, however, the Khmer people also *died* during the Pol Pot time, hundreds of thousands of people, making it forever impossible to see the Khmer Rouge period as an interesting but failed experiment in self-reliant development. The human cost of the incredible mass mobilization of Kampuchea from 1975 to 1978 was too great.[21] Thus, a food first development strategy, emphasizing self-reliance, may be rejected by many Kampucheans as being dangerously reminiscent of the Pol Pot approach. Discuss ''night soil'' and ''biofertilizer'' with a mid-level official in the Ministry of Agriculture and he will tell you that for nearly four years he collected and mixed human excrement with his bare hands as a forced laborer on an agricultural cooperative. Efforts to organize work teams to maintain irrigation works constructed under the Khmer Rouge fail amid fears that this is but the first step toward a reversion to the methods of Pol Pot cadre. Thus, peasants are either ignoring or resisting government efforts to organize and mobilize them, with the exception of work teams (*krom samaki*) that basically function as mutual assistance teams at peak work periods, a traditional part of rural life. The reestablishment of true cooperatives could only follow the Khmer Rouge period by many years.

Whether traditional tenancy patterns and creditor-debtor relationships are reemerging to fill the organizational vacuum is unknown, but definitely worthy of study. The current land surplus in Kampuchea would seem to work against the reestablishment of landlord-tenant relationships (known as *provas*) in which the landlord provided the land, seed, and implements in exchange for half of the resulting harvest.[22] The desperate shortage of seeds, implements, and draft animals, however, may have forced poor farmers into becoming tenants in relation to the minority with access to these resources. In this case, the ability of the government to supply needed inputs has been and will continue to be the crucial variable. Indeed, as a socialist government committed to agricultural development, the Heng Samrin regime must do so or be discredited in the eyes of the peasantry. Without government assistance, and without cooperatives able to function, inequality in rural areas may develop into a serious problem.[23]

The government's capacity to pay adequate prices for farm products, especially paddy, is also in question. To date the local authorities have relied on the provision of basic consumer goods such as cloth, sugar, and salt at very low subsidized prices as an incentive to peasants to sell to the state. The government, however, cannot provide cash during the "hungry period" before the harvest. Moneylenders are almost certainly filling this void, although, again, the extent of their role is not yet clear. Traditionally, the village merchant, either Chinese or Sino-Khmer, mediated virtually all of the Khmer peasantry's commercial relations with the outside world. Goods purchased from the local shop during the planting season were repaid with paddy at harvest time, at interest rates of 100 percent. The merchant milled the paddy and then sold it to the state or to rice merchants with trade neworks extending to Saigon, Hong Kong, or Singapore. While the export market does not yet exist, market prices are a powerful incentive for the reinstatement of the private marketing system so oppressive to farmers. Infrastructure development probably helps the government ease this threat by increasing the efficiency of its own internal transport and distribution systems. Pricing policies, however, outside the scope of development aid as such, will determine the extent to which the local merchant-moneylender again becomes a part of village life.

Thus, the peasantry of Kampuchea must address severe problems, some traditional and some a result of the human disaster of the past decade, as individuals, with little government or mutual support. In this sense, they face the problems of the 1980s with the resources of the 1950s. Little of a positive nature remains from the intervening decades. Their strength and capacity for survival lie in the practices upon which they now depend out of both tradition and necessity. Therefore, suggestions for developing the productive capacity of the Khmer peasantry discussed below presume that these practices represent a foundation to be built upon, not a legacy of backwardness to be immediately undermined and tossed aside.

Labor Power

The huge increase in land cultivated with rice from 1979 to 1980 seemed to promise rapid expansion of hectarage in subsequent plantings. In fact, rainy-season rice cultivation has been more or less frozen at 1.1–1.3 million ha through 1984, with the actual size of the crop being determined by climatic factors. In 1981, the Mekong basin flooded to the highest level in one hundred years of records, while simultaneously many of Kampuchea's central rice-growing provinces experienced severe drought, causing losses of about 200,000 ha. Total paddy production dropped over 400,000 mt from 1980. While a reliable monsoon in 1982 meant lower losses, in October 1982 the hectarage failed once again to exceed 1.3 million ha.[24] Even the total area planted during the 1984 monsoon season did not exceed 1.3 million ha.

This immediate stagnation of rice production results primarily from the severe labor shortage in the Kampuchean countryside. A glance at the statistical records from the 1950s confirms this. The average hectarage of rice production in Kampuchea from 1952 to 1957 was 1.1 million ha, slightly less than the 1980–82 figures. A population of 4.7 million owned 1.2 million draft animals.[25] Assuming a population evenly composed of males and females,[26] with about 40 percent less than fifteen years of age, the adult male population was about 1.4 million in the mid-1950s.

Population data for post-1970 Kampuchea are highly speculative, but FAO figures published in late 1981 yield remarkably similar results. While the present population is about 6 million, 30 percent is estimated to be under fifteen years of age, and as little as 25 percent of the adult population is male. Thus, there may be as few as 1.08 million adult males and certainly no more than the 1.4 million figure of the mid-1950s. Furthermore, the Ministry of Agriculture puts the total draft animal population at 1.24 million, almost exactly the figure Delvert gives for 1955.[27]

Table 2 summarizes the rudimentary statistical analysis.

Thus, while available labor power is at the level of the mid-1950s, as is the resulting rice production, the actual population of Kampuchea is greater by as much as 1.5 million people. Even allowing for exaggeration by a government desperately seeking outside aid, it is difficult to imagine the population being less than 5.2 million, a million less than figures submitted to FAO. This leaves Kampuchean farmers with the task of feeding 500,000 more people than their brethren of three decades before, in an atmosphere of insecurity that forces many

Table 2

Labor Inputs and Rice Production

	1954–55	1980–81	1984
Total population 000	4,700	6,200	N/A
Adult males 000	1,400	1,085	N/A
Adult females 000	1,400	3,255	N/A
Draft animals 000	1,200	1,240	1,180
Rice cultivation (rainy season) 000 ha	1,221	1,320	1,229

Sources: Jean Delvert, *Le Paysan Cambodgien*, for the 1954–55 data; FAO for the 1980–81 data; Jean-Claude Salle, "Final Report on the Food Situation in Kampuchea," *Kampuchean Humanitarian Assistance Programs* (United Nations, December 1984), for 1984 figures.

young men to join the fledgling Heng Samrin army rather than remain on the farm.

FAO, with the assistance of several voluntary agencies, aided the Kampuchean Veterinary Department in conducting a mass vaccination program for draft animals in 1981–82. The maintenance of a regular vaccination program over the next several years would ensure that the draft animal population continues to grow. Breeding programs are beyond the capacity of the Veterinary Department and, without external help, the Ministry of Agriculture as well. Khmer peasants themselves do not breed draft animals and do not show great concern for assuring their animals adequate pasturage.[28]

Whether as a result of government inefficiency in distribution and training or simple peasant resistance, peasants did not use imported power tillers and "iron buffalo." Occasionally peasants would hook up power tillers to a cart for local transport. Tractors, however, which were used extensively in Battambang in the late 1960s, would seem to have great potential in this land-rich province. The Agricultural Machinery Enterprise has 102 tractors at its disposal—58 are new and the remainder were repaired with spare parts collected in the Battambang area or imported from abroad. The number of hectares plowed by these tractors has risen from 712 in 1979 to 25,508 in 1981.

The tractors concentrate on floating rice land, which tends to be planted in very large plots suitable for mechanization, and villages with severe shortages of oxen and buffalo. The *krom samaki* contracts with the Machinery Enterprise at the rate of 200 kg of paddy for each hectare plowed, payable at the time of the harvest. This rice goes toward paying the 374 employees of the Machinery Enterprise.[29]

The director, Meas Phanna, a Berkeley-trained economist, estimates that an additional 100 tractors would be required to bring floating rice hectarage up to prewar levels. Battambang's current production of 286,000 mt of paddy is still less than half of peak production figures from the 1960s. The vastness of Battambang's rice plains and the relatively large size of the plots cry out for mechanized exploitation as a means of expanding production. Even the repair and maintenance capacity exists, thanks to the superb facility organized in Battambang town. With another shipment of spare parts, another twenty to thirty tractors could be repaired. Large shipments of new tractors are beyond the means of the international oganizations at this point, so new machines will have to be part of a bilateral aid package, from the Soviets or one of their allies.

A major constraint on the full use of even this tiny tractor fleet has been shortages of fuel. Indeed, unless bilateral donors agree to assure Kampuchea adequate fuel supplies in the coming years, no development that requires fossil fuel is going to take place. Thus, even a small-scale mechanization approach, limited to urgent needs in the "rice bowl" province of Kampuchea, may well be doomed to failure. Elsewhere in Kampuchea, the tiny size of the plots and the lack of repair facilities mitigate against efficient use of tractors. As the Khmer peasantry has emphatically rejected power tillers, "iron buffalo," and other labor-saving devices for working the land, the short-term prospects for expanding the amount of cultivated land in Kampuchea by increasing the available labor power are very bleak.

Rice Seed

Use of improved seed stocks to increase yields on currently available land is therefore mandatory. According to Delvert, Khmer peasants routinely select the best seeds at harvest time, grouping them by variety for the next planting season. Choosing the varieties best adapted to the local soil and water conditions is the key task, but the taste of the rice is also an important consideration. A peasant may plant as many as five or six rice varieties on different small plots, thereby significantly lessening the risks of a total crop failure.[30]

Recent experience confirms this analysis. Peasants in Kandal who participated in rice-seed trials organized by the Agronomy Department during the 1981 rainy season agreed to plant three international varieties—IR36, IR5, and Mashuri—in addition to several traditional ones. Interviewed just after the harvest in January 1982, they indicated that based on the results from the completed 1981 planting they expected to plant one-sixth of their land with IR36, one-sixth with Mashuri, and two-thirds with traditional varieties in 1982. (All peasants decided to plant the seeds in roughly the ratios given—this is not an average across peasants making radically different choices.) Although IR5 produced more than Mashuri per hectare, 6.6 mt to 5.7 mt, the peasants much preferred the eating quality of Mashuri. They paid it the highest possible compliment by comparing its taste to that of a traditional Khmer variety. IR36 was valued for its production—nearly 7 mt per hectare. On balance, however, despite acknowledging the virtues of the international varieties, which include yields *five times* those of traditional varieties, peasants in this village in Kandal were planning to rely mainly on traditional varieties in the following rainy season.

Further illustrating the subtleties of improving yields through rapid extension of international varieties is the way peasants tend to use their production. In the Kandal village just cited, Mashuri, a 130-day variety, matured six to eight weeks before the traditional varieties planted alongside it. The villagers, who were very short of food rice just before the harvest, ate the entire production of Mashuri without setting aside a small quantity (80 kg would more than suffice for one hectare) for seed. Thus, in October 1982, the villagers confirmed that despite their intention to plant this new variety again, they had failed to guard seed stocks, expecting that more seed would be made available from the provincial Agronomy Department. The Agronomy Department itself, however, had failed to maintain a reserve because of the previous year's seed shortages caused by flooding and drought. Thus, in this area of Kandal, the lack of food reserves in the immediate postfamine period and the unreliable 1981 monsoon rains prevented the extension of a promising international variety for local farmers.

The optimistic moral to draw, of course, is that peasants will make sensible seed selections, including appropriate international varieties. The Ministry of Agriculture has selected IR36 as the international variety to promote, but figures from the 1982 monsoon season indicate the slow pace of extension and acceptance. Some 66,000 ha of IR36 were planted. Even assuming excessive seeding at the rate of 100 kg per hectare, the 6,600 mt of IR36 seed employed represents but a fraction of the total seed imported into Kampuchea during the emergency period. The Oxfam/NGO Consortium alone imported 7,300 mt of IR36 or its equivalent during 1980. Having dug themselves out from the avalanche of international seed, the Ministry of Agriculture and the peasants of Kampuchea are proceeding only gradually (and sensibly) with the expansion of high-yielding rice production.[31]

The multiplication of traditional varieties is another top priority at the national level. Many pure traditional varieties were lost during the Pol Pot time due to extensive forced internal migrations of the peasantry. In late 1980, someone at the Ministry of Agriculture, probably Kong Som Ol, remembered that the International Rice Research Institute (IRRI) in the Philippines had sent missions to Kampuchea before 1975 to collect seed varieties for its Germplasm Bank, where over 30,000 of the world's rice varieties are stored in small quantities. IRRI has over 800 Khmer rice varieties in its Germplasm Bank; 140 of these have been returned to Kampuchea in the past two years under the auspices of Oxfam America. The Department of Agronomy's Research Division is multiplying these varieties on agricultural stations in Kompong Speu and Battambang. The ministry's goal is to be able to produce these traditional varieties in sufficient quantities to distribute to peasants and keep on hand at the national level for future indigenous seed development programs.

Proper seed storage facilities are a necessity if the long-term goals of this and other seed storage programs are to be realized. While actual breeding of locally adapted hybrids is a long-term prospect requiring years of training of Khmer rice scientists, at a minimum a centrally located, air-conditioned, dehumidified room in Phnom Penh is needed to prevent deterioration of important international and traditional varieties. Due to lack of twenty-four-hour electricity, cold storage at the provincial level, except perhaps in one or two locations using generators, is not practical. At the village and farm levels, traditional methods of storage from harvest to planting are apparently adequate, although improved designs using local materials might minimize losses due to insects, rats, and excessive heat.

Given the shortage of trained Kampuchean rice scientists and agronomists, investment in expansion of indigenous research capacity will not bring immediate benefit to peasant farmers. This very shortage, however, does have a positive side to the extent that farmers become part of the research process. In Kandal and Kompong Speu, in the area of the Prey Phdau Agricultural Station, farmers have participated in seed comparisons, fertilizer trials, and other experiments. Thus aid to Prey Phdau, at present a simple thatch hut surrounded by rice fields on the grounds of the 1960s UNDP project, would not only bolster the Agronomy Department's ability to conduct field research but also, in effect, support the training of more farmers in the use of modern seeds. The Tuol Samrong Research Station in Battambang is already receiving Soviet aid in the form of material and personnel.

Of all the aspects of Kampuchea's agricultural development, trusting the proverbial wisdom of the peasantry seems most appropriate in the area of seed improvement. Khmer peasants will make the correct choices. The Ministry of Agriculture's commitment to improving seed stocks in a methodical manner complements the conservatism of peasant farmers. International high-yielding varieties, however, demand good soils, high nutrient levels, excellent water control, and trained extension workers to produce optimally. These are lacking in Kampuchea today.

Fertilizers

Delvert develops at length the distinction between the rice farmer (*neak sré*) and the cultivator of field crops on riverbank land (*neak chamcar*). The poverty of the former derives from the difficulty of eking out an existence from a single crop planted on infertile soils. The *neak chamcar* works the most fertile land in Kampuchea, double-cropping lucrative cash crops sold traditionally to merchants who connected with urban and export markets. Urban markets have revived to such an extent, particularly in Phnom Penh, that the production of food crops for local consumption is probably as profitable as ever.[32]

Rice farmers predominate in Kampuchea, as do poor soils. Traditionally, Khmer peasants have not worked intensively to improve the soil through application of animal manure or nitrogenous plants. As noted above, the Khmer Rouge approach to agriculture included constant application of biofertilizers, including human wastes and corpses. These practices, even the acceptable ones, have had no noticeable carry-over into the post-Pol Pot period.

There may well be cultural factors at work here involving a taboo on handling waste materials. Khmer peasants have never used night soil to my knowledge. Furthermore, in the absence of intense population pressure on available land, the need to extract every last bit of fertility from the soil has not been great. Preparation and application of biofertilizers are very labor-intensive, and labor is very scarce in Kampuchea today. Kampuchea thus lags well behind its Asian neighbors in the application of biofertilizers on rice soils.

Several voluntary agencies have attempted to interest the national Agronomy Department in experimenting with blue-green algae, a nitrogen-fixing organism that multiplies rapidly in specially built tanks and then must be plowed into moist paddy soil. Dr. Goldwater of Oxfam hand-delivered some Burmese innoculant of blue-green algae to the Agronomy Department, but subsequent attempts to propagate it failed. The use of blue-green algae has proven successful, however, in climates similar to that of Kampuchea. Farmers use it extensively in southern India and Burma; India has a research center in Tamil Nadu devoted to study and extension work. Azolla, which is widely used as an alternative to chemical nitrogen fertilizers in northern Vietnam, cannot withstand the higher temperatures prevailing at certain times in southern Vietnam and Kampuchea.

Blue-green algae has promise for farm-level use, but only after a carefully orchestrated government program of trial and extension. At the moment, interest seems lacking at the national level, as the entire orientation of the Ministry of Agriculture is toward encouraging the import of chemical fertilizers on a grant basis. This policy is very short-sighted but understandable in the context of the relief effort, which placed unprecedented amounts of chemical fertilizers into the hands of Kampuchean farmers.

The last Sihanouk five-year plan envisaged construction of a urea factory in Kompong Som, using petroleum products from the local refinery.[33] Such grandiose dreams came to naught, of course; Kampuchea will be completely dependent on outside aid for chemical fertilizers for the foreseeable future. As aid from international organizations winds down, the Ministry of Agriculture probably wishes it had stockpiled some of the thousands of tons of chemical fertilizer it received during the emergency period. The scarcity of chemical fertilizers, which produce no lasting benefits for nutrient-poor soils in any case, will retard the expansion of the hectarage of fertilizer-dependent, high-yielding rice varieties.

What rice farmers have traditionally relied upon, however, and now miss greatly as demand far outpaces the partially revived production, is locally crushed ground rock phosphate. Kampuchea's soils are phosphate poor, particularly in the eastern provinces of Prey Veng and Svay Rieng. The Chinese-built plant at Tuk Meas in Kampot crushed 12,000 mt of phosphate per year, which met only a portion of the country's needs.[34] Illegal trade in superphosphate fertilizer, available in abundance in South Vietnam during the war, also supplied Khmer peasants with needed phosphate. In the west, farmers in Battambang and Siem Reap used crushed phosphate from a small plant at Phnom Thom in Battambang province as a seed mix, dipping their floating rice in the ground rock before seeding.

The plant at Phnom Thom has started production again, after a long repair process involving the import of spare parts from the American manufacturer of the crusher. While its capacity is 7,000 mt per year, initial production will almost certainly be much lower. Even operating at capacity, the plant will supply enough phosphate to fertilize about 100,000 ha, assuming the use of only 50 kg of crushed rock per hectare. The Soviets have rebuilt the larger factory at Tuk Meas. Assuming full exploitation of the phosphate reserves in both Kampot and Battambang, Kampuchea will exhaust these reserves within fifteen years.[35]

Thus, the immediate prospects for developing alternatives to chemical fertilizers are poor. However, Kampuchea has advisers from Vietnam and India, where some of the best work in the world in this area is being done. Advances, therefore, should result from bilateral cooperation, at the very least in the form of training Khmer technicians and the creation of demonstration farms. The transfer of these appropriate, self-reliant, but labor-intensive technologies to Kampuchean peasants will take many years, retarding the improvement of rice yields per hectare that remain among the lowest in the world.

Irrigation

The Pol Pot regime made solving Kampuchea's water problems its number one development priority. This fit in nicely with the fervent nationalism of the Khmer Rouge, who harked back to the greatness of the Angkor Wat civilization. Angkor's wealth derived from the harnessing of Kampuchea's water resources to produce abundant crops, and the destruction of the magnificent system of irrigation reservoirs and channels brought about the decline of the Khmer Empire and ushered in a complete dependence on uncertain rains and the rise and fall of the waters of the Mekong basin that continues to this day.

Today in Kampuchea long irrigation canals criss-cross the landscape, memorials to the thousands of people sacrificed to their construction. These canals hold rain and stream water in the rainy season, but they are usually as bone-dry as the surrounding fields during the dry season. Khmer hydrologists and Ministry of Agriculture officials say that most of the Khmer Rouge irrigation projects are worthless. A genuine assessment by Kampucheans in the service of the government is impossible, however, because the political line of the Heng Samrin regime discredits *a priori* the entire Pol Pot regime. Rodolfo Juarez, a Cuban hydrologist who worked in Kampuchea for almost two years for Church World Service, felt that some Pol Pot era projects were soundly designed and could provide a basis for further development of Kampuchea's irrigation infrastructure.

The Khmer Rouge constructed seven large pumping stations, which could irrigate in excess of 10,000 ha each. In these cases, they relied on machinery and expertise imported from North Korea. At Po Lors, for example, in Ba Phnom District in Prey Veng, seven diesel engines power seven large pumps that draw water from a large channel that has water in it year-round. The station is an impressive and imposing structure, but recent discussions with the one mechanic from the Pol Pot time who continues to work at the station revealed that it had only functioned during the 1978 dry season, and then on a limited basis. Of the seven pumps, three were run at a time for fourteen hours per day, three days per week. Presumably the lack of fuel prevented operating the station at capacity.

At Po Lors only two of the seven engines are in working condition, due to scavenging of parts by starving people to sell or barter for food and sabotage by Khmer Rouge sympathizers in the chaos following the Vietnamese invasion. At nearby Ksach Sar, however, shortage of fuel has prevented this large station from operating, even though four out of six engines are in working order.

That the Khmer Rouge opted for such large-scale systems wholly dependent on fossil fuels is rather surprising. And since Kampuchea today is equally short of fuel, the probability is that these stations will become utterly useless relics of the Pol Pot period. To prevent this terrible waste of potential, Oxfam America explored the possibility of an alternative fuels scheme on a trial basis to be located adjacent to one of these large irrigation stations. Consultants recommended producer-gas, but the necessary wood fuel resources did not exist in the project area in Prey Veng. On a large scale the costs would be prohibitive for a private voluntary agency, but the whole realm of alternative fuels development might offer attractive "prestige project" possibilities for a future bilateral donor.

The gravity-fed water transport systems built under the Khmer Rouge suffer from inadequate maintenance of dikes and channels. Kampuchean teams trained in the use of heavy earth-moving equipment by Mr. Juarez have performed major repair jobs in the past several years, notably saving 20,000 ha from annual flooding due to faulty design of a Khmer Rouge project in Prey Veng. These teams have also constructed several new earthwork systems themselves. On a daily basis, however, peasants have not been mobilized to shore up channel walls and keep them free of weeds and silt. As a result, water collects in the channels of these systems, but the systems themselves do not function as intended. This forces peasants to use traditional pumps to lift the water into their fields or to grow rice in the channels themselves.

Thus, the major irrigation need in Kampuchea is an efficient, labor-saving pump that would enable peasants to utilize water available in ponds, streams, and man-made channels during the rainy season. Small diesel-powered pumps have been useful for keeping seed beds watered or getting enough water into a paddy for plowing. But maintenance and fuel problems have plagued these pumps continually. In the dry season, when small diesel pumps would be most useful, peasants cannot depend on enough fuel being available for an entire season to risk putting labor into a crop that will not be irrigated through maturity. For small tasks, traditional pump designs such as the pedal-powered *rohat* and the *snach*, a basket suspended from a wooden tripod, are adequate and reliable but require large amounts of labor.

Wind- and solar-powered pumps should be tried in Kampuchea, as they are now being tried all over Asia. Winds may prove to be inadequate, except along the southern seacoast, but there is at least one functioning windmill of standard American farm design in the Phnom Penh area. Photo-voltaic systems are very expensive and work best when lifting water more than five meters. They are not optional in the low-lift situations most common in Kampuchea. These systems, however, have the crucial advantage of being easy to maintain and free from dependence on fossil fuels. The cost of solar alternatives will continue to drop as research continues.

Prototypes of promising small-pump designs would provide the Department of Hydrology with the opportunity to reach its own conclusions about alternatives to traditional designs. Like their counterparts in the Agronomy Department, the technicians in Hydrology were spoiled during the emergency period by the abundance of pumps and fuel to power them. Now they face the necessity of mobilizing farmers to maintain existing systems, gradually expanding their own capacity to construct new water management schemes, and adopting alternative pump designs for trial and replication. The former is a political task, but outside donors can make key contributions to their work by providing and maintaining equipment and facilitating the import of worthy small-scale pump designs.

Training

Kampuchea has been one vast training ground since January 1979. Young people have resumed studies interrupted on April 17, 1975, technicians have retrained themselves on the job, and the few survivors with advanced degrees have tried to reacquaint themselves with advances in their fields during the years they were completely isolated from the outside world. Every group has also received a healthy dose of political education to complement their technical learning.

It is precisely the political component of training that has prevented the voluntary agencies from participating to the extent that they would have wished. There has been no shortage of offers to the Ministry of Agriculture of foreign experts to help train cadres in a variety of fields. When Western agencies have participated in training, it has largely been in informal settings in which learning by watching and imitating is the rule. Examples of this type of training include instruction in the use of earth-moving equipment or in animal-vaccination techniques. The formal classroom has been closed off entirely.

Agricultural training is taking place abroad, particularly in Vietnam, at agricultural colleges in the Phnom Penh area, and in classes organized in the provinces by the local agricultural committees. Training abroad has removed numerous talented young people and even experienced technicians from the country for extended periods, up to five years in some cases. The Heng Samrin government has been willing to give up the contributions these people could make to present reconstruction efforts in Kampuchea for the skills and political attitudes they will bring back with them from abroad and apply to solving the country's long-term problems. Some have in fact begun to return, but it is still impossible, as with people trained locally, to assess their impact on development in rural areas. Just getting trained people working at the provincial level will be important. In Kompong Cham in 1980, *four* technicians were responsible for agricultural development in this province of one million people. Presumably the new generation of trained cadre is already beginning to ease such impossible work loads.

Inside Kampuchea, the Ministry of Agriculture has shown a refreshing pragmatism thus far. At Chamcar Daung Agricultural College, courses have been limited to three-month intensive training sessions in tractor repair, water management, veterinary science, and other subjects. Students are nominated and sent by the provinces; upon completion of the course they are expected to return immediately to the provinces and get back to work. While there is no way to judge the effectiveness of the crash courses, their practical work orientation is laudable.

Certain tensions within the government have a significant impact on training designs. Prior to 1975 many current government officials studied in Europe, the United States, or Cambodian institutions patterned after the French system. Their visions of appropriate models for training differ radically from those of long-time communist cadre with little formal education, or recent graduates of Vietnamese and Soviet institutions.

In agriculture, these conflicting approaches will have to work themselves out at the Agricultural College, from which a new generation of agricultural technicians will emerge over the next decade. While the present bent of the courses is decidedly practical, the one-time director of the college, now a mid-level official in the ministry, produced the 1969 catalogue of the Royal Agricultural College to give visitors an idea of what type of curriculum the college intends to offer. Whether such an elite model for education will be tolerated in revolutionary Kampuchea remains to be seen; however, it is probable that one reason for the slow start-up of the two-year, degree-granting institution has been lack of consensus regarding a return to the methods of the 1960s. Clearly, a formal, scientific curriculum taught in French would not meet Kampuchea's urgent needs in the agricultural sector. Starting a college that is giving courses in French is a pretty crazy idea in Kampuchea in the 1980s. The only way to make sense of the idea is to see it in the context of nationalism: the need to have a degree-granting Agricultural College to assert national sovereignty and independence in this key field.

Vietnam and India have crucial roles to play in training people to manage Kampuchea's agricultural development. Both are rice-producing countries that must obtain maximum yields at every planting to feed their populations. Both have experience with high-yielding varieties, indigenous seed development programs, and alternative technologies. The ability of their specialists and training institutes to communicate their knowledge effectively to their Khmer counterparts will be a crucial determinant of Kampuchea's agricultual development over the next decade.

Voluntary agencies should continue to contribute needed material aid to the ministry's training efforts, perhaps at the provincial level where needs are greater and the training is closer to village-level concerns. Shortages of school supplies will continue—both consumable materials such as pens and notebooks and items such as laboratory equipment and audio-visual materials which have more lasting value. Kampuchean agriculturalists are desperate for reference books, particularly in French, and descriptions of recent scientific research. In the present political situation, agencies can perform a useful function simply by supplying these resources. As the ministry develops its own training materials for extension work with peasant farmers, funds will be required to print these guides. Even while being excluded from contributing personnel to training institutions, international agencies have a variety of options to sustain the efforts of the Ministry of Agriculture at all levels.

Conclusion

In their book *Food First*, Lappé and Collins answer skeptics who protest that a focus on food production will not generate the capital surplus required to sustain development, by invoking the example of a commune in China. The Yuechi commune, by fully mobilizing its work force, was able to increase production and generate enough income to invest in farm machinery and other equipment. The following is a quote from the FAO report cited by the authors:

> What a commune of modest means does with its accumulation fund is illustrated by the example of Yuechi Commune in Wuxien Hsien, Kiangxu province, one of the less prosper-

> ous communes. Within the five years of the Third Plan, the collective bought two 35-HP tractors and 67 8-, 10-, and 12-HP hand tractors, set up 11 irrigation drainage stations, purchased 12 heavy transformers, 15 electric engines totalling 284 kw, 71 electric pumps, 18 harvesters, 183 small threshers, 193 sprayers, 11 milling sets, 56 rice transplanters, 19 4-ton ferro-cement transport barges, 520 other sampans, 12 road trailers, and 43 large handcarts.[36]

In this essay I have deliberately avoided defining the term "development." Perhaps Yuechi Commune best illustrates the sense of dynamism, of plunging headlong into a future of rapid change, that seems an essential aspect of the development process. Yet this type of development cannot take place in Kampuchea today. In Kampuchea development means the process of rebuilding, not only in a material sense, but in a cultural and spiritual one as well.

In March 1981 in Ba Phnom district in Prey Veng, villagers set up a thatch-roofed hut bedecked with flags beside National Route 1 to collect money to rebuild the local pagoda. Passersby, responding to the music and the amplified entreaties of the organizers, were giving one, two, five riels at a time toward the fund. Inside the pagoda compound, five young monks stood before the site of the former temple, which had been razed to its foundations.

I returned to Ba Phnom in October 1982. On the old foundations stood the new temple of simple design—brick construction covered with a thin layer of cement, whitewashed with colorful trim, covered with a thatch roof.

The existence of this new temple symbolizes the development process in Kampuchea, which is as much a process of recreating the past as pushing forward into a new world. Villagers in Ba Phnom have not invested in farm machinery, electric engines, threshers, or sprayers. Indeed, they cannot, even if they so desire. They have invested in new ox carts, draft animals, houses, clothes, weddings, and a new temple.

In this way, mindful as much of traditions as of the future, the people of Kampuchea have set about the task of healing and rebuilding. By emphasizing the creation of local food self-reliance, appropriate development aid contributes to the ongoing effort of the Khmer people to build a haven of security in a violent and capricious world.

Notes

1. Jean Delvert, *Le Paysan Cambodgien* (Paris: Mouton, 1961), pp. 651–55.

2. Oxfam UK created and led the consortium, which consisted of the following members in December 1980:

	Received £	Promised £	Total Commitment £
A.F.S.C.	46,522	—	46,522
Community Aid Abroad	829,748	—	829,748
Caritas International	49,500	—	49,500
Caritas Netherlands	40,353	—	40,353
CEBEMO	71,100	—	71,100
CNCD & Belg. FFH	89,466	—	89,466
Frère des Hommes	28,202	—	28,202
C.R.S.	22,568	—	22,568
Deutsche Caritas	1,826,595	121,600	1,948,195
DWHH	1,179,658	—	1,179,658
C.E.C.	1,513,005	—	1,513,005
Dutch Medical Committee	67,591	—	67,591
Frère de Nos Frères	28,319	—	28,319
Lutherans	2,254,832	688,000	2,942,832
Manitese	115,683	—	115,683
Mennonites	89,926	—	89,926
Miserior	934,425	240,000	1,174,425
Novib	545,618	323,000	868,618
OXFAM America	1,438,048	—	1,438,048
OXFAM Belgique	149,887	146,000	295,887
Help the Aged	77,689	—	77,689
OXFAM Quebec	115,475	—	115,475
OXFAM	6,511,918	—	6,511,918
Redd Barna Norway	144,930	—	144,930
Sol. Soc. de Belgique	78,907	109,000	187,907
Trocaire	82,250	—	82,250
World Relief	133,763	—	133,763
Total	£18,465,978	£1,627,600	£20,093,578

3. Discussion with the author, October 1981.

4. Food and Agriculture Organization, Office for Special Relief Operations, *Kampuchea: Report of the FAO Food Assessment Mission*, W/P0180, (Rome, November 1980), p. 11.

5. George Hildebrand and Gareth Porter, *Cambodia: Starvation and Revolution* (New York: Monthly Review Press, 1976), p. 20.

6. There is no reliable compilation of bilateral aid for the emergency period. I therefore rely almost exclusively on FAO figures for multilateral and private voluntary assistance to suggest the magnitude of the relief effort.

7. FAO Report, November, 1980, appendix c.

8. International Bank for Reconstruction and Development, International Development Association, *Report of Economic Mission to Cambodia—1969*, EAP-13a,

October 1970, vol. 1, pp. 13–4.

9. In 1980 I travelled to Kompong Cham, Svay Rieng, Kandal, Takeo, Kompong Speu, and Kompong Thom provinces; only once did I speak to a farmer who had planted an international variety, despite the obvious interest of our government guides in showing that relief aid was being used effectively.

10. This discussion owes a great deal to the work of John Dennis, Cornell University, who anticipated many of the problems I have presented here in his brilliant unpublished paper written just as the relief effort was getting under way, "Rice Production in Kampuchea." The thesis that peasants ate international rice seed also helps explain why there was no recurrence of famine conditions as harvest time approached in 1980, despite the fact that the World Food Program was late with its food rice deliveries to Phnom Penh and Kompong Som ports.

11. Discussion with the author, October 1982.

12. Food and Agriculture Organization, Office for Special Relief Operations, *Report of the Food and Agriculture Assessment Mission to Kampuchea, 23 October-4 November 1981*, W/P5662, (Rome, November 1981), p. 4.

13. Discussion with the author, July 1981.

14. FAO Report, November, 1981, p. 14.

15. Frances Moore Lappé and Joseph Collins, *Food First: Beyond the Myth of Scarcity* (Boston: Houghton Mifflin, 1977).

16. IBRD Report, vol. 1, p. 83.

17. Ibid., vol. 2, annex 2, pp. 28–30.

18. Discussion with an international consultant, October 1982.

19. This is not to say, however, that visits by World Bank consultants, ASEAN-based businessmen, and USAID experts—likely adjuncts of any "nonaligned" government in Kampuchea—would necessarily be an improvement from the standpoint of Kampuchea's economic independence.

20. I owe this ironic observation to Michael Vickery, from his article "Looking Back at Cambodia, 1942–76," in Ben Kiernan and Chantou Boua, eds., *Peasants and Politics in Kampuchea, 1942:1981* (London: Zed Press, 1982), p. 112.

21. See Kiernan and Boua, eds., *Peasants and Politics*, pp. 227–362.

22. Delvert, *Le Paysan Cambodgien*, pp. 503ff.

23. The data of Hu Nim and Huo Youn notwithstanding (see Kiernan and Boua, eds., *Peasants and Politics*, part 1, pp. 31–86), I agree with Delvert that "*la société paysanne cambodgienne est une Démocratie rurale presque parfaite*" (italics in original), particularly when compared with other countries in Asia. The key problem was that virtually all peasants were equally under the yoke of the local merchant. This problem is treated below and extensively in the Delvert study. See ibid., pp. 509, 510–13.

24. FAO Report, November, 1981, p. 12. Ministry of Agriculture figures presented in a discussion with the author. For 1984 figure, Jean-Claude Salle, "Final Report on the Food Situation in Kampuchea," Kampuchean Humanitarian Assistance Programs, United Nations, December 1984.

25. Delvert, *Le Paysan Cambodgien*, pp. 235, 305, 322.

26. An assumption confirmed by the 1962 census. See IBRD Report, vol. 3, table 1.1.

27. FAO Report, November 1981, pp. 2, 45.

28. Delvert, *Le Paysan Cambodgien*, pp. 242–43.

29. In this and the following discussions, information on present development efforts in Kampuchea is based on project work by the author during four visits to the country between July 1981 and October 1982.

30. Delvert, *Le Paysan Cambodgien*, pp. 337–38.

31. Peasants were making the right choice because none of the factors conducive to the successful planting of international varieties on a wide scale is present in Kampuchea today (no fertilizers, poor water control, etc). To go slowly with the expansion of international varieties is very sensible in this context.

32. Delvert, *Le Paysan Cambodgien*, pp. 322–424.

33. IDRB Report, vol. 2, p. 77.

34. Ibid., vol. 1, p. 28.

35. Ibid., vol. 2, p. 77.

36. Lappé and Collins, *Food First,* p. 389.

Orlin J. Scoville

REBUILDING KAMPUCHEA'S FOOD SUPPLY

Pol Pot's Legacy to the Peasantry

When Pol Pot fled Phnom Penh in January 1979, the Kampuchean peasantry was in the midst of a brutal transformation from a system of control dominated by moneylenders, landlords, the bureaucracy, and the military to a military regime guided by an extreme form of peasant socialism. Cities were emptied; there were perhaps 30,000 people left in Phnom Penh. The Central Bank had been blown up and money and credit abolished. Most schools were closed. Technicians, scientists, administrators, and merchants lived in fear. Many fled the country, lived in hiding, or were killed.

The years of war and disturbance had begun more than a decade earlier. The peak year for rice production was 1969–70; the highest level of per capita food production was reached in 1968; the last favorable balance of trade was in 1965.[1]

The Khmer Rouge often are stereotyped as bandits bent only on wanton destruction. They also, however, had a program of construction based on the realization of a socialist, peasant-dominated society. In a harsh way, they mobilized brigades of peasants and former urban dwellers to build roads, dams, and canals. Their goal was to increase sharply, through irrigation, double-cropping of rice. Most of the irrigation works, however, were never completed, and some agricultural officials say they were poorly engineered.

Private land ownership was abolished. The dikes or bunds that had outlined the boundaries of owners' rice paddies were eliminated, and fields were laid out in rectangular, one-hectare blocks. That arrangement was more efficient for irrigation and tillage. It has been continued by the present regime, and the resulting pattern can be seen everywhere as one flies over the rice fields.

As Pol Pot's National United Front of Kampuchea (NUFK) gained control of the countryside, peasants were organized into "groups of solidarity for increasing production" (*krom samaki bangkar bankaoeut phal*).[2] The production groups in turn were organized into larger cooperatives, of which there were said to be 30,000 by 1975. The entire labor force of the production groups, from children to the aged, was assigned to specific farm, land development, or village tasks, although each peasant retained use of a small garden plot. According to Hildebrand and Porter, the solidarity group system had become the basic unit of agricultural work by 1975.[3] The Heng Samrin regime has continued with public land ownership and the production solidarity groups, even including the name, *krom samaki*. They have developed their own regulations and have not attempted to revive the cooperative superstructure.

Around villages, one sees simple buildings constructed in the Pol Pot era that were intended for use as meeting halls and refectories to foster collective enterprise. There are numerous clusters of small houses built for the cadre and now referred to as "Pol Pot villages."

After years of border incidents, Vietnam invaded Kampuchea in December 1978. Kampuchea's troubles were not over when Pol Pot fled to the forest. War and unrest spread into the level rice-growing regions at harvest time and continued through much of the planting season in 1979. This turmoil generated further population movement, disrupting food production. The flood of refugees to the border reached its peak that year.[4]

In early 1979, the world took note of the plight of the Kampuchean people and some relief efforts were begun, especially at the Thai border. The UN Economic and Social Council Executive Board, at a May 1979 meeting in Mexico City, authorized UNICEF to cooperate with the International Committee of the Red Cross (ICRC) to give help in Kampuchea if it were possible. In October, the Heng Samrin administration in Phnom Penh agreed to let the ICRC and UNICEF conduct large-scale relief within Kampuchea.[5] Pol Pot similarly allowed relief operations in the areas controlled by the Khmer Rouge.[6]

To coordinate the relief work of UN agencies, the secretary general appointed a special representative, Sir Robert Jackson, and posted him in Bangkok. When the relief agencies began their work they reported that 700,000 hectares had been sown to paddy in the 1979 monsoon season, compared with 2,427,000 in 1969. Only 600,000 hectares would be harvested. Further, fish production in 1979 was 15,000 metric tons or less, compared with a pre-1975 level of around 200,000 mt. Only 850,000 draft cattle and buffalo were fit to work, compared with about 2.5 million before the war. There were 200 working tractors and very few trucks and other forms of road, rail, and water transport. Even simple farm tools were in short supply. Many bridges and irrigation works had been destroyed, and roads were in poor repair.

Training schools had closed, their buildings and equipment damaged. The network of animal health stations was destroyed and the veterinarians and vaccinators were gone. Foot and mouth, anthrax, rinderpest, black quarter, and other diseases were a growing menace. Trained staff for schools, ministries, laboratories, and factories were scattered or dead. Before 1975, Kampuchea had 1,600 technicians in the Ministry of Agriculture; by late 1980, there were 200, of whom not more than 10 had university degrees.

Agriculture-related industries were destroyed or severely damaged. Only a few of the 405 rice mills operating in 1969 were usable. The phosphate mines were closed, the equipment gone. Spinning and weaving mills were damaged or destroyed. Virtually no facilities remained for making fish nets or supplying fishing gear. Even blacksmiths, though surrounded by abundant scrap metal, found it hard to operate because they lacked equipment and supplies. Village markets were few and disorganized. Farmers could not obtain fertilizer and pesticides. Rice seed was scarce; peasants consumed what normally would have been kept for planting. Finally, there was a shortage of able-bodied men in the work force.

The Strategy of Relief

The relief program in Kampuchea has been conducted with the permission of the authorities in de facto control of the territory. For Phnom Penh and most of the countryside, that has meant dealing with the People's Republic of Kampuchea under the authority of Heng Samrin, although the government seated by the United Nations and recognized by many of the member countries has been Democratic Kampuchea, initially under the leadership of Pol Pot and currently under a coalition government with Prince Sihanouk as president and Son Sann as prime minister.

There are precedents for UN programs of relief to populations controlled by authorities not recognized by the United Nations or its member governments. Such relief programs require authorization by the General Assembly. For Kampuchea, annual resolutions since 1979 have authorized humanitarian assistance and relief. The scope of relief activities is dealt with in general terms. More specific ground rules are followed by donor governments, particularly the United States, the ASEAN countries, and some others. The U.S. State Department is constrained by provisions in the "Trading with the Enemy Act" (40 stat. 4.1, Oct. 6, 1917 as amended), the Export Administration Act of 1979, and the law that authorizes appropriations for assistance in Kampuchea (PL 96-110). According to that law, assistance "shall be for humanitarian purposes and limited to the civilian population, with emphasis on providing shelter, transportation for emergency supplies, and personnel, and similar assistance to save lives." Under these acts, the State Department tries to distinguish among relief, rehabilitation, and development. U.S. government grants to UN agencies, the International Red Cross, or private voluntary agencies cannot be used for purposes ruled to be developmental. Assistance with rehabilitation may be approved if urgently needed to meet relief objectives. For programs of American voluntary agencies, even where private funds are used, federal export licenses must be obtained for virtually every commodity sent to countries coming under the purview of the various laws. No similar constraints have been placed on technical personnel supplied by voluntary agencies.

There is a continuum from simple relief to development, and rule-making for Kampuchean programs is difficult. Provision of rice for food or seed has never been a problem. Simple farm tools have always been acceptable. Spare parts for tractors and other major equipment raise more questions, but usually have been approved. Repairs to small rice mills used to mill domestically consumed rice have been acceptable. Repairs for large mills, historically used to mill for urban consumers and also for export, would not qualify. Tractors and bulldozers, though needed to rehabilitate even a fraction of the prewar irrigation network, have not been allowed as contributions from U.S. public or private sources. Animal health programs have been acceptable, partly because of their importance to local food supplies and also because they help deter the international spread of diseases.

UN agencies have supplied fishing gear, boats, and boat motors to rebuild the fishing industry. Such aid is designated as relief. Ice-making machines and equipment to make fishing nets have also been supplied but have been criticized by some donor governments as being developmental.

The offer of relief food to people in exchange for labor on roads or dams could expedite Kampuchean rehabilitation. Provincial and central authorities in Kampuchea have expressed interest in such "food-for-work" schemes. But to carry out programs successfully, close cooperation between donor and host country in planning, administration, and monitoring would be necessary. Donor-country constraints on development projects and Kampuchean aloofness toward Western advisers discourage the establishment of "food for work" programs.

An example of a rather extreme application of the "no-development rule" occurred in 1981 when the Mennonite church sought to donate pencils and paper for use by school children. A U.S. export permit was denied because the goods would be handed out by employees of the People's Republic of Kampuchea and thus enhance the image of the administration. The ruling was subsequently reversed on the condition that the Mennonite agency oversee the distribution of the materials.

The programs of private voluntary agencies are restricted in varying degrees by their own policies. They raise their funds from their own constituents as well as receive grants from governments and the United Nations. Voluntary agencies vary in the "tough-mindedness" of their policies, but in general they tend to be less vigilant about the political aspects of their aid. A few seem to have adopted an attitude of "casting bread upon the waters." Nearly all voluntary agencies are more development-oriented than are UN agencies. For both the United Nations and voluntary agencies, the most frequently expressed frustrations have been the limited opportunity to work directly with people and to see enough of the results of their programs.

It is hard to identify a neat set of rules by which the Heng Samrin administration requests or uses foreign aid. One key principle seems to be that assistance with training comes almost entirely from socialist countries. Some help from the West has been requested and accepted for constructing or equipping buildings at training institutions. One American voluntary agency, for example, had a good reception for its technical assistance, including training, provided by Cuban experts. Another principle is that foreign commodities, at least from the West, are to be handed over to the authorities upon arrival in the country. Thereafter, they are the property of the central or provincial authorities. Some monitoring is allowed to insure that goods are used in accordance with overall relief programs. Sometimes the donors are well informed about end uses, sometimes not. A third Kampuchean principle sometimes detected but not as clearly evident is that there should be an egalitarian distribution of donated commodities. Early food-rice distributions tended to be handed out in small quantities to many districts and to many people in each one. Fertilizer, instead of being concentrated in optimal applications per hectare, tends to be applied over large areas, with inadequate attention to location of the high-yield, fertilizer-responsive rice varieties, and with applications at such low levels as to reduce effectiveness of the fertilizer.

Among food-relief programs, Kampuchean authorities have always clearly given top priority to gifts of rice. The reasons are readily apparent. First, rice has met an urgent food need. With some logistical help, it has been possible to move rice quickly to food-short areas. Second, rice has filled a gap as a medium of exchange. Emergency rice began to arrive in Kampuchea even before money was reintroduced. Rice was indispensable as a supplement to salaries for ministry staff and militia in an economy with nothing to sell. In October 1980, the mayor of Phnom Penh reported that adult volunteer workers and militia were given thirteen kilograms of rice free each month (if they also were farming, only six kg), and children were allowed three kg.[7]

Workers were also allowed to buy up to two kg of rice a month at state shops for one riel per kg. The market price at that time was 2.5 riels per kg. The same policy was reported in various provinces, with some variation in allowances.

A second high priority has been for inputs to increase crop production in the short run and to improve animal health. Rice seed, fertilizer, pesticides, simple farm tools, and tractor power have been important. Use of modern production techniques has put a heavy burden on the few trained agriculturists. Kampucheans are being trained in Vietnam, the Soviet Union, the Democratic Republic of Germany, and other socialist countries for short and long periods. Training has at times taken key officials from repair shops or other facilities just when they were needed to receive and assemble relief shipments of machines, spare parts, or other technical materials. Sometimes the training has seemed to be more political than technical.

The Kampuchean authorities would like to have much more help with development-oriented or growth-promoting programs such as irrigation systems, road and bridge building, animal breeding, and large rice mills.

Kampuchean development programming seems unconcerned with the rapidly growing private sector petty trading and artisan production. Whether that reflects overt policy or merely an inability to curb such activity is not clear.

To what extent has the "no-development" policy of donor governments restricted self-sustaining recovery? It has clearly added to costs by encouraging stop-gap relief at the expense of expanded capacity to put more land under irrigation or produce more work animals to farm more land. Because it has placed greater dependence on imported fertilizers and motor fuel it has made the economy more fragile. But even in those relief programs that have been implemented, production might have been enhanced with more cooperation from the Kampuchean authorities. The productivity of programs has been reduced because Kampuchea has restricted the entry of Western technicians and delayed or interfered with their ability to travel. Offers of Western-supplied training have not been accepted. The authorities have not even been willing to explore such neutral sources of technical help as the International Rice Research Institute (IRRI) at Los Banos, the Philippines.

On balance, the UN relief model might have been improved, but it has been highly effective and has performed within the constraints of *realpolitik*, both Western and socialist.

Food and Agricultural Relief in Kampuchea

The total relief effort for Kampuchea included programs within Kampuchea, at the border where refugees came for food and seed (the "landbridge"), in the border camps, in holding camps in Thailand, and in Thailand itself, for Thais affected by border disturbances. By the end of 1982, total UN relief to the people of Kampuchea reached $685 million. Half went to programs at the border; this was roughly divided between food, on the one hand, and health and housing, on the other. The other half went inside the country—two-thirds for food and agriculture, one-third for other programs, of which health was the most important. The remainder of this chapter discusses relief efforts within Kampuchea—mostly programs to restore production. This is not to say, however, that the food and health aspects were not of vital importance.

The relief programs of both the Food and Agriculture Organization (FAO) and voluntary organizations began early the effort to stimulate agricultural recovery. At the outset, there was dire need for almost every production requisite—whatever was brought in could be used. Initial emphasis was on seed rice, fertilizers, pesticides, fishing tackle, and small irrigation pumps for use in rice nurseries. Later programs became more sophisticated, with a gradual shift of emphasis toward tractor power, power tillers, fishing boats, motors, nets, ice-making equipment, rehabilitation of irrigation systems, animal health needs, and rice mills. By late 1981, work-animal health and tractors for land preparation became priorities. There was growing dissatisfaction with the importation of large quantities of seed rice, and schemes were launched to expand in-country seed multiplication and to import food rice to be exchanged with farmers for seed rice.

Rice Production Campaigns

Soon after coming to Kampuchea in late 1979, FAO worked with the Ministry of Agriculture to develop a pilot rice program for 10,000 ha in the dry season of 1979–80. Rice seed, fertilizers, pesticides, raticides, hoe heads, irrigation pumps, and a few other inputs were provided. By early March 1980, 4,117 ha of rice had been planted in five provinces. Yields varied from good to poor. Although results were less than planned, a beginning had been made and valuable experience gained.

A much more ambitious program was launched for the 1980 monsoon season. FAO undertook to supply rice seed for 377,000 ha, of which 100,000 would be an integrated production campaign, with the most crucial inputs provided. The integrated campaign was begun in six important rice-producing provinces expected to have food deficits. The campaign was planned in consultation with the Ministry of Agriculture and with the assistance of FAO experts on mechanization, plant protection, and rice cultivation.

There was some delay in the procurement and delivery of inputs, but once in the country, they were promptly distributed to the provinces. It was difficult to obtain enough suitable rice seed. A small quantity was readily available from the Philippines, but purchase from Thailand was complicated. The large rice-export industry in Thailand is limited to milled rice for food. The Royal Thai Government as a matter of policy does not permit export of unmilled rice, and government permission must be obtained for each order. Moreover, there were no experienced seed firms that could procure, process, and test rice seed.

Technical problems arose in Kampuchea over the proper use of pesticides and rodenticides because few officials and farmers were experienced with those chemicals and the identification of plant pests. Other problems arose in assembly, operation, and repair of tractors, power tillers, pumps, and sprayers. Fuel shortages were frequent and were not always a result of unavailability in the country. Sometimes they resulted from lack of transport, or problems of allocation among ministries.

Other rice-production assistance was given in 1980 by voluntary agencies including Oxfam, the Consortium for Agricultural Relief and Rehabilitation in Kampuchea (ARRK), World Council of Churches, World Vision International, International Council for Economic and Social Development (CIDSE), and American Friends Service Committee. Vietnam provided some rice seed, assistance with plowing, and technical help.

For the 1980–81 dry season, FAO supported a seed multiplication program in two provinces, with a goal of producing 3,600 to 4,000 mt of quality seed. The program supplied high-yield variety seeds, fertilizers, pesticides, seed-testing equipment, and other materials. The short-term planning services of a rice expert were provided. Most of the crop was grown on peasant farms, and the ministry expected to buy seeds from the farmers. The farmers, however, kept most of the production or sold it to their neighbors, so the ministry only procured 700 mt for distribution to others. The project was considered sufficiently successful to justify a project in 1981–82 on 4,000 ha intended to produce about 10,000 mt of seed. In April 1982, FAO reported that planting had reached 96 percent of the target, with an average yield of 4 mt per ha. Though 15,200 metric tons of seed were produced, the ministry was only able to procure 1,000 mt, the rest being sold or used by farmers.

Farmers have been very receptive to the rice seeds from the Philippine Bureau of Plant Industry, particularly IR36, a variety that matures in about 100 to 110 days and has considerable latitude with respect to date of planting. The Bureau of Plant Industry distributes seed for IRRI at Los Banos. Farmer experience with the IRRI varieties has been favorable, but realized yields of only about two or four mt per ha on the best-farmed fields were far below what could have been obtained with adequate use of fertilizers, pesticides, and proper control of water. Experience with seed obtained from countries other than the Philippines has been fair to poor. Adapted varieties of floating rice have posed especially difficult problems.[8]

Recognizing the need for Kampuchea to return to self-sufficiency in seed rice, FAO in 1981 launched an experimental program to provide the ministry with a supply of food rice that it could exchange with farmers for seed. The experiment was successful and was repeated on a larger scale in 1982, when the Ministry of Agriculture sold 16,559 mt of food rice to buy 11,038 mt of seed paddy from farmers. Most of those seeds were of floating or local main-season varieties.[9]

For the monsoon season of 1981, FAO provided inputs to support an integrated campaign on 200,000 ha in eight provinces. The components in the production packages remained about as before, but it was becoming evident that the land area in cultivation was approaching the limit of capacity for plowing because of the lack of oxen and buffalo and the widespread incidence of disease. Both FAO and the Ministry of Agriculture recognized the difficulties involved in further reliance on tractors and agreed to expand the animal-health program. One of the voluntary agencies then began a small vaccination effort.

Work-Animal Health

FAO surveyed the state of animal health and brought in veterinarians, vaccines, and necessary equipment. The Ministry of Agriculture set up short training courses for vaccinators, and FAO and a Church World Service veterinarian offered on-the-job training. By November 1981, it was common to see oxen and buffalo travelling the roads with red tags in their ears—a sign they had been vaccinated. Farmers eagerly brought their animals to the vaccination points. From July 1981 to November 1982, 1,234,000 vaccinations were given under FAO's program. Most of the work stock in infected areas had been vaccinated at least once. About 230 vaccinators had been trained in the ministry's school and on the job by veterinary advisers. By November 1982, FAO's veterinarian reported fairly good control of contagious diseases in workstock. But follow-up and booster vaccinations will be needed to maintain control.

Animal diseases are frequent among swine and poultry, and these animals are important to food production. Some voluntary agencies have given assistance, but donor governments, who readily give assistance to health protection for work animals, regard assistance to other species as too closely related to development.

A good argument can be made for increasing the supply of work animals through breeding programs. Possibilities would include importation of bulls or artificial insemination. Calves could be raised economically on underutilized forages. The skills needed to raise and manage oxen and buffalo are well known. Reliance on animal power would reduce the demands for imported fuels and spare parts. Donor governments have rejected proposals that they help to expand the number of work animals because of the time required to produce animals of working age and because animal breeding is viewed as development. Oxfam, Lutheran World Service, and perhaps one or two more voluntary agencies give a little assistance.

Mechanization

The readily apparent shortage of animal power needed for land preparation made it imperative that something be done to enhance plowing capacity. By 1980, the donor community had supplied about 75 field tractors, 1,300 power tillers, and some tractor plows. Other power equipment included several hundred power sprayers and 3,900 small irrigation pumps. Socialist countries, especially the USSR and Bulgaria, provided additional field tractors.

Futher equipment was given in 1981 and 1982, especially irrigation pumps. By 1982, the tractors and power tillers were used on only 34 percent of the 208,000 ha of land it was hoped they would plow. Fuel shortages, insufficient shop equipment, and poor administration of the tractor fleet and workshops were responsible. After 1980, FAO concentrated its field mechanization assistance on spare parts and maintenance facilities. By 1982, the annual demand for fuel for the agricultural machinery, including irrigation pumps, was about 8,000 mt (8 million litres). Fuel shortages had become so bothersome that by late 1982 UNICEF and some voluntary agencies had provided about 3,700 mt.

Irrigation Programs

With the shortage of work animals and manpower, irrigation to permit a second crop in the dry season offers a good opportunity to increase production. Neither the Ministry of Agriculture nor the major relief agencies gave high priority to irrigation in early programs except for irrigation of rice nurseries. The rehabilitation of irrigation works, repair of dams, and provision of large pumps had to await engineering studies. At least some key officials in the Ministry of Agriculture felt that the remaining dikes and dams constructed in the Pol Pot era were not properly engineered to make them usable—a judgment that may have been influenced by partisan sentiments. Major dams or dike repairs entail much earth-moving—a difficult feat without bulldozers—and the United States and some other donors did not wish to supply such equipment.

Church World Service initiated an irrigation program in 1980, recruiting an experienced agricultural engineer through the Ecumenical Council in Cuba. He was able to travel quite freely through the country, planning rehabilitation projects and training surveyors. Church World Service, after being denied U.S. permission to export a bulldozer, eventually obtained one through European contributions. By late 1982, the Church World Service program had made 13,000 ha irrigable in nine projects. FAO and three or four voluntary agencies have supplied irrigation pumps, including motors and hoses. The Australian Freedom from Hunger Campaign has made a major contribution of large pumps, together with engineering assistance.

The Ministry of Agriculture estimates the potentially irrigable crop area of Kampuchea to be 280,000 ha, including that irrigable by the traditional *decrue* or "flood retreat" method.[10] By the end of 1980, about 122,000 ha had been made to some extent irrigable, although most of that area still needed some additional land preparation or improvement in water supply. An additional 157,000 ha would be irrigable if pumps were supplied or works repaired. Although total irrigable area is only a little over 15 percent of the total area now planted to rice, the potential yields under irrigation are four or five times as high as can be realized with rainfed monsoon crops, and plowing is done when labor and work animals are readily available. The potential contribution of irrigation to production is thus great with the use of modern technology.

Fisheries

In the 1960s, Kampuchea produced up to 170,000 mt of fish from fresh waters and about 40,000 mt from marine fisheries. There were modest exports. By 1978–79, facilities for marine fishing had disappeared and freshwater fishing only yielded about 15,000 mt. Prior to 1975, the fishery branch of the Ministry of Agriculture had 350 technicians; by 1980, only 9 of these remained, and most of the new staff members were only partially trained.

FAO initiated in 1980 a project to repair barrages (weirs that project out into rivers and Tonle Sap), fishing tackle, motors, boat-building equipment, workshops, and other supplies. An ice-making plant was provided. Three or four voluntary agencies also gave fisheries equipment. FAO's equipment was slow in arriving, and confusion among Phnom Penh and provincial officials further delayed distribution of the equipment. The refrigerant for the ice plant was sent to the wrong province. The plant, needed to preserve fresh fish in the peak fishing season, stood idle for many months but was finally put to use in 1982.

Rice Mills

Early relief assessments for Kampuchea noted the virtually universal destruction or inoperability of rice mills. As a result, peasants and householders had to hand pound their rice, using time badly needed for other work. More importantly, the yields of milled rice were reduced. The conversion ratio for good rice mills is estimated at up to 70 percent—that is, 100 kg of paddy should yield 70 kg of milled rice. With hand pounding, the average yield would be around 60 kg or less. Rice milling also improves the availability of bran for livestock feeding.

In late 1980, FAO initiated a program to repair mills and add some small mills. Repair parts for old mills and 115 small mills, which arrived in late 1981 and early 1982, improved the milling situation. As large mills were not repaired, further investment will be needed as rice production increases.

Some 250,000 burlap rice bags were provided as a component of the recreation of a rice-milling enterprise. Kampuchea has also received a large quantity of burlap bags from relief shipments of food rice. They are a popular item of trade in all markets, and it seems likely that the Kampuchean authorities could have procured from local markets most of the necessary bags for rice milling if that had been necessary.

Results: Agricultural Production

Relief efforts in Kampuchea have been highly successful considering the haste with which aid had to be given and conditions in the country. The area planted to rice for the monsoon seasons since 1979 is listed in table 1.

Total paddy production from monsoon and dry-season crops in 1979–80 was only 637,000 mt; for 1980–81, 1,470,700 mt. In 1981–82 it was only 1,453,000 because of an adverse monsoon season. For 1982–83, a projection in November placed the monsoon crop at about 1,450,000 mt. In addition, about 257,000 mt should have been attainable in the 1983 dry season, assuming continued assistance at the modest level proposed in November 1982. So total paddy production for 1982–83 could have been about 1,707,000 mt.

Rice production available for consumption in relation to needs is shown in table 2.

The UN's worldwide standard for emergency food supplies is 400 grams of cereal per person per day (146 kg per year). For Kampucheans, the daily calorie requirement per person is 2,220, according to FAO.[11] The emergency ration provides 1,260 calories; other food crops and animal products are required to make up the remaining 40 percent of the calorie requirements. In Kampuchea, mung beans, cassava, corn, bananas, vegetables, fish, wild plants, game, chickens, and pigs have helped but have not completely closed the gap. Each FAO food

Table 1

Area Planted to Rice, Monsoon Seasons, 1979–1983

Year	Hectares
1979	700,000
1980	1,317,000
1981	1,325,000
1982	1,336,000
1983	1,450,000

Table 2

Rice Production Available for Consumption, 1980–1983[a]

		Milled Rice Production Available for Consumption	
Year	Population[b]	Total mt	Per Capita kg
1980	6,500,000	340,000	52
1981	6,591,000	757,000	115
1982	6,682,000	691,000	103
1983	7,033,000	914,300	130

a. After deducting for field losses, seed requirements, and an emergency reserve; paddy converted to rice at extraction rates of 60 percent for 1979 to 1981, 61 percent for 1982, and 63 percent for 1983 crops. Increasing rates of conversion are due to improvements in rice milling.
b. FAO Food Assessment Team estimates (population growth rate = 3 percent).

assessment has reported shortages in total food availability, particularly in the areas of lowest productivity.

Statistics on crop and livestock production are sketchy. Ministry of Agriculture estimates indicate that acreages in the major crops such as corn, manioc, and sweet potatoes rose sharply in 1980 but have since declined. Production of subsistence crops, especially bananas, on individual plots are believed to have increased steadily. Fish production has increased from 1,500 mt in 1979–80 to 67,500 mt in 1982, but even this provides an average of less than ten kg of fish per person per year, from which deductions must be made for heavy losses from spoilage.

On balance, it is believed the total nonrice supplies of food per capita may have declined somewhat since 1980, in part because of weather, but also because of the drop in the price of corn, manioc, and other food crops in relation to tobacco and other cash crops.

The November 1982 Food Assessment Mission was the first to include a nutritionist. That mission reported the existence of a considerable degree of undernutrition and malnutrition among children in several provinces.

The health of draft animals has been significantly improved through relief programs, and the Ministry of Agriculture is regaining its capacity, with some further assistance for veterinary supplies, to maintain levels of immunity and combat future outbreaks. At the end of August 1982, a voluntary agency was making plans with the ministry to establish a diagnostic laboratory for animal diseases.

On balance, with normal weather and no major pest or disease attacks, Kampuchea is approaching minimum self-sufficiency in food, but with no reserve capacity to cope with an adverse season. Whether on its own the country can find the foreign exchange for spare parts, fuel, fertilizer, and pesticides to maintain the momentum needed to keep up with population growth remains to be seen.

The level of coordination and cooperation among the UN agencies and with the voluntary agencies has been exceptional in Kampuchea and generally good in New York and Bangkok. In Rome there is coordination between FAO and the World Food Program (WFP). The fact that most agency leaders in Kampuchea live and work in two buildings has facilitated communication there. Loans of materials and vehicles are frequent among agencies. Program plans are discussed. At times, one agency has acted as buyer of relief supplies for another. Interagency meetings including UN and voluntary agency staff are regular and frequent. There is some communication between the UN and voluntary agency community and socialist country representatives, but not much. Results of FAO food assessment missions and similar surveys are made available to socialist country representatives. Information concerning Kampuchea's needs is passed on to the socialist country representatives who might be able to help or provide additional information, such as the need for motor fuel, seeds, or fertilizers.[12]

Alternative Donor Policies

What other policies could have been followed, and would they have been more productive?

Funding

Could the program have effectively used more money? No definitive answer can be given, but the level of funding seems to have been about right given the prevailing operating conditions. In some instances, earlier approval would have permitted more timely procurement. But more timely procurement could also have been achieved had the Kampuchean authorities, FAO, and other agencies been able to expedite their procedures. Decisions on major FAO purchases are made in Rome under procedures designed to ensure prudent buying but not intended for emergency procurement. Some voluntary agencies moved at dramatic speed, but in some instances there may have been countervailing disadvantages of higher costs or less-carefully selected seed varieties or pesticide formulae.

More funds could have been effectively used if Kampuchea had been willing to accept more technical help and if donors, particularly the United States, had not maintained so rigid a distinction between relief and development. The main constraints on effective use of more assistance were as much political as economic.

The Constraint on "Development"

The strict interpretation of the limits to relief had damaging effects, but most truly developmental programs would have been viewed with suspicion by Kampuchean authorities anyway because they would have entailed Western advisers and monitoring. The consequences of limiting aid to relief were many. First, all negotiations with Kampuchean coworkers tended to be conducted at arm's length. Joint planning of projects was discouraged. "Food for work" was a good concept, but application of such a program would require mutual planning, execution, and monitoring. As it was, nearly all planning had to be for very short-term operations.

Second, even if a project such as dam repair was approved, labor-saving equipment could not be obtained in the United States or with American funds because of developmental connotations (of earth-moving equipment, for example).

Third, such medium-term projects as work-stock breeding were not acceptable because of development inferences.

Fourth, any projects designed to stimulate the economy by encouraging internal trade, improving markets, or enhancing some export would have been flatly rejected by one or more donor countries.

Finally, projects to help build institutions could not be considered because they would be developmental and might strengthen the ties of the administration with the people. Bringing in hand-sickles may have strengthened the black market, but bringing in forges, blowers, anvils, files, and welders would have restored the blacksmith trade and thus been developmental.

Food Relief vs. Aid to Food Production

Donors and aid agencies struck a good balance between gifts of food and gifts of food-producing inputs, and the ratio between them has moved steadily toward more inputs. Experience has shown that it was much cheaper to stimulate in-country food production in Kampuchea than to ship in food, or to feed and support people at the border, or to cope with an increased flow of refugees. Table 3 illustrates this situation.

Table 3

Per Capita Relief Assistance Costs, 1982[a]

Production of a one-year supply of rice	$ 25
Purchase and shipment of a one-year supply of rice	$ 64
One year of support in border camp	$142 total/$100 food only
Receipt, placement, and transportation of one refugee to United States	$1,200 or more

a. Food costs refer to food purchased for consumption in 1982, while food production costs are for food produced in 1982, most of which would be eaten in 1983, thereby reducing needs for relief food purchases in that year.

To produce 1 mt of milled rice for consumption (which feeds seven persons) requires 1.5 ha at 1.3 mt paddy per ha. Externally supplied inputs include 120 kg of seed at $66; 225 kg of fertilizer at $62; $20 worth of pesticide and rodenticide; 40 liters of fuel at $8; and miscellaneous costs at 12 percent of the above, or $19. This totals $175 of external support, or $25 per person.

The UNICEF budget statement of March 6, 1981, estimated that $35 million was needed to provide 75,000 mt of food, or $467 per mt. Since some food would be of higher value than rice, an average of $450 per mt is used here, or $64 per capita. The UNICEF budget presentation of November 25, 1981, indicated the 1982 cost for Khmer assistance at the border at $30.6 million for an average of 215,000 beneficiaries, or $142 per capita, of which $100 was food.

Refugee costs are as estimated by the Office of the Coordinator for Refugee Affairs, U.S. Department of State. No estimate is available for additional public costs incurred on behalf of refugees after initial placement, but for many states and localities such costs are substantial.

It may seem inappropriate to compare the costs of handling refugees who come to the United States with merely supplying food, and increasing food production in Kampuchea will not entirely stem the flow of refugees to the border and beyond. But the correlation between food shortages and the exodus of refugees, along with the associated importance to the United States and other nations of an assured food supply in Kampuchea, are well established, as has been noted by the Marshall Green Panel.[13]

The programming needs of food production activities are far different than for food relief. If some emergency food stocks are judiciously prepositioned in anticipation of potential needs and combined with efficient delivery systems, food aid can be brought in at any season, as critical needs arise; however, aid to food production—seeds, fertilizer, pesticides, and other inputs—must be in the hands of peasants when needed or the chance to produce an adequate crop in that year will be lost.

Kampuchean Policies

Would different Kampuchean rules have improved results? It is difficult to conjecture about effects of Kampuchean rules without being certain what those rules were. In some instances one can only guess at alternative outcomes. Does denial of a donor representative's request to visit a project arise from bureaucratic fumbling and obstructionism or is one being protected from the threat of ambush? Some constraints seemed to flow from ideological considerations rather than from mere random or whimsical bureaucratic decisions.

The Basis of Production Organization

The key evidence of socialist principles of organization in agriculture is in the nature of resource allocation and control. The rules are laid out in a declaration of the Peoples' Revolutionary Council issued in 1980 (see appendix b). In general, peasants are grouped into production solidarity groups (*krom samaki*) of ten to fifteen families. Each family is assured of a house and a plot of from 1,500 to 2,000 square meters of land. Family plots not used may revert to the group for reassignment. Field cropland is assigned to each group on the basis of about one ha per family. Workstock is apparently left in the hands of individual owners, but they are encouraged to share their animals with others on the basis of "workpoints." Produce from the group land is allocated to families on the basis of workpoints earned. Rules are vague and imply that the fields are to be farmed collectively. Visits with a number of production groups leave the impression that most land is alloted to individual families to be farmed, but without firm continuing rights to any plot. Fruit trees and coconuts belong to occupants where they occur on family plots. Otherwise, they are under control of the group.

By 1980, there were about 80,000 production groups of farmers. There were similar groups in fishing, and for some trades and crafts. Farm production groups seem to function much like traditional Khmer clan or extended family groups, but there appear to be growing roles for the solidarity groups in statistical reporting, ascertaining member seed and relief needs, distributing production inputs, and handling collective production. There is no discernible effort to form cooperatives or collectives at higher levels.

Administrators' Attitudes Toward Peasants

The Heng Samrin authorities maintain a low profile toward peasants, interfering with their activities as little as possible, giving them inputs without much attempt to collect payments in cash or kind, and collecting no rents or taxes. Officials claim that villagers are not required to donate labor on public projects, but are paid in cash or rice or both. Pol Pot's heavy use of the corvée system made this a sensitive issue with peasants. As indicated, in the seed multiplication project, the Ministry of Agriculture had plans to take back some paddy in return for the inputs supplied, but it was able to make few collections.

Major farm equipment such as tractors, power tillers, sprayers, and pumps are kept at ministry or provincial offices and are usually operated by their staff. One adviser reported that the Soviet Union wanted its tractors to be sold to individual *krom samaki*, but Khmer officials did not do so, apparently feeling the local groups were not ready for such responsibility.

Central and provincial officials purchase rice at numerous buying points. Farmers sell there or at local private markets as they wish. Payment is in cash. Current policies toward peasants have aided production; if the administration tried to consolidate control of agriculture, farm production would certainly fall.

Private Commerce

As in agriculture, a traveller sees little evidence of the national or provincial influence on economic activity beyond the provision of some public market facilities. Small traders, merchants, and artisans are active at numerous roadside markets, apparently under little regulation or control. Thriving markets in consumer goods from Thailand or Vietnam exist at major towns. Gold buyers are found at least at Sisophon and Phnom Penh buying the rings and necklaces that still come out of hiding. In Phnom Penh there are numerous money changers, who in mid-1982 were offering twenty-seven riels for a dollar although the official rate remained at four riels. There appears to be competition between private restaurants in the major cities and others run by the municipal or central authorities. Some control of the inflow of nonessential goods and the accumulation by small traders of wealth that they are not likely to reinvest in productive enterprise is surely needed, but it its doubtful that any government could do much until a better supply of necessary consumer goods is available.

Training Policies

At national and provincial levels there is reluctance to have more than minimal technical presence from the Western countries and an adamant refusal to receive training from them or from other nonsocialist countries. But if more-intensive agricultural practices are to be assisted by donors, the level of technical competence will need to be expanded rapidly. A rice expert suggests that, as a rule of thumb, a simple peasant agriculture can yield good results with modern rice varieties and as little as 100 kg of fertilizer per hectare. Above that level, a balanced package of inputs and appropriate training of farmers are essential. FAO's recommendations for Kampuchea already exceed 100 kg of fertilizer per ha.

The policies of some donor governments have precluded development projects, but the policies of the Heng Samrin administration would have restricted the productivity of high-technology projects anyway. But good results would be obtained from more irrigation dikes and dams together with a few more large pumping stations, animal-breeding projects, and health programs for pigs and poultry.

Restrictions on Project Monitoring

Kampuchean central and provincial administrators have usually been amenable to some project monitoring. UN and voluntary agency staff have travelled frequently to major rice-growing regions and veterinarians from donor countries have usually been able to carry out vaccinations as needed. But there have been frequent periods when certain areas have been closed to outsiders, because of "bad roads" and "bandits." Some high officials in Phnom Penh are usually available for consultation, but in some ministries, notably the Ministry of Commerce, visits have been hard to arrange. There has been an annoying reluctance to give access to the rice storage depots at Kompong Som. WFP officials have seldom been able to construct a fully satisfactory account of the stocks and distribution of relief foods, some of which goes to public servants as a supplement to salary. Enquiries among peasants indicate that a great many receive relief rice, but in small amounts. How much actually gets to the destitute is unknown. Problems of relief-food distribution strengthen the argument for helping the peasantry reach self-sufficiency.

Prospects

It is not a vain boast for the United Nations, donor countries, and voluntary agencies to say that with their help, Kampuchea has recovered from disaster. In the short run, barring bad harvests, the country should be able to limp along at a near-subsistence food level with modest continued external aid from socialist or Western sources. Until Kampuchea develops an export market, it will continue to need fertilizer, fuel, spare parts for tractors, vehicles, and other equipment, pesticides, veterinary supplies, and some technical help from abroad. In the longer run, the rates of growth of population and food production will be running a close race with an uncertain outcome.

Moving upward from subsistence will require more aid in more sophisticated forms. The future provision of technical assistance is a matter to be decided by the authorities in Phnom Penh and the potential donors. Whether further aid would strengthen a communist regime I will leave to the political analysts. Agricultural assistance thus far has made the peasantry materially more self-reliant. A sturdy peasantry will be an element in shaping the institutions that eventually emerge in Kampuchea.

Notes

1. Data from Committee for Coordination of Investigations of Lower Mekong Basin, *Annual Statistical Bulletin* (Bangkok, 1970).

2. George C. Hildebrand and Gareth Porter, *Cambodia: Starvation and Revolution* (New York: Monthly Review Press, 1976), pp. 69–71.

3. Ibid., p. 72. The authors tended to accept uncritically reports made by NUFK on its accomplishments.

4. U.S. Committee for Refugees, *Cambodian Refugees in Thailand* (Washington, 1982), pp. 2–5.

5. ICRC, "Back from the Brink," Geneva, October 1981, and other sources.

6. U. S. Department of State, "Humanitarian Aid to Kampuchea," *Gist* (September 1979), p. 1.

7. Interview with Chan Ven, mayor of Phnom Penh and minister of education.

8. Rice-growing areas subject to deep flooding are planted to varieties that grow very rapidly and produce stems that may be five meters long so that the heads remain above the water. These "floating rices" were historically about 15 percent of the rice area.

9. Main-season varieties are planted from mid-May through July and require from four to eight months to mature. They were historically about 75 percent of Kampuchean rice acreage.

10. On a part of the flood plain with suitable soils, rice is transplanted as the flood waters recede. The crop then depends almost entirely on water stored in the soil.

11. FAO, *Fourth World Food Survey* (Rome, 1977), app. C, p. 77.

12. The periodic UN team reports going to higher levels, usually entitled "Humanitarian Operations Arising Out of Developments in Kampuchea," contain a brief summary of information supplied by the "bilateral donors."

13. U.S. Department of State, Special Refugee Advisory Panel, "The Indochina Refugee Situation: Report to the Secretary of State" (Washington, D.C., August 1981), p. 16.

Appendix A

Principal Items Supplied for Agricultural Rehabilitation in Kampuchea, 1979–1982, by Agency

Item \ Agency	FAO	CARE	CWS	Lutheran World Relief	ARRK	AFSC	OXFAM	WCC	WVI	CIDSE	Other	Total
Seed												
Rice	66,538*	7,620	—	—	1,652	1,144	2,100	4,200	4,825	7,000	400**	95,479
Vegetables & other	73	—	—	—	3	—	1,431	1	—	731	—	2,239
Food												
Rice (mt)	—	640	—	—	—	—	—	—	—	1,500	—	2,140
Fertilizers (mt)	99,133	300	3,500	—	—	2,445	15,000	4,400	6,700	6,600	200**	138,258
Pesticides & raticides (mt)	646	—	25	—	—	2	95	1,108	84	177	—	2,137
Tractors & power tillers	1,325	—	—	6	—	—	—	—	76	—	—	1,407
4-wheel vehicles	39	—	—	—	—	—	53	—	—	8	—	100
Motorcycles, bicylces	199	—	—	—	—	—	231	—	—	—	—	430
Irrigation pumps	1,485	—	Survey equipment	Well-drilling equipment	300	58	2,002	—	35	6	21†	3,907
Sprayers	13,625	—	2.500	—	—	400	7,000	1,000	—	1,500	—	26,025
Earth-moving equipment	—	—	bulldozers	—	—	—	—	—	—	Bull-dozers, exca-vators, roller	—	—

Diesel fuel (mt)	—	—	60	—	—	—	216	—	44	240	3,100˚	3,660
Tool kits	20,000	—	—	—	1,000	—	—	—	100	—	—	21,100
Hoes, plow points, & sickles	115,000	1,000	10,000	—	35,000	—	313,450	—	—	110,100	—	584,550
Workshops	6	—	Shop truck & equipment	—	—	Survey equipment	6	—	some	—	—	—
Crop processing equipment	115 rice mills, 40 winnowers, 250,000 jute bags	—	—	—	—	—	Composting equipment, rice mill repairs, jute factory, 2 cotton gins repaired	—	—	Repairs to rice mills, rubber mill, and phosphate mill, equipment for rice seed storage	—	—
Fishing gear, motors, equipment, nets, lines	1 acre plot, several motors, shop equipment	—	—	—	Nets 6T.	Various	Various	Motors, nets	Motors, nets	—	—	—
Animal hlth. vaccines, doses	2.3 mill. Units	—	250,000 Units	Equipment for A.I., piggery, & poultry	—	Vaccines, Vet. supplies	Vaccines, cold boxes	Equipment	Vet supplies, mechanical feeds	—	15 refrigerators	—
cold boxes, misc.	2.1 2 generators, 115,000 syringes		Motor bikes, vet supplies									
Education Rehab. of schools and teaching supplies	—	Various	—	—	Various	—	Various	—	—	Tools, survey equipment	—	—

Source: Compiled by the author from tables in the reports of the Food Assessment Missions in Kampuchea, FAO, Rome, November 1980, February 1981, November 1981.

Notes: *Includes 11,038 mt seed acquired by Ministry of Agriculture using 16,559 mt food rice given by FAO for seed acquisition. **EEC. †Australian Freedom from Hunger Campaign. ˚Through July 1981, UNICEF.

Appendix B

Krom Samaki
(''Increasing Production Solidarity Groups'')

Introduction

Reasons for Establishment of Krom Samaki

To rehabilitate land devastated by war, to organize work forces to replace those killed, and to reconstitute the destroyed or lost means of production. *Krom samaki* should collect and assemble what remains of human and material resources in order to reestablish agricultural production by utilizing all available facilities to the maximum extent possible.

Size of Groups

Ten to fifteen families are considered a general average number of families per group. In mountainous or remote areas the number would be reduced to five to seven families.

In describing the functions of this new group, emphasis is laid on ''family economy'' and ''cooperative economy'' leading to ''cooperative production,'' and animated nationally and locally through the group.

Incentives

Krom samaki's overall aim is to create incentives for farmers to increase agricultural production. They must be assured of land and a house. An average area of land for one family is 1,500–2,000 square meters, with variations according to the population of the district. Farmers holding more land, however, should be allowed to retain it.

Each family in a group should have its own plot of about 1,500–2,000 square meters on which it is free to do the production and to dispose of the products. It may also be able to ''borrow'' more land from *krom samaki.* This will be the case mainly in the dry season so that they can transplant rice and grow vegetables. Following harvesting of the dry-season product the land will be returned to the group, but if the latter is unable to use and administer it, the families or groups may continue to work on it if they can do so profitably. The land operated under control of a group is about one ha per family.

Forest Clearance

Families and groups will also clear forest land where required for planting of crops for periods of three to five years or more. For such purpose the State will advance funds for equipment, animals, and materials, i.e., oxcarts, rice seed, and food for the families. This will be against guarantee that the borrowers of land and funds really will carry out their tasks effectively.

Trees

Fruit-bearing trees grown on family plots are the property of the family. On borrowed land the produce of the trees will be administered by the village or subdistrict, or handed over by these to the families to administer.

Animal Production and Fishery

Animal production and fishery resources should be developed by families. Agricultural waste should be used to feed fish and part of the family's income should be set aside for animal feed. The State will set up an "Animal Breeding Base" under which "animal breeding regions" and, within these, "animal breeding centers" will be formed.

Handicrafts

Traditional handicrafts are to be rehabilitated. These include silkworm breeding, cloth weaving (sarongs, mats), production of goods from bamboo and beans, plate making, tile making, carpentry, and ironwork. Locally available materials should be sold or exchanged to obtain the material needed to expand craftsmanship. The State will endeavor to provide such materials and will purchase the artisan products from the families or groups.

Technical Instruction

The *krom samaki* also have an important role to play in the technical instruction of family and group members.

Produce Sharing

The family keeps its own produce which it may consume, sell to the State, or sell on the free market. No taxes will be raised on "family economy" and no pressure is to be exerted on families to hand over any part of their produce to the group, at subdistrict or village level.

Instructions for *Krom Samaki*

Work Forces

A series of simple instructions to *krom samaki* aims at avoiding some of the pitfalls experienced in the trial period 1980. For instance, each group will divide members into two work forces, i.e., a main and an auxiliary force, the former consisting of fully able-bodied persons, and the latter of people able to carry out only lighter work. Remuneration will be based on a point system as will be also the hire of cows, oxen, or buffaloes. The actual application of the point system will be decided on by the families themselves at local level.

Seeds

Groups must retain 100 kg of floating rice, 70 kg of light or heavy rice, and 30–35 kg of corn per hectare for the next season and year. If possible, they should, however, try to retain an additional 10 percent of rice seed, i.e., a total of 110 kg per hectare. Emphasis is placed on the need for proper storage, and the dangers of worms, insects, and water are clearly indicated.

Monitoring

Groups will maintain lists of families' names, numbers of dependents, etc. Workers themselves will record days worked and calculate points due to them.

Harvest Estimates

Value and quality of the coming harvest is to be estimated by groups or families with a view to tentative sharing and distribution plans. After harvest, this estimate will serve as a guideline for actual sharing of the produce. Before general harvesting, specimen fields will be harvested to compare weight of rice before and after drying, thereby indicating the approximate percentage loss to be expected.

Source: "Declaration of Decision Number 2 38-8," People's Revolutionary Council of Kampuchea, summary drawn from a rough unofficial translation. FAO, Rome, Food Assessment Mission in Kampuchea, November 1980. August 30, 1980.

REFUGEES

Michael Vickery

REFUGEE POLITICS: THE KHMER CAMP SYSTEM IN THAILAND

When in late 1980 United Nations High Commissioner for Refugees (UNHCR) coordinator for Southeast Asia Zia Rizvi announced plans for a $14 million aid program for Cambodian refugees returning home, that is, aid money to be spent within Cambodia and which would attract people back from the Thai border, it was greeted with a lack of enthusiasm by some potential donors, who objected that UNHCR "was being politically used."[1] Squadron Leader Prasong Soonsiri, the newly appointed secretary general of the National Security Council of Thailand, in an interview with two Singapore journalists said it was a "UNHCR programme I don't want to be concerned with," in spite of the regularly expressed Thai view that the Cambodian refugees are "illegal immigrants" who should leave Thailand as soon as possible.[2]

A more controversial assertion that the UNHCR Cambodian refugee program was politically used was made by journalist John Pilger, who claimed that the refugee system was manipulated for the destabilization of Cambodia and Vietnam.[3] At various times Thailand and Vietnam have each claimed that the other was making political use of the refugees.

Even if one wished to dismiss those assertions as unfounded or exaggerated, the circumstance that within little more than half a year (June 1979-January 1980) Thai authorities pushed potential refugees back over the border, then opened up refugee camps ("holding centers") to all comers, only to close them when barely one-third of the expected number of arrivals had been reached, would indicate that political concerns rather than purely humanitarian considerations may have been paramount. This suspicion is further strengthened when we find that Thai officials have at different times both prohibited resettlement from the camps to third countries and complained that resettlement was proceeding too slowly, and have both threatened to force repatriation back across the border into Cambodia and blocked a UNHCR plan to achieve the same thing.

Thus whatever else the Cambodian refugee system is, it is also eminently political. Moreover, there are solid material reasons behind some of the politics.

From their beginning the Cambodian camps along the Thai border were cross-border trading centers. At about the same time as Rizvi was presenting his plan for aid to returning Cambodians it was estimated that "about 50 million baht (U.S. $2.5 million) of goods in the black market is daily transacted [with the Cambodians grouped along the border] and in the banks in Aranyaprathet [the main Thai town near the refugee camps], about 30–40 million baht of money is deposited every day . . . mostly from the black market operation."[4] Much of the profit deposited in the Thai banks originated with the gold and other valuables Cambodians brought with them to the border.[5]

It was revealed after two years that since October 1979, $350 million had been spent in Thailand in refugee relief efforts, and an equal amount had "served to bring the population in Kampuchea back from the verge of extinction,"[6] presumably via cross-border aid relief work within Cambodia, which would also have involved expenditure within Thailand. Clearly the Cambodian refugee operations have been big business with very profitable spin-offs.

It should be obvious that I take a critical view of the refugee operations, and in fact I shall argue that most of those people should never have become refugees; but it must be understood that in so doing I do not deny that those Cambodians who chose to become refugees were suffering to some degree—some in fact were severely malnourished and dangerously ill—or that they sincerely believed that they were making a legitimate choice, or that Cambodians and Cambodia deserved some form of international aid. One of the spin-offs of the refugee operations, however, has been a revelation that much of the rural Thai population normally lives in conditions approximating those the Cambodian refugees wished to escape.[7] If the same incentives and facilities were offered by the wealthy Western nations, tens or hundreds of thousands of Thais and probably also Filipinos, Indonesians, Mexicans, and Brazilians would also choose to become "refugees." They cannot because it would be politically and economically intolerable for the potential hosts, and because the special refugee facilities were offered to the Cambodians for reasons other than their perceived suffering. In any case, mass population movements are not the way to solve the problems of the world's poor, and if special political interests had not been at work, that would not have been an option for hundreds of thousands of Cambodians either.

The Cambodian Refugee System—the Volag Archipelago[8]

"So there you were ready to set up Khao-I-Dang and you had to go out and find people to fill it." The young UNHCR official to whom I addressed this remark one day in August 1980, after listening to his explanation of the genesis of the largest Khmer refugee camp, looked somewhat startled, but then concurred that my observation was not entirely inapt. And, on the following day, while I was talking to some refugees about the circumstances of their arrival at Khao-I-Dang (KID), one of them spontaneously came out with very nearly the same words.

The expression "refugee camp" conjures up images of temporary emergency shelter organized out of humanitarian motives to help people who are victims of some kind of natural catastrophe, war, or political persecution. Furthermore, to qualify under international law as a refugee from other than natural catastrophe, a person's situation must meet certain conditions of racial, religious, national, or political persecution, broadened in some countries to include "persecution by economic proscription"—denial of the right to work.[9]

The Khmer refugee camp system was ostensibly begun for those reasons, but it has become much more than that. At its greatest extent in mid-1980 it consisted of at least ten camps near or on the Thai-Cambodian border, led by the huge KID complex, plus several transit centers in Bangkok. The total population was over 200,000, and the very existence of such a system affected the internal politics of both Cambodia and Thailand as well as their relations with the rest of the world.

Since at least late 1979, in addition to its humanitarian work, the camp system has also, intentionally or not, functioned as a magnet drawing off tens of thousands of people who would otherwise have remained to work productively in Cambodia. Through this drain of personnel, plus the gold, cash, and other valuables which they bring, it has served to destabilize the already fragile Cambodian economy. Although the system is run jointly by the Thai government and UNHCR, most of the implementation of aid efforts and the actual work of running the programs day by day have been turned over to several dozen voluntary agencies (Volags) employing several hundred people from at least twenty, mostly Western, countries.

We may accept that the Volags are all genuine humanitarian organizations; but most of them are also bureaucracies with at least a skeleton staff of professionals who wish to expand the area of operation of their organizations wherever possible. They also have theories and projects they wish to test in operation, and the refugees provide an excellent captive population among whom to work. In saying this I do not mean to denigrate either the Volag personnel or their projects, most of which in themselves are laudable. Nevertheless, instead of temporary emergency shelters the Khmer refugee system and its Volag armature has grown into a network of communities increasingly permanent in nature and increasingly open to political exploitation, with the potential for "creating a Palestinian factor," as was noted at the very beginning of KID's existence.[10]

The Volag archipelago in 1980 consisted, at a minimum, of the following camps:

Khao-I-Dang, by far the largest camp with a July 1980 population of 136,000. It is situated on a barren plain below a long, low-lying mountain of the same name 30 km north of the town of Aranyaprathet and about 15 km from the border. It is officially a "holding center" for illegal entrants to Thailand pending their resettlement in third countries or return to Cambodia. It is run by the

Supreme Command of the Royal Thai Armed Forces and the UNHCR with the various services contracted out to Volags.

Nong Samet (also "New Camp," "007," "Rithisen") and Nong Chan, about 30 km north of Aranyaprathet in a poorly demarcated zone that straddles the border and may even be mostly on the Cambodian side. The conventional names of the camps are those of the adjacent Thai villages. Their populations have fluctuated between 40,000 and 100,000, with the higher figures reported in early 1980. They are run internally by Khmer military and civilian committees. Access to them from the west is controlled by units of the Thai Supreme Command, who also exercise a large measure of control over the Khmer administration. An international presence has been maintained by UNICEF, the International Committee of the Red Cross, and various Volags, who provide medical care, supplementary food, and educational and social services.

The "old" Aranyaprathet camp located about 5 km north of the town. This was set up before 1979 for Khmer fleeing Democratic Kampuchea, and most of its inhabitants had come out during that earlier period.

Sakeo I and II (the latter also known as Ban Kaeng, from a nearby village), about 50 km westward from the border and the town of Aranyaprathet near the district seat of Sakeo. Sakeo I was set up in October 1979 for Democratic Kampuchea (DK) military, supporters, and camp followers, and in July 1980 it had a population of about 25,000. These were moved to a new site (Sakeo II) nearby, and the sociological composition of the population was altered through transfers between Sakeo and KID. Like KID, Sakeo is run by the Thai Supreme Command and UNHCR with services provided by Volags.

Kamput in Chanthaburi province and Mairut (closed in 1981) in Trat, two smaller holding centers under the Surpreme Command and UNHCR, originally set up to receive mostly DK refugees who crossed the border in those areas.

Surin and Buriram holding centers, dating from the pre-1979 period.

A new processing and transit center at Phanatnikom in Chonburi province for people chosen to be resettled in other countries.

A newly constructed holding center at Kap Choeung, north of KID.

I should probably also include the DK fortified camps just across the Cambodian border in Phnom Chat and Phnom Malai, since rice, water, and medical aid have regularly been supplied to them by one or another of the Volags or the international agencies.

Since 1980 the number of centers within Thailand and their population have decreased. Kap Choeung, Buriram, Mairut, Surin, and the old Aranyaprathet camps were closed during 1981 and early 1982, and by the middle of 1982 KID was down to around 40,000 inhabitants, while Sakeo II, due for closure in 1982, had 30,000. On the border, however, although Nong Samet and Nong Chan were down to about 44,000 each, at least twenty camp locations were recognized with a total population of about 300,000. Since many are mainly military camps with little relevance for refugee operations, I will not discuss them here. At Kamput most of the early DK residents were transferred to KID and a new resettlement processing center was established which was still active at the end of 1982.[11]

The comparison with another famous "archipelago" has been drawn because once an inmate had entered he was no longer a free agent. He could not just turn around, walk out the gate, and return whence he came. This is still true. His life may be circumscribed by authorities—the Thai military and the Volags—whose purposes he finds strange and whose ideas of discipline and organization may seem vaguely and disconcertingly reminiscent of what he wished to escape in Cambodia.[12] Opportunities to earn his own living may be arbitrarily limited. At twenty-four hours' notice he may be transported, in buses fictitiously marked as though for ordinary public transportation, from one island of the archipelago to another without being told his destination in advance—a circumstance which given the purpose of the first such convoy in June 1979 is always somewhat terrifying.[13]

Normal channels of communication with the outside world (postal service, telephone), where he may have friends or relatives, are circumscribed if not totally blocked, and the official channels open to him do not inspire trust. Finally, his term of assigned residence is indefinite, to be decided by people he does not know and may not trust.

Important differences from the other "archipelago" are, first, that entrance is usually a free choice—one can at that point say "no." Also, the administration, even if dictatorial, is generally benign: one is adequately fed, is not required to work, is offered better medical care than one has known for at least five years, perhaps ever, and efforts, albeit slow, are made to locate and reunite family members. Finally, if one falls into a small minority—in 1980 perhaps somewhere between one-fifth and one-tenth of the total (now a larger proportion but a smaller number)—who fulfill certain special conditions, he will eventually be taken out to a new life in one of those Western paradises about which he has always dreamed. Pending that, however, and for the other four-fifths to nine-tenths of the original refugee population for an unforseeably distant future, one will most likely remain a prisoner, although a well-treated one, among the Volags. If one tries to just walk out, the benign treatment suddenly ends. At KID, at least, one would probably be shot on sight by the Thai military guards. In spite of the expressed Thai desire to rid themselves of refugees, such unorganized departures are not permitted. It is possible to be granted permission to return, under escort, to the Cambodian border; and in fact permission has always been granted to the few who have made that choice. But they were special cases: cadres of the DK forces who wished to return to fight the Vietnamese, or a few thousand ordinary peasants who realized they had no future in the camp system, or, after Sihanouk's visit to KID in July 1982, several more thousand who in the first flush of enthusiasm volunteered for a move to the Sihanoukist base of O Smach in northern Cambodia. It always seemed clear to those working in the holding centers that permission would not be so easily granted if suddenly 20,000 of the educated bourgeois, or even peasants, of KID asked to go home to live under the present Phnom Penh government, and since 1981 such doubts have proven accurate.[14]

Genesis of the System

It would not be fair to say that the journalistic coverage of the Thai-Cambodian border and the refugees in 1979–80 was deliberately obfuscatory; but I nevertheless found, on arriving in the Aranyaprathet area in April 1980, that even with a good deal of knowledge of Cambodian problems and previous trips to the then existing refugee camps in 1975 and 1977, I had been misled by press accounts of the situation. An accurate picture must be even less accessible for nonspecialists.

Cambodians began fleeing their country even before the end of the war on April 17, 1975, and the first refugees proper crossed the Thai border the next day, mostly in the Aranyaprathet and Pailin areas. A camp was set up for them in Aranyaprathet, not far from the center of town. Although the Thai government has never signed the international convention on refugees, it was recognized that escape from Democratic Kampuchea could be well motivated and refugee treatment, at least temporary, was accorded. At that time the refugees in Aranyaprathet were free to leave the camp to move around town, visit acquaintances, and find work if possible. Most of them were more or less well-educated town dwellers with contacts, friends, or relatives abroad, and most eventually made their way to Western countries, principally France and the United States. By 1977 that first camp had been superseded by another situated five miles outside town. The refugees were no longer allowed to move around freely, and their condition was more like that of prisoners. Up to January 1979 the total number of Khmer who had fled to Thailand was about 35,000.[15]

With the destruction of Democratic Kampuchea in early 1979 and the ensuing freedom of movement, many people began moving toward the border. Just like the refugees of the 1975–79 period, this new movement involved mostly former urban residents who rejected peasant life and sought a way of life like the one they had known before April 1975. Unlike the pre-1979 movement, these new "refugees" were not fleeing from political oppression, which, for them, had ended with the destruction of the DK administration in their districts. It cannot be too strongly emphasized that the only people in danger of political persecution in 1979 were DK cadres and combatants. Members of the former educated class, or the urban bourgeoisie, could return home and were offered integration into the new administration, thus a privileged status; if they did not want that, they were free to try to support themselves by petty trading. Neither were they, at least in the first half of 1979 and often longer, fleeing from starvation, since the stocks of rice left by the old regime together with the rice in the fields ready for harvest meant that, for several months, there was adequate food in most parts of the country *for those who stayed in place.*

Thus the vast majority of those who were to become refugees in 1979 and

1980 fit neither the 1951 Convention definition of a refugee nor even the expanded category of refugee by economic proscription.

The principal reasons for the new movement were to make contact with the outside world for the purpose of going abroad or contacting friends and relatives already abroad; to trade across the border for commercial purposes; and to join, or organize, one of the paramilitary or bandit groups loosely called Khmer Serei—"Free Khmer."[16]

The first people who tried to go abroad, or even to contact relatives, were mostly from the former wealthy, well-educated groups who had had some earlier experience abroad and who spoke French or English. In the beginning, when they were few in number, it was relatively easy, particularly if they still had some currency or gold, to cross the border, contact a foreign embassy, and get out to some other country. They would then write back to family and friends in Cambodia about the ease with which they had managed their departure, thus encouraging more and more to attempt it.[17] However, as numbers increased, so did the Thai border controls, and such immediate departure became virtually impossible.

Many more people came to trade. Few had been peasants before 1975, and they considered petty commerce both higher in status and more remunerative than farming. They came to the border with currency, jewels, gold, or other valuable objects hidden since 1975 and bought Thai products to take back and sell at a profit to finance another journey. Throughout 1979 there was a constant procession of thousands or tens of thousands of such people on the road from Battambang and Siem Reap to the border. Some, having started as border traders, then decided to attempt emigration, which might involve several months waiting at the border for the right occasion to cross. Others decided to remain at the border as middlemen in the growing volume of trade, or they joined a Khmer Serei organization, which also lived off the trade, and plotted the reconquest of Cambodia.[18]

The third main group of border arrivals were the "politicals," again mostly former urbanites or military men who had been victimized by the DK regime, but who were equally opposed to its successor on grounds of its socialism and dependence on Vietnam. These people wanted the restoration of a system like that of Sihanouk's Sangkum or Lon Nol's Republic, and to a greater or lesser extent they were willing to fight for that goal—in contrast to people who had given up on Cambodia and thought only of going abroad.

They came to the border to organize their resistance both because it was impossible to do so within Cambodia and because they hoped for external aid, in particular from the United States and Thailand. Any doubts about the reality of such aid may be dispelled by noting that the rehabilitation from scratch of the three resistance forces that have now formed a coalition could not have occurred without foreign aid, which could only have been transported through Thailand. Much of their military supplies are from China, while food and medicine have been channeled through "refugee" supplies.

In spite of their proclaimed goals, most of the Khmer Serei in 1979–80 seemed less rather than more eager to fight, could not in any case agree on leaders or organization, and found their true vocation in the control of cross-border trade and refugee traffic—activities in which most of them degenerated to the level of bandits and racketeers.[19]

The places along the border to which these people came were clandestine border-crossing points known to smugglers, bandits, and various "politicals" long before 1979, or even 1975. The original Khmer Serei, some of whose members are now with Son Sann's KPNLF, had operated along the border in opposition to Prince Sihanouk in the 1950s and '60s; before them, Issaraks ("freedom fighters") had used the same forest clearings and border trails in the 1940s. Then, just as at present, they were hoping for Thai aid against a government in Phnom Penh, that of the French. After 1975 there were still "Khmer Serei operating from bases inside Thailand"[20] and a lively cross-border trade between Thai merchants and representatives of the new Cambodian authorities, which on one occasion led to a murderous incident very close to the location of the present agglomerations.[21] One of the 1979 Khmer Serei leaders was reported to have been a teak smuggler based at Phnom Malai throughout the DK period.[22] The first people who came in 1979 knew, or could easily find out, the best border points for their purposes.

Several of those border points gradually turned into large camps during and since 1979. Often, because they were founded on trade, they are located opposite Thai villages that have given their names to the refugee agglomerations. The first, in terms of initial importance in 1979 (but which was largely destroyed in early 1980) was opposite the village of Non Mak Mun. Five miles to the north is the "new camp," opposite the village of Nong Samet; and three miles to the south, near the village of the same name, is the Nong Chan camp. By 1982 another center which had become very important, both politically and in terms of Volag involvement, was Ban Sangae, a KPNLF military and civilian base across the border from the Thai village of that name north of KID. Equivalent DK centers were Nong Prue and Tap Prik, on the border south of Aranyaprathet.

Since the refugee operations have straddled the border and have been so closely interwoven with border diplomacy, it is necessary to devote some attention to peculiarities of the Thai-Cambodian border north of Aranyaprathet. Along that section of the border the line that appears on modern maps does not correspond to specifications laid down in the 1970 treaty between Thailand and France, and on the ground it is impossible to determine precisely where either line should be. Existing border markers are far apart and may have been surreptitiously moved during the past several years. Thus no one knows precisely where the border is, and driving into border camps, refugee operation officials and embassy personnel like to joke that "somewhere here we are crossing into Cambodia, but we don't know where."

In practice, since the beginning of the refugee problems, the Thai authorities have always, with one exception, treated the border agglomerations as being either in Cambodian territory or in an imprecise border-straddling zone that they did not yet wish to dispute. Thus in a Thai document given to the UNHCR and dated February 1–15, 1980, KID is described in the following terms: "This center is prepared for Kampucheans expected to flee into Thailand after Vietnamese attack to [sic] Khmer Serei at Ban Non Mak Mun, Nong Samet and Nong Chan."[23] This implicitly places the three camps in Cambodia. This document was accompanied in the file by a map of "Concentration of Kampucheans," dated December 25, 1979, which showed a number of "refugee" camps along the border, including Nong Samet, Non Mak Mun, and Nong Chan, designated by arrows as across the border in Cambodia.

Even after the June 23–24, 1980, Vietnamese "incursion," when the location of the border was a more sensitive issue, local news reports spoke of "two Khmer refugee encampments straddling the border at Nong Chan and . . . opposite this village (Mak Mun)."[24] The same paper also published maps, without any subsequent rectification from the government, showing Nong Samet, Mak Mun, and Nong Chan right on the border, and quoted a spokesman of the Supreme Command as very delicately announcing that "foreign forces"—mostly Vietnamese—"had attacked a Khmer Serei unit at border mark no. 44 *near* [my emphasis] the Thai border. . . [and] the fighting *spilled over* [my emphasis] . . . into the Thai village of Non Mak Mun."[25]

Three weeks later there was a coup among the Khmer Serei and the DK forces centering on Nong Samet, which, of the big border camps, stands the best chance of really being within Thai territory. Even then, Thai sources were extremely careful. They spoke of fighting "near Camp 007 [Nong Samet] opposite the Thai village of Nong Samet," with the result that "thousands of Kampucheans fled screaming into Thailand" coming from "120,000 Khmer civilians at the border encampment." It was reported that "Thai military forces were dispatched to the area to *prevent a spillover of the fighting onto Thai territory*" (my emphasis). Furthermore, "Camp 007 was pounded with Vietnamese artillery shells [and] . . . Thai forces . . . saw. . . huge flames . . . from the direction of the Kampuchean encampment."[26] Two days later the same newspaper reported a pushback of refugees who fled into Thailand from the fighting at "Camp 007, straddling the Thai-Kampuchean border."[27]

The Thai position, then, except for propaganda following the June 24 "incursion," which went far beyond the first official statement, was to treat the three large border camps as Cambodian and their inhabitants not only as nonrefugees, which legally they really were, but not even as "illegal immigrants" confined to the holding center of KID. They were Cambodian, under one or more Cambodian administrations, and subject to Thai control only to the extent that border security demanded it.

This attitude toward the border is apparent in subsequent statements. In a retrospective comment the *CCSDPT (Committee for Coordination of Services to*

Encampments Along the Thai-Kampuchean Border

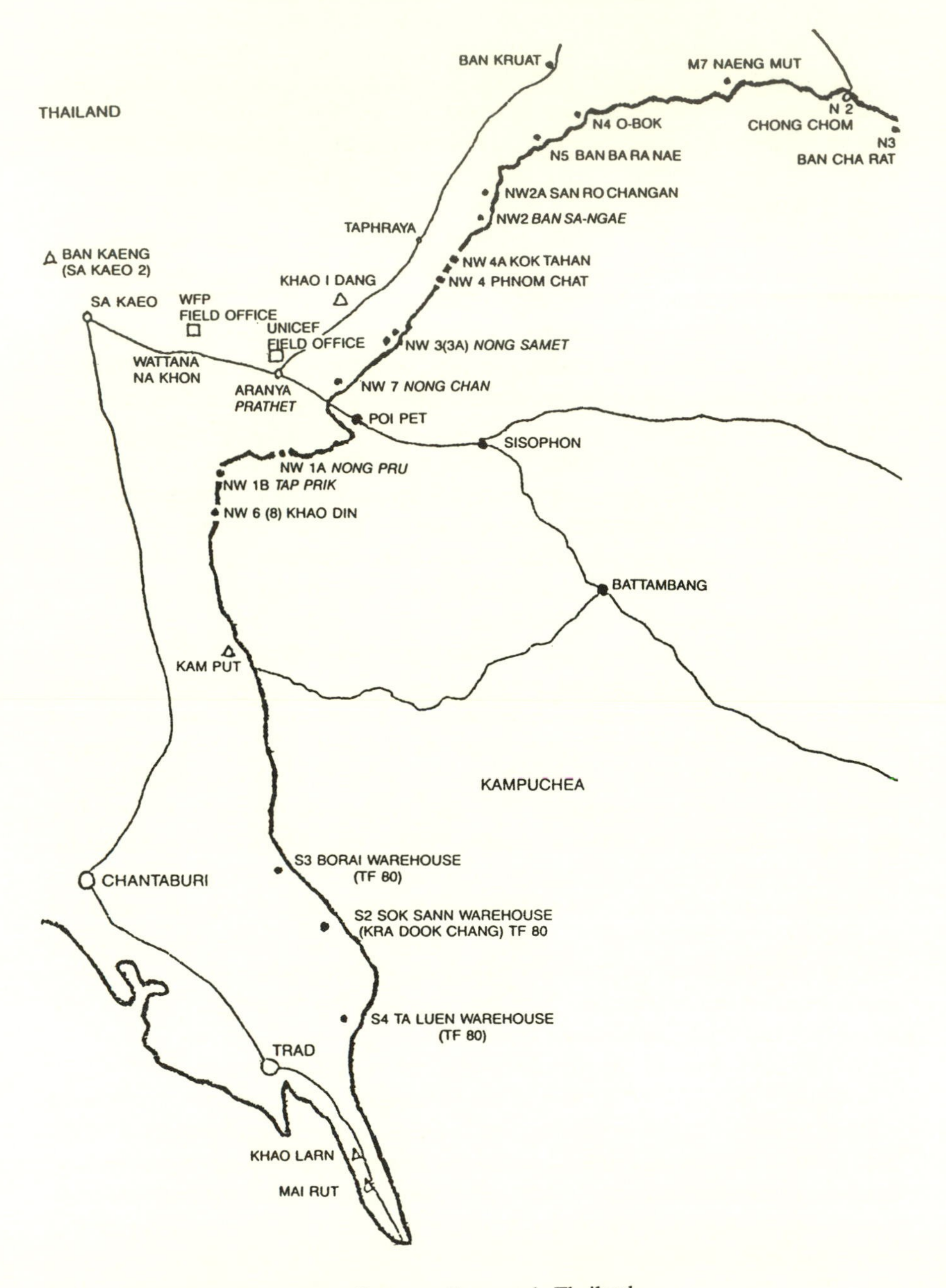

Source: Joint Operations Center, Supreme Command, Thailand.

Displaced Persons in Thailand) Handbook said that "as an additional consequence of the Vietnamese incursion, altogether more than 50,000 Kampucheans spilled over the border into Thailand," presumably from Cambodia. This could only refer to the people who on that occasion moved out of the Nong Chan camp toward the west.[28] The same source also stated that the border encampments "are generally connected with the main Khmer political and military groups in the area," and that the largest, "Ban Sangae, Nong Chan, Nong Samet, . . . are located mostly in Kampuchea."[29]

Thailand's foreign minister, perhaps unintentionally, seems to have gone even further. In July 1982 he was reported to have "ridiculed a statement by Hun Sen [Cambodian Foreign Minister] . . . denied a charge by Hun Sen that Thailand was providing bases for the Khmer Rouge and other resistance forces." He said it was untrue that Thailand was providing shelter for Khmer Rouge guerillas, and he also scoffed at a Phnom Penh claim that the new tripartite coalition has no control over Cambodian territory. "I think they control a sizeable area," he said.[30] Implicitly, then, the bases those Cambodian resistance forces undeniably possess, which include Ban Sangae, Nong Samet, and Nong Chan, are to be considered in Cambodian territory.

The Cambodian inhabitants of those border areas have always held that they were in Cambodia, and most of them had no intention of crossing the Thai border to become refugees. They were there, as indicated above, to engage in Cambodian politics and trade. Those camps had always attracted anticommunist fugitives, and by June 1979 there were, in addition to the "Khmer Serei" nucleus, over 40,000 unorganized, non-DK people massed along the border north of Aranyaprathet near Non Mak Mun, Nong Samet, and Nong Chan.

Remnants of the DK forces were also arriving at the border. Near the end of April 1979, attention was drawn to a group of 50–80,000 DK soldiers and accompanying civilians who were allowed to cross Thai territory to avoid attacking Vietnamese and reenter Cambodia at another point.[31]

Although most of the noncommunists, and nearly all of the DK soldiers, were content to remain on the Cambodian side of the border, increasing numbers of people were hoping to cross into Thailand and proceed to other countries. Unlike the first few hundred who had earlier succeeded in such plans, there was no possibility of the new, large numbers being accepted abroad. The Thai, furthermore, did not consider people who came over after January 7, 1979, the date the Salvation Front (SF) and Vietnamese forces captured Phnom Penh, as genuine refugees. They were "displaced persons" on the Cambodian side of the border, and "illegal immigrants" on the Thai side. The people nevertheless kept coming, and the Thai government, possibly intent at that time to remain neutral in the Indochina conflicts, feared embarrassment or danger from refugee politics and financial difficulties if stuck with their support.

In the third week of June 1979, about 42,000 of the non-DK Khmer north of Aranyaprathet were loaded onto buses on the Thai side and taken on a long journey around the northern borders of the two countries to a point south of Srisaket and forced down narrow mountain trails in the Preah Vihear area, sometimes across minefields, back into Cambodia. Perhaps thousands died.[32] Some of the survivors were indeed discouraged and decided to make their peace with the new government. Many others, perhaps most, drifted back to the border and can be found again today in all the refugee camps.

The Thai move was effective in drawing attention to the problem. There was a wave of international protest, and some pressure was exerted on Thailand to institute more humane policies.[33] Their action, if planned as a measure to rid themselves of the pseudorefugee problem at the border, proved in the end to have been counterproductive. It called attention to the Khmer massed along the border north of Aranyaprathet and eventually resulted in programs that would attract even more of them, bringing them across the border semi-permanently. It is arguable that had Thailand left these people alone, tolerating the trading that had supported most of them and taking severe action only against the few who might have tried to force their way into Thailand, the stated goal of discouraging the refugee exodus might have been readily achieved.

It was suggested at the time that another 30,000 Khmer in Chanthaburi and Trat provinces, who may have been part of the group that had crossed Thai territory just south of Aranyaprathet in April,[34] might get the same treatment, but they were DK forces and the different treatment accorded the two groups is perhaps a clue to certain unexpressed goals of Thai policy. One Thai official was quoted to the effect that although the world was accusing Thailand of lack of humanitarian feeling, ''when we help them, they say we're not neutral,'' and Bangkok is accused of helping Pol Pot ''merely because refugees had been permitted to enter.'' This disingenuous explanation ignored the differences between the two groups. Those sent back were all anti-Democratic Kampuchea, as everyone well knew, whereas those whose asylum might conceivably help Democratic Kampuchea were given special treatment. ''Thai officials were never able to explain satisfactorily why they had not chosen instead (for expulsion) the 40,000 Khmer Rouge soldiers and civilian suppliers who had taken refuge further south in Chanthaburi and Trat provinces,'' instead of the unorganized and helpless civilians north of Aranyaprathet.[35]

Moreover, a position of neutrality, under international law, would have demanded the disarming of those DK military who had entered Thai territory and their internment for the duration of the war. Thus the Vietnamese and Heng Samrin accusations of Thai complicity in DK operations were not without foundation.

The enforced return to Cambodia of the 42,000 coincided with increasing discussion of conditions within Cambodia and the need for aid to all Cambodians, whether on the border or in the interior. There were numerous reports of countrywide starvation and epidemics. The U.S. State Department, whose analysts had not considered that Cambodia was in a crisis, was pressured by ''American charities and their own embassy staff in Bangkok'' to change its views, even though it had ''serious second thoughts about the [embassy's] data.''[36] The genesis of this change in U.S. attitude is interesting in view of later information suggesting the crisis was exaggerated. Whatever the true situation, one of the stumbling blocks to increased aid was the question of whether it should be delivered directly to Phnom Penh or pushed across the Thai border. Another, related to the first, was the question of aiding both sides. In general the United States and Thailand favored a cross-border operation and no overt political discrimination; the Swedes were also pressing to send aid across the northern border to the 42,000 believed stranded in northern Cambodia.[37] Phnom Penh insisted that all aid should go through its hands and none to the Pol Pot remnants near the border. In the end aid went both to Phnom Penh, by plane or through the port of Kompong Som, and across the north and south of Aranyaprathet, where the Thai continued to supply DK remnants as they had done in the past,[38] and where the international and voluntary organizations gave help both to the DK groups and to the anticommunist Khmer Serei camps, from which food was then transported inland to the northwestern provinces.

While attention has been fixed primarily on conditions within Cambodia and on the anticommunist refugees north of Aranyaprathet, an entirely different group of people was slowly proceeding toward the border. Their appearance, beginning in September 1979, was to be the catalyst for a new system of refugee organization.

When the Salvation Front/Vietnamese forces invaded Cambodia in early 1979 and defeat for the DK regime was imminent, most of the DK military and political forces, together with as many ordinary people as they could gather, withdrew gradually from the towns and rice plains into the forests and mountains of western and northwesten Cambodia. From there, taking up to seven months, they moved slowly away from the attacking Vietnamese toward the Thai border, where some of their comrades had already appeared in April. During this long march through inhospitable, malaria-ridden country, with dwindling food supplies and no medicine, they were also wracked by internal tensions left over from the factional disputes and purges of the Pol Pot years.[39]

By the time they erupted onto the Thai border—not at the points discussed previously, but to the south of Aranyaprathet—they were in the pitiful condition shown to the world by the press in September and October 1979, dying by the scores from illness and hunger. Other groups of these DK refugees also crossed into the southeast in Chanthaburi and Trat, where there had already been a large border concentration at Ban Laem as early as July. But the group that attracted the most attention was that which came out about 20 km south of Aranyaprathet. The pity their condition aroused caused people to forget or ignore their leaders' political past, and emergency aid was rushed in from all quarters.[40]

In mid-September Thai officials led by Air Marshal Siddhi Sawetsila, then secretary general of the National Security Council, visited the border where the new exodus was taking place and announced that 60,000 people were massing there, moving closer to escape Vietnamese pressure. In early October they crossed. Under this pressure Thai policy gradually changed. Thailand accepted that aid must be given, but not by Thailand alone. Help was requested from Western countries and international agencies on condition that aid going through Thailand must be nonpolitical and must go to all sides of the Cambodian conflict.[41]

By the end of October, Prime Minister Kriangsak had announced an open-door policy "allowing all Khmer refugees who wished to come to Thailand to do so."[42] This was not meant to be a change of strategy; "there must be some people alive in order to oppose the Vietnamese in Kampuchea . . . it will just take longer," meaning apparently longer than the earlier tactic of forcing everyone back into the country as soon as they reached the border.

This statement at least demonstrated that any observer who had seen covert objectives, beyond concerns for Thai security, in earlier Thai policy toward the refugees was not entirely wrong. The covert objective was now clearly to use the refugee situation to influence future political developments within Cambodia.

About 30,000 of the newly arrived DK refugees were settled in a camp near the town of Sakeo, about 50 km from the border,[43] and another large group established itself on and around the fortified base of Phnom Malai, an old Khmer Serei hideout just inside the Cambodian border about 20 km south of Aranyaprathet. Still a third group of these DK remnants set up a base at Phnom Chhat, inside Cambodia north of Nong Samet, and in the southeast the camps at Kamput and Mairut welcomed those who crossed over in that region.

The focus of the press on these DK refugees had several important effects. First, international attention was directed to the Cambodian refugees and relief efforts were intensified. Second, the Thai government reversed its policy and agreed to open its borders and establish "holding centers" to care for the refugees until such time as they could either return home or go on to "third countries." Third, a belief grew both abroad and in Thailand that *all* Cambodian refugees were in the same pitiful shape and that they were fleeing starvation at home. Finally, this supposed evidence of administrative failure served as propaganda ammunition against the SF regime and the Vietnamese efforts to support it.

From few press accounts would the reader have been able to distinguish between these DK refugees and the "refugees" of the Mak Mun, Nong Chan, and Nong Samet camps who were an entirely different group of people, were at the border for different reasons, had not been victims of persecution since the end of the DK regime several months earlier, and were only in rare cases near starvation.

Some hints of the different situation did come through from a close reading of the *Far Eastern Economic Review* (*FEER*), which reported accurately that north of Aranyaprathet conditions were different. At the Nong Samet camp, whose estimated population of 80,000 was believed to be the largest concentration of Cambodians in the world, "most people [were] in relatively good health"—in fact there were attempts to conceal the quantities of food in the camp—and a brisk trade across the border into Cambodia was observed.[44]

The journalists nevertheless believed things were changing. Newly arrived refugees reported tighter travel restrictions, and three had stories of Vietnamese firing on people to keep them from reaching the border.[45] Thus the reports of increasing starvation within Cambodia and the new rumors of Vietnamese brutality served to convince outside observers that the refugee exodus would increase at all points and that they would all be in increasingly poor physical condition. It should have been recognized as significant, though—and the significance increases in the light of what happened in November and December—that even when Prime Minister Kriangsak opened the door in October, there was no large-scale movement of the 80,000 people at Nong Samet or the other thousands in Mak Mun and Nong Chan, to take advantage of it.

With the door open and a massive exodus expected, a place had to be prepared to receive them. There was already a plan for one huge holding center for 200,000–300,000 at Mairut in Trat province and four or five smaller centers elsewhere. Then, in late October or early November, it was decided to build the large center at Khao-I-Dang and leave Mairut as one of the smaller camps.[46] All the reasons for the change may not have been made explicit. Officially it was lack of water at Mairut, but the KID site, as indeed nearly all of the holding centers, suffers from the same defect. Mairut, however, was in the area of the first large exodus of DK refugees, and the originally planned giant camp would have served as a relief and rehabilitation center for them; but by September or October it was clear that the main anti-Phnom Penh operations were going to be farther north near Aranyaprathet.[47]

More important, probably, was a belief that vast numbers of new refugees were being pushed out of the Northwest by famine within Cambodia and Vietnamese harassment. The *FEER* wrote on November 16, 1979, that 180,000 people had already crossed the border north of Aranyaprathet and cited diplomatic sources as placing another 130–150,000 within striking distance. It added that within the next two months Thailand could receive up to 750,000 people. By November 30, the estimate was 600,000 on the border, meaning that 14 percent of Cambodia's reputed 4 million surviving population was either in Thailand or ready to enter. By December 7, as the ''plight of the Khmers daily grows more desperate,'' ''the survival of the Khmer race [might] depend on the exodus into Thailand.'' It was foreseen that in the coming months a million Khmer, or up to a quarter of the estimated population, could be under Thai control.[48] In these circumstances, Thailand, which would not consider accepting 40,000 in June, now agreed to take several hundred thousand. Certain more astute observers did not fail to note the political advantages that might accrue to Bangkok from the effective control of such a large part of the total Cambodian population.

As a result of the new developments and predictions, the Thai Supreme Command chose Khao-I-Dang as the site for the new major holding center and gave UNHCR the green light to set it up in the expectation that 300,000 or so miserable Khmer would rush across to settle there. On November 21, 1979, the first small team of UNHCR officials waited on the bleak landscape for the buses and trucks sent out to bring the people in.[49]

To their astonishment, in the first week after the opening of KID, only 28,000 people took the opportunity to enter, and they were in fairly good condition. Many of these had cash or gold and hoped to set up businesses in the new campsite. In the second and third weeks 16,500 and 29,800 respectively arrived; the numbers then dropped to under 4,000 for each of the following three weeks. In the seventh week, the first week of January 1980, the total jumped again to just over 21,000, because of fighting among Khmer Serei factions in the border camps; but immediately afterward it fell to 2,800 for the eighth week, under 2,000 in the ninth week, and then fewer than 1,000 per week. On January 24, when the total population was about 111,000, just over one-third of what had been expected, Thai authorities ordered KID closed to further entry.

It appeared that the UNHCR might have been misled. Although there may have been half a million people at the border, the number prepared to become refugees was only a fraction of what had been estimated, and most were hardly in circumstances justifying refugee treatment. Indeed, many of those who did come required persuasion, or they came to KID like the mountain climber, ''because it was there.'' Otherwise they would have continued to trade between the border and the interior, and as conditions at home improved, gradually returned.[50]

In view of Thai complaints about the political and economic dangers posed by the refugees, and their frequently restated view that the remaining refugees should all return to Cambodia, the reader may find a suggestion that people were persuaded to become refugees scarcely credible, yet something of the sort did indeed happen and is one of the reasons for the immense growth of the Volag archipelago.

The evidence for persuasion also supports the allegations that the refugee exodus was in part an artificially created movement, designed to destabilize Cambodian society by drawing off people—administrators, technicians, doctors, teachers—who are needed in Cambodia and who would not have left if the refugee centers had not been created.

Discussion of this should begin with the fact that neither Thailand nor the United States (and the U.S. embassy in Bangkok played a very active part in the development of the refugee system) approved of the change of government in Cambodia. DK brutality, as such, had never unduly disturbed the Thai government, and as *FEER* wrote, ''although no one in Bangkok was willing to say it . . . Thais would prefer a savage but independent Cambodia to a more humane regime under the thumb of Hanoi.''[51] By 1977 probably, and at least by 1978, some U.S. analysts shared the same view, realizing that Democratic Kampuchea did not represent what they had earlier feared (a socialist regime attractive to workers and peasants of Thailand or Malaysia) and could be used profitably against Vietnam.[52]

All considerations concerning the refugee situation, beyond the immediate problem of Thai security, were directed toward the weakening of the PRK regime, even if it meant the return of Pol Pot. An exodus of refugees seems at first prima facie evidence against any regime, and the United States had long encouraged people to vote with their feet against communism. Forcing refugees to return to Cambodia would, on the contrary, indicate support of, or indifference to, the new regime, and would at least convince potential refugees that they had no choice but to make peace with their government and work for it. Since many of the refugees were people who could be put usefully to work in the new Salvation Front administration, their flight could weaken a regime that both the Thai and U.S. governments opposed.

What had to be devised was a system in which the flow of refugees from Cambodia could be maintained, in order to make the government look bad and to deprive it of needed skills, and yet which would not be a burden on the host country of Thailand. The solution would be an open-door policy funneling people into a huge camp, or holding center, run by UNHCR and the Volags, where people could come out of Cambodia yet be isolated from Thai society. The change in location of this planned center from Trat to KID reflects a realization that most of the people who wanted to leave Cambodia were coming out in that area, and that they were not ''politicals''; they were irrevocably opposed to Pol Pot and also against the SF regime, however benign it might be. The difference of course was that if they were not allowed to leave Cambodia they *could* work for the SF regime, which wanted them, and strengthen it with their abilities.

As for the politicals, DK or Khmer Serei, they could be kept on the border, where they wanted to stay anyway, and supplied overtly or clandestinely as part of the general refugee relief—thus the U.S. and Thai insistence on cross-border aid.

In the end, in spite of the predictions about numbers up to one-quarter of the Cambodian population massed to rush over into Thailand, the most difficult problem proved to be to persuade enough people to become refugees to create an impressive holding center.

During visits to the Khmer Serei border camps between July and September 1979, international aid officials were surprised to find that few of the Khmer were interested in coming over to Thailand. They were already rather well supplied with food, and the main type of aid they requested was arms.[53] The same thing occurred when the DK refugees appeared in September 1979, south of Aranyaprathet. Even though they were in much worse condition, their first interest was not asylum, and when transportation arrived to take them to holding centers many, perhaps half or more, just disappeared back into the forest.

A suggestion was then made by one of the responsible aid officials to take the needed medical and food aid right to the border; but the UNHCR rejected it because people in the ill-defined border zone were not technically refugees and therefore UNHCR by the terms of its charter could not help them. Thus refugees had to be created out of people who were still only displaced persons in their own country.

When the Sakeo camp was first established for the DK refugees, those who went were persuaded with information that a camp had already been built and the sick would be in greater comfort and have access to better facilities for treatment. In fact, nothing had been built and they were deposited on bare ground in the rainy season. It is thus arguable that part of the terrible death tolls at Sakeo during the first few weeks might have been avoided if the refugees had not been moved.

Even then, half or more of the DK people at the border retreated into Cambodia, setting up a base in Phnom Malai which has impressed subsequent visitors as being better run than any of the camps on the Thai side. There the Thai rush to move people might have been for reasons of security. The DK group were objects of Vietnamese attack, and a new attack was believed imminent. Still, since half of them escaped the refugee dragnet, that problem was not eliminated.

The greatest surprise, and the most persuasion, was probably at KID, where only a trickle of those expected showed interest in coming. The first persuasion was arm twisting of the Khmer Serei leaders, telling them that if they did not allow people to go to KID, their supplies, of which they were getting enough, would be cut off by the Thai army. When this still did not produce enough people, word-of-mouth information went around that life at KID would be more comfortable and secure, which was true, and that people would get free food and would not have to work, which even if true was not the way to turn displaced persons into useful citizens. The Voice of America began Khmer-language broadcasts about the formation of KID, a place to which thousands of Khmer were going to "seek freedom," which influenced many people still within Cambodia to come to the border. Since a main activity of Nong Samet was trading, traders were assured that KID would have a free market where they could do better business in more secure conditions than on the border. Some of the "politicals" were led to believe that they could go to KID for a month or so of regroupment and then be returned better equipped to the border. Finally, rumors began that people who went to KID would be able to emigrate quickly to other countries. There was an element of truth in the last. Those refugees who met the requirements for resettlement could more easily be reached by foreign embassies at KID than at the border.[54]

In spite of these efforts, Khao-I-Dang, in terms of its initial plan, was a fiasco. Most of the people at the border did not want to become refugees. Moreover, by the end of December 1979 it was clear that conditions within Cambodia were not so bad as had been imagined, indicating that the U.S. State Department had been correct months earlier in resisting the "data" from the Bangkok embassy, and that nothing like one million, or even half a million, Khmer were going to rush across the border and put themselves under Thai control.

Although there had been a reasonable and justifiable concern that people on the border were too close to a war zone and should be moved out of artillery range, many more were of their own choice left on the border than were brought out to holding centers. Their preference for remaining where they were cuts the ground from under the claim that they had to be moved because of danger from the military situation. In fact, there was probably more danger from misdirected Thai fire than from the Vietnamese artillery.[55]

What Thailand got in the end was not one-quarter of the Khmer population, which could perhaps be used politically, but 100,000 or so of those Khmer who wanted nothing more to do with Khmer politics, whose only goal was resettlement in the West, and who, pending such a decision, were quite content to remain as welfare refugees in the hands of the Thai and international community. This was the last thing Thailand wanted, and it therefore closed KID to further immigration on January 24, 1980.

The analogy suggested by ''Volag archipelago'' is thus not just facetiousness, for there was no need to have such a large refugee problem at all. It was created for political reasons by attracting people who would not otherwise have come, and the initial expectation was that it would attract even more. For if UNHCR was unable to go right up to the border to aid those who needed medical care and food, other agencies not only could, but did, and still are doing so in cooperation with UNHCR. The International Red Cross and Catholic Relief Services, to name two, were already working on the border before the KID and Sakeo camps were set up; along with other Volags, they have continued to provide at the border, both for Khmer Serei and DK bases, most of the important services provided in KID.

By mid-1981, and even more clearly in 1982, more people were being supported by international aid at the border than had ever been in the holding centers, and supported more comfortably than in the first months at KID or Sakeo.

Although KID had been closed to new refugees, that was not the end of it. Once created, a place where people could sit indefinitely in security, on welfare, it inevitably became a magnet drawing more people out of Cambodia. The magnet effect was operative because the guards could be bribed to let new people in at night. The population thus rose from 111,000 in January 1980 to 136,000 in July. The increase represented almost entirely middle class former town dwellers whose goal was resettlement in another country. They were also often people possessing skills needed within Cambodia, who had been offered employment by the new regime, and who would not have attempted to come out if there had not already been a place like KID. The magnet was kept charged in various ways. Messages could easily be sent back to Cambodia via the same underground railway that brought out clandestine new arrivals, and they told friends and relatives of the good deal at KID. The Voice of America kept up its news of Cambodians finding freedom across the border. Some people even left KID clandestinely and went all the way back to Phnom Penh to lead out relatives who were too timid, or lacked funds, to make the trip alone. All of this traffic was facilitated by the nearly absolute freedom of movement permitted by the new

Cambodian authorities.

Until July 1980, when some major changes began to be carried out, the Volag archipelago remained in what we can call its original form. Khao-I-Dang, the most important holding center, was in general a refuge for members of the former bourgeoisie. According to official UNHCR statistics, over 70 percent were of this category, the rest being peasants. This estimate of peasants could be too high; certainly one had the impression that they were far fewer, and many people arriving in KID had listed as their previous occupation their enforced peasant status of 1975–79.

The camp at Sakeo retained its 25,000-plus population and its character of a DK sanctuary, although an indeterminable number of the people there had come out with the DK forces under duress and would have been happier in the milieu of KID.

On the border Mak Mun lost its importance after a factional Khmer Serei fight in April 1980 resulted in the expulsion of its leaders and most of its population moved to Nong Chan or Nong Samet. The latter two gradually developed quite different functions and characters. Nong Chan became, until early 1981, the site of the "land bridge" where rice, for both seed and food, was distributed to peasants who came from Cambodia and then returned.[56] The population fluctuated widely and the number of permanent residents was relatively small. Both the transients and the permanent residents were mostly peasants who realized the situation was temporary and who were not interested in becoming real refugees or trying to resettle abroad. Nong Samet, after the destruction of Mak Mun, had been the main center of Khmer Serei political activity and cross-border black-market trade. Before July 1980 the camp leadership pretended to keep population statistics showing a total of about 180,000 people, a figure which was accepted by the press and perhaps even by UNHCR in Bangkok. The foreign relief workers directly involved in food distribution and medical care considered, however, that the true figure was no more than half that, perhaps even less, and that the books were being cooked in an attempt to obtain extra supplies which the camp administration could then sell or distribute selectively for political support.[57]

Within the holding centers, in particular KID, conditions of life improved far beyond what should have been expected in a temporary camp for displaced persons who were supposed to return home once the emergency was over—the professed goal of both the UNHCR and the Thai government. Had such an objective been the whole story we should have expected KID to be maintained at the minimum level of comfort consonant with basic human needs, no encouragement or aid in developing special programs to make camp life attractive, and full information to be provided about developments within Cambodia, all effectuated to persuade refugees that return was preferable to stagnation in miserable holding centers. Voluntary return of individuals or small groups, because of the porous quality of the border, could have been effected without objections from Phnom

Penh, probably even without its knowledge. In fact, there always was a constant traffic into, as well as out of, Cambodia.

Instead of that, Khao-I-Dang, within a few months of its establishment, had all the accoutrements of a permanent settlement—schools, some adult education, special nutritional programs for mothers and children, even a Montessori kindergarten project—much of it, together with the high standard of medical care, superior not only to what is available in Cambodia now, but to what most of the camp's residents could have expected before 1975. The only aspects of camp life definitely inferior to prewar Cambodian circumstances, abstracting from the lack of freedom to leave the camp, were the schools, which could not yet, in 1980, offer a full syllabus or school day for all children, and the housing, very primitive at first, but steadily improving, with the newest units built in late 1980 suitable for long-term, if not permanent, residence. Since 1980, with the rapid decrease in the KID population to around 40,000 in 1982, conditions have still further improved. Interesting to those informed of developments within Cambodia was that the steady improvement there was paralleled by the equally steady improvement in camp life, almost as though the purpose was to make certain that refugee life remained more attractive. Moreover, instead of disseminating accurate information on progress within Cambodia, the Thai authorities, whose lead the UNHCR had to follow, insisted on blocking news that might have given a positive view of the Heng Samrin government. Short-wave radios were confiscated, and there were even sporadic attempts to prevent international news magazines and the Bangkok English-language press from reaching the refugees.

Furthermore, the actual operation of the camp programs was turned over to voluntary agencies, mainly Western, which as bureaucracies, however laudable their intent, wished to expand their areas of responsibility and test their own theories and projects. Their goals were generally to provide programs and services approximating as closely as possible normal conditions of middle-class existence, and they were not at all concerned about the policy of eventual voluntary return of the refugees. In fact, most of the personnel were ideologically opposed to the new Cambodian government as much as to Democratic Kampuchea, an attitude congruent with that of the refugees, and they were quite willing to encourage the latter in their insistence on resettlement. Thus, founded on a misapprehension, to which the anti-Heng Samrin policies of Thailand and the United States contributed, and allowed to grow without any overall policy control, Khao-I-Dang by late 1980 showed a very real potential for the Palestine-type situation that had been foreseen by some observers in the beginning.

Dismantling the Archipelago: Repatriation and Resettlement

When it was fully realized that residents of KID thought only of resettlement in third countries, that most of the politically utilizable refugees preferred to stay at the border, and that suitable aid could in fact be delivered to them, major efforts began in 1980 to resettle abroad or repatriate the refugees at KID and other camps within Thailand.

Emphasis was placed on the "holding center" definition of the camps as places for people fleeing temporary difficulties who, with a small number of exceptions, should return to Cambodia. Whether or not it was realized from the beginning, however, virtually none of those who volunteered to come to KID had any intention of returning, and the "temporary holding centers" assumption was valid only for the DK cadre among the Sakeo population, plus a very small number of peasants at KID.

The attitude of the KID population was clear once the camp was in operation, but in spite of this the Thai authorities began pressing in mid-1980 for a "voluntary repatriation," which everyone knew would end in a fiasco if it were to be really voluntary. UNHCR had to go along with it in principle, since its policy too, quite correctly, is that Cambodians should go home once the immediate danger has passed. They do insist, though, on the *voluntary* aspect of repatriation, and their concern was to assure that any such operation avoid a repetition of June 1979.

The publicity attendant on the Thai government's announcement of the projected repatriation was all else than reassuring for those aware of refugee mentality. On June 11, Air Marshal Siddhi Sawetsila announced that "mass repatriation" had been planned for some time, and that about 3,000 from KID were expected to volunteer at first and would perhaps then "create an impetus for a larger movement," since it was "likely that many of the would-be returnees would want to go back to fight against the Vietnamese."[58] The same source noted that the move would coincide with the start of the rainy season with an increase in anti-Vietnamese activity, which revealed an important motive in the Thai effort. It was also acknowledged that it would coincide with a deterioration of the food situation inside Cambodia,[59] which makes the repatriation scheme appear as a gigantic effort to undermine the Cambodian economy, thus subverting the aid efforts that had been initiated and leading eventually to a larger refugee exodus.

At the same time the public was informed that the move would start on June 16 with the 24,000 refugees from Sakeo, who were stronger and healthier than those of the other camps (and thus better fighters against the Vietnamese), after which that camp would be closed.[60]

Both Bangkok newspapers spoke of new developments at "Sarokkok," a name which meant nothing to anyone not directly familiar with the border. The *Post* said land was being cleared there to make "a safe haven zone"; while the *Nation* reported that the 40,000 people remaining at Nong Samet were being moved to Sarokkok, "about 2 km. away and about 3 km. inside Thai territory." Sarokkok, in fact, is a ruined Angkor-period temple within the area of the Nong Samet agglomeration—like the entire area, of dubious location with respect to the border. Up until early June most of the residents of the camp had lived to the southwest of it, but then the Thai military forced them to move back beyond it, which meant a shift of 2–3 km toward, or farther into, Cambodia. No explanation was given for the relocation order, which caused considerable hardship and was complied with under threat of armed attack.

A day later Singapore's deputy prime minister in charge of foreign affairs, Sinnathamby Rajaratnam, on the eve of leaving for a meeting in Bangkok, stated that "all should be sent home." "Thailand should send home all the Kampuchean refugees living in its territory," which "would enable them to fight for their country's freedom and independence."[61] During the following days the tension was cranked up still further. General Prem approved the mass repatriation, and Air Marshal Siddhi added that it would "involve more than 100,000 Khmer refugees initially at major holding centers"—that is, nearly the entire population of KID and Sakeo—a statement rebroadcast in Khmer by Voice of America that panicked the refugees at KID, in spite of the assurance that the repatriation would of course be "voluntary."[62]

On the same day, the deputy secretary general of the National Security Council predicted that "widespread hunger is expected [in Cambodia] for the whole of 1980 and the first part of 1981,"[63] which again makes one wonder what Thailand had in mind in trying to send back 100,000 people.

As border-crossing points for the returnees the newspapers publicized Nong Chan, the "land bridge"; Tap Prik, the gateway to the DK fortress of Phnom Malai; Sarokkok, with a population that was then estimated at 140,000, all moved from Nong Samet; and, mysteriously, Mak Mun.[64] I say "mysteriously" with all due deliberation. Among the crossing points Tap Prik was the obvious route for refugees, while Nong Chan and Sarokkok-Nong Samet would handle the "bourgeois," or non-DK peasants, from KID. There was no reason to use Mak Mun, unimportant in the "archipelago" since its destruction in April, unless it was part of some special plan.

Although Mak Mun camp had not been important as a refugee center since April, it still had some residents and was heavily guarded by the Thai military, who prohibited visitors and journalists, an unusual circumstance at that time on that stretch of the border. As for nearby Mak Mun village, which I had visited a couple of times, it seemed to be the headquarters for a Khmer Serei armed unit, and its mysterious atmosphere was compounded by the warning given to an International Rescue Committee (IRC) colleague by an "American diplomat" that "Mak Mun village was strictly off-limits."[65] Since Mak Mun figured as a major objective in the Vietnamese "incursion" a week later, even though no, or very few, refugees had been sent back there, it seems that Thailand may have hoped to use it to transfer a Khmer Serei force of comparable importance to the DK force that went from Sakeo through Tap Prik. The Vietnamese could already have had some inkling of its special function.

To add to confusion in the public mind, there was also speculation that the camps, which would have been virtually emptied by the projected moves, "may be permanent.'' Previous policy was said to have been reversed with plans ''to house more than 100,000 Kampucheans in Thailand in conditions which can only be described as permanent.''[66]

The anti-Phnom Penh purpose of the whole exercise was obvious, and from that quarter came a threat to crush the scheme, perhaps in the belief that 100,000 or so would really be pushed across. The *Post* responded piously that ''threat will not deter'' the operation.[67]

When the exercise was finally carried out, during the week of June 16–23, 7,000 hard-core DK cadre and military left Sakeo, crossing the border at Tap Prik to rejoin their comrades at Phnom Malai. At KID it was a total fiasco. Less than 2,000 of the camp's 136,000 people, mostly poor peasants who were out of place there, volunteered and were taken to the border either at Nong Chan or Nong Samet (Sarokkok). UNHCR, somewhat to the chagrin of the Thai military, assured the voluntary character of the operation by insisting that all volunteers be interviewed singly in the presence of a UNHCR official. This may have reduced the total below that which the Thais had expected.

If the numbers volunteering to go back and fight the Vietnamese proved a disappointment, the political goal that may be inferred from the Thai statements—a confrontation with the Vietnamese—was a smashing success. On the night of June 23–24, the latter occupied Nong Chan camp and Mak Mun, bringing down on themselves the censure of the entire ''free world'' for their ''invasion'' of Thai territory and the death and destruction among the refugees.

There were several things peculiar about this incident. The invasion aspect involves discussion of the location of the Thai border, some of the complexities of which have been noted previously. It is sufficient here to reiterate that the border camps had always been treated as though they were in Khmer territory or in an undefined no-man's land. Furthermore, the first official Thai reaction was very low key, even conciliatory. The Supreme Command's information office announced an attack on a Khmer Serei unit ''*near* the Thai border [my emphasis],'' which then ''spilled over . . . into the Thai village of Non Mak Mun.'' General Prem added that the ''Vietnamese troops could have entered Thai territory while pursuing the Khmer Rouge.''[68] For Air Marshal Siddhi the ''news reports of the border clashes were confusing,'' and in his first statement he seemed mainly concerned to emphasize that the attackers were Vietnamese, not Heng Samrin Khmer.[69]

This moderate reaction was fully in accord with reports of independent observers close to the action. An American Volag employee who worked daily at Nong Chan and returned there even before the fighting was over to help care for the wounded said that the Vietnamese entered Nong Chan and Mak Mun camps peacefully on the night of June 23–24. They did not come in shooting or brutalizing people. What they wanted was an evacuation of the camp areas, and they gave people a choice of moving either eastward into Cambodia or westward toward, or farther into, Thai territory. Thai artillery opened up on this mass of people, killing about 40 at Nong Chan and 150 at Mak Mun. The Volag employee was convinced that no civilians, either Khmer refugees or Thai, were deliberately killed by the Vietnamese, and he believed that all civilian deaths were either due to Thai artillery or accidental, in the crossfire between Thai and Vietnamese forces that followed.[70] Other foreign observers with quick access to the border, including some ''American diplomats,'' agreed that the Vietnamese had initially entered the camps in a relatively gentlemanly manner, and some noted that they even seemed to be trying to respect a line of demarcation, albeit not the one the Thai side would accept as a definitive boundary.

In the days immediately following the first news, and for weeks thereafter, both official sources and the press emitted increasingly shrill noises about a Vietnamese invasion and the ensuing menace to peace in Southeast Asia. The Thai foreign minister denied that it had only been a ''spillover in hot pursuit,''[71] and one journalist wrote two weeks later that ''in the early hours of . . . June 23, Vietnamese artillery and mortar fire pounded into Nong Chan and . . . Camp 204 at Ban Non Mak Mun,'' which seems to have been quite contrary to fact. He also referred to reports of Vietnamese atrocities, saying one man saw over 300 bodies.[72] Those closer to the action denied Vietnamese atrocities at all and attributed the largest death toll, whatever the number, to Thai artillery.

The Vietnamese claim was that the camps they tried to disorganize were on the Cambodian side of the border and thus a legitimate field of action, and that the immediate reason for closing them down was the Thai attempt to send back anti-Heng Samrin combat forces through those camps. It is noteworthy that they did not try to move on Nong Samet, whose position on the Thai side of the border is less doubtful than that of Nong Chan or Mak Mun. In spite of this, they really did intrude on Thai teritory, in the *village* of Mak Mun, but in a way that suggested they did not know precisely where they were. As it was reported in the press, as they occupied Mak Mun village they kept asking the inhabitants whether they were Khmer or Thai. The Thais interpreted their comments to mean that they were only after Khmer and did not wish to hurt the Thai villagers. But Khmer who had been close to the events and who spoke to me a few days later on the road near the border interpreted the Vietnamese actions to mean that they were attacking only the Thai and did not want to hurt the Khmer. Since my American informant understood that the Vietnamese had not attacked civilians at all, we should, it seems, interpret their questions in Mak Mun village as an effort to discover whether they were in Thai or Khmer territory.[73]

The Vietnamese interpretation of the Thai purpose in the repatriation exercise is not at all overdrawn. A return of fighters was certainly one of the Thai objectives. What seems at first glance strange in the Vietnamese action was that their move in June was against the Khmer Serei camps, where the Thai objective was not realized, and not against the Tap Prik-Phnom Malai area, where Thailand had sent back about 7,000 seasoned DK fighters.

This anomaly, though, is only apparent. The Phnom Malai area had been under Vietnamese attack for a year, and a new offensive, obviously in preparation, started a few weeks later. There would have been no purpose in speeding it up just to meet the returnees from Sakeo. The Khmer Serei camps, on the contrary, had been virtually ignored by the Vietnamese, and their decision to move was probably prompted not by the small number of people who finally came, but by the Thai propaganda that implied that 100,000 or so might be sent back. This threat of a massive return of potential enemies came at a time when there were increasingly strong rumors of alliance, or at least cooperation, among the various Khmer Serei factions and the DK forces at Phnom Chat, which if realized could have meant a serious threat to efforts from Phnom Penh to reorganize northwestern Cambodia. It also presented an occasion to protest physically the Nong Chan land bridge, which Phnom Penh believed was unnecessarily drawing people away from their fields to the border, as well as serving as a supply point for the anti-Phnom Penh military forces.

There is some evidence that the Vietnamese move was not an exercise prepared long in advance, but rather a hasty, makeshift operation, and that the intention was certainly not an invasion of Thailand. Khmer Serei intelligence discovered in their interrogation of Vietnamese prisoners that the incident occurred just when a major rotation of Vietnamese units was scheduled, indicating that the bogey of "Vietnamese invasion" may have been for the public only.[74]

Thus, a U.S. embassy border team member was able to reassure a Volag medical group who had been panicked by sensational press accounts and refugee rumors, telling them that the Vietnamese were definitely not about to invade Thailand and that the action at the border was a strictly localized incident. This is what clearer heads have always maintained, at least when they are not speaking for propaganda effect. Less than two weeks after the event General Saiyud Kerdphol told the *Washington Post* "It would take a 10-year Vietnamese buildup to create a serious invasion force for conquest of Thailand." He did not see even Vietnamese-fostered insurgency as a major threat.[75]

The Vietnamese military threat was similarly minimized two years later by the commander of the 9th Army Division in charge of the border in Prachinburi, Major General Somkid Chongpayuha. He reported that the Vietnamese troops have suffered high casualties and illness, that their offensive this year is less intense than earlier, and that most of them have withdrawn about 10 km from the border, from where they lack the capacity to strike into Thai territory.[76]

The incursion of June 1980 demonstrated that any attempt to reinforce DK or Khmer Serei troops under cover of refugee repatriation would be resisted, and it proved that repatriation could not be carried out by simply pushing a mass of people across the border without consulting the PRK authorities. No further operations of that nature have been attempted, but the original structure of the refugee camp populations has been drastically changed through large-scale population transfers among the camps beginning in July 1980. Some of those moves were a plan developed some time earlier by the Thai authorities and UNHCR to reduce the size of KID by spreading its population among the smaller holding centers and to obliterate the sociological difference between ''Pol Pot'' and ''bourgeois'' camps. Apparently once the original political purpose of a giant holding center had proved unviable—that is, Thailand was not going to control one-fourth of living Cambodians, and those who came were generally not politically utilizable—it was felt that a center so large was too difficult to administer and control. As for mixing the two different groups, it was believed that this would smooth out tensions between them and make it easier for all of them to cooperate in Cambodian society in the future.

The first of the several moves was a shift of about 5,000 from Sakeo to KID. The ideological climate of the former had already been attenuated by the departure of 7,000 cadre in June, leaving the remaining people much more freedom to declare their disassociation from DK policies, if they so desired.

One effect soon noticed was that the Sakeo population, who had formerly thought of eventual return to Cambodia rather than resettlement, began to adopt the bourgeois attitudes of the KID population, renounced their intentions of going home, and began to look for ways to get on the resettlement lists.

Other shifts took thousands more from KID to Mairut, Kamput, and a new transit center at Phanatnikom, and many of those people were designated for resettlement, which reached its peak in 1980–81. Thus the total Khmer refugee population within Thailand was rather rapidly reduced from around 200,000 in mid-1980 to something over 90,000 in mid-1981, with the most dramatic shift at KID, down to 42,000, less than one-third of its former size.

The magnet effect on the remaining middle-class Cambodians within Cambodia, however, may have increased as conditions at KID were upgraded to keep pace with improvement in Cambodia. New, livable dwellings were constructed, the space opened up as people departed was filled with lush vegetable gardens, and the remaining inhabitants whose chances of resettlement were small or nonexistent settled in for an indefinite stay.

In spite of the official closure of the camp to new arrivals, they kept coming clandestinely, were integrated into the population, and often, if they had good connections, were able to emigrate rapidly. Because of the relative ease of communication across the border, some of those who came to KID in 1981–82 and had families or well-connected friends abroad have been able to prepare everything for crossing the border, entry to KID, and resettlement processing even before they leave Phnom Penh. They are in fact making use of the camp system for their own convenience.

It is finally clear that the camps had become much more than temporary refugee shelters and that predictions of a ''Palestine factor'' had not been out of place. Permanent settlement within Thailand had always been ruled out, most of those who came to KID refused to consider return to Cambodia, and under existing rules applied by the important third countries, several tens of thousands would never qualify for resettlement abroad. Simply relaxing those rules would not alleviate the problem either, for so long as the border camps and KID existed, any increase in resettlement could entice more people out of Cambodia to take advantage of it. Although that could further serve to destabilize Cambodia, one of the original goals, Thai authorities were increasingly worried about being stuck with people who could not be moved out. Thai preference was for those remaining in the holding centers to be "repatriated," and that was also seen by UNHCR as a desirable solution, but it was contrary to the preferences of the refugees themselves and could not be carried out voluntarily.

One suggestion by Thai Foreign Minister Siddhi Sawetsila was that a UN observer team be stationed on the Thai side of the border, while on the other side, in western Cambodia, safe areas under UN supervision would be established for the uprooted civilian Cambodians encamped near the border and those in Thailand who wished to return home.[77] This was of course in part a desire to establish a buffer between Thailand and Vietnam and a counter to a Vietnamese demand for a demilitarized zone on both sides of the border. It would also mean UN protection for the DK remnants and other anti-Phnom Penh armed forces who had been reinvigorated since 1979, aid to whom had been threatened by the international organizations' attempted cut-off following the Vietnamese attack of June 23–24, 1980.

For UNHCR mere relocation to the border and UN supervision of a safe zone for DK and Khmer Serei fighters was not a desirable solution. The repatriation it supported meant safe transit to original homes within Cambodia and security upon return monitored by the United Nations. UNHCR also realized that part of the reluctance of refugees to consider repatriation was a nearly total lack of information about changing conditions within the country. Since crossing the border into Thailand, their only sources of news had been the Voice of America, rumors, and stories of more recently arrived refugees. Even the last hardly brought the earlier arrivals up to date with any accuracy since, in order to justify flight, those who had recently left had to make the most negative possible statements about life within the country.

Thus, to find out if a significant number of refugees, realizing that third-country resettlement was extremely unlikely, would choose to return home under favorable circumstances, UNHCR in March 1981 engaged Dr. Milton Osborne, then senior research fellow in international relations at the Australian National University, to conduct a survey on "Attitudes Towards Voluntary Repatriation" among Cambodian refugees in Thailand.[78]

Osborne found that "there is currently a substantial degree of interest in returning," particularly among those with peasant and low-level urban worker backgrounds, groups which by that time, after the exodus of many of the bourgeoisie for resettlement, may have made up nearly half the remaining population of KID. He found that nearly 5 percent of that group would accept immediate return, another 41 percent would return if UNHCR said it was safe, and 24.4 percent would go with further UN guarantees, or after reestablishment of "peace" in Cambodia.[79]

Osborne also reported on the "profound lack of information about contemporary developments in Kampuchea among refugees," and he recommended a program of voluntary repatriation "preceded and accompanied by a major campaign to make information on the conditions prevailing inside Kampuchea available to refugees."[80] At the same time Osborne deplored the effect of an accelerated American effort to move large numbers, about 31,000, out of KID in early 1981, since it strengthened the desire for resettlement among refugees who had little chance of acceptance abroad. "More than incidentally," his report stated, "the current accelerated American programme is quite clearly having the effect of encouraging a further outflow of former middle class Kampucheans from inside Kampuchea in the hope of being resettled."[81]

Osborne noted as well that "difficulties . . . might be raised by the Thai authorities when faced with a program of action that could be seen as giving an increased degree of legitimacy to the Vietnamese-backed regime in Phnom Penh and as undercutting their effort to encourage refugees to become recruits for Khmer Rouge [DK] and Khmer Sereika [KPNLF-MOULINAKA] forces."[82] Indeed, Osborne was proven correct: the proposed information program and any kind of mass repatriation to the Cambodian interior were blocked by Thailand.[83]

This explains the seemingly contradictory news reports that Thailand was simultaneously threatening to force 100,000 refugees back to Cambodia and rejecting a UN plan to fly 20,000 back to Phnom Penh.[84] Since, for public consumption, it has all along called the refugees an unwanted burden, it is difficult for the Thai authorities openly to block repatriation and admit that they would prefer to keep them around a while longer for the politico-military purpose noted in Osborne's UNHCR report. Instead of that they have argued that sending refugees back across the border would be too dangerous, ignoring that such a move would only take place under joint UNHCR-Cambodian government agreement and would be less dangerous than the Thai plans to relocate refugees in quasi-battle zones. Then they assert that nothing can be done that implies recognition of the People's Republic of Kampuchea; as a possible alternative they have suggested repatriation via a third country, such as Burma, patently unfeasible if only because of the logistics.[85]

Later in the year there were indications of an apparently more conciliatory position. Zia Rizvi was reported to have said that most of the 130,000 people in the camps in Thailand would prefer to return to their villages if their safety could be assured (repeating the conclusions of Osborne's report), and that Thailand "had not ruled out allowing refugees to go back across the border to areas of Kampuchea controlled by the Vietnamese-supported administration if that was where they wanted to go."[86]

Then, at a meeting of aid organizations a Thai Supreme Command representative said that the 92,000 people in KID, Sakeo, and two other small camps had no chance of going to a third country and that as the situation improved they would have to go home.[87] At the same meeting the chief of the Foreign Relations Section, Ministry of the Interior, although not of policy-making status, said that "if in the future, the national [sic] organizations assisting the Kampucheans in Cambodia help them to help themselves, these people will want to live there and not want to escape to Thailand," indicating that more aid sent directly to Cambodia would be desirable.[88]

What the Thai authorities really wanted, however, was spelled out by National Security Council Secretary General Prasong Soonsiri in an address to the Foreign Correspondents Club in Bangkok on October 2, 1981. He said that there were areas of Cambodia controlled by one side or the other in which the population can exist in relative security, and it was reasonable that some refugees could return. Thus the DK, KPNLF, and MOULINAKA border camps were considered a part of Cambodia and acceptable places for repatriation.[89]

Prasong followed this up a week later in Geneva by requesting international support for a Thai-UNHCR plan to repatriate 100,000 Cambodian illegal immigrants and deter future emigration by neutralizing any motive for emigration (and this ignored that there was not yet Thai-UNHCR agreement on the modalities of repatriation). But Prasong also urged increased resettlement, which had always been recognized as an inducement to emigration from Cambodia and a deterrence to repatriation.[90]

The same positions were maintained in 1982—Thai efforts to get Cambodians from the holding centers to the border and UN insistence that such is an unacceptable substitute for genuine repatriation. A Thai Supreme Command report of June 1982 stressed the "Voluntary Relocation [border] Program . . . pending . . . the larger Voluntary Repatriation Program."[91] Interestingly, this report was issued in connection with the transfer of border relief operations from UNICEF to the United Nations Border Relief Operations (UNBRO), about which more is said below.

In July 1982 at the annual conference on Indochinese displaced persons, Secretary General Prasong complained that third countries had not been resettling refugees fast enough, and that Thailand might have to take drastic action, perhaps meaning forcible relocation, if other countries did not honor their commitments.[92] At the same time a senior U.S. refugee official was reported to have said that the United States planned to take 40,000 Indochinese refugees in 1982, down from 88,000 the previous year, and that this "served to discourage would-be refugees from fleeing their countries," thus tacitly admitting the magnet effect of the refugee system and the fact that most were not legitimate refugees.[93]

Prasong had admitted the same thing when he said the Thai authorities had not allowed subsequent groups [1981–82] of displaced persons to depart for third countries, a "tactical" plan "aimed at neutralizing the Indochinese people's motive for leaving their homeland," and he thereby undercut his own previous statement that Cambodian displaced persons were "increasing in number *as a consequence of the fighting and starvation in their troubled land*" [my emphasis].[94]

Another maneuver in the efforts both to get rid of the refugees and to retain them for political use was prompted by the new coalition agreement among the DK, KPNLF, and Sihanouk forces signed on June 22, 1982. Following that Sihanouk visited the border and on July 7 went to KID. Although the KID population had been apathetic toward his expected visit, several hours of exhortation over the camp loudspeakers managed to bring out a respectable crowd to greet him when he arrived just after 3 pm.[95] In his speech he invited people to join him in the anti-Vietnamese struggle, and in the following weeks several thousand signed up for transfer to the O Smach base of Sihanoukist forces in northern Cambodia.

About 2,000 had actually left when the movement was called off at Sihanouk's request because the camp was not in fact prepared to receive them.[96] Among the volunteers left in KID enthusiasm quickly waned. By the end of August the move to O Smach was an unwelcome subject of conversation, and most who were willing to comment on it at all said they were comfortable in KID and would think again about moving to O Smach when the international organizations and Volags had prepared proper facilities and could guarantee their safety.[97]

Since then some of those who went to O Smach have drifted back to KID, disillusioned with what they found—a circumstance which is likely to undermine future efforts at relocation to the border. Prasong has firmly reiterated that "Thailand has ruled out any direct repatriation of the Kampucheans to Phnom Penh or into the Vietnamese hands."[98]

One new development that should give cause for concern about the ultimate objective of some of the refugee aid programs is the increase in Volag activity in the border camps, now more tightly organized under the new tripartite anti-Phnom Penh coalition, and in camps across the border in Cambodian territory, such as Ban Sangae, all under the umbrella of UNBRO, which, under the World Food Program (WFP), took over border relief operations from UNICEF in June 1982.[99]

The formation of the Khmer coalition and the expansion of UNBRO activities seems in fact to represent the creation of the UN-supervised safe zone on the eastern side of the border that Thailand has long desired. It is not, however, just a safe zone for displaced persons, but a series of military bases, formally united, at war with the authorities in the interior of Cambodia. The United Nations might well find itself sponsoring and giving protection to a pro-Thai buffer zone, which, even if a legitimate desire for the Thai government, is not something the international community, at least in the present circumstances, should guarantee.

It is also appropriate to raise the question of the status of foreign Volag personnel working across the border in areas for which they have no proper documentation, in fact supporting a guerilla operation against a government in power. Their legal station would be particularly relevant if they were to be caught in hostilities.

The View from the Other Side

The above discussion has treated the refugee problem almost exclusively in relation to the politics of Thailand and those other powers that support or encourage the Thai position. It is thus not the whole story, which must include the "refugee politics" of the People's Republic of Kampuchea and Vietnam as well.

As emphasized above, most of the Cambodian refugees, and probably all of those still remaining in Thailand, left their country after the overthrow of Democratic Kampuchea, and thus at a time when they no longer qualified for refugee status. Moreover, a large number of them, perhaps the majority, have been people of sufficient education and training to qualify for state employment in the new, post-January 1979 regime, which had promised to reconstruct and maintain normal town life and infrastructure; and in fact such people were offered positions throughout 1979. Had they accepted they would have qualified for state support, which, although inferior to their prerevolutionary living standard, would have provided basic necessities and was probably equivalent to refugee fare at the time. They were not driven out of the country by persecution, but found flight easy because the new PRK regime kept the promise made by the Salvation Front in December 1978 to restore freedom of movement.

Having refused to participate in the reconstruction of their country and fled abroad, they have been considered disloyal by the PRK government, and in the words of a Foreign Ministry official in 1981, "they are of the last priority" among Cambodians abroad whom the new regime might like to attract homeward. By September 1982, however, the PRK position seemed to have softened somewhat, with Foreign Minister Hun Sen declaring that even members of the tripartite anti-Phnom Penh coalition, if they abandoned that body, could return home and even participate in elections. This offer of amnesty, however, would not stand forever, and if coalition members wait until their forces have dwindled away, "they will be treated as prisoners of war."[100]

Whatever categories of people the People's Republic of Kampuchea might choose to readmit in principle, they are likely to insist on formal negotiations with the UNHCR and a screening process over which they would have control. Just as in 1980 when relocation of several thousand DK cadres and soldiers to the border provoked a military response, there is no question of the remaining refugees simply walking en masse back to the border.

Nevertheless, most of the refugees who have already been resettled in third countries were of the educated bourgeoisie, who were offered places in the new state organizations and whose refusal implied disloyalty. With each passing year an increasing proportion of the remainder consists of former peasants and urban workers whose position is much less sensitive politically. Such people already go back and forth constantly, although clandestinely, between KID and the border, and between the border and the Cambodian interior. During the Cambodian New Year in April 1982, for example, many left the border to visit their home villages for the celebrations. Such people could easily fit again into their communities, and at worst, if apprehended on return, might be subject to a week or two of "reeducation." There is probably no serious difficulty for that portion of the remaining refugee population to be accepted again within Cambodia if they were to return in an organized fashion, with preliminary UNHCR-PRK negotiations and the possibility for screening by the Cambodian side. It thus would imply some measure of recognition of the existing government.

Some Conclusions

The refugee problem is inextricably involved with the total Indochina political situation and cannot be solved separately without an unacceptable degree of violence, for example forcing people back to, or across, the border.

So long as the border camps and KID exist, they will attract Cambodians who are dissatisfied with life at home, and so long as anti-Phnom Penh forces are maintained at the border the attraction will be compounded by the fear that Pol Pot, supported by the United States, China, ASEAN, and the United Nations (as it appears from Phnom Penh) may well return to power.

Although many, perhaps a majority, of those remaining in KID and other camps within Thailand would choose repatriation over resettlement, they will not do so until they feel the country is safe, which means until they believe that neither Pol Pot nor a coalition in which DK elements are important is likely to return to power.

Even in the best of circumstances, there will remain several thousands, perhaps 10–20,000, who will adamantly refuse repatriation or relocation, and who wish only to be resettled in third countries. Since many of them are of the educated or political groups who rejected employment by the People's Republic of Kampuchea, or who worked against it, their fears of returning are justifiable. They should in the end be airlifted out and resettled en masse. After all, some of the Western countries, in particular the United States, bear a large part of the responsibility for the destruction of the society those people knew.

Such resettlement, however, would have to be accompanied by a total closure of the refugee camp system and a termination of all aid to the border, to prevent the mass resettlement from enticing more people to try their luck. This would of course mean a nearly complete reversal of the policies now pursued by the United States, China, and Thailand, but it would contribute to the political health of the last-named country and is a necessary step in the reconstruction of Indochina, which has been too long delayed.[101]

With respect to the question of the impact of refugee repatriation on Cambodia and its effect on the rebuilding of its "social structure," the answer is "very little." Those refugees whose return would have had some effect, the well-educated professionals, the skilled businessmen, the former administrators, the teachers, have already been resettled abroad, are not welcome in Cambodia, at least not immediately, and probably would not wish to return to work under prevailing conditions or any conditions likely to obtain in the near future.

In the meantime their places are being filled with surprising rapidity by survivors of some education and talent who now rise more swiftly, and to higher positions, than they could ever have hoped for before 1970. For them the new regime has brought a real improvement in status, if not in material prosperity, and they will probably now work hard to maintain what they have gained. In addition to these survivors, there are numerous, possibly hundreds, of young Khmer abroad in the socialist countries for technical education who will gradually fill the places of those who died under Democratic Kampuchea or fled across the border. Cambodian social structure is reforming without the refugees, and the latter will be increasingly redundant. The new social structure will not be just what they knew before the revolution, but that is hardly a matter for lament.

Notes

1. *Bangkok Post* (*BP*), September 2, 1980; *The Nation Review* (*NR*) (Bangkok), September 3, 1980.

2. *BP*, October 2, 1980.

3. John Pilger, "America's Second War in Indochina," *New Statesman*, August 1, 1980.

4. *NR*, September 12, 1980.

5. See description of the various camps below.

6. Statement by John R. Kelly of the United Nations, *Annual Conference on Indochinese Displaced Persons in Thailand 1982* (*Conference*), p. 53.

7. In connection with refugee operations several Volags organized medical teams to visit Thai border villages, and some of the Western members of those teams, whom I met in Aranyaprathet during May-September 1980, were shocked at the health conditions in those villages, which resembled the situation among the Cambodian refugees. It is also significant that from 1980 through 1982 the Bangkok press gave unusual attention to the parlous state of much of the country's health. The suspicion that the new concern might have some connection with the refugee situation seemed confirmed by a remark of the then minister of public health that about 1 percent of Thai infants "look just like the Kampuchean children who arrived in Thailand during the start of the refugee influx" (*BP*, October 18, 1981). In other contexts it was stated that "many school children especially in rural areas were starved and suffering malnutrition" (*NR*, September 19, 1980), that "eighty percent of Thai children have been under the malnutrition classification" (*NR*, August 16, 1981), or that over half the Thai children suffer malnutrition (*BP*, August 12, 1982). As countermeasures the public health minister and two Magsaysay Award-winning Thai doctors advocated a series of measures not unlike the health care system designed under Pol Pot in Democratic Kampuchea. Because of the scarcity of doctors, particularly outside the towns, rural villagers would have to be taught "simple primary health methods to take care of themselves," and a system of quickly taught, unpaid "health communica-

tors'' (five days of training) and ''health volunteers'' (two weeks of training) should be organized in the villages. Traditional herbal medicine should also be supported, and instead of more large hospitals, many simple clinics within easy reach of villages should be constructed (*NR*, August 6, 1981; *BP*, October 18, 1981; *NR*, June 5, 1982; *NR*, June 17, 1982).

8. Credit for coining this phrase goes to my Khao-I-Dang colleague and fellow International Rescue Committee agent, Daniel Steinbock.

9. The international definition of ''refugee'' is established by the Convention Relating to the Status of Refugees, signed in 1951 in Geneva and extended by the 1967 Protocol Relating to the Status of Refugees, neither of which, incidentally, have been signed by Thailand. See Office of the United Nations High Commissioner for Refugees, *Handbook on Procedures and Criteria for Determining Refugee Status* (Geneva, September 1979). See also James C. Hathaway and Michael S. Schelew, ''Persecution by Economic Proscription: A New Refugee Dilemma,'' *Chitty's Law Journal*, vol. 28 (May 1980).

It should be emphasized that the mere preference of a person for settling in a country other than his own does not qualify him for refugee status. Obviously, most of the Khmer who crossed the border to KID wished to settle in the advanced Western countries, and for quite understandable reasons—life was perceived as easier: urbanites believed they would not have to resort to agricultural or menial labor, desired consumer goods were more readily available, children might have access to better education, and there would be less danger of being caught in war or revolution. There was, however, no more reason to consider the personal preferences of those Khmer than the preferences of the Thai or other third world people, who live in miserable conditions and would prefer to live in the United States, Canada, or Europe. One of the tasks of U.S. consular officers is to screen visa applications for individuals who may be seeking a way to reach the United States to satisfy their personal preferences. As Zia Rizvi pertinently remarked, refugee movements are most likely to assume importance when the flow is from socialist to capitalist countries. And that is not necessarily because living conditions are worse in the socialist countries, but because the major capitalist governments are reluctant to recognize the poor and dissatisfied of other capitalist countries as legitimate refugees.

10. *Far Eastern Economic Review* (*FEER*), November 2, 1979, p. 13.

11. ''Thai Supreme Command Headquarters, Border Relief Operations,'' June 28, 1982 (SC-6/28/82); Committee for Coordination of Services to Displaced Persons in Thailand, *The CCSDPT Handbook, Refugee Services in Thailand* (1982).

12. Very often the bourgeois refugees of KID, who were preponderant there in 1980, resisted to some extent all organizational discipline that went counter to economic and social laissez faire, saying it resembled ''Pol Pot.'' Among instances I observed were unceasing efforts to circumvent camp medical authorities' efforts to limit or prohibit useless or noxious medicines; objections to suggestions that where possible they should supplement their own food supply by growing vegetables on empty land; refusal to accept, until continuation of the program was threatened, that a special course of English-language instruction, due to the numbers of people involved, had to be limited to school children; and most serious of all, persistence in black-market dealings with surrounding Thai villages that could have endangered the status of the camp. In September 1980 it was discovered that certain enterprising refugees had been collecting the surplus from the camp population's daily rice rations and clandestinely selling it to Thais outside the camp at below market prices (having received the rice free, any price was profit for the

refugees), something which could pose a political threat to Thai authorities by calling attention to the "privileged" status of Khmer refugees in comparison to poor Thai villagers. In this case UNHCR had to resort to Pol Pot-type discipline—a threat to turn the guilty over to KID's Thai guards, who were perceived to be as brutal as DK cadres.

13. Reference is made to the expulsion in June 1979 of 42,000 Khmer from the border north of Aranyaprathet. According to survivors, they were enticed farther into Thailand, then invited to board buses which they believed would transport them to a safer place, but instead they were forced back across the border in northern Cambodia in a very dangerous area.

14. See below for evidence that the supposition was correct.

15. Supreme Command (SC), June 28, 1982.

16. This is derived from my own conversations with refugees during May-September 1980. See also Milton Osborne, "The Kampuchean Refugee Situation, a Survey and Commentary," report prepared for the United Nations High Commissioner for Refugees (UNHCR), Bangkok, April 23, 1980.

17. Such messages could be sent via acquaintances in Thailand to the Khmer Serei camps on the border, and thence into Cambodia. Such clandestine mail routes continue to operate.

18. See Osborne, "Kampuchean Refugee Situation," pp. 54–55, for examples.

19. The political degeneration of the border camps during 1979–80 discouraged some of the more talented potential "politicals" who left Phnom Penh for the border and who then, disillusioned, continued on to KID and resettlement. Others held out at the border until better discipline was established. Some "politicals" have been on the border since 1975, while some of the better known leaders, such as Son Sann, Dien Del, and In Tam, have been imported from exile in a political aid program bearing some resemblance to the return of the "Khmer Hanoi" to Phnom Penh since January 1979.

20. *FEER Yearbook 1979*, p. 313; and in July 1982, at Nong Chan, I met KPNLF officers who claimed to have been on the border since 1975.

21. I am referring to the affair of January 28, 1977, in which thirty people were at first reported killed in an unprovoked DK incursion. Eventually it turned out that the incident was probably provoked by smuggling in the area, and in June-July three Thai were executed and others imprisoned for having provoked it. In spite of that no light was shed on the responsibility for the killing, and there was even some doubt about the identity of the killers. The best attempt at a serious analysis is Larry Palmer, "Thailand's Kampuchea Incidents," *News from Kampuchea I* 4 (October 1977):1–31. In *FEER*, February 11, 1977, pp. 8–10, Norman Peagam wrote of the incident expressing doubt about DK involvement. Less than a month later he was expelled from Thailand (*FEER*, March 4, 1977, pp. 9–10).

22. *FEER*, December 7, 1979, p. 15.

23. This document was in a confidential UNHCR file which accidentally fell into my hands in July 1980. English as in the original.

24. *NR*, June 25, 1980, p. 10.

25. *NR*, June 24, 1980.

26. *BP*, July 12, 1980.

27. *BP*, July 14, 1980.

28. *Handbook*, p. 5.

29. Ibid., p. 43.

30. *BP*, July 12, 1982; *NR*, July 13, 1982.

31. Michael Battye, Reuters, *New Straits Times* (Kuala Lumpur), May 2, 1979; *FEER Yearbook 1980*, p. 293.

32. Such are the figures given by survivors, but like all such estimates they are subject to caution. See also *FEER*, August 3, 1979, p. 19; August 17, 1979, p. 17; *Asiaweek*, June 22, 1979, pp. 12–13.

33. See *Asiaweek*, June 22, 1979, pp. 12–13; *FEER*, August 2, 1979, p. 19.

34. *FEER Yearbook 1980*, p. 293.

35. *Asiaweek*, June 22, 1979, pp. 12–13; *FEER Yearbook 1980*, p. 293.

36. *FEER*, July 20, 1979, p. 39.

37. *FEER*, August 3, 1979, p. 19; August 17, 1979, p. 17.

38. *FEER*, September 7, 1979, p. 10.

39. Stephen R. Heder, *Kampuchea Occupation and Resistance*, Asian Studies Monographs no. 027, (Institute of Asian Studies, Chulalongkorn University, January 1980).

40. *FEER*, July 6, 1979; November 2, 1979; November 9, 1979.

41. *FEER*, September 21, 1979, p. 7; November 9, 1979, p. 40; September 28, 1979, p. 12.

42. *FEER*, November 2, 1979, pp. 12–13.

43. *FEER*, November 9, 1979, p. 29.

44. *FEER*, November 2, 1979, p. 13; November 9, 1979, p. 42.

45. *FEER*, November 2, 1979, p. 13; November 9, 1979, p. 41.

46. *Asiaweek*, November 16, 1979, p. 46; UN report at KID.

47. Because the DK forces were establishing bases at Phnom Malai and Phnom Chat. See *Asiaweek*, November 9, 1979, pp. 20–21; November 16, 1979, p. 46; December 7, 1979, p. 35; *FEER*, November 11, 1979, p. 29; November 30, 1979, p. 15.

48. *FEER*, November 16, 1979, p. 25; *Asiaweek*, November 30, 1979, p. 16; *FEER*, December 7, 1979, pp. 5, 14; William Shawcross, "The End of Cambodia?" *New York Review of Books*, January 24, 1980, p. 25.

49. Conversations with UNHCR personnel at KID.

50. Conversations with refugees who had been part of that first movement to KID. John Pilger, "America's Second War in Indochina," wrote that much of the refugee operation was a deliberate effort to draw off population and destabilize Cambodia and Vietnam. I believe he was generally correct, but the evidence to justify as definite a statement as his is lacking and unlikely to be discovered.

51. *FEER Yearbook 1979*, p. 313.

52. Michael Vickery "Democratic Kampuchea—CIA to the Rescue," *Bulletin of Concerned Asian Scholars*, April 14, 1982. One may not, of course, exclude the possibility that the worst peasant excesses against the bourgeoisie in Democratic Kampuchea aroused the envy of peasants in other countries.

53. This and all unattributed information below is from conversations with UNHCR and other aid organization personnel and refugees who were involved.

54. The only officially acknowledged pressure was that on the Khmer Serei leaders to allow those who wished, to take advantage of the opening of KID. The Voice of America broadcasts were also overt, and whatever their conscious objective, *many* refugees said that those broadcasts had been the most important impetus to their decision to cross the border to KID. For the rest, it is impossible to separate spontaneously generated rumor from deliberate suggestions by foreign personnel. The other instances of persuasion noted

here were strictly unofficial and in fact supposedly prohibited; but both refugees and foreign personnel told me they had occurred. The UNHCR people generally blamed overenthusiastic Volags or American Embassy agents, and one of the latter, who regretted what KID had become, blamed UNHCR employees.

55. See *FEER*, November 23, 1979, on an unprovoked shelling of the Nong Chan camp by Thai artillery on November 8, 1979.

56. Since then distribution has been intended only for noncombatant residents of the border area and is effected by handing 7 kg of rice weekly to every female ten years of age (120 cm tall) or older, who is calculated to represent 2.5 persons receiving a daily ration of 400 g.

57. The new Nong Samet administration of July 1980 in fact claimed no more than some 40,000 people, and the difference in the pre- and post-July figures may have been the origin of the UNHCR claim later in the year that around 130,000 Cambodians had returned from the border to the interior of the country—a claim that took everyone familiar with the situation by surprise.

58. *NR*, June 11, 1979.

59. Ibid.

60. *BP*, June 11, 1979.

61. *NR*, June 12, 1979.

62. *NR*, June 13, 1980, and personal observations.

63. *BP*, June 13, 1980.

64. *NR*, June 14, 1980.

65. We had gone to the village to locate the son of a refugee in KID. A number of "diplomats" patrolled the border as members of the Kampuchea Emergency Group, the Joint Voluntary Agencies, or openly as Foreign Service officers.

66. *BP*, June 14, 1980.

67. *BP*, June 15, 1980.

68. *NR*, June 24, 1980.

69. *BP*, June 25, 1980.

70. Conversation, end of June 1980.

71. *NR*, July 5, 1980.

72. Jim Gerrand, in *BP*, July 6, 1980, p. 6.

73. The language of Mak Mun is in fact Lao, but most of the villagers also speak Khmer, and if they have been to school, standard Thai. Khmer from the border camps were constantly moving through the village.

74. Information provided by a Western researcher in close contact with all of the border groups.

75. *Washington Post*, July 5, 1980; quoted in Hans H. Indorf and Astri Suhrke, "Indochina: the Nemesis of ASEAN?" *Southeast Asian Affairs, 1981* (Singapore), p. 67.

76. *BP*, July 12, 1982, p. 2.

77. Address to the United Nations, reported in *BP*, October 2, 1980.

78. Milton Osborne, "Kampuchean Refugees in Thailand, Attitudes Towards Voluntary Repatriation," ms., March 30, 1981.

79. Ibid., p. 18 and table E.

80. Ibid., p. 18.

81. Ibid., pp. 4–5; proof of the ill-considered nature of that move is that many of the refugees affected did not meet the criteria for resettlement, suggesting that the political

aspect of the move was uppermost in the minds of its organizers. Some of those refugees were subsequently returned to KID, which should tend to dampen the hopes for resettlement, but about 20,000 of them, still in the Kamput camp where they had been transferred from KID, remained, in late 1982, a sore point between the U.S. State Department and the Immigration and Naturalization Service as well as between American and Thai authorities. See *NR* and *BP*, November 1–2, 1982.

82. Osborne, "Kampuchean Refugees," p. 17.

83. See *NR*, April 15, 1981, for mention of the UNHCR survey, the UNHCR request to repatriate directly more than 20,000 refugees who wished to go home, and the Thai rejection by the "National Security Council . . . which has ruled against it." There was further commentary in *NR*, May 28, 1981, and in an *NR* editorial, April 16, 1981, which questioned the validity and integrity of the survey, and to which A.J.F. Simmance, UNHCR regional representative for Western South Asia, responded with a letter dated April 16, 1981, defending the results of the survey. My own information is also based on conversations with various UNHCR employees.

84. Rod Nordland, *Philadelphia Inquirer*, May 26, 1981.

85. *FEER*, May 1, 1981, pp. 22–23.

86. *The Age* (Melbourne), June 6, 1981.

87. Col. Sanan Kajornklam, JOC/SC, CCSDPT report, August 7, 1981.

88. Mr. Pranai Suwanratn, in ibid.

89. *BP*, October 3, 1981.

90. *BP*, October 15, 1981.

91. SC-6/28/82.

92. *Conference*, pp. 25–26.

93. *NR*, July 16, 1982.

94. *Conference*, pp. 25–26.

95. Personal observation; see also *BP*, July 8, 1982.

96. *BP*, August 11, 1982.

97. Conversations at KID.

98. *NR*, November 1, 1982, p. 6; apparently in response to a Japanese offer of financial assistance to construct airfields along the border to facilitate mass repatriation.

99. SC-6/28/72; *CCSDPT Handbook*. In principle the transfer occurred on January 1, but UNICEF staff remained on duty for another six months to aid in the orientation of their replacements.

100. *FEER*, March 3, 1983, p. 12.

101. Such a change of policy is not even being discussed. At an early 1983 meeting of donors to the "UN relief operation inside Cambodia [sic] and along its border with Thailand," the donors, including the United States, Japan, Australia, Germany, Singapore, Norway, and Sweden, earmarked all but $1.2 million out of $14.2 million for "refugee camps, holding centers and affected Thai villagers." The sole contribution to aid for *Cambodia* was Sweden's. Fortunately the Soviet Union has been providing large amounts of bilateral aid in sectors important for Cambodia's recovery. See *FEER*, February 10, 1983, p. 12.

J. D. Kinzie

THE "CONCENTRATION CAMP SYNDROME" AMONG CAMBODIAN REFUGEES

> The patient is a widowed, thirty-five-year-old Khmer woman from an urban middle class background, referred because of suicidal thoughts and the potential for child abuse. She is an attractive woman and appears social, proper, and deferential to myself and to Ben, a Cambodian mental health worker who is translating. She speaks softly at times, sits still, and shows little expression except a slight tear. I take the history in the usual slow, low-keyed manner. She is clearly very depressed and suicidal. "If I must die, my children should go with me," she says. These symptoms are not uncommon for refugees, but her past history is. Without expression, she tells of being placed in a "labor camp" after Pol Pot took over Cambodia. In the next few years she witnessed the execution of her husband and the death by starvation of one of her children and her parents. She was separated from her brothers and sister, who are presumed dead. Two remaining children were separated from her during each sixteen-hour work day. She describes symptoms of starvation and malnutrition, and frequent arbitrary executions in the camps. At the end of my interview, I feel at a loss for words. I sit back and say to Ben, "It is very sad." "Yes," he says, and interprets it to her. She nods and says no more. My feelings range from profound empathy to outrage at this inhumanity.

I began to wonder about the clinical implications for this patient. Should not trauma this profound have clinical effects besides depression? Should not the effects be long term and difficult to change? How could anyone return to normal living after these experiences? These questions opened our investigation into the effects of the Pol Pot regime on those who survived it.

Since 1978 the Department of Psychiatry at the Oregon Health Sciences University has held a weekly psychiatric clinic for Indochinese refugees. In the first five years more than 250 patients were evaluated, resulting in over

3,000 clinic visits.[1] With the help of trained mental health counselors and interpreters from each of the cultures represented, we have developed an approach to the evaluation and psychotherapy of these patients that seems compatible with their cultural values and yet acceptable to their needs in this country.[2] The contact with patients has been relatively successful and has received widespread community and refugee support. About 15 percent of the refugees are from Cambodia.

In the last several years we have gradually become aware that there is something different about the more recent Cambodian refugees. We were used to hearing horror stories and almost unspeakable human tragedies, but somehow the Cambodians seemed unique. The amount of violence reported in their lives was massive. The deaths and murders that they had seen seemed both more numerous and more wanton than those reported by other refugees. In addition they spoke of forced labor, migration, separation from family, starvation, illness, and hardship that other refugees did not encounter. We realized, however, that what separated Cambodians more than anything else was not that their stories were more horrible, but that their manner of speaking was more passive, unemotional, and almost uninvolved in the events they described. The manner in which they told their stories was so detached as to be almost unconvincing. Often they did not elicit the same degree of emotional anguish within us that other refugee stories had. Gradually, however, we became aware of the intensity of the experiences and their contribution to the patients' symptoms and style of presentation.

The patients' horror stories are largely confirmed by other reports of the situation in Cambodia under the Pol Pot regime from 1975 to 1979. Recent publications describe the true dimensions of the terror during this time. There were political and nonpolitical executions, forced migrations from the cities, starvation, forced labor, separation of families, lack of health facilities and health care.[3] The killing of Cambodians by other Cambodians has been described as autogenocide.[4] The experiences of the Cambodian refugees reminded us of the Nazi concentration camps.

The parallels were so striking, in fact, that we began to wonder if the Cambodians suffered from some of the same symptoms—referred to collectively as the concentration camp syndrome—as Nazi victims. In 1961 Eitinger published a preliminary report on the pathology of the concentration camp syndrome.[5] He believed that the symptoms were organic in nature and were caused by physical shock and malnutrition. Others who studied the same prisoners also noted similar symptoms but thought the problems were caused by the massive, prolonged psychological traumas of internment in a concentration camp.[6] They felt the persistent reaction was due to the separation and loss of relatives, frequent and arbitrary executions, forced labor, deprivation of food and shelter, illness, and constant fear of death. Remarkably consistent behavior has been found in all victims of concentration camps. This includes chronic depression and anxiety, sleep disorders, fatigue, recurring nightmares, isolation, rumination, irritability, and a hyperactive startle reaction. Some survivors of severe persecution have a chronic reactive aggression with persistent anger and hate.[7] Most victims of these camps have had persistent symptoms. And even those who apparently are well adjusted twenty years later can be dissatisfied and continually depressed.[8] It has been proposed that these problems result from the massive threat to the person's existence which remains ever present in one's mind long after the reality of the threat has passed.

Post-traumatic Stress Syndrome

Combat and prisoner-of-war experiences have also led to major symptoms not unlike those observed in concentration camp victims, though POWs usually exhibit a lesser degree of psychopathology. As a result of the Vietnam war experience there has been a renewed interest in the post-traumatic stress syndrome (PTSS). The postwar difficulties of adjustment of Vietnam veterans and POWs rekindled interest in this delayed stress syndrome.[9]

In 1980 the American Psychiatric Association issued the third edition of the *Diagnostic and Statistical Manual of Mental Disorders.*[10] This developed and made operational new criteria for major psychiatric disorders. After extensive field testing these new definitions proved to be very reliable. A new diagnosis was described in this manual that was not in previous editions—a post-traumatic stress disorder of both acute and chronic forms. The criteria for diagnosis are listed in table 1. Among the significant findings is the fact that recognizable stress evokes significant symptoms in almost everyone. Such symptoms include reexperiencing trauma through intrusive recollections of the effect, recurrent dreams, numbness to or reduced involvement in the external world, hyperalertness, sleep disturbance, survivor guilt, trouble with concentration, avoidance of activities that arouse recollection, and intensification of the symptoms due to exposure to stimuli that symbolize or resemble the traumatic events. Clearly, many of the symptoms were derived from the literature on the concentration camp syndrome as well as combat veterans. However, post-traumatic stress symptoms have resulted from a variety of situations: a fire at the Coconut Grove nightclub in Boston in 1942,[11] the collapse of a dam in Buffalo Creek in 1972,[12] the kidnapping of children in Chowchilla,[13] and even car accidents.[14] A wide variety of disasters or catastrophic events can cause symptoms that are remarkably similar and persistent. A recent work by Atkinson et al. has shown that the post-traumatic stress syndrome does seem to be a specific disorder distinguishable from other disorders and from nonspecific adjustment disorders.[15]

Cambodian Refugee Patients

With such a widespread description of symptoms in the Western literature, we asked why Asian patients show so few of these symptoms. None of our patients came to the clinic complaining or seeking treatment for their Pol Pot experiences.

Table 1

Post-Traumatic Stress Disorder

Differential diagnosis

Diagnostic criteria

A. Existence of a recognizable stressor that would evoke significant symptoms of distress in almost anyone.
B. Reexperiencing of the trauma as evidenced by at least one of the following:
  1) Recurrent and intrusive recollections of the event
  2) Recurrent dreams of the event
  3) Sudden acting or feeling as if the traumatic event were reoccurring, because of an association with an environmental or ideational stimulus
C. Numbing of responsiveness to or reduced involvement with the external world, beginning some time after the trauma, as shown by at least one of the following:
  *1) Markedly diminished interest in one or more significant activities
  2) Feeling of detachment or estrangement from others
  *3) Constricted affect
D. At least two of the following symptoms that were not present before the trauma:
  1) Hyperalertness or exaggerated startle response
  *2) Sleep disturbance
  *3) Guilt about surviving when others have not, or about behavior required for survival
  *4) Memory impairment or trouble concentrating
  5) Avoidance of activities that arouse recollection of the traumatic event
  6) Intensification of symptoms by exposure to events that symbolize or resemble the traumatic event.

Acute Form: Onset of symptoms within six months of trauma and duration of less than six months.
Chronic Form: Duration of six months or more.
Delayed Form: Onset of symptoms at least six months after trauma.

*May also be depressive symptoms.

Initially we did not identify symptoms as stemming from the Cambodian tragedy. Most patients suffered from depression, but other symptoms such as conversion reactions or psychosomatic illness were also observed. Were the reactions of Asians or Cambodians to massive stress somehow different or were we simply unable to identify these reactions in our patients? Was there anything different about Cambodians who had a thorough, prolonged "concentration camp experience" that separated them from other refugees? Until 1981 we did not know the answers to these questions. We are just beginning to look at the problem today.

In 1982 we undertook a more in-depth psychiatric review of our Cambodian patients. Recently the post-traumatic stress syndrome has been incorporated in the Diagnostic Interview Schedule.[16] This structured interview allows us to ask detailed questions about the trauma, the symptoms, and their duration, to determine specifically the presence of post-traumatic stress symptoms and to arrive at a diagnosis. The interviews were done with the help of a Cambodian mental health worker with three years of mental health work experience in an American university who served as a translator. We then began to use this interview on all Cambodian patients who had been in Cambodia during the Pol Pot regime. None of these patients had complained of symptoms relating to their experiences during that period. They all had come for other reasons, predominantly depression. The patients were both returning patients who had been in treatment for some time and new patients. Our original work with thirteen patients formed the basis of the first published study on this syndrome.[17] Currently we have reviewed in detail the symptoms of twenty Cambodian refugees, nineteen of whom have all the major symptoms of the post-traumatic stress syndrome.

Description of the Syndrome

It is significant that none of our patients sought treatment or was referred specifically for PTSS. None came to discuss Cambodian experiences or the symptoms that came from them. Most complained of major depressive symptoms, though three patients had conversion reactions and one each suffered from alcoholism, a severe ulcer disease, and schizophrenic symptoms. The depression symptoms, which originally confused the issue because they overlap with but are not the decisive aspect of PTSS, were similar to the primary complaints of other refugees: sleep disorders, insomnia, poor concentration, and poor appetite. The patients suffered from two independent psychiatric disorders, usually major depressive disorder and PTSS. The coexistence of PTSS with other major psychiatric illnesses in the same patient has been described elsewhere.[18] We now believe that most victims of Pol Pot's regime suffer from PTSS. However, unless there are other incapacitating symptoms these people will not seek medical attention solely for PTSS. The reasons for this are probably related to avoidance of thoughts of the past, as will be discussed below.

The summary of PTSS symptoms is shown in table 2, the summary of patients in table 3. The most common symptom, which also was a separate diagnosis in fourteen patients, was major depression. The most striking feature of all nineteen patients was their active avoidance or minimization of the events that happened under the Pol Pot regime. They typically rushed through the first discussion of that period of their life, showing very little emotion and offering no details. None of our patients elaborated on these events, and none volunteered any personal reaction to the situation. Only with the structured interview approach and the establishment of a relationship were patients willing to go into detail. Then the full history and range of symptoms emerged. Nineteen out of twenty patients qualified for post-traumatic stress syndrome, delayed type. Although this diagnosis is accurate, it tends to minimize the problems they faced: it is quite clear that their symptoms were as pervasive as those of the survivors of the Nazi concentration camps. Furthermore, their symptoms had been present for two to four years, usually beginning one or two years after the Vietnamese invasion of Cambodia in 1979. All of the patients stated they had not told their stories before, either to Cambodians or to Americans. Nevertheless, it is possible that they were able to and did talk about their experiences previously, perhaps early in the refugee camp experience. It seemed clear, however, that once the syndrome developed they stopped talking about their past under Pol Pot. On direct questioning most would reluctantly begin to describe the catastrophic horror they experienced—forced migration from their home, forced long labor, separation from family, starvation, little sleep, random beatings, serious illnesses, and the sight of frequent, indiscriminate killings. For some it was even longer before they could tell of their personal losses—deaths of members of their own families, and their own personal terror during that time. A unique aspect of these symptoms was avoidance of any events, even in the interview, that reminded them of that part of their life in Cambodia. Patients did not now discuss these experiences with friends or others, nor did they elaborate upon them in initial meetings in which we began taking psychiatric histories, and when mentioned they were often played down. They similarly avoided all events in their life that reminded them of Cambodia, such as news on television and newspapers. Most also avoided all news or movies that involved violence or natural disasters.

Even the one patient who technically did not qualify for the diagnosis of post-traumatic stress syndrome described appalling events during the Pol Pot regime and then rapidly denied they had any effect on her. She said that she did not want to discuss them anymore, and finally that she did not even want to be a patient in the clinic. This denial of problems was probably a frantic attempt to avoid any further stimulation by the thoughts of what had happened in Cambodia.

The second most common symptom among our patients was strong periodic intrusive thoughts about the past. Despite their inability to discuss them, most experienced intrusive thoughts and nightmares, occuring nightly to weekly, regarding these disturbing events. These usually involved the death of family members. Our interviews often stimulated this process. Once we asked about the events and their reactions patients commonly gave vent to a full flood of emotions which could not be controlled. These outpourings, however, did not reduce the intrusive thoughts. Most patients described emotional reactions of the same intensity that they had had during the actual events. It took a great deal of patience and support during the interview to help them control their feelings. Most of our patients on follow-up visits found no relief in describing these events, and many, at least temporarily, became worse.

Many of the patients described an exaggerated startle reaction or hyperactivity to stimuli that reminded them of events in Cambodia. A startle reaction is a sudden, unexpected response with involuntary movements, increased pulse rate, sweating, and a sensation of anxiety. For example, a knock on the door brought a

Table 2

Symptoms of Cambodian Post-Traumatic Stress Syndrome (N = 19)

A. Present in over 75% of patients
 1) Depressive symptoms with disturbances in sleep, appetite, and concentration
 2) Avoidance of memories or activities and events that arouse memories
 3) Recurrent dreams (nightmares)
 4) Recurrent and intrusive thoughts
B. Present in over 50% of patients
 1) Detachment or estrangement from others (emotional numbing)
 2) Exaggerated startle response
 3) Intensification of symptoms by exposure to events that symbolize or resemble the trauma
 4) Intensification of symptoms with minimal academic, social, or vocational stress
C. Present in over 25% of patients
 1) Family violence, anger, or severe irritability
 2) Guilt or shame about surviving
D. Other: suicidal thoughts or thoughts about death

reaction like a knock on the jail cell before taking a prisoner for execution. They reacted to sudden sounds like explosions. Several patients began to have thoughts about mutilations after they underwent surgery. Other stimuli seemingly unrelated to events in Cambodia—certain noises and unfamiliar sounds—caused immediate startle reaction and hyperactivity. Some patients experienced a sense of detachment, emotional numbing, and loss of interest in their environment.

Six patients had difficulty relating to their spouses or children and were sometimes very irritable or aggressive. The majority said that they were afraid of, did not feel attached to, or did not want to be with anyone outside of their immediate surviving family.

Guilt or shame was present in less than half of the patients. Some had marked ambivalence about leaving relatives behind, and some felt ashamed for having food, as some of their relatives had starved to death. Some patients, however, stated that relatives told them to get out quickly, even though the relatives knew they themselves were dying.

Poor concentration and memory and sleep disorders were almost universal. These problems seemed more subjective since they were not apparent in the interview. Nor were these symptoms specific to the post-traumatic syndrome since they are common among all depressed Asians. Only a few ever displayed any active suicidal tendencies, but thoughts of death, and even a certain resignation to it, were present in some of the patients' minds. Only one patient, an American-trained military officer, displayed anger toward his aggressors of the Pol Pot era and a desire for vengeance. None of the other patients ever mentioned this.

On the whole, the patients were extremely impaired by their symptoms. One was marginally employed, doing odd jobs. The rest were unemployed. Most even found the stress of attending school, with deadlines for class performance, more than they could handle. One patient who attempted to work as a dishwasher found that the pressure of the job reminded him of forced labor such that he became intensely anxious and unable to perform. When their government financial assistance was delayed or the assistance process was not understandable, several panicked and stated they were going to die of starvation, as they almost did in Cambodia. They were an extremely vulnerable group of patients. Their impairment was much greater than was apparent in the earlier interviews, and the threat of death seemed to be just beneath the surface of their consciousness.

Case Examples

Case #1

The patient is a fifty-six-year-old widowed Cambodian refugee. She was originally referred because of tiredness, sleeping problems, some depression, and headaches, which she said had been increasing in frequency for several months. She attributed the onset of her present symptoms of fatigue and swelling legs to the forced labor in the farm system of the Pol Pot regime in Cambodia. Nevertheless, the problems continued when she came to the United States. She developed problems sleeping, and her weight dropped by about eighteen pounds. Her immediate worry was survival in the face of loss of welfare assistance that she had been receiving for eighteen months. A further problem that emerged during the interview was the relationship with her two sons, whom she felt were quite Americanized and did not provide for her or were not interested in her. She said that at times she felt life was not worth living. She was afraid of not being able to make it on her own, and of being put in a nursing home, despite an apparent willingness on the part of her sons to take care of her.

She was from a royal family in Cambodia and apparently was distantly related to a past ruler. She did not attend school. She was married to her husband for thirty-one years. She had a very close relationship with her family, which included four sisters and one brother. Her husband was a successful businessman and politician and rose high in the Cambodian government. He apparently was

Table 3

Patient	DSM III diagnosis	Major stressors	Major PTSS
1. Male age 24	Major Depression	4 years forced labor; starvation; death of friends when escaping; whereabout of family unknown	conscious avoidance of past memories; general ahedonia; panic feelings; startle reaction
2. Female age 35	Major depression	4 years forced labor; child & parents starved; husband tortured & executed; starvation	avoidance of activity, of thoughts of Cambodia; nightmares; uncontrollable intrusive thoughts; ahedonia; suidical thoughts; homicidal thoughts to children
3. Female age 53	Major depression	4 years forced labor; husband & several family members executed; starvation	avoidance of talking about past; instrusive thoughts of husband's death; recurrent dreams that husband still alive; isolation from others; hyperactive response to noises
4. Female age 57	Major depression	4 years forced labor; 2 sons-in-law & 1 son executed in front of; husband died of starvation	"I tried to forget about it" nightmares & recurrent intrusive thoughts of children's execution; guilt of leaving other family members behind; hyperactive startle response
5. Female	Adjustment reaction; non-specific somatic complaints	3 years forced labor; 3 years wandering in jungle; malnutrition	recurrent thoughts & nightmares; emotional numbing & avoidance; hyperalertness; marked intensification of symptoms when recalling her past
6. Female age 56	Major depression	4 years forced labor; siblings executed; husband died of untreated disease	"I don't want to think about it" intrusive thoughts; emotional numbing; poor concentration; avoidance of events similar to past because of intensification of symptoms
7. Male age 58	Major depression, agitation	4 years forced labor; son drowned while escaping	marked difficulty discerning past events; agitation & anger; family violence; marked nightmares & intrusive thoughts of son's death; very poor sleep & concentration; separated from family & lived alone
8. Male age 30	Major depression, agitation	4 years forced labor; hid identity; constant fear of execution; parents & 2 siblings executed; 1 sibling died of starvation	agitation; violent episode; unsuccessful attempts to avoid thoughts of past; desire to be left alone; not interested in any activities; thought he should have died in Cambodia than starve here
9. Males age 44	Alcohol abuse	2 years forced labor; all members of family disappeared; starvation; frequent executions	intrusive thoughts of past; active avoidance of memory of past; startle reaction; recurrent dreams of being chased by Pol Pot
10. Female age 30	Atypical dissociative disorder	4 years forced labor; husband & father executed; 1 child died of starvation	avoidance of any reminders of past; frequent moves because something in neighborhood reminds her of Cambodia; always felt on guard that past was going to repeat itself; emotional numbing & social withdrawal

11.	Female age 40	Mixed dissociative & conversion symptoms	4 years forced labor; husband tortured for 10 months before killed; patient beaten; 1 child died when escaping from Cambodia	hospitalized with suicide attempt; numbing; minimal social responses to family & children
12.	Male age 27	Major depression	4 years forced labor; parents executed, separated from family; members whose whereabouts unknown	hospitalized with suicide attempt; avoidance of thinking of aything related to Cambodia; periodic rage reaction, guilty about being alive; intrusive thoughts & nightmares; some symptoms reduction when received letter from sister informing him of several siblings alive
13.	Male age 30	Major depression, agitation; peptic ulcer	4 years forced labor; parents, 3 brothers & 2 sisters executed; changed identity—constant fear of being exposed as former military officer	hospitalized for depression; irritable & angry episodes towards family; avoidance of thoughts of past, "It's beyond people's understanding." nightmares; easily startled; feels as if Pol Pot's people still here; feels ashamed at being alone; angry at Pol Pot—wants to get revenge
14.	Female age 27	Major depression	4 years forced labor; parents died of starvation; 32 of 36 relatives died; saw cousin beaten to death, dug own grave day before Vietnamese invaded	nightmares of hearing voices of people being executed; easily startled by small noises; avoid "at all costs" memories or reminders of past
15.	Female age 26	Asthma; Major depression	4 years forced labor; 8 of 10 of her siblings killed or starved to death; father killed	recurrent dreams & nightmares; shame & guilt; intrusive thoughts; avoidance of situations & memories of Cambodia
16.	Female age 40	Conversion disorder of body weakness; major depression	4 years forced labor; malnutrition; 2 siblings & father executed; 4 other siblings disappeared or died	continual nightmares, startles easily; intrusive thoughts; avoids thinking about past; easily tearful; suicidal thoughts
17.	Female age 48	Major depression; multiple physical symptoms	4 years forced labor; with malnutrition; husband & 3 children executed; 2 children starved to death	nightmares & intrusive thoughts; startles easily; much avoidance of of thoughts and memory; detached to situations around her; resigned to death
18.	Male age 30	Major depression; gastric ulcer requiring 2 major surgeries; marked weight loss	4 years forced labor; lost track of 4 siblings, 1 presumed dead; father executed; hid identity of being in military	recurrent nightmares; always avoids memories of past guilt about being alive; avoids emotional problems; irritability
19.	Female age 31	Schizophrenia; seizure disorder	4 years forced labor; starvation conditions; threat of execution; uncle & aunt were killed	nightmares; tries to avoid thinking about past but unable to do so; startles easily, irritability & anger; outbursts towards family

quite sophisticated and traveled frequently. Her two sons were in the United States studying at the time of the revolution in Cambodia. Both are now doing well in the United States. When the communists took over, she and her husband had to leave Phnom Penh immediately and were forced to travel through the country working, receiving little food. Her husband, apparently suffering from malnutrition, began to "swell up" and died after two years. When her royal background was exposed, all her siblings were executed. She had no contact with any family member after the communists took over. She described no other details and was obviously reluctant to talk about her experience.

She is a petite, poised woman who appeared tired and older than her age. She had a great deal of social presence and skills and managed some laughter, but fought to hold back her tears. We began by treating her for depression. It was not until six months later that we did a more structured interview on the symptoms to determine her degree of impairment and suffering. Since arriving in the United States she has had frequent nightmares about what happened in Cambodia. These occur weekly and she often awakes in terror at night. She frequently becomes jumpy and is easily startled by small noises. Almost any noise startles her, although she cannot identify the stimuli further. She has to stay on guard and will not let herself get involved in events or with people. She is especially guarded when alone, or when listening to bad news on television about violent crime. Since arriving in the United States she has had marked difficulty sleeping, which was alleviated by antidepressant medicine. She also has had marked difficulty concentrating since 1979. She says, "I have lost all my memories. I cannot study well and can hardly concentrate." The problem had intensified during the preceding year. She says that she has lost feelings for others around her, people she normally cares about. "I only think of myself, I don't have much feelings for other people." She states that she still has some interest in a few activities that relate to her Buddhist religion. Otherwise she shows little interest in anything. She misses those she left behind and feels a little ashamed about being alive. She would like to go back if life were a little better. "I don't want to stay in the United States. Sometimes I see people here that are not friendly." She avoids at all costs anything that reminds her of the evil "blood bath" of Pol Pot. As this information was being described the patient became visibly more agitated and developed a full flood of emotions of pressured tearfulness. Each time the events of Cambodia were raised she showed the same degree of frightful, tearful anxiety that she had shown previously.

Case #2

The patient is a twenty-seven-year-old single Khmer male who was referred to our clinic by a hospital after a serious suicide attempt. He had taken an overdose of medication because he was very demoralized about his future, his inability to get a job, lack of financial resources, and a feeling of separation from his family. He came from an urban middle class family, attended a university in Cambodia, and suffered a great deal of social disruption following the Pol Pot takeover. He was sent to the countryside and his parents were executed. He was separated from his siblings and has heard nothing from them. He came to the United States three years ago and has been increasingly demoralized and desperate. He has had other problems as well. Every since he escaped from the camps he has had periodic rage attacks during which he became very irritable and angry. These would occur about every three days. He will feel "wild" or "crazy," and then he will calm down and feel normal again. He also has problems with poor sleep, decreased energy, and poor concentration. He feels guilty and shameful about being alive and has frequent suicidal thoughts, which culminated in the overdose. He recently ran out of the money he had saved and is unable to find work. He has marked intrusive nightmares about past experiences, as well as about what happened to his parents. He avoids thinking about the past whatsoever. He avoids any reading or television that would expose him to violence or war scenes, and he frequently has a startle reaction to minor noise. He appears desperate and sad during interviews and even shows some startle reactions with brief hyperactivity as he describes his past. He particularly avoids any mention of Cambodia and is very troubled when he sees television scenes of people in other countries begging for food. As he describes the past he becomes visibly pressured and says he hates the word Cambodia. He lives in constant fear and only wishes that "a foreigner had done it to us.

Cambodians did it to themselves, we do not know what to say. It would be better had it been done by the enemy but it was done by our own countrymen. Cannot be angry about that, better not to think about it.''

Case #3

The patient is a forty-eight-year-old widowed Khmer refugee referred by a health clinic for a large number of physical symptoms for which no medical cause could be found. Besides her physical symptoms, she has poor concentration, headaches, nightmares, poor sleep, poor appetite, weight loss, persistent fatigue, and almost no interest in anything around her. She shows avoidance behavior regarding all the events in Cambodia and feels very guilty about having left. She has startle reactions and intrusive thoughts and nightmares regarding the horror she suffered. She says that she is sad much of the time, and that she has resigned herself to death should it come.

The patient was the uneducated wife of a soldier and the mother of eight children in Cambodia when the Pol Pot regime took over in 1975. When she and her family were separated, she performed forced labor and suffered malnutrition. Her husband was executed in 1977. One son and two daughters were executed; one son and another daughter died of starvation. She was forced to do five hours of labor a day, ate little, and very frequently became sick. This ended only in 1979 when the Vietnamese took over. She and three of her children left Thailand in 1982 and came to the United States. One daughter is married and lives in another state, two daughters live with her. She states that she had a very happy life in Cambodia until the communists took over and feels guilty about leaving Cambodia and some of her relatives behind. She tells her story in a rather detached, unemotional way. She tries very hard to hold back tears—or any reaction—and appears very estranged from others around her. She actually minimizes what she has undergone. At times she seems sad and detached, estranged and indifferent. She seems actively to avoid all memories, detaching and separating herself from events around her. She appears to be quite numb, stating that she has interest in only one daughter at this time. She has no interest in other activities, feels

> ashamed for being alive when some friends are starving. She actively avoids anything that reminds her of Cambodia (e.g., movies and television) or that would expose her to any war situation. She had never told the events that happened to her to anyone before and seems to receive no relief from telling them to us.

These patients told their stories in a straightforward manner. But the amount of stress they experienced is truly catastrophic. Their anguish became evident only after persistent interviewing. It was also difficult for us to listen to their stories without either retreating to a detached "professionalism," with a tendency to deny the real human suffering, or conversely, overreacting and losing our ability to listen compassionately and intensely to their anguish. They are some of the most difficult, if not the very most difficult, stories and feelings a psychiatrist can witness.

The Psychiatric Literature and Its Application to Cambodian Patients

It is clear that the concentration camp syndrome is a rather severe manifestation for what in generic terms is described as the post-traumatic stress syndrome, chronic type. It is also clear that the syndrome that has been described in the West is evident among victims of the Pol Pot regime, thus giving cross-cultural evidence for the syndrome. The most unusual aspects of the syndrome include its duration for as long as four years, the presence of depressive symptoms, and the predominant pattern of avoidance that characterizes thoughts, behavior, and activity. Victims consciously structure their lives to avoid thinking about what happened to them, making detection of the syndrome difficult. But this avoidance notwithstanding, intrusive thoughts, nightmares, and other bad memories are common. They cannot forget. The clinic interview even served as a stimulus for further intrusive thoughts, which intensified the existing symptoms. There was no catharsis or healing effect.

Eitinger, in his earliest work studying concentration camp victims in World War II, argued that this syndrome could be due to an organic brain disease brought on by the physical trauma and malnutrition the victims endured.[19] Most of our patients also described multiple physical traumas, tortures, and malnutrition and showed symptoms of vitamin deficiency. Although these physical disorders were probably not the sole cause, they certainly contributed to the syndrome by weakening the victims and leaving them biologically vulnerable to the psychological trauma.

Kolb and Mutalibassi reported that when exposed to combat sound-stimuli, some PTSS Vietnam veterans developed an immediate physiological hyperactivity in the cardiovascular and neuromuscular systems. The authors proposed that some veterans had an abnormal potential for fright arousal.[20] Such patients may have an on-going perceptual abnormality which results in an inability to distinguish between past and current perceptions. This can cause a primitive startle reaction, through emotional conditioning to a "current" stimulus, which cannot be distinguished in the patient's mind from a previously threatening stimuli. The response patterns to stimuli of many Cambodians, such as startle reactions, seem to resemble a primitive conditioned response that is very difficult to extinguish.

The psychodynamics of the concentration camp syndrome have been the subject of numerous investigations. Early psychoanalytic thinkers described traumatic neuroses as the reexperiencing of a childhood trauma. They concluded, therefore, that adult traumatic neurosis could not exist unless a childhood trauma had preceded it. Chadoff criticizes this view on the basis of the universality of adult traumatic neurosis in victims of concentration camps and concludes that the traumatic neurosis of the concentration camp syndrome is almost entirely the result of the severe trauma itself. This could affect a previously healthy personality. On the other hand, Grauer suggests that the condition really is closer to a psychosis or a borderline syndrome because of the chronicity of the symptom and the poor response to therapy. He feels that severe trauma results in permanent ego changes: ego exhaustion, similar to what occurs in the elderly, and changes in the ego-superego boundaries. The punishing superego continues to attack the impoverished guilt-ridden ego. The ego does not have the capacity to neutralize the superego; there is no incorporation of the superego within the ego. The ego cannot be gratified, so it becomes drained. These superego disturbances cause shame and guilt which result in repetitive nightmares, daily rumination, and phobias.[21] The consensus among psychoanalysts is that the most usual pattern seen in concentration camp victims is a massive repression of the aggressive-like impulses that affect survivors. This repression, necessary in the concentration camp, became permanently fixed in the personality, causing guilt and depression. It seems clear that trauma in adult life adequately explains the syndrome. It is difficult, however, to know if the concept of the superego-ego imbalance is applicable to Asians in view of the relatively infrequent findings of guilt. Certainly the massive repression of aggression seems to be common among our patients.

Lifton explores the importance of the "death imprint" in the post-traumatic stress syndrome. He contends that the survivor "is one who has come in contact with death in some bodily or psychiatric fashion and has remained alive." In his view this death imprint consists of an intrusion into the mind of an image associated with death. This image is accompanied by the feeling that a premature or an absurd death is unacceptable. The death image is powerful, and an individual faces anxiety when unable to assimilate it into life. This may be followed by an incapacity to respond and a resulting sense of guilt. Lifton feels that the traumatic syndrome is a state of being in which one is exhausted by images that can neither be assimilated nor cast aside. The suffering is associated with being "stuck." This guilt is static. The guilt and the loss tend to be overwhelming in relationship to an unacceptable death. This results in a psychic numbing and a death-linked image of denial. The psychic numbing may be a form of symbolic death in order to avoid permanent physical or psychological death. A further concept that Lifton explores is the conflict between the joy of having survived and the quality of one's life afterward.[22] Although the concepts Lifton uses tend to be abstract, he certainly has extended our view of the traumatic stress syndrome, especially concerning those who were near death. Although not formally tested in our interviews, our patients had a rather severe preoccupation with death images, commonly associated with the meaninglessness of life. It is as Lifton suggests: they faced death and came back from it with an image that left them numb, vulnerable, and struggling for a sense of meaning.

Horowitz and his colleagues describe two common stress reactions, a frozen avoidance response and persistent intrusive thoughts.[23] The major response may relate to the previous personality style of the individual. Our Cambodian patients predominantly had a frozen avoidance response, oscillating with persistent intrusive thoughts. There was a conscious effort to avoid the persistent intrusive thoughts; sometimes it was successful and sometimes it was not. The struggle tended to consume most of the psychological energy of our patients.

In combination, these concepts may help explain PTSS. The biological trauma, malnutrition, and illnesses certainly left the patients weakened and vulnerable to psychological stress. In the case of many of the symptoms, there appears to be a conditioned response to a stimulus, especially in a state of hyperarousal. It is also clear that the syndrome can occur in previously healthy adults, and that probably the most common defense mechanism has been denial of aggressive feelings previously adaptive but now fixed and maladaptive. The concept of a death image has contributed to the psychology of survival and has advanced our understanding a great deal.

Distinctive Features of the Cambodian Case

Beyond an almost universal brutality, the Cambodian refugees were set apart from other victims of persecution in significant ways. First, the brutality was perpetuated by other members of the same ethnic group—that is, by other Cambodians. Second, the very social and cultural fabric of Cambodian life was destroyed. Family life was shattered, past beliefs and traditions were demolished, religions were completely suppressed, and people of high status were systematically exterminated. Our patients were left with an overwhelming sense of powerlessness. Their reaction was surely influenced by a tradition of nonconfrontation or withdrawal, which contributed to demoralization and emotional numbness.

The withdrawing and avoidance behavior finds cultural support in the traditional Cambodian belief that personal bad fortunes stem from dishonorable events in a previous life. Cambodians under Pol Pot were destroyed by other Cambodians, a shameful event which some perceived as the result of a previous dishonorable or tragic event in Cambodian history. Therefore, Cambodians believe that they experienced shame on behalf of the whole nation due to transgressions by their ancestors. Such collective shame may have influenced avoidance behavior and contributed to the feeling that the suffering should be borne alone and in silence.

A third problem the Cambodians faced was that their difficulties did not end with the fall of the Pol Pot regime. They still had to endure a series of adjustments and traumas: avoiding invading Vietnamese soldiers and Thai bandits; trying to find their way across Cambodia; and adjusting to refugee status in Thailand for an indeterminate time, unsure of their future and with no knowledge of their relatives. Typically they were in refugee camps from one to three years before they were relocated in the United States, where they confronted a new culture, a new language, a limited social network, and precarious financial status. They also faced unemployment, an uncertain future, and continuing stress, usually without friends and in an unsympathetic environment. They often bore these burdens alone. Clearly their problems and symptoms did not end with escape from Cambodia.

The pathological states induced by the concentration camp should not cause us to overlook some of the coping and survival tactics used by survivors. Because of a reluctance to talk about their experience and because little has been written by the victims themselves, we are only beginning to understand how people cope. This is in contrast to the voluminous literature on and by victims of the Nazi concentration camps.[24] All survivors we saw in the clinic attempted to disguise their past identity during the Khmer Rouge period, especially if they had been associated with the government or the military, or if they were well educated. They pretended to be farmers, peasants, or people with little contact with or knowledge of the former government or outside world. One university graduate pretended to be a bicycle repairman. A former government official pretended for three years to be deaf and dumb.

They all tried not to stand out but to disappear in the crowd. To be singled out often meant death. They also tried to live one day at a time. Denial of death did not seem to be a strong impulse since many recognized they could die at any time, especially if their true identity were exposed. Still, coping one day at a time was not easy. Finding enough to eat—bugs, leaves, and insects often meant survival—was a major preoccupation.

Until more information is available and former labor camp prisoners feel freer to remember and describe their experiences, we can only marvel at the human ingenuity they used to cope with adversity.

Psychotherapeutic Considerations

It has been questioned in the past whether psychoanalysis or dynamic psychotherapy effectively treats victims of severe persecution. DeWind reviews a great deal of controversy among psychoanalysts on this subject.[25] It is the experience of most that there are many obstacles to the successful therapy of these victims.

The reasons for this are many. The massive traumatic experiences are extremely painful for the patient to work through. The patient simply cannot stand to relive the experience. The therapist also may not be able to tolerate the terrible stresses of the survivors and may react with anxiety and guilt. This can result in a whole range of defensive maneuvers on the part of the psychiatrist, ranging from over-identification at one extreme to reaction formation and denial disguised as objectivity at the other. Perhaps the simplest explanation for the difficulty of psychotherapy is that the capacity of persecution victims to trust other human beings is so impaired that they are not able to enter into reciprocal relationships.[26]

It is clear that the goal of therapy for the massively traumatized is not "cure" in the sense of being rid of painful memories.[27] In the end the sad experience of persecution can never be forgotten. Still, some therapeutic approaches can greatly alleviate the pathogenic influences and suffering. A limited, supported psychotherapeutic approach, oriented toward symptomatic relief rather than personality reconstruction, is most useful. This minimizes the need continually to reexperience the trauma, does not produce such strains on the therapist, and can allow trust to develop slowly as symptoms are reduced. It also focuses on the future and plans instead of the past and memories—a focus which helps to limit intrusive thoughts.

A promising technique advocated for Vietnam veterans with post-traumatic syndrome is group psychotherapy.[28] Various psychotherapeutic drugs can also offer some relief for PTSS victims.[29]

The therapy of Cambodian refugees presents special problems. First, there is a singular unwillingness to talk about the trauma. This makes any type of verbal, interactive psychotherapy extremely difficult. Also there is no historical or cultural analog of individual psychotherapy in Cambodia. Our usual means of dealing with these experiences—verbal interaction—seems inappropriate and strange to the Khmer. They are also reluctant to talk in groups about personal matters, which makes group therapy an unlikely mode of treatment.

We have found some medicines to be effective in reducing some of the biological symptoms of depression as well as some of the aggressive and hyperarousal symptoms. On a few occasions they have produced dramatic benefits. In our psychotherapy we have aimed to be consistent and warm without appearing too intrusive regarding their symptoms and the past. This seems to be necessary since further inquiry often intensifies the symptoms. It is important to take an interest in their personal needs and to offer assistance in securing benefits, whether financial or medical. They are often so vulnerable that loss of these basic supports results in intense insecurity and increased symptoms. It is also important for many refugees to document their impairment. They often feel guilty about not working but are so impaired that the pressure of work increases their symptoms. Giving formal sanction to their impairment has often reduced their symptoms a great deal. Of great importance is the psychiatrist's recognition that this is a chronic illness and a long-term therapeutic approach and commitment are needed. To know that the therapist is consistent, is committed to a long-term relationship, and is not demanding a "cure" provides much security for these patients.

Preliminary data at this time indicate we can perhaps be more optimistic about the outcome of these symptoms than we were in the past. Certainly at least some of the biological symptoms can be reduced and the increased arousal, startle reflex, and even rage reactions can be modified. The consistent concern and interest of the therapist is helpful, particularly if the amount of intrusiveness that the patient tolerates can be accurately judged. There are times when it may be helpful for the patient to talk about the past and times when it is contraindicated.

Afterthoughts

How could another holocaust have happened in the twentieth century? After being shocked by the Nazi concentration camps, one somehow has the feeling that our civilized world could not let it occur again, at least not on such a scale and so systematically. One can ask an even more naive question: What could make these nice, quiet-appearing Asians kill each other?

Psychiatrists are used to hearing about personal violence. We hear of our patients' hatreds, their desires, and fantasies of killing others. We interview those who have been extremely aggressive or have committed homicide. Doctors have witnessed personal violence, which has become a large part of our current life. In Vietnam I witnessed the almost casual cruelty of both Americans and Vietnamese. Even so, the individual psychiatric patient case does not explain behavior in such a holocaust.

The question remains: how did a particular long-standing stable culture go through such a frenzy of brutal and homicidal behavior? There is no clear answer. Perhaps the Cambodians were carrying to a logical extreme the violence that they had witnessed around them. The contending powers in Indochina—backed by the military might of East or West—unleashed massive destructive power. When fanatical revolutionary leaders pursued the destruction of their enemies out of the world's view and awareness, the violence became uncontrolled and even genocidal. Finally, once the leaders and carriers of traditional culture were eliminated, there were no cultural values or traditions to check the destructive process. There was neither an outside voice nor an inner voice telling them to save themselves. As history has shown, out of chaos can come systematic violence. Cambodians are no different than other people—capable of good and evil at different times. But Cambodian experience from 1975 to 1979 is among the most grim reminders of the human potential for evil. This poem written by a Khmer refugee seeks to express the Cambodian experience:

Hell and Evil Society

The year was 1975, I remember it very well.
One regime had been dissolved.
A new regime came full of thirst for blood.
Killing in cold blood all over the nation.

This new regime, is definitely called communist.
Nothing to hide, nothing to cover.
Broadcasting throughout at the provinces.
They said, this new regime is *equality*.

OH! My God! It is opposite to what we thought!
They are the new oppressors on us.
They have enough to eat for them—both men and women,
But for us—people are starving to death from this oppressive regime.

Work hard labor all day and night with no breaks.
Living with fear, which no one could ever imagine.
One dip of rice porridge per day.
Then pushed tight and held in one place.
Then beaten and executed us.

On the one hand, no food. We are all starving to death.
On the other hand frightened, fearful, afraid of cadres.
Killing with no mercy. Butchering and executing like barbarians.
Each person is starving to death, their bodies are so skinny, look like vines.
Each day, all day, all you can hear is their voices from the grave crying for help.
Others are sick, laying there in the sewer, crying for help, but I cannot open or close my eyes.
Skinny bodies, skeleton-like, all over the place under the heat of the sun.
No medication, nor medical doctors for help. None.
Not even a little. None whatsoever.

Corpses lay, death all over, filling the ground.
If people see, it will tremble their heart, but none are left to see.

You could see your own corpse which the butcher axes cut with scars.
Marks all over your body, with both hands and feet tied to your back.
No clothes. Everyone is a corpse.
Young, old, males, and females dead on top of each other—creating a mountain of corpses.
All the hair fallen off of the corpses—rotting to the ground.

Some lost their wives, some lost their husbands.
Even the rice paddies and the food we produced is killed.
This new society is very blood thirsty, killing the young, the old, with no mercy.
This is the real hell and the evil society.
Pol Pot is the real and the most evil of all mankind in the world.

Po, Bunthan—A former student of philosophy
Phnom Doun-Penh, Phnom Penh

Translated by Rath Ben

Notes

1. J. D. Kinzie and S. Manson, "Five Years Experience with Indochinese Refugee Patients: What Have We Learned," *Journal of Operational Psychiatry* 14 (1983):103–11.
2. J. D. Kinzie, "Evaluation and Psychotherapy of Indochinese Refugees," *American Journal of Psychotherapy* 35 (1981):251–61.
3. *Political Killings by Governments, an Amnesty International Report* (London: Amnesty International Publications, 1983).
4. D. Hawk, "The Killing of Cambodia," *New Republic* 187 (1982):17–21.
5. L. Eitinger, "Pathology of the Concentration Camp Syndrome," *Arch. General Psychiatry* 5 (1961):371–79.
6. P. Chodoff, "Psychiatric Aspects of Nazi Persecution, *American Handbook of Psychiatry,* 2d ed., 6 (1965):932–46; E. C. Trautman, "Fear and Panic in Nazi Concentration Camps; A Biosocial Evaluation of the Chronic Anxiety Syndrome," *International Journal of Social Psychiatry* 10 (1964):134–41.
7. K. D. Hoppe, "Chronic Reaction Aggression in Survivors of Severe Persecution," *Comp. Psychiat.* 12 (1971):230–37.
8. P. Ostwald and E. Bittner, "Life Adjustments After Severe Persecution," *American Journal of Psychiatry* 124 (1968):1393–1400.
9. N. E. Archibald and R. D. Tuddenham, "Persistent Stress Reaction Following Combat: A Twenty Year Follow-up," *Arch. General Psychiatry* 12 (1965):475–81; O. K. Buck and J. I. Walker, "Post-traumatic Stress Disorder in Vietnam Veterans: A Review," *South Med. Journal* 75 (1982):704–706; J. S. Frye and R. A. Stockman, "Discriminant Analysis of Post-traumatic Stress Disorder Among a Group of Vietnam Veterans," *American Journal of Psychiatry* 139 (1982):52–56; H. A. Wilmer, "Post-traumatic Stress Disorder," *Psychiatric Annals* 12 (1982):995–1003; F. T. Corcoran, "The Concentration

Camp Syndrome and USAF Vietnam," *Psychiatric Annals* 12 (1982):991–94; W. H. Sledge, J. A. Boydstun, and A. J. Rabe, "Self-concept Changes Related to War Captivity," *Arch. General Psychiatry* 37 (1980):430–43.

10. American Psychiatric Association, *Diagnostic and Statistical Manual of Mental Disorders*, 3d ed. (Washington, D.C.: APA, 1980).

11. A. Adler, "Neuropsychiatric Complications in Victims of Boston's Coconut Grove Disaster," *JAMA* 123 (1943):1098–1101.

12. J. L. Titchener and F. T. Kapp, "Family and Character Change at Buffalo Creek," *American Journal of Psychiatry* 133 (1976):295–99.

13. L. Terr, "Chowchilla Revisited; The Effects of Psychic Trauma Focus," *American Journal of Psychiatry* 140 (1983):1543–50.

14. J. I. Walker, "Post-traumatic Stress Disorder After a Car Accident," *Post Grad Med* 69 (1981):83–86.

15. R. M. Atkinson et al., "Diagnosis of Post-traumatic Stress Disorder in Vietnam Veterans," preliminary findings presented at 136th annual meeting of American Psychiatric Association, New York, 1983.

16. L. N. Robbins et al., "NIMH Diagnostic Interview Schedule, Wave II," 1981.

17. J. D. Kinzie et al., "Post-traumatic Stress Syndrome Among Survivors of Cambodian Concentration Camps," *American Journal of Psychiatry,* forthcoming.

18. F. S. Sierles et al., "Post-traumatic Stress Disorder and Concurrent Psychiatric Illness: A Preliminary Report," *American Journal of Psychiatry* 140 (1983):1177–83.

19. L. Eitinger, "Concentration Camp Syndrome."

20. L. C. Kolb and L. R. Mutalipassi, "The Conditioned Emotional Response: A Sub-class of the Chronic and Delayed Post-traumatic Stress Disorder," *Psychiatric Annals* 12 (1982):979–87.

21. H. Grauer, "Psychodynamics of the Survivor Syndrome," *Canada Psychiatric Association Journal* 14 (1969): 617:22.

22. R. J. Lifton, "The Psychology of the Survivor and the Death Imprint," *Psychiatric Annals* 12 (1982):1011–20.

23. M. H. Horowitz, "Stress Response Syndromes: Character Style and Psychotherapy," *Arch. General Psychiatry* 31 (1974):768–81; N. H. Zilberg, D. S. Weiss, and M. J. Horowitz, "Impact of Events Scale: A Cross-validation Study and Some Empirical Evidence Supporting a Conceptual Model of Stress Response Syndromes," *Jounal of Consult. and Clin. Psychology* 50 (1982):407–14.

24. V. E. Frankl, *Man's Search for Meaning: An Introduction to Logotherapy* (New York: Washington Square Press, 1959, 1963); T. Radel-Weiss, "Men in Extreme Conditions; Some Medical and Psychological Aspects of the Auschwitz Concentration Camp," *Psychiatry* 46 (1983):259–69.

25. E. Dewind, "Psychotherapy After Traumatization Caused by Persecution," *International Psychiatric Clinic* 8 (1971):93–114.

26. *Political Killings by Governments,* an Amnesty International report.

27. E. Dewind, "Psychotherapy After Traumatization."

28. J. I. Walker, "Group Therapy in the Treatment of Vietnam Combat Veterans," *International Journal of Group Psychotherapy* 31 (1981):379–89.

29. V. A. van der Kalk, "Psychopharmacological Issues in Post-traumatic Stress Disorder," *Hospital and Community Psychiatry* 34 (1983):683–91; J. I. Walker, "Chemotherapy of Traumatic War Stress," *Military Medicine* 147 (1982):1029–33.

Cynthia M. Coleman

CAMBODIANS IN THE UNITED STATES

From the United States to Democratic Kampuchea: Returning to the Homeland

At the time of the fall of the Lon Nol regime to the Khmer Rouge in 1975, a number of Cambodian military men were in the United States to study military tactics. Some had arrived as recently as three weeks before the fall of their government. They had come on student visas, jointly sponsored by the U.S. and Cambodian governments, and had left their families behind in Cambodia. In late May or early June 1975, the U.S. government cancelled their visas and scholarships and told them to report to Camp Pendleton, California, where they would be processed and resettled in American society as refugees.

Some of those who reported to Camp Pendleton decided to attempt to return home to Cambodia instead of becoming refugees in the United States, primarily because all had wives and families in Cambodia. They felt responsible for having left their families behind at a time when they knew that the old regime was near collapse. There was a strong sense of both guilt and duty toward those they loved. They were joined in the camp by several families who had fled Cambodia during April 1975, leaving behind children and spouses. The families, after several weeks in a U.S. refugee camp, were obsessed with the fate of their loved ones, and they came to the conclusion that their flight had been precipitate. Camp Pendleton was rife with rumors about the Khmer Rouge takeover—rumors fed by the refugees, U.S. military and civilian personnel, and U.S. resident Cambodians who went to the camp to seek information about their relatives. The families and ex-military men heard the stories, believed (or feared) that many of the stories were true, and banded together in their resolve to return home, to protect their families. Barring that, they hoped to gain admission to a third country, perhaps a third world nation in Africa or Latin America, from which they felt it would be easier to be accepted by Cambodia for repatriation. Since the new communist regime in Cambodia was hostile to the U.S. government, the group was worried that the Khmer Rouge would not accept them back if they came from the United States.

When Fort Chaffee, Arkansas, opened as a refugee camp and Camp Pendleton began to wind down, this small group was moved to Chaffee, where it was joined by other Cambodians seeking repatriation. By November, there were 114 Cambodians—81 men and 11 families—at the camp. The entire group still refused refugee status or sponsorship, afraid that any legal recognition by the U.S. government would jeopardize their chances of returning to Cambodia. They appealed instead to the United Nations High Commissioner for Refugees (UNHCR) for UN status.

In early December 1975, while still at Fort Chaffee, the group was informed by UNHCR that they would shortly be sent to Philadelphia, where they would be allowed to wait for three months at the YMCA for Cambodia to accept them back. After that, they would have to accept refugee status in the United States if no decision on their repatriation had been reached. UNHCR refused to give them UN status or to treat them as visitors; it wanted them to take jobs.

While still at Fort Chaffee, the group was also turned down on a number of small requests: an office for their leader, UN flags, badges, ID cards, and armbands. At this point, the group announced that they would starve themselves if they had to enter U.S. society, and they talked of the possibility of self-immolation. They said that they would refuse to leave Fort Chaffee. They told the UN representative that the United Nations was untrustworthy and that the U.S. government was responsible for their difficulties. They demanded to be parachuted into Cambodia, or taken to the Thai-Cambodian border and released to walk into Cambodia, or provided passage on a ship—anything to facilitate their return.

After several days of intense negotiations, the Cambodian group said that they were willing to go to Philadelphia. They had no intention of causing trouble. If they could not go home within three months, they would quietly starve themselves.

A private, voluntary agency—one of the agencies which contracts with the U.S. Department of State to resettle refugees in the United States—signed a contract with UNHCR to provide care and maintenance for the group during their stay in the Philadelphia YMCA. Before the group arrived, agency representatives met with representatives of UNHCR and the U.S. Department of State. UNHCR suggested that the group had little hope of repatriation. It was, in fact, unclear who outside of Cambodia was empowered to speak for the Cambodian government—formal channels of communication had apparently not been established, thereby making it almost impossible to repatriate the group officially. The State Department's representative, on the other hand, expressed the hope that the private agency would be able to erode the group's solidarity, so that they could be individually sponsored and integrated into American society as refugees. The State Department had spent months attempting to talk the group out of repatriation. By the time the group left Fort Chaffee, the department estimated that the agency had, at best, a 10 percent chance of success.

The group of 114 Cambodian would-be repatriates arrived in Philadelphia on December 11, 1975. They organized themselves quickly, setting up a system of internal control: check-in procedures were instituted, and a 10:30 pm curfew established. No single men were allowed on the twentieth floor, where the families and women were housed. The group began to keep a close watch on the voluntary agency's staff, following them around the building, clocking them in and out of the office set up for them, even on weekends.

The group as a whole decided to study Cambodian revolutionary theory in preparation for their eventual repatriation. Two or three, however, went to the staff secretly to ask for English books, explaining that studying English had been proscribed. Many of the military men refused to speak to the agency staff, referring any questions back to the group leader.

Sam, the group's spokesperson, talked cryptically to the staff about U.S. Central Intelligence Agency (CIA) spies within the group who needed to be weeded out before the group could be repatriated. He said that he was in daily contact with the Cambodian mission in New York. It took the staff quite a while to discover that the "mission" actually consisted of several Cambodian permanent residents who, while they had no official status in the United States, maintained contact with the Royal Government of National Union of Kampuchea (GRUNK) mission in Paris through the Chinese courier at UN Headquarters in New York.

Sam talked a good deal about the need for the group members to purify and discipline themselves, to study Cambodian nationalism, and publicly to confess impurities and receive "condemnation" from the group in order to be allowed to return home to Cambodia. The group began to hold evening meetings with mandatory attendance for these purposes. Agency staff members were specifically not invited.

Within a month of the group's arrival in Philadelphia, it was clear that the leadership was receiving specific instructions on how to comport themselves, although at the time it was not clear to agency staff where the information was coming from. Staff members were getting conflicting stories from different sources. Sam was still talking about his "boss" in New York at the Cambodian "mission." In mid-January, that man went to see UNHCR's representatives. After their conversation, UNHCR telephoned agency staff to report that it felt all of this was getting a bit out of hand. The Cambodian who had seen them was not a "contact" at all but a professor of economics in New York who intended to return home to Cambodia at some point and felt that he himself was not "tainted" by having lived or worked in the United States. The next weekend the man arrived in Philadelphia and reportedly told the group that they should go to work. None did.

Sam was beginning to use a new vocabulary—one day describing Cambodia as a "nonaligned, neutral power," the next day calling a staff member "comrade." The group formed a "Supreme Committee" of seven top-echelon leaders, six second-echelon leaders, seven "facilitators," and an overall chairman. The self-criticism sessions continued at night and on weekends, long meetings lasting several hours, where group members "confessed" to impure thoughts and deeds. Sometimes they confessed to being "weak" in their thoughts only, doubting the goodness of the new society or wanting to commit adultery. Other times they confessed to actions—sneaking out at night with a woman, studying English or Christianity alone in their rooms, or talking to agency staff members about the last meeting. Group members then stood up, one by one, and condemned the indiviudal, which assisted the purification process. By the end of the season, they felt cleansed but were extolled to continue their struggle against their own weaknesses, in order to be able to return home with the rest of the group when the time came.

While the group was officially under UNHCR auspices, the money to maintain it was sent by the U.S. State Department—through the United Nations—to the private agency in Philadelphia. Policy decisions that affected the budget, therefore, were referred to the State Department, while questions about the process of negotiations for the group's return to Cambodia were referred to UNHCR.

By mid-January 1976, it seemed that UNHCR had still not contacted the GRUNK mission in Paris about the group's desire to repatriate, nor had any representatives of Democratic Kampuchea, as the new regime called itself, been officially consulted about this group. In fact, on January 14, UNHCR representatives reported that it appeared, contrary to what agency staff members heard from Sam and others in the group, that there was not just one "mission" in Paris consisting of three or four persons, but perhaps several groups operating as representatives of Democratic Kampuchea.

On January 19, the agency's staff met with representatives of UNHCR and the State Department to explain certain changes in the group: an increase in physical and psychological symptoms of anxiety and depression, as well as the reorganization that involved the emergence of the Supreme Committee. The UNHCR representative agreed to cable his headquarters in Geneva to ask that a representative be sent to Paris to open discussions with the GRUNK mission or whoever was representing Democratic Kampuchea. He anticipated, however, that this would take some time, since the GRUNK mission would then need to contact the government in Phnom Penh and wait for an answer.

On February 2, 1976, having received an answer from the GRUNK mission, the UNHCR representatives came to Philadelphia to speak to the group. They reported that the situation looked bleak. They had been informed that persons could apply for repatriation at the mission, but that cases would be decided only on an individual basis. Most of the group felt that their chances of acceptance were better as a group, and that it would be safer for them all to return together. The UN representatives appeared to be telling the Cambodians that they were now on their own and that there was not much else that the United Nations could do for them.

Sam began saying that he had been told that the group was not intended to take the message from the GRUNK mission seriously, and that the mission was simply using it as a ploy to get the United Nations out of the discussions. After all, he suggested, Cambodians should only be dealing with Cambodians, without outside intervention. It was not clear to agency staff where this information had come from. Apparently, some members of the group did not believe him and accepted the UNHCR message as fact.

A week after this meeting, several members of the group, including the chairman of the Supreme Committee, traveled to New York to meet privately with the UNHCR representative. The meeting lasted three hours. The chairman later told staff members that he had requested that UNHCR supply the group with airline tickets to Paris so they could be interviewed by the GRUNK mission. The representative promised to confer with UNHCR headquarters in Geneva, but gave no indication of whether he supported this plan. The chairman said that much of the conversation focused on the agency's staff members and the representative told him to work with the staff and to trust their advice—they could help. The chairman would not be more specific.

That week was particularly tense for the group. More and more members complained of physical illnesses. Many of the men developed almost daily debilitating headaches, while a surprising number of women in the group complained of stomach pain or uterine cramps. The staff consulted a psychiatrist, who expressed the opinion that some of the women were experiencing a classic symptom of hysteria. Several women sought information about tubal ligation, and one woman had the operation performed at a local hospital.

Emotional stress became more directly evident. The Cambodian teacher, a member of the group who taught the school children Cambodian history, stopped teaching because he was not being paid by the group. A young mother who had fled Cambodia in April 1975, leaving her husband behind, talked to staff members for six hours one night about killing herself and her children, or returning home and leaving her children behind with the agency. One man, after saying that he feared the group and could not talk openly, told the staff to stay out of the group's problems: the Cambodians' problems would drive the Americans insane. He said that if the staff thought carefully about who he was and why he could not return to Cambodia—the first time anyone in the group had said this—then they would understand the magnitude of his situation. He spoke cryptically, in symbols, of the group being too powerful to fight, of preferring death in Cambodia to life in any country outside of his beloved homeland.

Several men told the staff that their mail from Paris was being opened by the group and asked staff to hand deliver mail to their rooms before it was intercepted. Many were clearly depressed by the GRUNK message and felt that since they were ex-members of Lon Nol's military who may have committed "war crimes" in the heat of battle, passing the scrutiny of the GRUNK mission on an individual basis would be extremely difficult. One man talked about having cut the liver out of a dead Vietnamese soldier, and eating it to dispel his fear of the enemy. Another felt that he was currently in danger of physical harm from unnamed members of the group—there were "accusers" all around him, people who said that he was collaborating with the imperialists. He said his only option, if he could not get to Paris by himself to talk to the GRUNK representatives, was self-immolation.

On February 11, a meeting was held between agency staff and the State Department representative in Washington, D.C. The purpose of the meeting was to outline what was happening to the group and to relay the group's request to be supplied with airline tickets to Paris to present themselves to the GRUNK mission. The State Department representative was interested that some of the group members were hopeful about the message from GRUNK, and he suggested that those persons should be encouraged, on an individual basis, to contact the mission. They would still have the option of becoming refugees in the United States, he added, if they were denied repatriation to Cambodia by the Paris mission. From the U.S. government's point of view, the only alternative was indefinite detention at an Immigration and Naturalization Service (INS) facility, particularly if violence broke out within the group. The State Department was still interested in keeping UNHCR involved if possible. As the money to maintain the repatriates in Philadelphia came from the U.S. government, the State Department would discuss with UNHCR whether a similar arrangement for the purchase of airline tickets was possible.

Two days later, the State Department agreed to provide airfare and a refugee travel document for anyone who wanted to go to Paris. It would provide the money directly to UNHCR. Apparently, the State Department felt that UNHCR was lukewarm, at best, on continued involvement with the group. State felt that UNHCR had not yet cabled Geneva with the group's request for airline tickets, and that too much time would elapse while UNHCR headquarters was consulted.

By February 18, after a series of secret meetings between some group members and the staff, and a rash of telephone calls between New York, Washington, and Philadelphia, a number of the details had been worked out. The State Department would pay for a one-way excursion ticket to Paris for each person and a plane ticket from Paris to Beijing, which at that time was the only official point of entry to Cambodia. It was assumed that China would provide the repatriates with passage from Beijing to Phnom Penh.

Three days later, one of the group members—the same man who had spoken of being accused by the group of collaborating with the imperialists—held a secret meeting with agency staff. He had received a message from GRUNK that included a statement of costs for the trip, should he be accepted as a repatriate. Among the costs that he would have to bear was the airline ticket on the Chinese carrier from Beijing to Phnom Penh.

When agency staff members left the YMCA to go to the Pan Am office in the city, three of the group followed them, waiting in the bookstore across the street from the office, watching the transaction. The airline ticket for the Chinese carrier presented something of a problem. Pan Am finally issued a Multilateral Carrier Order, which resembles a money order made out for a certain amount in the name of the carrier.

After having secured refugee travel documents at INS in Philadelphia, several of the group secretly went to New York for a day to apply for tourist visas from the French consulate.

On February 26, 1976, five Cambodians left Philadelphia for Paris, where they applied to GRUNK for admission to Cambodia as repatriates. All were ex-Lon Nol military men. They had not told the rest of the group that they were leaving until the morning of their departure. The leadership called them traitors and cowards because they had broken the solidarity of the group.

Three days later, after a weekend of private and public meetings between the group's leadership and agency staff, a plan was worked out to move the rest of the group, in subgroups of five or six, to New York to get tourist visas, and from there on to Paris to make contact with the GRUNK mission.

A general panic ensued. Some were afraid of the group—particularly those who had been close to one of the five who had left for Paris. Others cited fear of the group as a reason for wanting to be among the next to leave for France. Although the leadership tried to form small travel groups, everyone was making their own deals, contacting staff secretly—on the back stairs, in alleys near the YMCA, or by telephone late at night.

The contract with UNHCR was extended on March 8 to allow time for the Cambodians either to leave in small groups for Paris or to resettle as refugees here. It had become clear that the Cambodians going to Paris would need more than forty-five days in France, as negotiations between GRUNK and the repatriates were moving slowly. The State Department authorized living allowances in Paris for three months and a round-trip ticket (New York-Paris-New York) should they not be accepted by GRUNK or should they decide against going to Cambodia after all.

By mid-March, one of the Cambodians had received a written message from GRUNK stating that it was acceptable for the host government—in this case the United States—to pay for the repatriation of all former students living outside of Cambodia. The State Department was pleased with the GRUNK message, as were the repatriates. The staff felt that there was much going on that they were not privy to. However, they saw their own task as assisting the Khmer in their desire to return home.

On March 30, one of the original five called collect from Paris to say that his visa was due to expire in ten days and that he could not have it renewed because his airline ticket was only valid for forty-five days and he was almost out of money. Though his contacts with the mission encouraged him to believe that he had been accepted for repatriation, no date had been set yet. A dozen or so of the others were in a similar position. He said he had been told by GRUNK that his travel documents were in order. It was decided that staff would travel to Paris to arrange for ticket extensions and a system for paying the additional forty-five-day living allowance in the event that the Cambodians were not repatriated during the first forty-five days. And arrangements were made to meet the Cambodian who had telephoned from Paris on April 5 at the Air France bus terminal in Paris—his choice for a meeting place—as he did not want the GRUNK mission to see him meeting with an American, should they be watching.

In France, the UNHCR staff met with the repatriates, although not in large groups and almost always on the street. Meeting times and places were passed by word of mouth. Agency staff were instructed not to telephone or come to the hotels where the repatriates were staying. In fact, the repatriates were loathe to give out the names of the hotels, saying that they moved every few days. They had been attending twice-weekly orientation sessions, lasting from two to four hours, at GRUNK headquarters. They watched movies about industrialization and agricultural development and the equality of women in the work force in the new Khmer society. After a series of individual interviews, GRUNK had issued them Cambodian passports. During these interviews they were asked about their life in Paris, the United States, and Cambodia before April 1975. They were asked how they felt about France and the United States, and who had paid their transportation to France and living expenses in France. They were also asked about their family members in Cambodia.

GRUNK told them that they would have to wait several months in Paris while the government in Phnom Penh located their families and verified their stories. They were to spend their time in Paris in the reeducation program at the Paris mission. Once in Phnom Penh, they were told, they would be reunited with their families, joining them in rice farming. They would be separated from their wives and children during the day, but would be able to be with them at night. It was their understanding that the reeducation would not continue once they reached Cambodia.

Returning to Philadelphia, staff brought back letters from the Cambodians in Paris, as well as the news that the original five were to leave Paris for Phnom Penh—via Air France and the Chinese carrier—on May 25, 1976. The rest of the group, including all of the families who had arrived in the United States as refugees a year before, were anxious to leave for Paris as quickly as possible. The mood of the group had changed. Most seemed excited and happy. The long group sessions continued, but the content changed. There were fewer purification sessions. Instead, the sessions were educational, and more relaxed. At the group's request, the American Friends Service Committee supplied several movies about industrialization in North Korea and China. Staff who had gone to Paris were greeted almost as comrades-in-arms upon return, and they were invited to a weekend session. It lasted for hours. The group crowded into a long hallway, sat or sprawled on the floor, while Sam stood in front of the dark movie screen, waving his arms and extolling the group to study the industrialization of "neutral powers." The movies seemed to go on forever, and some members of the group got bored and restless, particularly during one movie about workers in a small-parts factory. But they stayed, whispering quietly to their neighbors from time to time, drinking gin out of ginger ale cans.

On April 9, 1976, the State Department informed the staff privately that there was a rumor that the Khmer Rouge government in Phnom Penh was about to recall all of its mission representatives around the world. If true, this meant that there would soon be no place for Cambodians to apply for repatriation. On April 13, the last small group of Cambodian repatriates left Philadelphia for Paris to contact the mission.

Groups of repatriates from the United States left Paris for Phnom Penh—via Beijing—on May 25, June 8, and June 22, 1976. The night before they left Paris, several telephoned agency staff in Philadelphia. They felt sure that it was safe for them to return home; they had called to say good-bye. On the flight from Paris to Beijing, they mailed post cards to staff members from the refueling stop in Athens. The day of their flight from Beijing to Phnom Penh, one Cambodian sent a post card that said simply, "the city is beautiful and silent." He was writing of Beijing, but he might have been writing of Phnom Penh, indeed, of his country.

In the four years that followed, there was silence. Then, in April 1980, a partial list of those tortured and killed at Toul Sleng prison in Phnom Penh arrived in the United States. Nineteen of the repatriates were on the list. The first of the group to die in the prison was the man accused of being a collaborator with the imperialists. The last had been the chairman of the Supreme Committee. Of Sam, there was no word. None of the 114 has ever been heard from since the plane carrying them back home touched down on the tarmack in Phnom Penh.

Most of the Cambodians now living the United States fought hard to escape precisely what this group of 114 succeeded in returning to: Democratic Kampuchea. One cannot help but wonder if some of these repatriates might not have avoided their fate had they been in Cambodia, instead of the United States, on April 17, 1975. None of them, to our knowledge, was among the hundreds of thousands of refugees who found their way out of Cambodia into Thailand, and from there into resettlement nations. Certainly none came back to the United States, which currently has a Cambodian population of over 90,000. What follows is a description of the lives of those Khmer who have tried to establish new roots in the United States, especially since 1975.

From Democratic Kampuchea to the United States: Fleeing the Homeland

In the 1950s and 1960s, there were perhaps no more than 200 Cambodians living in the United States. In 1963, for example, 138 Cambodian students were studying in American universities, a number of them here on Cambodian government scholarships. France, of course, with its historical link to Cambodia, had far more students enrolled in its universities and was the preferred place to study. Perhaps an additional 50 Cambodians lived in the United States, mostly members of the Cambodian diplomatic missions and employees of the Foreign Language Institute and the Voice of America. Few others had joined this group until the early 1970s, when members of Lon Nol's military began to arrive for short periods to receive military training at U.S. installations. Nevertheless, by April 1975, probably fewer than 1,000 Cambodians were living in the United States. For the most part they were apolitical professionals or students; a few, however, were involved in the politics of Cambodia and remained here in exile, either self-imposed or as a result of being declared political agitators by one of the regimes in Phnom Penh. In fact, before 1975 there was no real Cambodian community in exile in the United States, unlike other national groups that have gone into exile here, and there was no true sense of community, as they were spread out across the country—in Washington, D.C., Dayton, Des Moines, Iowa, Los Angeles, etc.

In April 1975, Americans watched on television as the government in Saigon changed and the evacuation of South Vietnam began, but there was far less press coverage of the change in Phnom Penh, and it passed almost unnoticed by most Americans, except as an adjunct to the war in Vietnam. When the refugees from Vietnam and Cambodia arrived at Camp Pendleton in California, and later at Fort Chaffee, Arkansas, Eglin Air Force Base in Florida, and Indiantown Gap, Pennsylvania, the Vietnamese refugees vastly outnumbered Cambodians. The Cambodians, again, passed almost unnoticed.

The U.S. government, mindful of the experience of resettling large numbers of Cuban refugees into South Florida during the 1960s and the social effects that that resettlement program initially had on Miami, concluded that resettlement of the "Indochinese" would have to be done differently. The original idea was to find individual sponsors all over the country who would make a moral commitment to provide housing and a job to a refugee until the person could "get on his feet." Sponsors would be found by private, voluntary agencies (Volags). The agencies contracted with the Department of State and received a small sum of money per refugee to find a sponsor and to help with the resettlement in case the individual sponsorship did not work out. The Department of Health, Education and Welfare (HEW) joined a task force to work with the State Department, the Volags, and the refugees to move the refugees out of the four U.S. camps and into American society.

In those first weeks of May and June 1975, there was an amazing spirit of camaraderie in the camps. The Volags, most of them church affiliated, interviewed refugees, searched for sponsors, and worked through the various federal bureaucracies, including intelligence checks on each adult refugee. For the most part, the refugees sat patiently and somewhat humorously watched the swirl around them. With almost no information to go on, and no past experience of America, some refugees went "Volag shopping," trying to get the best deal for themselves and their families. Where was Iron Mountain, Michigan? What was it like?

Most of the Cambodians who left Cambodia in April 1975 were urban, middle and upper class, educated city dwellers. They came from Phnom Penh and from Battambang and other provincial cities with easy access to the sea or the Thai border. A few had known for some time that the war was not going well for them and had been traveling to Bangkok or Hong Kong periodically "on vacation," taking out a family member and valuables on each trip. One man brought slides of each of the prize-winning flowers he had grown at home in the capital.

Although unable to plan ahead, some brought all of their immediate family with them when they left, but most did not. Many were military men, on assignment away from home at the time of the fall of Phnom Penh, who made the decision to leave within an hour or two of hearing the news about the capital. Some were on patrol ships at the time, and the captain made the decision for them.

Everyone was waiting for news from home. Many of the Vietnamese thought at first that they were in the United States for a short vacation and would soon be able to go back. The Cambodians were hoping for a quick collapse of the new regime. But they heard nothing. From the first, it was as though Cambodia had fallen into a black hole. Almost no one was getting out. In May and June 1975, a few refugees straggled over the border from western Cambodia into Thailand; a caravan of foreigners made it to the Thai border after they were released from asylum in the French embassy in Phnom Penh, but they had little news, as they had been sequestered almost immediately after the change of government. Phnom Penh had been evacuated: to where? No one knew.

The Cambodians in the refugee camps found sponsors and left, family by family, to become refugees, as we thought of it, and to exist, as they thought of it. They would wait for news.

As a nation, the United States had no idea how to deal with Cambodians, and no idea what it was dealing with. Only a handful of Americans, mostly from the State Department, the CIA, or ex-missionaries, had ever worked with Cambodians. Now, HEW, Volags, private sponsors, and church groups were trying. Those few Cambodians who had made it out of Cambodia and to the United States were spread out all across the country. Even in large cities, like Philadelphia, only six families were resettled during 1975. The Volags and sponsors found jobs and housing and began to teach them English. The upper class Cambodians, accustomed to French society in Phnom Penh, learned English quickly. The others did not. In large U.S. cities, the housing found was generally below the standard the refugees had known back home. Inner city apartment dwellings, cold in winter and hot in summer, were a long way from the breezy, fan-cooled, spacious houses where one could grow prized orchids. Employment, too, was a problem. Entry-level jobs—a euphemism for factory work, floor cleaning on the night shift, and pumping gas—are a long way from being a general in the Cambodian army, or a teacher of history, or an engineer. Sponsors brought clothes to the refugees' homes because they needed them, and they were grateful, but they were ashamed, and they hated the shame of it.

Professional Cambodian women, who had been high school principals at home while their husbands worked as chiefs of surgery in the country's hospitals, were being retrained in antipoverty program classes as clerks, while their husbands remained unemployed. In many cases the women became the financial supporters of the family while the men took care of the children. It was not that the men would not work, or were lazy, but simply that the indignity was too great. Some tried to wash floors but the assault on their identity overcame them. Their sponsors did not understand, and in many cases, they did not understand either.

They were isolated and alone, cut off from home and family, from their culture and heritage, from other Cambodians in the United States, from the things that had always given them their identity. Americans seemed cold and emotionless, Americans all looked alike. Americans were rude and incomprehensible.

Finally, there was the terrible silence from Cambodia. Between 1975 and 1978, what news there was coming out of Cambodia was awful. The 14,000 Cambodians in the United States before 1979 had an informal network that spread across the country and to France—and into the refugee camps along the Thai border. But the network did not extend far into Cambodia, and almost no one was getting out. Each new rumor, of the evacuation of the cities, of forced labor camps, of starvation and pandemic malaria, of the razing of the medical college and the burning of all books, was fed along the network and deepened the refugees' sense of aloneness and helplessness. Every Cambodian in the United States had family members still at home in Cambodia—parents, children, wives, brothers—and they had heard nothing. There was no mail service, there was no underground network that extended beyond the borders. There was only a secret organization running the country, and, for two years, no one was quite sure who was even in charge.

While in the first few months there had been wild, unbelievable rumors of executions, now the rumors were more vague, and worse. Everyone believed them, and no one did. It was not possible. Cambodians were, after all, Cambodians. They would never do such things to their own people. They would wait for more news.

In the meantime, many began to leave their sponsors and move, mostly to Southern California, where the "ambience" was better. Southern California had a large Asian population, in addition to a significant proportion of the U.S. Cambodian population. The waterfront market in Redondo Beach feels like an Asian city market. It is not the climate, although that helps, but the familiar "feel" of Southern California, which makes it a more "comfortable" place to live. The sense of being different is not so great as in other places in the United States. While there is prejudice toward Asians, a refugee is more shielded from the prejudice by the sheer number of Asian faces. Of course, other Asians are often prejudiced against the refugees as well, but they do not really have to mix, and the prejudice is of a different nature, more familiar and therefore more acceptable.

Many of the Cambodians who moved within the United States were searching for extended family members or an ethnic community. Others, however, moved to get away from sponsors or their ethnic community. Some Cambodians, particularly young, ex-military men with wives and children back home in Cambodia, moved away from other Cambodians to try to block out the past. They wanted to cut themselves off from pain and to pretend they were Americans, not Cambodians at all. They bought stereos and cars and television sets, packed up their possessions, and headed off to a different city in search of jobs and themselves. They worked during the day and drank at night. When loneliness overcame them they sat in their apartments, alone, played Cambodian music on their tape recorders, and cried. Soon, they moved on again, to another city and another job. After several years, many of these young men ended up in Texas or Southern California or Oregon and reentered the Cambodian communities there.

These were the ones who were "acting out" their depression, and they were a minority. Most Cambodians projected a semblance of stability, going to jobs, raising children, acquiring material goods, in fact, functioning in an American context. But the news from home, and the lack of it, had forced a depression upon them which few people in the world—except holocaust victims and victims of massive natural disasters—had ever experienced: The sense of inexplicably having survived. Why had others died and why had they escaped? They felt that they alone did not deserve to escape—and the guilt of having survived was almost unbearable. If the reports and rumors were true, they would continue to exist and wait for an early release from this life, for a life that could produce this much suffering was not worth living, the burden had become too great. Cambodians began referring to themselves as "unlucky," but it carried with it a sense of a permanent state of being and something they could not fight against. They were unlucky to be alive, to be here in America, to be Cambodian, and they were unlucky as a people and as a nation. Nonexistence, the Buddhist principle, was the answer, but for that they would have to wait: They could not commit suicide, for to do so would condemn them to be reincarnated 500 times—to commit suicide in each of those lives.

Some Cambodians, although not many, had turned away from Buddhism in anger and despair, saying that Buddhism was to blame for what had happened to their beloved Cambodia. Perhaps the Christian God, working through the church groups trying to help them, was a better choice. Of course, some Americans tried to convert the refugees, but those who did not want to be converted simply moved away. The few who became Christians for the most part did so because they were searching for answers to unthinkable questions.

Besides, there were only three Cambodian Buddhist monks in the United States—one in Washington, D.C. and two in California—so that for many Cambodians, it had been a long time since they had been able to seek solace and counsel from a monk, once the center of cultural and religious life.

Approximately 14,000 Cambodians were residing in the United States in

November 1978, when Congress passed a Sense of Congress Resolution that Cambodian refugees living in several camps inside the Thai border should finally be allowed to attempt to qualify for the U.S. refugee program. Just under 15,000 refugees from Cambodia had lived in these camps for several years, awaiting resettlement, or working with the resistance, or just waiting for something to happen.

In late December 1978, it did. Vietnam pushed across the border into Cambodia at Parrot's Beak and within two weeks had captured Phnom Penh.

Cambodians in the United States went wild. It was as though the years of depression and guilt suddenly turned outward and unleashed a buried fury. They were excited and agitated. The rumor network went into full swing, operating day and night for weeks. The genocidal maniacs had been overthrown, but they had been overthrown by the hated Vietnamese. Surely all of the stories about Pol Pot could not be true, and the Vietnamese, enemies for thousands of years, would be worse. But at least there might be some word from loved ones. Young Cambodian men threatened to get guns and shoot the first Vietnamese they saw on American streets. Prince Sihanouk was under house arrest but might get out. Communism had been overthrown and soon everyone could go home. The rumors, and their logic, were out of control.

Oxfam outfitted a barge with rice, for a population rumored to be starving, and started into Cambodia. Finally, for the first time since April 1975, accurate, factual reports began to come out of this country with an impregnable wall around it, and many of the Cambodian refugees did not believe the reports. The Vietnamese were lying to the international community, they said. Cambodians met together all over the United States. Some met with Americans who had friends working in the few international organizations with access to information from inside Cambodia. They received the information in silence.

Some Cambodians believed all of it, some believed none. Most accepted the devastation of Phnom Penh, of Kompong Cham, of Takeo as true but felt that the torture chambers and mass execution sites were inventions of the Vietnamese to justify the invasion of Cambodia.

In March 1979, the first large group of refugees, fleeing with Pol Pot's forces, fleeing from Pol Pot's forces, fleeing the Vietnamese invaders, maybe just fleeing—crossed the border into Thailand. There were almost 1,000 of them, dressed in rags, bone-thin and rock-hard, looking like they were in a trance. The Cambodians in the United States saw pictures of them, and cried for their people, and began to believe what had happened. Some Cambodians quit their jobs. Most stayed at work during the week and huddled together on weekends, to gather news and despair together.

Rumors were more concrete that spring, and more accurate. They were coming not so much from France and other parts of the United States, as they had before, but from the Thai border, where refugees were crossing almost daily. The rumors were of the Pol Pot atrocities and, increasingly, of a starving population ranging across the land in search of food and family.

There were so many refugees that spring, in fact, that the Thai government must have felt that there was no longer a border at all. In any event, in June Thailand forcibly repatriated 40,000 Cambodian refugees back across the border into Cambodia and into a mine field. Some 20,000 died during the repatriation. Cambodians in the United States greeted this news more with despair than anger. After all that had happened to their people, they were being pushed back once again into fear, darkness, death.

Very little resettlement took place that year. Few of the 15,000 in the established refugee camps who had been there for years actually came to the United States. Cambodians already in the United States were oriented toward Cambodia and the Thai border, still waiting.

Weekly, the news got worse. Thousands upon thousands were carried, straggled, or fell across the border into Thailand. Others died on the way, or just inside the border. By the fall of 1979, mass starvation was upon Cambodia in full force. People made their way toward Thailand, traveling by foot for weeks at a time, eating roots and bark and leaves, trying to stay alive. By December 1979, there were 300,000 Cambodians in holding centers inside Thailand and an additional half million people living along the Thai-Cambodian border, looking for food and medicine. Finally there was news. There were almost no children under the age of five left alive. Conditions inside Cambodia had been too terrible for them to survive, or for pregnancies to occur, for that matter. It seemed that the entire population was starving. Those who had not been executed or died of hunger and disease were now starving to death.

Cambodians in the United States, as in France and elsewhere, became desperate. The Khmer race was about to become extinct. Cambodian women in the United States began getting pregant—they owed it to their race. Women in the camps who could get pregnant did the same thing. They did what they could to set up a tracing system to find relatives, or least to find out what had happened to them. They besieged the American Red Cross, the Volags, their own newsletters and networks, the State Department, the United Nations, anyone they could think of. They wrote letters to the camps and holding centers in Thailand, to be posted on boards and fences—"My name is____, I come from Kandal, I am looking for my mother and father. If you know of them, please write to me at this address"—and slowly, the word began coming back. "Do you remember me? My name is____. I saw your younger brother last month at the border. The rest of your family is dead. I want to come to the United States, but I need a sponsor. Will you sponsor me?" Almost every Cambodian in the United States had lost family members but a surprising number found family members too, and they began desperately to try to get them to the United States, before something else happened.

Cambodians in America spent the first half of 1980 writing letters, sending telegrams, and telephoning other Cambodians, both in the United States and overseas, trying to track down relatives and friends, and trying to get them to the United States, where they would be safe. The need to do everything possible to assure safety, perhaps born out of the guilt they felt for having survived, consumed their days and nights. Anxiety and depression were things that most U.S. Cambodians lived with constantly that spring and summer, but there was also a new sense of hope. More family members were alive than they had thought possible six months before. As for those who were dead, at least now they knew for sure. The curtain of uncertainty had begun to lift, and it was a release from the years of waiting in a void.

The Cambodia Crisis Center, a national fund-raising organization to collect money for international relief organizations feeding the Cambodians along the border, in the Thai holding centers, and inside Cambodia itself, was in full swing by the spring of 1980. Cambodian communities across the country formed groups, often with American churches, to raise money to donate to the crisis center. Here was a way to help. They gave speeches at universities, showed films and slides of the border, held church suppers to raise money. Three Cambodians made a fund-raising spot commercial for the crisis center, which was used by television stations all across America.

The first list of names of those interned and executed at Tuol Sleng prison, the Khmer Rouge's torture center in Phnom Penh, was brought to the United States that spring by Ben Kiernan, an Australian scholar, and distributed to Cambodian communities all over the country. It seemed that everyone knew someone on the list. Photocopies of the list passed from person to person, and the number of check marks in the margins next to the names grew.

By the summer of 1980, a few Cambodians with close relatives already in the United States began to arrive. Many initially stayed with their relatives, but for a number, problems began almost immediately. Cambodians who had been here since 1975 and had not lived through the Pol Pot era said that their relatives had changed almost beyond recognition. The new arrivals were hard, both physically and emotionally. They stole food and valuables from their relatives and hid them in their rooms. Families felt that a newly arrived brother or sister took everything and gave nothing in return. The resident Cambodians began to say that the newly arrived "had seen too much" and were cruel and hardened. They had lost their humanity and gentleness, they claimed.

Many of the new arrivals said that it was as though they had come to family they no longer knew, family who had not lived through the experiences that had engulfed their lives for five years and who could not possibly understand them now. Many felt that their families owed them more than they were willing to give. They were surrounded by fear. The shock of coming from Cambodia to America overwhelmed many. One woman became terrified when her husband entered an elevator and the doors closed—she thought he was gone forever. A man panicked when an automatic door opened in front of him—who had done it? These people who had walked 600 kilometers to escape Cambodia were afraid to go downtown to shop for fear of getting lost and never returning to their families. They were disoriented and bewildered. They were also deeply depressed for, in truth, they *had* seen too much.

Many left their families' homes and moved into Cambodian urban ghettoes that had begun to spring up in a number of U.S. cities. They sought other Cambodians who had not become refugees until 1979 or later. They sought the comradeship of those who could understand. A large percentage of these newer arrivals did not want to work, did not really want to study English, although this was the reason they gave for not seeking employment. They really wanted to sit and to try to heal. Many did just that, never venturing far from their inner-city apartments, except to the welfare office or the grocery store. When they went to Cambodian parties or get togethers, they looked out of place and were silent, watching their countrymen enjoy themselves.

During the summer of 1980, the U.S. Department of State announced that 20,000 Cambodian refugees from the camps in Thailand would be admitted into the United States in the next twelve months. This was a significant shift in U.S. refugee policy, as only a total of 20,000 Cambodians had come to the United States between 1975 and 1980. The federal government was in fact doubling the U.S. Cambodian population in this country in one year.

There were other implications as well. The Cambodians already here felt that 20,000 was far too small a number, given the fact that approximately 600,000 Cambodians remained along the Thai-Cambodian border, many of whom were seeking unsuccessfully to slip quietly into the holding centers, thereby becoming eligible for asylum in a third country. The holding centers and refugee camps held over 160,000 Cambodians at that time, almost all of whom were seeking refugee status. Cambodians here compared the number of 20,000 against the huge number of Vietnamese refugees being admitted into the United States and felt that the U.S. government was discriminating against Cambodians as a group.

From a social policy point of view, there was another implication as well. Of the 20,000 Cambodians residing in the United States by 1980, nearly half lived in Southern California between Los Angeles and San Diego, with approximately 7,000 settled in Long Beach, California. The city was inundated with refugees, many of whom were secondary migrants from other sections of the country. Furthermore, the voluntary agencies had traditionally resettled new refugees in the same city as their friends and relatives already here. Clearly, with a new wave of Cambodian refugees that would double the U.S. Cambodian population, resettlement agencies could not continue to place large numbers of people in Long Beach—or even in Southern California—where jobs were scarce, housing was even more scarce, and community tensions had become a serious problem. Refugees were finding their tires slashed and their cars spray painted. It was rumored that the Ku Klux Klan was planning a march through the streets of Long Beach to protest the presence of refugees. Cambodian leaders and voluntary agencies alike in Southern California were concerned that an additional influx of refugees into the area would only make matters worse. Of course, those with immediate family already there—mothers, fathers, wives, husbands, children—could not and should not be kept from their families. But the State Department was estimating that perhaps one-half of the 20,000 to be resettled would not have immediate family in the United States and therefore might be willing to resettle elsewhere in the United States.

It was in this atmosphere, then, that planning for the Cambodian Cluster Project, or Khmer Guided Placement (KGP) Project, as it was later named by the State Department, began in the summer of 1980. The program was designed to identify "nonimpacted" areas of the country where newly arriving Cambodian refugees could resettle successfully and to strengthen existing Cambodian communities in those areas in order to provide the newly arriving refugees with an ethnic community support base. It was hoped that Cambodian refugees would feel comfortable and not have to move.

Tentative sites for the program were selected by the Cambodian Association of America, a mutual assistance association which had begun in 1958 as the Cambodian Student Association of America and incorporated as a nonprofit organization in 1975 under its current name. The association, headquartered in Long Beach, was both political and social in nature.

Politically, U.S. Cambodian groups were split into several factions at that point—all noncommunist and all supporting the resistance activities along the Thai-Cambodian border and just inside Cambodia. The major split was between those groups that supported Son Sann, the sixty-seven-year-old ex-banker and prime minister under Prince Sihanouk, who left the comfort of Parisian cafes to lead a Cambodian resistance force in the border area in October 1979, and those groups that supported Prince Sihanouk. General In Tam was aligned with the Prince, although he had a significant following in his own right. A Federation of Cambodian Associations had been incorporated in the United States in August 1979, in an attempt to pull together the factions, but by 1980 the federation was coming unglued, as was the Confederation of Cambodian Associations, headquartered in Paris. Nearly all mutual assistance associations in the United States, at least those with a significant membership and leadership, were partly political in nature. The Cambodian Association of America had a significant following, supported In Tam and Sihanouk, and had the backing of the president of the federation during the summer of 1980. Headquartered in Long Beach, they were the most powerful association in Southern California. They were, in fact, almost the only choice among existing Cambodian associations to launch such a project.

The site selection for the Cambodian Cluster Project was based upon considerations of available jobs, housing, social services, and the community "atmosphere," e.g., the degree of receptivity to Cambodians. In addition, the association looked at how successfully other Cambodians had resettled in that area in an attempt to provide role models for the new arrivals. Cambodians everywhere in the United States initially put aside political factionalism and agreed to support the project. They felt it necessary to unite for the "salvation" of their people by helping to insure the safety of some of those tenuously living along the Thai border. They felt the need to help in order to assuage their own guilt for having survived and for being healthy, well fed, and comfortable. While this feeling lasted no more than six months at most—and among some Cambodians just six weeks—it allowed the cluster project to get off the ground before political in-fighting threatened its effectiveness to any serious degree.

The project was funded by the Department of Health and Human Services' (formerly HEW) Office of Refugee Resettlement (ORR) to provide social services to the new Cambodian arrivals in the twelve sites finally selected for participation in the project.

Once preliminary data had been collected on potential sites and Cambodians residing in those sites had been contacted by the association to ascertain their willingness to participate, the association entered into a period of negotiation with the thirteen national agencies engaged in the resettlement of refugees in the United States.

The Resource Center set up a schedule for KGP arrivals into sites, based upon Volag affiliates' estimates of local agency capacity in any given month. In addition, the Department of State set up a separate schedule of movement of Cambodians from holding centers into the processing center in Thailand, and from there either directly to the United States or to the Refugee Processing Centers in the Philippines (Bataan) or Indonesia (Galang). This included a painstaking effort to reschedule and restructure English language and cultural orientation programs, involving the scheduling and rerouting of thousands of Cambodian refugees within the refugee system in Southeast Asia.

During the spring of 1981, representatives from the Department of State, the American Council of Voluntary Agencies (ACVA), and the Cambodian Association of America traveled to Southeast Asia to do project and refugee orientation with the American embassy and voluntary agency staff working in the refugee program and, more importantly, with the Cambodian refugees who were to be sent to the cluster sites. The three representatives estimated that they were able to provide adequate orientation about the project to over 3,000 Cambodians during their six-week visit to Thailand, Indonesia, and the Philippines. Their orientation was particularly emphatic on the issues of seeking employment quickly and remaining in nonimpacted areas of the United States.

Rumors among the refugees in the camps and back in the United States abounded. This was a resettlement program to take care of widows and children only; men were expected to return to the border to fight in the resistance, preferably with the small Sihanouk force being established; it was a special program for widows, who would be provided for forever by the U.S. government; KGP refugees were obligated to stay in the cluster sites for three months but were then free to move wherever they chose. In fact, none of the rumors was true. No refugee was legally bound to remain in a site for any length of time, and a large number of unattached men entered the United States under the Cluster Project, in addition to widows and complete families.

At this point the project ran into its first major nonpolitical difficulty. In the spring of 1981, the INS ruled that Southeast Asian refugees should be interviewed for admission to the United States on a case-by-case basis, instead of continuing to be interviewed as though the refugees were "presumptively eligible," as had been the case since the beginning of the Southeast Asian refugee program in 1975. The result of this change in INS policy was that during one week in the spring, over 60 percent of the Cambodians who were interviewed were deferred by INS. The next week, the deferral rate dropped to 58 percent. This impaired the ability of the Department of State to move refugees either directly to the United States or to the refugee processing center to begin intensive language and cultural orientation before coming to the United States.

Another significant factor contributing to the slowdown in processing was that the INS interviewing staff consisted of only three officers in Thailand. At about the same time, the Songkhla refugee camp in Thailand, where Vietnamese refugees were interned, became overcrowded and ran short on water. A potential health hazard was created. The INS officers were moved to that camp to interview so that refugees could be moved quickly, thereby averting a potential health crisis. The side-effect of this emergency was that for several weeks no INS officers were available to interview in the processing camp where the KGP Cambodians were being held.

Cambodians who knew about this holdup took it as one more example of the U.S. government's discrimination against Cambodians in favor of Vietnamese refugees, processing the Vietnamese first while Cambodians waited. Movement of Cambodian refugees to the U.S. cluster sites, delayed to December 1, 1980, began as a trickle and did not accelerate until May and June 1981. The project was scheduled to end on September 30, 1981.

All of these events—the scheduling of Cambodians into processing centers for language and cultural orientation, the INS deferral rate during the spring, and the health emergency in the Songkhla camp—created early delays in the processing of Cambodians and resulted in a large number of KGP arrivals in August and September 1981.

During site visits to potential cluster sites, local Volag and Cambodian affiliates had been assured that a "manageable" number of Cambodians would arrive in the site over a period of six months, taking into account local site conditions. In fact, over 5,000 KGP refugees arrived in twelve sites during the month of September alone, creating an enormous burden for all concerned. For example, during two weeks in September 1981, 250 Cambodians arrived in Richmond, Virginia. Over 50 Cambodian refugees arrived at the airport in Houston, Texas on one night alone. The August and September rate of arrivals almost sank the project. As it was, it took months for the Volags and Cambodian associations to dig themselves out from under the deluge.

In the end, even with political factionalism inhibiting the project and the fits and starts inherent in the process, the project managed to resettle over 10,000 Cambodian refugees in twelve nonimpacted cities across the country.

In March 1982, the Indochina Refugee Action Center completed a study of the refugees at the twelve Cambodian cluster sites. They looked at the rate of secondary migration out of the sites and the financial status of the newly arrived Cambodian refugees. For the most part, secondary migration was low. Although it varied from site to site, the overall rate was 11.7 percent. When that figure was adjusted by the number of Cambodians known to have moved into a cluster site, the rate dropped to 9.2 percent. And of those who left, 23 percent had immediate family in other parts of the United States and should not have been included in the project in the beginning.

Their preliminary findings on the employment of those refugees were also encouraging. Forty-seven percent were fully employed and financially independent, even though the majority of those studied had arrived within the previous six months.

Between the summer of 1980 and the end of 1981, over 38,000 Cambodians were resettled in the United States. Since that time, perhaps another 25,000 have come. Statistically, they have done well. Emotionally, they have fared less well. Some have now moved on, trying to forget the past, trying to rebuild their lives, bereft of family and friends. A few still wait for news of loved ones, having heard nothing all these years. Many actively supported the resistance factions in 1979, 1980, and 1981. By now, some have dropped out, as the hope for a quick settlement of the problems of their beleaguered nation grows more distant. Most Cambodians, however, continue to see themselves as a people in exile, waiting for the time when they can return home, to Cambodia.

CHRONOLOGY OF CAMBODIAN HISTORY WITH EMPHASIS ON RECENT EVENTS*

Summary of the Khmer Polity

3d to 6th century	Funan Period
535–802	Chenla Period
802–1432	Angkor Period
1432–1864	Disorder and Contraction
1864–1940	French Protectorate
1940–1945	Vichy Authority, Japanese Power
1945	Interlude of Full Independence
1945–1954	Gradual Achievement of Independence
1954–1970	Sihanouk's Kingdom of Cambodia
1970–1975	Khmer Republic and Civil War
1975–1979	Democratic Kampuchea
1979–	People's Republic of Kampuchea and Resistance

Summary of the Khmer Communist Movement

1930–1951	Indochina Communist Party
1951–1954	Khmer People's Revolutionary Party
1954–1960	Pracheachon (Peoples Group)/KPRP
1960–1966	Worker's Party of Kampuchea
1966–1979	Communist Party of Kampuchea

Pol Pot Resistance	Heng Samrin Regime
1979–1981 Communist Party of Kampuchea	1979— Kampuchean People's Revolutionary Party
1981—Party Announces Its Dissolution	

*This chronology is deliberately short on dates, so that it can be read with greater ease. If there are more than two events in a year, the month is indicated; more than two events in a month, the day.

3d to 6th centuries, A.D.

Funan Period. Indianized kingdom of Funan (Sinicized name) dominates peninsular Southeast Asia. Khmer culture and political traditions in Angkor period developed directly from heritage of Funan.

535–802

Chenla Period. Funan attacked by its vassal state farther north on the Mekong, Chenla, and overcome. Funan eventually forcibly merged into Chenla. The term "Cambodge," and from that "Cambodia," is derived from Kambuja, the name of the supposed founder of the main ethnic group in Chenla, the Khmer.

802–1432

Angkor Period.

9th century

God-King Jayavarman II establishes Khmer kingdom of Angkor.

1177

Angkor temporarily invaded and sacked by Kingdom of Champa.

1181

Jayavarman VII captures throne and extends Khmer Empire to unprecedented reaches; replaces Hindu deities with Mahayana Buddhism.

13th century

In a popular surge, Theravada Buddhism displaces Mahayana in Cambodia. It remains a central part of Cambodian life at least until the Pol Pot regime attempts its elimination beginning in 1975.

1431

Angkor Thom, capital of Angkor, is abandoned to the invading Siamese (Thai). The Khmer Empire has collapsed.

1434

The Khmer court reestablishes itself on the site of the modern capital, Phnom Penh.

1470–1600

Capital relocated several times as a result of Siamese invasions and Khmer resurgences. At the end of this period, the Khmer royal family went into exile in Laos and a Khmer prince was enthroned as a vassal by the Siamese.

1620

A royal marriage cements an alliance between Cambodia and Vietnam, temporarily blocking Thai suzerainty over Cambodia.

17th and 18th centuries

Progressive encroachment on Khmer territory by Vietnam and Thailand, culminating in annexation of Battambang and Siemreap provinces by Thailand in 1794.

1864

Cambodia becomes French Protectorate. Rebellious activity against the French begins immediately.

1867

France appeases Thailand by officially ceding Battambang and Siemreap.

1887

Cambodia enters the French Union Indochinois.

1904

Khmer King Norodom dies. French block his son's accession to the throne, arranging instead for his half-brother Prince Sisowath to be crowned.

1907

French secure reintegration of Battambang and Siemreap into Cambodia.

1930

Vietnamese communist movements unite as the Indochina Communist Party (ICP). Supranational organization is in response to contemporary Comintern policy stressing internationalism. There are no Cambodian members until 1932.

5/40

France falls to Germany.

12/40

First local Khmer Issarak (Freedom) committee, dedicated to Cambodian independence, started. Rapid growth of the movement through the time of the 1950 convention. See below.

4/41

King Sisowath Monivong dies. French decide to revert to the Norodom branch of the royal family and enthrone eighteen-year-old Norodom Sihanouk.

5/41

First Japanese troops arrive in Cambodia. French Vichy administration in Cambodia tolerated by the Japanese.

3–10/45

Japanese dislodge the French administration and oversee the establishment of an "independent" Cambodian government with Son Ngoc Thanh as prime minister.

8/45

Japan surrenders to the Allies.

9/45

Ho Chi Minh proclaims the Democratic Republic of Vietnam in Hanoi.

10/45

Reimposition of French colonial rule. Thanh arrested and held in France.

1/46

Sihanouk signs agreement with France recognizing Cambodia as an autonomous state within the French Union. However, authority in Cambodia remains in French hands.

3/46

First Khmer political party formed, the Democratic Party, representing Cambodian nationalism and the desire for modernization.

1947

Under limited autonomy, new Cambodian parliament promulgates constitution.

1949

France grants Cambodia qualified independence, retaining some authority in some aspects of foreign relations and military policy. Following this, Cambodia is recognized as an independent state by the Western powers.

1950

United Issarak (Freedom) Front (UIF) founded at a conference. The president of its Central Committee is Son Ngoc Minh, a key figure in the communist movement.

9/11/51

Democratic Party, calling for immediate independence and a merely ceremonial kingship, wins overwhelming victory in National Assembly elections. Struggle between Prince Sihanouk and parliament ensues.

9/30/51

In the wake of the dissolution of the Indochina Communist Party, the Khmer People's Revolutionary Party (KPRP) is formed.

1953 & 1954

Left-wing and communist students return from study in France. Some go to Vietnam after the Geneva Accords, but most enter the Communist underground in Cambodia. Among them are Pol Pot, Ieng Sary, Hu Nim, Hou Youn, and, in 1967, Khieu Samphan.

Spring and Summer 1953

Sihanouk dissolves Democratic government as nationalist and communist antigovernment activity increases. He goes abroad on his *Croisade Royale pour l'Indépendance.*

10/53

France transfers remaining attributes of de facto independence to Cambodian government.

4/54

Geneva conference opens.

5/54

Dien Bien Phu falls to the Viet Minh.

7/54

Geneva Accords call for removal of Viet Minh forces from Cambodia and recognize Cambodia's independence and neutrality.

9/8/54

At U.S. instigation, the South East Asia Treaty Organization (SEATO) is formed. Sihanouk refuses to join.

9/30/54

From the jungle, where he had been a leader of the anti-French resistance, Thanh pledges his loyalty to the newly sovereign government and requests an audience with the king. Sihanouk refuses to see him and Thanh stays in the jungle to lead the Khmer Serei (Free Khmer), an anti-Sihanouk resistance group which is later funded by the U.S. CIA. Thanh returns to legal politics after Sihanouk is overthrown, holding positions, including prime minister, under Lon Nol.

10/54

As a result of the Geneva Accords, approximately 1,000 UIF members go to Hanoi, including a majority of KPRP. Those communists remaining form the legal Pracheachon (People's Group) to pursue political struggle in national elections.

3/55

Sihanouk abdicates in favor of his father, Norodom Suramarit, so that he might legally participate in national politics. Sihanouk forms political party, Sangkum Reastr Niyum (People's Socialist Community), which sweeps elections later in the year. Sangkum, a very broad and changeful coalition, wins every seat in every National Assembly election until Sihanouk's overthrow in 1970. However, the elections are far from fully free.

4/55

Sihanouk attends Bandung conference of nonaligned nations where he agrees on peaceful coexistence with China and denounces SEATO.

1956 & 1958

Sihanouk recognizes the USSR and PRC.

6/60

King Suramarit dies. National Assembly revises constitution and elects Sihanouk head of state. He holds this position until his overthrow in 1970. No new king is named. The Khmer monarchy ends.

9/60

Cambodian communist movement reorganizes as the Worker's Party of Kampuchea (WPK) at "2d General Assembly" of the party. Pol Pot appointed to the Central Committee.

6/62

Sihanouk choses several leftists to run as candidates of the Sangkum, which wins all seats. Sangkum MP Khieu Samphan, later head of state of Democratic Kampuchea, appointed secretary of state for commerce. He is forced out a year later.

7/62

WPK Secretary-General Tou Samouth disappears, perhaps murdered, perhaps by Pol Pot.

5/63

Third General Assembly of the WPK. Pol Pot given permanent appointment as secretary general. Central party operations moved from Phnom Penh to the Northeast.

11/2/63

Ngo Dinh Diem assassinated.

11/6/63

Sihanouk terminates all U.S. economic and military aid to Cambodia in protest of U.S. support for the Khmer Serei operating against his regime. Deteriorating relations with the U.S. over the next several years also involve Sihanouk's fear of U.S. influence in Cambodia, U.S. refusal officially to recognize Cambodia's borders against the claims of Thailand and Vietnam, and growing U.S. involvement in what Sihanouk sees as a losing cause in Vietnam.

11/10/63

To compensate for U.S. aid and make Cambodia generally more self-reliant, Sihanouk partially restructures the economy in a policy he refers to as "Buddhist socialism," which involves the nationalization of trade, banking, and some industry.

3/65

U.S. combat troops arrive at Da Nang.

5/65

Sihanouk severs diplomatic relations with U.S.

Summer 1965

Sihanouk first allows North Vietnam to use Cambodia to transport materiel and base troops fighting in the South.

9/66

Name change from Worker's Party to Communist Party of Kampuchea (CPK), signifying the leadership's militancy and assessment that the revolutionary struggle had reached an advanced stage.

9/11/66

Conservative National Assembly elected. Lon Nol becomes prime minister. The beginning, perhaps, of the polarization of national politics lasting for thirteen years.

4/67

Peasant uprisings in Battambang province. Savagely suppressed by Lon Nol. Khieu Samphan and other prominent leftists still in the government flee to the countryside to join the *maquis* after a frightened Sihanouk blames them for instigating the disturbances.

8/67

Association of South East Asian Nations (ASEAN) formed.

1/68

Tet Offensive in South Vietnam.

3/68

Amid at least the perception of economic crisis, Sihanouk is forced to begin abandoning Buddhist socialism, encouraging private industry and foreign investment.

1/69

Sihanouk is the only head of state to attend Ho Chi Minh's funeral.

3/69

First U.S. bombing of Cambodia. Directed at Viet Minh sanctuaries. Secret and illegal.

6/69

Cambodia restores full diplomatic relations with the U.S.

3/18/70

Prime Minister Lon Nol and Deputy Prime Minister Sirik Matak stage successful coup against Sihanouk while he is abroad travelling. In the following days, coup prompts rejoicing in Phnom Penh and pro-Sihanouk demonstrations in the countryside.

3/23/70

From Beijing Sihanouk announces formation of National United Front of Kampuchea (NUFK) to overthrow the Lon Nol regime.

3–4/70

The manipulation of anti-Vietnamese sentiment to rally support for the coup d'etat triggers spontaneous anti-Vietnamese attacks. Several hundred Vietnamese civilians living in Cambodia are slaughtered in racist outbursts in scattered locations.

4/30/70

U.S. and South Vietnamese armed force (50,000) invade Cambodia to attack Vietnamese bases and bolster faltering Lon Nol regime. Prime Minister Lon Nol finds out about the invasion after it has begun.

5/13/70

Royal Government of National Union of Kampuchea (RGNUK) formed in Beijing with Sihanouk as chief of state. Most major positions go to

Sihanoukists. With the passage of time, Sihanoukists lose their positions to Khmer communists.

5/13 & 5/19/70

Lon Nol regime restores relations with Thailand and South Vietnam which had been broken in the early 1960s.

6/70

U.S. ground troops leave Cambodia; South Vietnamese troops remain for another year, greatly antagonizing the population.

10/70

Khmer Republic established.

11/70

"Regroupees" (Cambodian communists who left Cambodia after the 1954 Geneva Accords) return from Hanoi to Cambodia to aid the NUFK cause.

7–11/71

Lon Nol army defeated in its last and most important offensive of the war, the Chenla 2 campaign.

10/71

Lon Nol declares a State of Emergency. Constitutional rights are suspended.

2/72

U.S. President Nixon visits China.

3/72

Large student demonstrations against Lon Nol.

6/72

Lon Nol wins fraudulent election as president of the Khmer Republic.

1/73

Paris Peace Accords signed.

2–8/73

U.S. bombing of Cambodia intensifies.

Early 1973

Intensification of drive by communist party center to purge NUFK of Sihanoukists, and communists with ties to Vietnam. Intensification of CPK drive to force North Vietnamese troops out of the country.

5/73

Beginning of ''Democratic Revolution'' in regions controlled by CPK. Traditional village life is upset as land is collectivized, people are forcibly relocated, political study sessions are required, and traditional customs and religious rites are proscribed. Poor peasants are given leadership positions.

8/73

U.S. bombing of Cambodia ends.

1974

U.S. President Nixon resigns.

1/75

Khmer Rouge launch major (and final) offensive against the capital, Phnom Penh.

Early 1975

U.S. President Ford unsuccessfully appeals to Congress for additional aid for the Lon Nol and Thieu regimes, ostensibly so that the U.S. could negotiate from strength in the creation of coalition governments.

2/75

NUFK congress decides on evacuation of cities once victory is achieved.

4/17/75

Khmer Rouge gain final victory after surrounding Lon Nol's last stronghold, Phnom Penh. In the following weeks, the Khmer Rouge forcibly evacuate Phnom Penh and the provincial capitals. Lon Nol government officials and military officers are executed. Labor camps are established in the countryside.

4/19/75

DK leadership moves quickly to get control of territory claimed by both Cambodia and Vietnam while North Vietnamese troops are still focused on defeating South Vietnam.

4/30/75

South Vietnamese government falls as communist troops take Saigon.

5/75

An American container ship, the *Mayaguez*, is seized by the new Cambodian government. As a show of force, the U.S. bombs numerous targets on the Cambodian mainland.

8/75

Beijing grants economic and technical aid to the new regime in Cambodia.

9/75

Sihanouk returns to Cambodia for the first time since the government he nominally heads took power.

1/76

New constitution is adopted, renaming the state Democratic Kampuchea (DK).

4/76

Sihanouk forced to resign as head of state. Khieu Samphan named president; Pol Pot, prime minister. Sihanouk placed under house arrest until Khmer Rouge need him again in 1978.

9/76

Radio Phnom Penh announces that Pol Pot has stepped down as prime minister for "health reasons."

10/6/76

Radical leadership in China, the "Gang of Four," arrested.

10/15/76

Pol Pot reinstated as prime minister.

4/77

Cambodia escalates border raids against Vietnam.

6/77

SEATO dissolved.

8/77

Cambodia mounts armed attack deep into Tay Ninh province in Vietnam.

9/77

Pol Pot visits Beijing. Upon his return, he delivers major public speech and reveals the existence of the CPK as well as his position as secretary-general.

10/77

Vietnam responds to Cambodian border incursions by launching a major offensive into Cambodia.

12/77

Cambodia announces it is temporarily breaking relations with Vietnam.

1/3/78

Cambodia rejects Vietnamese proposal that both sides withdraw their forces five kilometers from the border and negotiate a truce.

1/78

China increases arms shipments to Cambodia.

2/78

Leadership in Vietnam apparently decides to support an ouster of Pol Pot by his opponents inside Cambodia, and perhaps, if that fails, direct military action to bring down the regime.

4/78

Massive numbers of ethnic Chinese flee Vietnam by boat and across the border into China.

5/78

Revolt of CPK cadres from the Eastern Zone against the Center. Tens of thousands of people in the region are slaughtered in a general purge by Pol Pot forces; some escape to Vietnam.

6/78

After long refusing, Vietnam joins COMECON.

7/78

After relations have spiralled downward dramatically since February, China cancels its aid to Vietnam.

8/5/78

DK Defense Minister Son Sen returns from Beijing claiming assurances of Chinese support and assistance in Cambodia's ''just struggle'' against Vietnam.

8/21/78

U.S. Senator George McGovern calls for an international force to overthrow the Democratic Kampuchean regime.

11/78

Vietnam and USSR conclude Treaty of Friendship and Cooperation.

12/3/78

Announcement in Vietnam of the founding of the Kampuchean National United Front for National Salvation (KNUFNS), led by Heng Samrin.

12/25/78

Vietnam launches full-scale invasion of Kampuchea; 120,000 troops enter the country and meet little resistance. The Vietnamese force in Cambodia reaches 200,000 before declining to between 120,000 and 160,000 for most of the period since 1978.

1/7/79

Vietnamese forces occupy Phnom Penh. Khmer Rouge forces retreat in disarray with army of approximately 40,000 to Thai border where they set up resistance bases.

1/8/79

Formation of the People's Revolutionary Council of Kampuchea, headed by Heng Samrin.

1/10/79

Proclamation of the People's Republic of Kampuchea (PRK).

1/13/79

Foreign ministers of ASEAN jointly condemn the Vietnamese intervention.

Spring/Summer/Fall 1979

Massive travel throughout Cambodia as people return to their pre–1975 homes and search for relatives. Little planting occurs and a severe food shortage follows.

2/17/79

Chinese launch major invasion of northern Vietnam as "punishment" for its actions in Cambodia. Perhaps 50,000 die before China withdraws on March 16.

2/18/79

Vietnam and Cambodia sign a twenty-five-year Treaty of Peace, Friendship, and Cooperation during visit to Phnom Penh of Vietnamese delegation led by Prime Minister Pham Van Dong. Treaty pledges support to Cambodia in case of attack but also promises friendship to ASEAN nations.

4/79

Beginning of large-scale influx of Cambodian refugees into Thailand.

6/79

Many thousands die as Thailand forcibly repatriates 43,000 Cambodian refugees by pushing them across the border without food in a remote and deserted area littered with land mines. No effective effort is made by any party to save lives.

7/79

International Committee of the Red Cross (ICRC) and UNICEF representatives visit Phnom Penh to assess the food shortage and establish an aid program. Agreement on aid is delayed until October because of conditions put down by PRK and Vietnam.

8/18/79

Pol Pot and Ieng Sary tried in absentia on charges of genocide; sentenced to death by a government tribunal.

8/21/79

Khmer Rouge forms new Patriotic and Democratic Front of the Great National Union of Kampuchea to attract broader support. Sihanouk declines post as president of Front.

10/9/79

Former Cambodian Prime Minister Son Sann establishes the Khmer People's National Liberation Front (KPNLF) to oppose the Vietnamese occupation as well as the possible return to power of the Pol Pot forces.

10/19/79

UN secretary general launches a $110 million aid appeal for Cambodia on behalf of ICRC and UNICEF. First of many appeals over several years by high UN officials for relief and reconstruction aid for Cambodia, nearly all of which are coupled with warnings of imminent catastrophe. However, none of the appeals raise the amounts from donor nations that UN and relief agency officials ask for. Some fall far short. Subsequently, doubt is raised about the real extent of the food shortage.

10/21/79

First of several annual UN General Assembly votes on attempts by allies of Vietnam to deny accreditation to the exiled Democratic Kampuchean regime. Democratic Kampuchea prevails 79 votes to 35, with 34 abstentions and 12 absences.

11/14/79

UN General Assembly approves by 91 to 21, with 29 abstentions, resolution spearheaded by the noncommunist nations of Southeast Asia calling for the withdrawal of foreign (i.e., Vietnamese) forces from Cambodia. A similar resolution is approved every year hence by increasing margins.

11/21/79

Khao-I-Dang refugee camp opens along the Thai-Cambodian border. At its height, in the range of 200,000 Cambodian refugees are living there, making it the second largest Cambodian city in the world, after Phnom Penh.

12/79

Khmer Rouge Congress approves leadership reshuffle in which Pol Pot gives up post of prime minister but retains leadership of army and party.

3/18/80

Tenth anniversary of the overthrow of Prince Sihanouk. Cambodia is on its fourth radically dissimilar regime in ten years.

3/25/80

PRK government employees paid in cash, the first time currency has been used for nearly five years.

6/11/80

Following his failure to make much progress in uniting Cambodian exiles around him, Prince Sihanouk announces that he is giving up all political activity.

6/22/80

Vietnamese troops are accused of making a limited incursion into Thai territory in pursuit of anti-PRK guerrillas. Hanoi's standing is seriously injured within ASEAN and among nonaligned nations.

7/80

India recognizes the PRK, one of a few non-Soviet-bloc states to do so by 1985.

10/14/80

Despite particularly vigorous Vietnamese and Soviet lobbying, UN General Assembly votes 74–35 with 32 abstentions to continue recognizing the exiled regime of Democratic Kampuchea.

10/22/80

Annual ASEAN resolution approved by the General Assembly contains special provision calling for the convening of an international conference on Cambodia.

2/81

Sihanouk says he is prepared to cooperate with the Khmer Rouge and lead a national front against the Vietnamese occupation, given Chinese support.

3/1/81

Economic agreement signed in Phnom Penh by PRK, USSR, and Vietnam.

3/25/81

Prince Sihanouk announces formation of new organization to oppose the Vietnamese-backed government in Phnom Penh, National United Front for a Cooperative, Independent, Neutral, and Peaceful Cambodia (FUNCINPEC).

4/81

Major delivery of Chinese arms to the KPNLF forces on the Thai border.

5/81

The Kampuchean People's Revolutionary Party opens its first congress in Phnom Penh.

6/81

Meeting of National Assembly in Phnom Penh approves draft constitution and chooses members of new Council of State and Council of Ministers. Heng Samrin elected chairman of the Council of State and party leader Pen Sovann is made prime minister.

7/81

The International Conference on Kampuchea, convened by the UN, meets in New York. Attended by representatives of 79 countries as full participants, and 15 countries with observer status, the event is boycotted by Vietnam and its allies. Conference adopts declaration calling *inter alia* for a ceasefire in Cambodia, withdrawal of all foreign forces under supervision of UN troops, and free elections under UN supervision. China is successful in blocking provision that would call for Democratic Kampuchean forces to be disarmed if Vietnamese forces withdraw from Cambodia.

9/4/81

Three major resistance leaders—Sihanouk, Son Sann, and Khieu Samphan—agree in Singapore on joint action against Vietnamese-backed regime in Phnom Penh.

9/18/81

UN General Assembly accepts the credentials of the DK regime, 77–37–31.

12/2/81

Pen Sovann, secretary-general of Kampuchean People's Revolutionary Party, is removed from his posts and disappears.

12/7/81

Khmer Rouge announces abolition of the Communist Party of Kampuchea. Few observers believe this has actually occurred.

6/82

Under pressure from the PRC and ASEAN states, the two anticommunist Cambodian resistance groups and the Khmer Rouge sign an agreement in Kuala Lumpur establishing a tripartite Coalition Government of Democratic Kampuchea (CGDK) to oppose the Vietnamese-backed regime, with Prince Sihanouk as its president, Son Sann as prime minister and Khieu Samphan as vice-president.

10/82

Attempt to deny credentials to the CGDK at the UN is defeated 90–29–26.

12/82

Vietnamese create a cabinet-level position in their government to handle matters pertaining to Laos and Cambodia.

2/83

Kampuchean representation is debated at the Seventh Nonaligned Summit at New Delhi. The seat remains vacant, another victory in this forum for the PRK.

3/6/83

Nguyen Co Thach proposes unconditional talks between the Indochinese states and ASEAN nations.

3/23/83

ASEAN foreign ministers reject Vietnamese proposal for "bilateral" negotiations in special ad hoc meeting in Bangkok.

4/83

Sihanouk, Son Sann, and Khieu Samphan hold CGDK cabinet meeting on Cambodian soil for the first time. Ambassadors from China, North Korea, Bangladesh, and Mauritania present their credentials to Sihanouk.

5/3/83

PRK holds ceremony to accompany the withdrawal of an estimated 10,000 Vietnamese troops from Cambodia. Thai authorities claim that troop withdrawal is part of a rotation.

5/6/83

Two PRK documents, dated September 13 and October 9, 1982, establishing immigration policy for Vietnamese settlers in Cambodia, surface in the French press. Leaked by the U.S. embassy in Bangkok, these documents suggest that Vietnamese settlers be given special treatment and that Hanoi determines the policy governing their status. Resistance estimates of Vietnamese immigrant population are as high as 1,000,000. PRK foreign minister places figure at 60,000.

5/11/83

A team from the UN Food and Agriculture Organization reports that half of Cambodia's young suffer moderate or severe malnutrition. An independent UNICEF report corroborates these findings.

5/20/83

PRK marks a "National Day of Hatred" to commemorate the atrocities of the Pol Pot government.

Late 5/83

Vietnamese arrest and purge some 300 PRK officials, including members of the security police, in an attempt to control subversion.

6/19/83

The PRK Armed Forces commemorates its founding as June 19, 1951, when the Issarak army was formed. Foreign estimates of the size of the PRKAF are 20,000–30,000.

6/23–25/83

Sixteenth ASEAN foreign ministers meeting in Bangkok.

7/83

PRK Foreign Minister Hun Sen criticizes communique issued by 16th ASEAN foreign ministers conference, especially Thai demand that the People's Army of Viet Nam (PAVN) troops withdraw 30 kilometers as a condition for talks.

8/7/83

Indonesian Foreign Minister Mochtar Kusumaatmadja rules out new ASEAN diplomatic initiative after Vietnam rejects ASEAN proposal that PAVN troops be withdrawn 30 kilometers from border as condition for talks.

8/16/83

PRK National Assembly issues report asserting that more than 2.7 million Khmer perished under the DK regime. Figure based on interviews.

10/83

CGDK is seated as the representative of Cambodia at the UN. For the first time since Democratic Kampuchea was ousted from power in Cambodia, Vietnam does not force a vote on a credentials challenge. It has also declined to do so since.

11/83

PRK news agency reports that there are 20 groups of Soviet specialists assisting in economic development and organization work, and that more than 1,000 Khmer students are studying in the USSR.

1/84

Hun Sen, at press conference, rejects any kind of reconciliation with Sihanouk or Son Sann.

3/84

Data from USSR embassy in Phnom Penh indicates that USSR has supplied some $450 million in nonmilitary aid to PRK since 1979.

7/84

Regular meeting of ASEAN foreign ministers drafts communique on Cambodia stating its position: (1) national conciliation would permit a role for Heng Samrin rank and file; (2) self-determination for the Khmer people through supervised elections; (3) withdrawal of PAVN forces.

9/1/84

Vietnamese Foreign Minister Nguyen Co Thach says that PAVN troops will withdraw from Cambodia when: (1) China stops threatening Vietnam; (2) Thai sanctuary for the resistance forces ends; (3) PRK armed forces are sufficiently developed to defend the country.

9/23/84

President Reagan meets with Sihanouk and Son Sann in New York after his address to UN General Assembly.

1/85

Eighth PRK National Assembly ends in Phnom Penh. Hun Sen is elected prime minister.

3/85

Hun Sen outlines the requirements for settlement of Cambodian situation: (1) elimination of Pol Pot forces; (2) withdrawal of the Vietnamese forces after the elimination of the Pol Pot forces; (3) free and supervised elections; (4) creation of peaceful coexistence and stability among Southeast Asian countries; (5) cessation of all interference by countries outside the region; (6) creation of international guarantees and supervision of above agreements.

4/3/85

At the urging of ASEAN, the U.S. House of Representatives approves $5 million in aid to the noncommunist resistance under the leadership of Sihanouk and Son Sann. Initiative was spearheaded by Democratic

Representative Solarz and has bipartisan support. Despite initial opposition, President Reagan is expected to sign the legislation.

4/17/85

Tenth anniversary of the victory of the revolution.

BIOGRAPHIES OF KEY FIGURES

Heng Samrin

Khmer. Born in Kompong Cham province, Cambodia, 1934. Joined KPRP, 1959; officer in KPRP armed forces; deputy commander and political commissar, Eastern Zone of DK, 1977; fled to Vietnam, 1978; chairman, Central Committee of KNUFNS and president, People's Revolutionary Council of Kampuchea (head of state of PRK), 1/79- ; chairman, State Council, PRK, 1981- ; secretary-general, PRPK, 12/81- .

Hun Sen

Khmer. Born in 1952. Joined GRUNK army, 1970; company commander, 1973; regimental commander, 1976; fled to Vietnam, 1978; minister of foreign affairs, PRK, 1/79- ; Vice Premier, PRK, 1/79- ; member, Politburo, People's Revolutionary Party, 12/81; Chairman, Council of ministers, PRK, 1/85- .

Ieng Sary

Sino-Khmer. Born in Vinh Binh province, southern Vietnam, 1925. Primary education in Vietnam; secondary education Lycée Sisowath in Phnom Penh; lived in France while earning doctorate from Institut d'Etudes Politiques in Paris, 1950–56; joined PCF, 1951; president, Union of Cambodian Students in France (leftist), 1955–56; returned to Cambodia and joined KPRP, 1957; member, Central Committee of WPK, 1960–63; teacher in Phnom Penh, 1958–63; went underground, 1963–71; member, Politburo CPK, 1966- ; special envoy of CPK to Sihanouk and outside world, 1971–73; member, Politburo, FUNK, 1970–75; special adviser to Office of RGNVK deputy prime minister (Khieu Samphan), 1973–75; second deputy prime minister in charge of foreign affairs, DK, 1975- ; sentenced to death in absentia, Phnom Penh Tribunal, 8/79; member, CGDK Coordinating Committee for Economy and Finance, 7/82- ; brother-in-law of Pol Pot.

Khieu Samphan

Khmer. Born in Svay Rieng province, Cambodia, 1932. Lived in France while earning doctorate in economics from University of Paris, 1954–59; dissertation: "L'Economie du Cambodge et ses Problèmes d'Industrialisation," decried impact of dependent development on Cambodia while advocating moderate structural reforms; secretary, Federation of Cambodian Students, 1955–59; founded French-language leftist journal *L'Observateur* after returning to Phnom Penh, 1959; self-declared "liberal Marxist," arrested and immediately released by Sihanouk, 1962; Sangkum MP, 1962–67; appointed by Sihanouk as secretary of state for commerce, 1962; resigned under pressure, 1963; fled to countryside and joined Khmer Rouge, 1967; member, Politburo, FUNK, 1970–75; deputy prime minister and minister of defense, GRUNK, 1970–75; commander-in-chief, KPNLAF, 1972–75; president of State Presidium (titular head of state), DK, 1976–79; prime minister of DK, 12/79– ; vice-president in charge of foreign affairs of CGDK, 1982– .

Lon Nol

Khmer, with some Chinese ancestry. Born in Preyveng province, Cambodia, 1913. Secondary schooling in southern Vietnam; graduated Royal Military Academy, Cambodia; government official, 1937–52; governor, Kratie province, 1945; head of national administrative services, 1949; chief of national police, 1951; governor, Battambang province, 1954; commander-in-chief of the Khmer Royal Armed Forces, 1959; appointed defense minister, 1960; prime minister, 1966–67 and 1969–70; suppressed peasant revolts, 1967–68; led a successful coup against Prince Sihanouk, 3/70; most important figure in noncommunist Cambodia, 1970–75; suffered severe stroke, 1971; prime minister, 1971–72; first president of Khmer Republic, 6/72–5/75; went into exile, 1975; currently in California, USA. Author: *The New Khmer Way* (1972).

Norodom Sihanouk

Khmer. Born in 1922. Educated in Saigon and at military academy in France; enthroned as king of Cambodia by French colonial administration, 1941; dissolved National Assembly, 1/53; embarked on *Croisade Royale pour l'Indépendance* to secure international support for Cambodian independence, 1953; most prominent and powerful Cambodian from Geneva Accords in 1954 to 1970; abdicated kingship in favor of his father to enter political arena legally, 1955; founded and headed Sangkum, leading political party in Cambodia, 1955–70; prime minister 1955–60; held various posts, including foreign minister, permanent representative to UN, 1955–60; recognized Soviet Union, 1956, and PRC, 1958; head of state 1960–70; cut off all U.S. aid, 1963; severed diplomatic relations with U.S., 1965; reestablished ties with U.S., 1969; overthrown in coup while in the Soviet Union, 3/70; announced alliance with Khmer Rouge and formation of FUNK and GRUNK from Beijing, 1970; titular head of resistance to Lon Nol regime while residing in Beijing and Pyongyang, 1970–75; titular head of state, DK, 1975; returned to Cambodia, 9/75; forced to resign as head of state, 1976; under house arrest, 1976–1/79; special envoy of Khmer Rouge to UN and member of People's Representative Assembly, 1976–79; founder and president, FUNCINPEC, 1981– ; head of MOULINAKA; president, CGDK, 8/82– . Author: *L'Indochine vu de Pekin* (1972); *My War with the C.I.A.* (1973); *War and Hope: The Case for Cambodia* (1980); *Bittersweet Memories* (1981).

Pen Sovann

Khmer. Born in Takeo province, Cambodia, 1936. Joined Viet Minh-led anticolonial struggle in Cambodia, 1950; joined KPRP, early 1950s; studied at Nguyen Ai Quoc School for party cadres in Hanoi, late 1960s; deputy to division commander in Eastern Zone, 1970; claims to have left Pol Pot's CPK and Cambodia in 1974, though also confirms working in Hanoi with FUNK news agency from 1970 to 1975; commander-in-chief, Revolutionary Armed Forces (PRK), 1/79–12/81; vice-president of People's Revolutionary Council of Kampuchea, and minister of national defense, PRK, 1979–12/81; secretary-general of PRPK, 5/81–12/81; member, Politburo, and Secretariat, PRPK, 5/81–12/81; apparently purged from all party and government posts, 12/81; reputed to be in Moscow and Hanoi, 1982– .

Pol Pot (an alias; given name: Saloth Sar)

Khmer. Born in Kompong Thom province, Cambodia, 1928. Lived in France, pursued but did not complete degree at Ecole Francaise de Radio-électricité, 1949–53; joined pro-Viet Minh communist group and PCF while in France; joined Unified Issarak Front, 1953; arrived Unified Issarak Front headquarters on border of Kompong Cham (Cambodia) and Tay Ninh (Vietnam) provinces for training, 1953; returned to Phnom Penh and joined Phnom Penh committee of KPRP, 1954; lecturer in private school, 1954–63; arrested for writing and editing left wing journal, *Solidarity,* 1955; appointed to committee to elaborate party line after major defections and loss of membership, late 1950s; member, Politburo WPK, 1960–63; secretary general, WPK, 1963–66; fled Phnom Penh and went underground, to Northeast, where he also became regional party secretary, 1963–70; six-month trip to Hanoi and Beijing, 1965–66; upon return to Cambodia renamed party CPK and remained secretary general, 1966–81; moved base area to Kompong Thom province, 1970–75; signatory to FUNK appeal to Lon Nol, 1971; vice-chairman in charge of military, KPNLAF, 3/72–?; visited PRC, 1975; announced on state-run radio to be prime minister of DK, 4/76; unseated as prime minister, 9/27/76, but reinstated 10/22/76; delivered first public speech (five hours long) and revealed existence of CPK and his position as secretary-general, 9/77; made trip to Beijing, 9/77; overthrown by Vietnamese invasion, 12/78; maintained Khmer Rouge resistance organization on Thai/Cambodian border region, 1979– ; sentenced to death in absentia at Phnom Penh Tribunal, 8/79; relinquished prime ministership but retained leadership of army and CPK, 12/79; reported in poor health, early 1985.

Son Ngoc Thanh

Khmer-Vietnamese. Born in southern Vietnam, 1908. Secondary schooling in France; librarian at Buddhist Institute in Phnom Penh during early 1930s; early leader of Khmer nationalist movement and cofounder of the first Khmer-language newspaper, *Nagara Vatta,* est. in 1936; after 1942 demonstration against

French, spent three years in Japan; served briefly as prime minister at Japanese behest, 1945; jailed by French at end of war; released in 1951, went to jungle where he worked for Cambodian independence; attempted to make peace with Sihanouk, who rebuffed him, 1954; returned to jungle and formed anti-Sihanouk Khmer Serei with assistance from the U.S. CIA, late 1950s and 1960s; adviser to head of government, June 1970; prime minister of Khmer Republic, 1972; after dispute with Lon Nol, returned to South Vietnam where he reportedly died in 1976.

Son Sann

Khmer. Born in Phnom Penh, Cambodia, 1911. Educated Ecole des Hautes Etudes Commerciales de Paris; deputy governor, Battambang and Prey Veng provinces, 1935–39; minister of finance, 1946–47; vice-president, Council of Ministers, 1949; minister of foreign affairs, 1950; MP for Phnom Penh and president of Cambodian National Assembly, 1951–52; governor, National Bank of Cambodia, 1954–68; minister of state (finance and national economy), 1961–62; vice-president in charge of economy, finance, and planning, 1965–67; president, Council of Ministers, 5/67–12/67; first vice-president in charge of economic and financial affairs, 1968; resided in Paris, 1970–79; founder and president, KPNLF, 10/79– ; prime minister, CGDK, 8/82– .

Son Sen

Khmer. Born in Tra Vinh, southern Vietnam, 1930. Advanced schooling and anti-Sihanouk leftist activism in France, 1953–56; teacher, 1956–63; went underground, 1963; chief of General Staff, KPNLAF high command, 1972–75; deputy prime minister in charge of national defense affairs, DK, 1975– ; secretary general of NADK Supreme Commission, 9/79– ; member, Coordinating Committee for National Defense, CGDK, 9/82– .

GLOSSARY

Angkar "The Organization." Term used by Cambodian communists and others to refer to the ruling authorities in communist-controlled areas before 1975, and everywhere after the victory of the revolution in 1975. The Communist Party of Kampuchea first publicly acknowledged its existence in September 1977.

Angkor Wat Massive temple complex built by Khmer King Suryavarman II in the twelfth century, the remains of which are in Siem Reap province. It is the largest single religious monument known to exist anywhere in the world. Now the most potent symbol of Khmer ethnic identity and Cambodian nationalism.

April 17 People *See* New People.

ANS Armée Nationale Sihanoukienne (National Army of Sihanouk). The umbrella organization of the military forces loyal to Prince Sihanouk. Formed in June 1981, it is based at various points along the Thai-Cambodian border. Its total forces, including MOULINAKA, consisted of 5,000 armed soldiers as of 1983. A widely respected former prime minister and senior Cambodian politician, In Tam, is ANS commander-in-chief.

ASEAN Association of South East Asian Nations. An organization founded in 1967 to promote economic cooperation and security among member nations. Current members are Thailand, Malaysia, the Philippines, Indonesia, Singapore, and Brunei. ASEAN has spearheaded the opposition to the Vietnamese occupation of Cambodia.

Base People *See* Old People.

Beng "Basin"; "reservoir."

Buddhist Institute Founded in 1930 by French scholar Suzanne Karpeles to promote study in Khmer Buddhism. Became center of anticolonial activity. Destroyed by the Khmer Rouge in 1975.

CGDK Coalition Government of Democratic Kampuchea. Technically the CGDK is the ruling coalition in the state of Democratic Kampuchea. The term is also used as if the CGDK were the state itself. In practice, the CGDK is a loose political and military coalition of the three forces operating on the Thai border to oppose the Vietnamese occupation of Cambodia. The three divergent groups were pressured by ASEAN and the PRC to form a coalition in August 1982 so as to increase the international visibility of the resistance and to help prevent the Cambodia seat at the United Nations from being declared vacant or being given to the People's Republic of Kampuchea. Also, the coalition appears to have reduced the number of military clashes among the resistance forces themselves. The coalition is composed of the forces of its president, Prince Sihanouk; its prime minister, Son Sann; and the remnants of the Khmer Rouge, nominally under the leadership of Khieu Samphan, who is vice-president of the CGDK. The CGDK is recognized as the official government of Cambodia by the United Nations and numerous countries.

Cham *Also called* "Khmer Islam" *and* "Cham-Malay." Islamic Malayo-Polynesian group in northern Cambodia and southern Vietnam. Official Khmer Rouge policy from 1975 to 1978 was to "Khmerize" them and destroy their separate communities. This often involved large-scale massacres. Remaining population estimated at 200,000.

CPK Communist Party of Kampuchea. The Worker's Party of Kampuchea changed its name to the CPK in 1966 as part of the adoption of a more militant line spearheaded by its secretary-general, Pol Pot. The party dominated the resistance forces to the Khmer Republic from 1970 to 1975 and ruled Cambodia from 1975 to 1979. Party dissidents cooperated with the Vietnamese in their December 1978 invasion of Cambodia and now lead the People's Republic of Kampuchea. They have formed a new communist organization, the Kampuchean People's Revolutionary Party, which sees itself as the true heir to the pre-Pol Pot communist movement. After being ousted from power, the main branch of the CPK operated from the jungle as a resistance movement from 1979 to 1981. In 1981 they announced the dissolution of the CPK, but whether this actually occurred is doubtful.

Democratic Kampuchea The name of the new state proclaimed by the leadership of the Communist Party of Kampuchea on January 8, 1976, eight months after they defeated Lon Nol's Khmer Republic and acceded to power in Phnom Penh. Technically, Cambodia was ruled by the Royal Government of National Union of Kampuchea from April 17, 1975, until January 1976, but "Democratic Kampuchea" is widely used to refer to the entire period of CPK rule, from April 17, 1975, to January 7, 1979. The DK leadership was routed by the Vietnamese invasion of December 1978 but maintains a government-in-exile and a resistance movement on the Thai border. In August 1982 DK became an autonomous part of the Coalition Government of Democratic Kampuchea.

FANK Forces Armées Nationales Khmeres (Khmer National Armed Forces). The military of Lon Nol's Khmer Republic.

FAO Food and Agriculture Organization. Based in Rome, a branch of the United Nations that researches and acts on problems in agriculture and food production, especially in third world countries. Several FAO missions have gone to Cambodia on fact-finding tours.

First Indochina War The war between colonial France and the communist-dominated independence movement in Vietnam which ended in July 1954 with the signing of the Geneva Accords, which mandated the temporary partition of Vietnam at the 17th parallel, elections within three years to determine the terms of reunification, and the withdrawal of all French administration from Indochina. A significant armed resistance also operated against the French in Cambodia (*see* Khmer Issarak), but the French gradually negotiated away sovereignty over Cambodia between 1946 and 1953.

Front for the Construction of the Motherland of Kampuchea The mass organization of the People's Republic of Kampuchea in the latter half of 1981 only. It succeeded the Kampuchean National United Front for National Salvation and was succeeded by the Kampuchean United Front for National Construction and Defense in December 1981.

FUNCINPEC Front Uni National pour un Cambodge Independant, Neutre, Pacifique, et Cooperatif (National United Front for an Independent, Neutral, Peaceful, and Cooperative Cambodia). The main political organization of Prince Sihanouk. Formed in March 1981 and based along the Thai-Cambodian border. It is an autonomous component of the Coalition Government of Democratic Kampuchea.

FUNK or **NUFK** Front Uni National du Kampuchea; National United Front of Kampuchea. A political and military front dedicated to the destruction of the Lon Nol regime. It was established by Prince Sihanouk in Beijing shortly after the Lon Nol coup ousted him from power in 1970, but nearly from the beginning it was controlled by his allies in the Front, the Communist Party of Kampuchea. This was the first time Prince Sihanouk aligned himself with the Khmer Rouge. It was in the name of NUFK, with Sihanouk as nominal head, that the revolution was won in April 1975.

Geneva Accords, Geneva Conference International conference on Indochina attended by France, the United States, the United Kingdom, the People's Republic of China, the Soviet Union, communist and noncommunist Vietnamese factions, the Royal Cambodian Government, and Laos from April through July 1954. The signing of the Geneva Accords marked the end of the First Indochina War and specified a framework for geopolitical and military development thereafter. The Accords stipulated (1) temporary division of Vietnam; (2) holding of general elections in Vietnam within two years to inaugurate reunification; (3) neutrality of Cambodia and Laos; (4) on-the-spot demobilization of anti-French and anti-government rebels in Cambodia. The Accords were signed by all parties except the United States and South Vietnam.

GRUNK or **RGNUK** Gouvernement Royal d'Union Nationale du Kampuchea; Royal Government of National Union of Kampuchea. Government-in-exile formed and headed by Prince Sihanouk after his overthrow in 1970. Within a short time of its formation, GRUNK's high-level positions were relinquished by Sihanoukists in favor of officials of the Communist Party of Kampuchea. Technically in power April 17, 1975, to January 8, 1976, when it was replaced by the state of Democratic Kampuchea.

Hanoi Khmer A term used to refer to the 1,000 or so Cambodian communists who went to live in Vietnam after the Geneva Accords in 1954. Most were killed by Pol Pot's wing of the communist party between 1970 and 1975, after they had returned to Cambodia to participate in the revolution. Some who escaped and returned to Vietnam went back to Cambodia in 1979 as officials in the People's Republic of Kampuchea.

ICK International Conference on Kampuchea. Conference sponsored by the United Nations in compliance with an October 1980 General Assembly resolution sponsored by ASEAN. Advertised as an attempt to resolve the Cambodia crisis through negotiation and with international guarantees for all parties, the July 1981 Conference was attended by seventy-nine nations as full participants but boycotted by Vietnam and its allies. It approved a resolution calling inter alia for a cease-fire in Cambodia, withdrawal of all foreign forces, and UN-supervised elections. The conference highlighted strains between the PRC and the noncommunist states of Southeast Asia over the extent to which the Democratic Kampuchea forces should be delegitimized to encourage Vietnamese concessions. The conference also voted to maintain a standing committee to work toward a settlement of the Cambodia issue.

ICP Indo-China Communist Party. The unification of several Vietnamese parties in Hong Kong in 1930 under the leadership of Ho Chi Minh. Little significant Khmer participation until after the Second World War. Dissolved into three national parties in 1951, including the Vietnam Worker's Party and the Khmer People's Revolutionary Party.

ICRC International Committee of the Red Cross. Voluntary relief organization active inside the People's Republic of Kampuchea and along the Thai-Cambodian border since 1979 in the distribution of food aid and medical services.

Kampuchea Khmer-language word for Cambodia.

Kampuchean People's Revolutionary Council Also translated as Revolutionary Council of the People of Kampuchea. The governing body of the People's Republic of Kampuchea from January 8, 1979, to June 1981. The Council was superceded by a National Assembly, which is the highest organ of government according to the constitution adopted in 1981. The executive powers of the state lie with committees that emanate from the Assembly.

KCP Kampuchean Communist Party. *See* CPK.

KGP Khmer Guided Placement Project. Also called the Cambodian Cluster Project. A major multidepartmental effort by the U.S. government to cluster new Cambodian refugees in a limited number of sites so as to create viable and stable communities. Undertaken in the early 1980s when there was a large number of incoming refugees, the intention was to prevent secondary migration and thereby lessen the burden on a few cites to which Cambodians were moving after arriving in the United States.

Khmer Islam *See* Cham.

Khmer Issarak "Independent Khmer." Independent committees that sprung up throughout Cambodia starting in the early 1940s to fight for independence from France. Often armed resistance was used. United Issarak Front was founded in 1950. The Khmer Issarak forces were disarmed and demobilized in 1954 as stipulated by the Geneva Accords. The Khmer Issarak movement was the vehicle for the first significant growth of the communist movement.

Khmer Krom Ethnic Khmer born and raised in southern Vietnam, portions of which were once in the Khmer state.

Khmer Loeu "Hill tribes." The ethnic minorities that live in the mountainous regions of Cambodia.

Khmer Republic or **Republic of Cambodia** The government established by

Lon Nol after the ouster of Prince Sihanouk in March 1970. It was officially proclaimed October 9, 1970, ending almost 2,000 years of uninterrupted monarchial government. Backed by the United States, it fell to a coalition of Prince Sihanouk and the Communist Party of Kampuchea, backed by the PRC and North Vietnam, on April 17, 1975.

Khmer Rouge "Red Cambodians." A derogatory term coined by Prince Sihanouk to refer to Cambodian communists. Sometimes the term is used specifically to refer to those communists who stayed in Cambodia when most of their colleagues went to North Vietnam after the Geneva Accords in 1954, though it is also used as an umbrella designation for all Cambodian communists. *See also* Hanoi Khmer and Khmer Viet Minh.

Khmer Serei "Free Khmer." Right-wing, anti-Sihanouk guerrilla movement covertly supported by the U.S. CIA and Thailand during the 1950s and 1960s. Cambodian nationalist Son Ngoc Thanh was one of its leaders. In mid-May 1970 Thanh dissolved his group, as Sihanouk had been ousted from power.

Khmer Viet Minh A derogatory term used by Sihanouk to refer to Cambodian leftists, especially while they were organizing pro-independence agitation in the Cambodian countryside in alliance with the Vietnamese independence movement in the 1940s and early 1950s.

Khum "Subdistrict."

KNAF Khmer National Armed Forces. *See* FANK.

KNUFNS Kampuchean National United Front for National Salvation. Also commonly referred to as the Kampuchean United Front for National Salvation, as well as simply the Salvation Front. The first incarnation of what has remained the main political organization in the People's Republic of Kampuchea besides the Kampuchean People's Revolutionary Party. First established on Vietnamese soil in December 1978 to provide a Cambodian structure to help legitimize the Vietnamese invasion and the ouster of Democratic Kampuchea. The Front has numerous noncommunists in its leadership, including Buddhist clergy, though it is by and large controlled by the communist leadership. Its name was changed in mid-1981 to the Front for the Construction of the Motherland of Kampuchea.

KPNLA Khmer People's National Liberation Army. *See* KPNLAF.

KPNLAF *Can refer to either of two distinct entities*: Kampuchean People's National Liberation Armed Forces (*also known as* People's Armed Forces for the National Liberation of Kampuchea) or the Khmer People's National Liberation Armed Forces (also known as Khmer People's National Liberation Army). The Kampuchean People's National Liberation Armed Forces were the army of the Sihanouk/Communist Party resistance to Lon Nol's Khmer Republic from March 1972 to April 1975 (*see also* RAK). The Khmer People's National Liberation Armed Forces are the military force of Son Sann's anti-Vietnamese resistance, the Khmer People's National Liberation Front.

KPNLF Khmer People's National Liberation Front. A political and military organization founded and led by former prime minister Son Sann. Its avowed aims are to end the Vietnamese occupation and to establish Western-style political and economic structures in Cambodia. The KPNLF is an autonomous part of the Coalition Government of Democratic Kampuchea, which has Son Sann in its number two position, prime minister. In 1983, the KPNLF had approximately 15,000 troops under arms in the Thai-Cambodian border region. It is the Cambodian political faction most favored by U.S. government policy.

KPRAF Kampuchean People's Revolutionary Armed Forces. The military of the People's Republic of Kampuchea.

KPRP *Can refer to either of two distinct entities*: Khmer People's Revolutionary Party *or* Kampuchean People's Revolutionary Party. The Khmer People's Revolutionary Party was founded in September 1951 as the Indochina Communist Party dissolved into three national parties. Its leadership and policies were aligned with those of the Vietnamese communist movement. The name of the party was changed in 1960 to the Worker's Party of Kampuchea, and again in 1966 to the Communist Party of Kampuchea. The Kampuchean People's Revolutionary Party is the communist party that functions in the People's Republic of Kampuchea. The date of its founding is unclear. Many sources list May 1981, when the first public party congress was held, but the present form of the party may have come into being at any point after mid-1978. The party itself refers to the May 1981 event as the Fourth Party Congress, the third being in January 1979 and the second being in 1960. This reflects the fact that the party sees itself as the legitimate heir to the Cambodian communist movement that existed before Pol Pot "subverted" it.

Krom Samaki "Solidarity Teams." A term used in both DK and PRK to designate small agricultural cooperatives. In the PRK cooperative farming has largely been restricted to labor-intensive peaks in the production cycle, such as sowing and harvesting.

Kuantan Formula A 1980 diplomatic initiative by President Suharto of Indonesia and Prime Minister Datuk Hussein Onn of Malaysia which suggested that a solution to the Cambodia conflict involve recognition of Vietnam's special security interests in Cambodia in exchange for reducing the pressure on Thailand. It emphasized the importance of reducing the high level of ongoing conflict in the region, which was seen as benefiting only the PRC. The initiative was promptly rejected by both Thailand and Vietnam.

KUFNCD Kampuchean United Front for National Construction and Defense. The mass organization of the People's Republic of Kampuchea. It succeeded the Front for the Construction of the Motherland of Kampuchea in December 1981.

KUFNS Kampuchean United Front for National Salvation. *See* KNUFNS.

Land bridge The points of passage between the Cambodian interior controlled by the People's Republic of Kampuchea and the areas near the Thai border held by the resistance groups or the Thai military. These points were used by some Cambodians after 1978 to leave for Western countries and were used by Western relief agencies to distribute food and supplies to be taken into the PRK.

March 18 People *See* Old People.

MOULINAKA Mouvement pour la Liberation Nationale du Kampuchea (Movement for the National Liberation of Kampuchea). Military organization formed and led by Prince Sihanouk and based among the large civilian camps on the Thai-Cambodian border. Founded in 1979, its troop strength in 1983 was approximately 3,000. *See also* ANS.

NADK National Army of Democratic Kampuchea. Also known as National Army of Kampuchea. The army of the Democratic Kampuchea resistance. Created in 1979, it succeeded the Revolutionary Army of Kampuchea.

National Army of Kampuchea *See* NADK.

New People or **April 17 People** Terms commonly used by local authorities in Democratic Kampuchea to refer generally to urbanites or to Cambodians who were in Phnom Penh when it surrendered to the revolutionary forces on April 17, 1975. New people were seen as antagonistic to the revolution because of their political beliefs and social backgrounds and were discriminated against in terms of allocations of food and privileges. A high proportion of those so designated died during Democratic Kampuchea rule.

Nixon Doctrine A statement by President Nixon in July 1969 that future U.S. policy would be to provide military and technical aid to third world allies while denying American troops for use in their conflicts. Nixon later called Cambodia "the Nixon doctrine in its purest form."

NUFK National United Front of Kampuchea. *See* FUNK.

Old People or **Base People** or **March 18 People** A designation used by authorities in Democratic Kampuchea to refer to Cambodians who: (1) had been living in the areas under Khmer Rouge control after the March 18, 1970, coup against Sihanouk; (2) had been living in areas under Khmer Rouge control prior to the victory of the revolution on April 17, 1975; or (3) were of peasant background. This included a majority of the Cambodian population. These people were politically and socially privileged as compared to the New or April 17 People whose loyalty to DK was suspect in the eyes of its leadership.

Oxfam A voluntary relief organization that since 1979 has emphasized the reconstruction and development of Cambodia's agriculture and economy, as opposed to immediate relief or assistance to refugees. Primarily active within the People's Republic of Kampuchea.

PAFNLK People's Armed Forces for the National Liberation of Kampuchea. *See* KPNLAF.

Paris Peace Accords Agreements between the United States, North Vietnam, South Vietnam, and the Viet Minh to end the war and provide a postwar political settlement in Vietnam. Signed on January 27, 1973, the Accords called for withdrawal of all U.S. troops within sixty days, reunification of Vietnam through peaceful means, and recognition of Cambodia's and Laos's neutrality.

PAVN People's Army of Viet Nam. The armed forces of the Socialist Republic of Vietnam. As of 1983, there were approximately 150,000 PAVN troops inside Cambodia.

PDFGNUK Patriotic and Democratic Front of Great National Union of Kampuchea. Mass organization established by the exiled Communist Party of Kampuchea leadership on the Thai border in September 1979 for the declared purpose of ousting the Vietnamese from Cambodia. Khieu Samphan appointed as "provisional chairman."

Phum "Village."

Pracheachon "People's Group." Legal political party formed by the communist movement after the Geneva Accords to contest national elections and pursue political struggle. It failed to win any seats in the National Assembly, as was true for all political parties except Sihanouk's Sangkum.

Prek "Small river"; "canal."

PRK People's Republic of Kampuchea. The state established by Vietnam and its Cambodian allies immediately after Vietnamese troops invaded Cambodia and ousted the Democratic Kampuchea government in January 1979. The highest ranking official is the president of the People's Revolutionary Council, Heng Samrin, a Khmer communist who broke with the Pol Pot leadership and fled to Vietnam in 1978. The PRK is recognized by most Soviet-bloc countries and a few others, such as India. It has insecure control of most Cambodian territory, though there is a significant armed resistance under the rubric of the Coalition Government of Democratic Kampuchea.

PRPK People's Revolutionary Party of Kampuchea. *See* KPRP.

RAK Revolutionary Army of Kampuchea. The armed forces of the Communist Party of Kampuchea and the state of Democratic Kampuchea. Founded in 1968. In 1979, after the Pol Pot regime was forced out of power, the RAK was renamed the National Army of Democratic Kampuchea, sometimes also referred to as the National Army of Kampuchea. It operates as a resistance army from the Thai border area.

RCGNU Royal Cambodian Government of National Union. *See* GRUNK.

Republic of Cambodia *See* Khmer Republic.

RGNUK Royal Government of National Union of Kampuchea. *See* GRUNK.

Sangha Refers to either the Buddhist monkhood or the national organization of same.

Sangkum Reaster Niyum "People's Socialist Community." A political party formed by Prince Sihanouk in 1955 and led by him until he was overthrown in 1970. It soon absorbed most opposition parties and its candidates won every seat in every National Assembly election until 1970. It was an unusually diverse coalition, ranging from the aristocracy and right wing to prominent leftists who later emerged as leaders of the Communist Party.

Second Indochina War The war for control of South Vietnam and Laos after the departure of France in 1954, and for control of Cambodia after the overthrow of the Sihanouk regime in 1970. U.S. involvement escalated gradually throughout the late 1950s and 1960s until the United States had over half a million troops in Vietnam by 1969. The Paris Peace Accords ended U.S. troop involvement in 1973. Communist victories in all three Indochinese countries in the spring of 1975 ended the war.

SF Salvation Front. *See* KNUFNS.

Srok "District."

SRV Socialist Republic of Vietnam. Proclaimed in June 1976 after the reunification of North and South Vietnam.

Theravada Buddhism "Lesser Vehicle" Buddhism. The dominant religion in Cambodia, Thailand, and Burma since the thirteenth century. It was nearly completely banned by the Khmer Rouge from 1975 to 1979. The People's Republic of Kampuchea has encouraged the restoration of the temples and the rebuilding of the monkhood, within limits, since 1979.

Third Indochina War Refers to the Vietnamese invasion of Cambodia in December 1978 and the ensuing Chinese invasion of Vietnam in February 1979. This war is, in effect, still being waged since Vietnamese troops continue to engage the forces of Democratic Kampuchea in Cambodia, and since China and Vietnam clash regularly along their border.

Tonle Sap "Great Lake." Dominates the central basin of Cambodia and is replenished by the Mekong and several Cambodian waterways. An extremely abundant source of fish and a critical part of Cambodia's agricultural cycle.

Toul Sleng Formerly a school in Phnom Penh, the main prison and execution center for Pol Pot's enemies within the communist movement during Democratic Kampuchea. Of some 17,000 brought there, 7 survived. The Khmer Rouge fled Phnom Penh in such haste at the end of 1978 that detailed prison records were left behind.

UNICEF United Nations International Children's Emergency Fund. A UN organization that played a major role in alleviating the 1979 famine and refugee crises inside Cambodia and along the Thai-Cambodian border.

Wat A Buddhist temple.

WPK Worker's Party of Kampuchea. The name of the communist organization in Cambodia from 1960 to 1966. It succeeded the Khmer People's Revolutionary Party and was succeeded by the Communist Party of Kampuchea. The leadership of the WPK gradually shifted from Vietnam-affiliated Khmer Issarak veterans to French-educated, urban-based intellectuals such as Pol Pot, Ieng Sary, and Son Sen.

RECOMMENDED SECONDARY SOURCES ON POST–1970 CAMBODIA

Barnett, Anthony, and John Pilger. *Aftermath: The Struggle of Cambodia and Vietnam.* London: New Statesman, 1982.

Carney, Timothy M. *Communist Party Power in Kampuchea (Cambodia): Documents and Discussion.* Data Paper no. 106, Southeast Asia Program, Cornell University, 1977.

Chandler, David P., and Ben Kiernan, eds. *Revolution and Its Aftermath in Kampuchea: Eight Essays.* Monograph Series no. 25, Yale University Southeast Asia Studies, 1983.

Elliott, David W. P., ed. *The Third Indochina Conflict.* Boulder: Westview Press, 1981.

Heder, Stephen R. *Kampuchean Occupation and Resistance.* Asian Studies Monographs no. 027, Institute of Asian Studies, Chulalongkorn University, 1980.

———. "Kampuchea, October 1979 to August 1980, the Democratic Kampuchea Resistance, the Kampuchean Countryside, and the Sereikar." Ms.

Honda Katuiti. *Journey to Cambodia. Investigation into Massacre by Pol Pot Regime.* Tokyo: Privately published, 1981.

Kampuchea in the Seventies. Helsinki: Kampuchean Inquiry Commission, 1982.

Kiernan, Benedict Francis. "How Pol Pot Came to Power: A History of Communism in Kampuchea, 1930–1975." Ph.D. dissertation, Monash University, 1983.

Kiernan, Ben, and Chanthou Boua, eds. *Peasants and Politics in Kampuchea, 1942–1981.* Armonk, N.Y.: M. E. Sharpe, 1982.

Ponchaud, Francois. *Cambodia Year Zero.* New York: Holt, Rinehart and Winston, 1977.

Poole, Peter A. *The Expansion of Vietnam in Cambodia: Action and Response by the Governments of North Vietnam, South Vietnam, Cambodia and the United States.* Athens: Center for International Studies, Ohio University, 1970.

Shawcross, William. *Sideshow: Kissinger, Nixon and the Destruction of Cambodia.* New York: Simon and Schuster, 1979.

———. *The Quality of Mercy: Cambodia, Holocaust and Modern Conscience.* New York: Simon and Schuster, 1984.

Thion, Serge, and Ben Kiernan. *Khmers Rouges! Materiaux pour l'histoire du communisme au Cambodge.* Paris: J. E. Hallier/Albin Michel, 1981.

Vickery, Michael. *Cambodia: 1975–1982.* Boston: South End Press, 1984.

Whitaker, Donald P. et al. *Area Handbook for the Khmer Republic (Cambodia).* Washington, D.C.: Foreign Area Studies, The American University, 1973.

CONTRIBUTORS

Anthony Barnett is an associate of the Trans-National Institute in Amsterdam and has been a visiting fellow at the Southeast Asia Program, Cornell University. He is an editor of *New Left Review* and a frequent contributor to *New Statesman*. He is the author of *Iron Britannia: Why Parliament Waged Its Falklands War* (1982), and co-author of *Aftermath: The Struggle of Cambodia & Vietnam* (1982).

Timothy Carney is a Foreign Service Officer with Southeast Asia, and especially Cambodia, as his area of particular interest. He has served in Vietnam, Cambodia, and Thailand. Born in 1944, Mr. Carney graduated from MIT in 1966 and later attended Cornell University for a year of advanced Southeast Asian area studies. He speaks Khmer, Thai, and French. Among his published works are a Cornell monograph on the Communist Party of Kampuchea and a photo-essay on the first two years of the Cambodian refugee and relief emergency.

David P. Chandler was posted to Phnom Penh as a U.S. Foreign Service officer in 1960–1962. After resigning from the Foreign Service in 1966, he took degrees at Yale and the University of Michigan. Since 1978 he has been research director of the Centre of Southeast Asian Studies at Monash University in Australia. His major publications include *The Land and People of Cambodia* (1972), *Cambodia Before the French: Politics in a Tributary Kingdom, 1794–1847* (1974); (tr.) *Favorite Stories from Cambodia* (1978); *A History of Cambodia* (1983), and (ed., with Ben Kiernan) *Revolution and Its Aftermath in Kampuchea: Eight Essays* (1983).

Joel R. Charny, after attending Brown University, served for two years as a Peace Corps volunteer in the Central African Republic. He initially worked for Oxfam America as a volunteer while obtaining an Ed.M. degree from the Harvard Graduate School of Education. He worked as administrative officer for the OXFAM Consortium team in Phnom Penh, Kampuchea, during the emergency period. Since 1981 he has worked based in Boston as the Southeast Asia Projects Officer for Oxfam America.

Cynthia M. Coleman was project director of the Cambodian Repatriation Project funded by the United Nations High Commissioner for Refugees in 1975–1976. Much of her paper in this volume was based on this experience. From 1978 to 1980 she was director of a project for Southeast Asian refugees sponsored by the Pennsylvania Office of Mental Health. Between 1980 and 1982, Ms. Coleman was a senior staff member of the Indochinese Refugee Action Center and the Cambodia Crisis Center and served as program officer of the Cambodian Cluster Project of the U.S. Department of Health and Human Services. She has been director of the Migration and Refugee Service of the Lutheran Children and Family Service since 1982.

John Dennis has worked for four years in rice agriculture in Thailand and is currently a doctoral candidate at Cornell University with a major in Development Sociology and minors in Soil Science and Asian Studies. In 1980 he identified sources of rice seed in Thailand for relief agencies and administered a program for Oxfam America supplying subsistence agriculture inputs to Kampuchean farmers coming to the Thai border. Subsequently, he has worked part-time as an agricultural consultant to Oxfam America for various projects in Kampuchea.

Meng-Try Ea was born in Phnom Penh (Kampuchea) where he studied at the Faculty of Law and Economics Sciences (1960–1964). He went to France in 1973 and got a Ph.D. (in demography) in 1980 from the University of Paris. He worked as a lecturer in geography from 1965 to 1971 in Phnom-Penh when he was a member of the National Assembly in 1972–1973. In 1981–1982 he was visiting fellow for the Department of Demography at the Australian National University with a fellowship of The Population Council. On returning from Australia he worked for the Laboratory of Historical Demography, Paris, under the guidance of Professor Jacques Dupâquier.

May Ebihara received her Ph.D. from Columbia University and is presently associate professor of Anthropology at Lehman College and The Graduate Center of the City University of New York. She has also taught at Bard College, Mount Holyoke College, and Columbia University. She is the only American anthro pologist to have conducted field research in a Cambodian village and has written a number of articles on aspects of Khmer peasant culture. She has served on the Southeast Asia Council of the Association for Asian Studies and is presently on the executive committee of the Thai/Laos/Cambodia Studies Group of the AAS.

David W. P. Elliott is associate professor of Government and International Relations at Pomona College. He spent a total of six years in Vietnam between 1963 and 1973 with the U.S. Army, the Rand Corporation, and as a private scholar. He received his B.A. from Yale, and a Ph.D. from Cornell University. Elliott was a member of the US-ASEAN dialogue group of scholars that spent two months in the ASEAN countries in 1980. His most recent trip to Southeast Asia was a visit to Vietnam and Thailand in 1982. Elliott is the editor of *The Third Indochina Conflict* (1981).

David R. Hawk is adjunct lecturer at Hunter College, City University of New York, and seminar associate of the Columbia University Seminar on Human Rights. In 1980 and 1981 he was based in Bangkok as director of the Khmer Program of the World Conference on Religion and Peace. From 1974 to 1980 he worked for Amnesty International, U.S.A., as its executive director and then as a consultant on International Human Rights Treaties. His articles on Asia, international affairs, and human rights have appeared in a variety of books and periodicals including *The New Republic* and *Worldview*. Mr. Hawk received his Master's of Divinity degree from the Union Theological Seminary and has done advanced degree work at Oxford.

J. D. Kinzie graduated from the University of Washington School of Medicine and did a psychiatric residency at the same institution. He was a fellow in transcultural psychiatry at the University of Hawaii, John Burns School of Medicine. Dr. Kinzie spent two years as a general practitioner in Vietnam and Malaysia and taught at the Faculty of Medicine in the Department of Psychological Medicine at the University of Malaya. Previously, he was on the faculty in the Department of Psychiatry at the John Burns School of Medicine, Honolulu. He is now associated with the Oregon Health Sciences University where he is professor of Psychiatry and director of Clinical Services for the Department of Psychiatry. His primary research interests include psychiatric education and transcultural psychiatry, with a special emphasis on the treatment approaches of Indochinese refugees.

Orlin J. Scoville, now a private consultant, is professor emeritus at Kansas State University, where he taught from 1966 to 1976. From 1962 to 1966 he led a large agricultural development program as deputy assistant director of U.S. A.I.D./Thailand. Between 1939 and 1962 he was an agricultural economist for the United States Department of Agriculture. He received his Ph.D. in agricultural economics from Harvard in 1949. In 1980–81, Dr. Scoville served as a member of three United Nations Food Assessment Missions to Kampuchea.

Serge Thion studied sociology at the Sorbonne. Since 1971 he has been a research fellow in the Centre National de la Recherche Scientifique in Paris. He visited Cambodia first in December 1967 and most recently in August 1981. He taught in a Phnom Penh high school in 1968–69 and, in 1972, was the only Western observer ever invited to visit the Khmer Rouge areas. He reported on his visit in *Le Monde*. His articles on Cambodia have appeared in *Indochina Chronicle*, *Liberation* (NY), *Cahiers Internationaux de Sociologie*, *Les Temps*

Modernes and the *Bulletin of Concerned Asian Scholars*. He is the author, with Ben Kiernan, of *Khmers Rouges! Matériaux pour l'histoire du communisme au Cambodge* (1981) and, with J. C. Pomonti, of *Des Courtisans aux partisans* (1971). He has also written on Vietnam.

Michael Vickery was born in the U.S. He received his BA in Russian and Far Eastern Studies from the University of Washington, in 1952; and his Ph.D. in History from Yale in 1977. He lived in Cambodia, as an English language teacher, from 1960 to 1964, and in Laos from 1964 to 1967. After three years in residence at Yale, he did dissertation research in Cambodia and Thailand, 1970–72. He was lecturer in Southeast Asian History, Universiti Sains Malaysia, Penang, 1973–1979; research fellow, Australian National University, 1979–82; research associate, University of Adelaide, 1982 to present. During May–September 1980 he worked in Khmer refugee camps and holding centers in Thailand and made further visits to them in 1981 and 1982. In 1981 he spent three weeks in Cambodia, for the first visit since 1974. He wrote his dissertation and published articles on the premodern history of Cambodia and Thailand and has published articles and a book on contemporary Cambodia.

David Ablin graduated from Princeton University in June 1986 and is working toward a Ph.D. in political science at Yale. **Marlowe Hood** is a graduate student in the Politics Department at Princeton. He is currently writing a dissertation on the local press in the People's Republic of China. Together they are the authors of articles on Southeast Asia which have appeared in the *New York Times*, *Chicago Tribune*, *Christian Science Monitor*, *New York Review of Books*, and elsewhere.